FUNCTIONAL ANALYSIS

FRIGYES RIESZ
and
BÉLA SZ.-NAGY

Translated from
the 2nd French edition by
LEO F. BORON

DOVER PUBLICATIONS, INC.
NEW YORK

This Dover edition, first published in 1990, is an unabridged republication of the work first published by Frederick Ungar Publishing Co., New York, in 1955. The Appendix, *Extensions of Linear Transformations in Hilbert Space Which Extend Beyond This Space,* separately published by Ungar in 1960, has been added to this edition.

Manufactured in the United States of America
Dover Publications, Inc.,
31 East 2nd Street,
Mineola, N.Y. 11501

Library of Congress Cataloging-in-Publication Data

Riesz, Frigyes, 1880–1956.
[Leçons d'analyse fonctionelle. English]
Functional analysis / Frigyes Riesz and Béla Sz.-Nagy ; translated from the 2nd French edition by Leo F. Boron.
p. cm.
Translation of: Leçons d'analyse fonctionelle.
"An unabridged republication of the work first published by Frederick Ungar Publishing Co., New York, in 1955. The appendix, Extensions of linear transformations in Hilbert space which extend beyond this space, separately published by Ungar in 1960, has been added to this edition"—T.p. verso
ISBN 0-486-66289-6
1. Functional analysis. I. Szökefalvi-Nagy, Béla, 1913– II. Title.
QA320.R513 1990
515'.7—dc20 90-33977
CIP

PREFACE TO THE FIRST EDITION

This book has developed from courses entitled "Real Functions," "Integral Equations," "Hilbert Space," etc., which the two authors have taught for several years at the Universities of Szeged and Budapest. As the printing was delayed by technical difficulties we have in the meantime added some paragraphs dealing with recent results.

The first part, on modern theories of differentiation and integration, serves as introduction to the second, which treats integral equations and linear functionals and transformations. This division into two parts corresponds to the division of the work by the two authors; although they have worked together, the first part was written principally by the first, and the second by the second author.

The two parts form an organic unit centered about the concept of linear operator. This concept is reflected in the method by which we have constructed the Lebesgue integral; this method, which seems to us to be simpler and clearer than that based on the theory of measure, has been used by the first author in his lectures for more than twenty years although it has not been published in definitive form.

The first part begins with a direct proof of the Lebesgue theorem on the differentiation of monotonic functions and its application to the study of the relations between the derivatives and the integrals of interval functions. After this we construct the theory of the Lebesgue integral and study the spaces L_2 and L_p and their linear functionals. The Stieltjes integral and its generalizations are introduced in terms of linear operations on the space of continuous functions.

The second part begins with a chapter on integral equations, the subject which prepared the way for the general theory of linear transformations. We present several methods for arriving at the Fredholm alternative, and in the succeeding chapter we apply them to completely continuous functional equations of general type on either a Hilbert space or a Banach space. *Symmetric* completely continuous linear transformations are studied in a separate chapter.

We then develop the spectral theory of self-adjoint transformations, either bounded or unbounded, of Hilbert space. We also consider the problem of the extensions of unbounded symmetric transformations. A special chapter is devoted to functions of a self-adjoint transformation, as well as to the study of the spectrum and its perturbations. Stone's theorem on groups of unitary transformations and some related theorems, as well as certain ergodic theorems, are the subject of another chapter.

The last chapter surveys the beginnings, as yet fragmentary, of the spectral theory of linear transformations which are not necessarily normal; we present the method based on the calculus of residues, and we also include a study of the very recent results of J. von Neumann on spectral sets.

In the exposition we have not attempted to study in detail all possible generalizations; rather we have sought to present the principal problems and the methods for handling them. At times we have presented several methods for attaining the same goal and have compared them and discussed their scopes.

We wish to express here our profound gratitude to the Hungarian Academy of Science for publishing our book in French, thus assuring for it an international public. Our hearty thanks are also due Á. Császár, who read the manuscript and whose critical remarks helped us to improve the text. We also acknowledge our gratitude to H. Grenet and K. Tandori for the care they gave to the correction of the proofs.

Budapest and Szeged, February 1952

PREFACE TO THE SECOND EDITION

The favorable reception which this book has received has necessitated a new edition. In it we have tried to eliminate the misprints of the first edition and to improve some passages. There are major changes in Chapters X and XI, mainly in the sections dealing with semi-groups of general type, the relations between the spectrum of a linear transformation and the norms of the iterated transformations, and the spectral sets of von Neumann.

We wish to express our thanks to all those, and particularly to Á. Császár, who by their criticisms have facilitated our task of improving the text.

Budapest and Szeged, May 1953 F. R. and B. Sz.-N.

CONTENTS

Part One

MODERN THEORIES OF DIFFERENTIATION AND INTEGRATION

CHAPTER I: DIFFERENTIATION

CHAPTER II: THE LEBESGUE INTEGRAL

Part Two

INTEGRAL EQUATIONS. LINEAR TRANSFORMATIONS

CHAPTER IV: INTEGRAL EQUATIONS

CHAPTER V: HILBERT AND BANACH SPACES

CHAPTER VI: COMPLETELY CONTINUOUS SYMMETRIC TRANSFORMATIONS OF HILBERT SPACE

CHAPTER VII: BOUNDED SYMMETRIC, UNITARY, AND NORMAL TRANSFORMATIONS OF HILBERT SPACE

CHAPTER VIII: UNBOUNDED LINEAR TRANSFORMATIONS OF HILBERT SPACE

CHAPTER IX: SELF-ADJOINT TRANSFORMATIONS. FUNCTIONAL CALCULUS, SPECTRUM, PERTURBATIONS

CHAPTER X: GROUPS AND SEMIGROUPS OF TRANSFORMATIONS

PART ONE

MODERN THEORIES OF DIFFERENTIATION AND INTEGRATION

CHAPTER I

DIFFERENTIATION

LEBESGUE'S THEOREM ON THE DERIVATIVE OF A MONOTONIC FUNCTION

1. Example of a Nondifferentiable Continuous Function

In classical analysis it is generally assumed that the functions considered possess derivatives, indeed even continuous derivatives up to some order, although it is true that occasional exceptions are allowed. In spite of this, up to the beginning of the present century mathematicians rarely asked whether the functions belonging to a particular category, for example the continuous or the monotonic functions, necessarily possess derivatives; or, if they are known not to possess derivatives everywhere, then at least whether they possess them on the complement of a set whose nature can be made precise. The results in this direction were limited to several almost obvious facts, for example that a convex function necessarily has a left and a right derivative at each point and therefore is differentiable at every value of x except at most for a denumerable set of exceptional values.

The first serious consideration of these problems came in 1806, when in a paper entitled "Sur la théorie des fonctions dérivées" the great scholar AMPÈRE [**1**][1] tried without success to establish the differentiability of an "arbitrary" function except at certain "particular and isolated" values of the variable. Taking into account the evolution of the idea of function one is led to believe—although the original text says nothing positive on this point—that the efforts of Ampère could hardly have been directed beyond functions consisting of monotonic components.

During the entire nineteenth century, as fruitful as this century was for the development of analysis, the solution of the problem did not advance; it even seems at first glance that mathematicians were farther from it. In fact, the first important result, after more than half a century, was furnished by the critique of WEIERSTRASS,[2] who put an end to the repeated attempts

[1] The numbers and asterisks in square brackets refer to the bibliography at the end of the book.

[2] Published by DU BOIS-REYMOND [**1**].

to establish the differentiability of an arbitrary continuous function by constructing *a continuous function without a derivative.* Afterwards these examples multiplied, simpler and simpler ones were invented, and their mutual relationships and relations to other problems were carefully investigated. This was a work in which almost all the great masters of analysis of the second half of the century took part and which has continued up to our time. Here is such an example, perhaps the most elementary, which is due to VAN DER WAERDEN [**1**] and is based on the obvious fact that an infinite sequence of integers can be convergent only if its terms remain equal from some point on.

Let us agree to say, as usual, that the function $f(x)$ possesses a derivative at the point x when the ratio

$$\frac{f(x+h)-f(x)}{h}$$

tends to a finite limit as $h \to 0$ and $x+h$ runs through values for which $f(x+h)$ has a meaning. Denote by $\{x\}$ the distance from x to the nearest integer. Let us form the function

(1) $$f(x) = \sum_{n=0}^{\infty} \frac{\{10^n x\}}{10^n};$$

since the terms of this series represent continuous functions and, furthermore, since the series is majorized by the geometric series $\Sigma\, 10^{-n}$, the function $f(x)$ is obviously continuous. But in trying to calculate the derivative at the point x we shall encounter a contradiction.

Let us observe that we can obviously restrict ourselves to the case where $0 \leq x < 1$ and let us write x in the form

$$x = 0.a_1 a_2 \ldots a_n \ldots,$$

with the agreement that when the option arises we shall write x in the form of a finite decimal fraction completed with zeros. We distinguish two cases according as

$$0.a_{n+1} a_{n+2} \ldots \leq \tfrac{1}{2} \text{ or } > \tfrac{1}{2}.$$

In the first case,

$$\{10^n x\} = 0.a_{n+1} a_{n+2} \ldots,$$

whereas in the second,

$$\{10^n x\} = 1 - 0.a_{n+1} a_{n+2} \ldots.$$

We set $h_m = -10^{-m}$ when a_m equals 4 or 9 and $h_m = 10^{-m}$ otherwise. Consider the ratio

(2) $$\frac{f(x+h_m)-f(x)}{h_m};$$

by formula (1) this ratio can be expressed by a series of the form

$$10^m \sum_{n=0}^{\infty} \pm \frac{\{10^n (x \pm 10^{-m})\} - \{10^n x\}}{10^n}.$$

But it is clear that the numerators are zero starting with $n = m$ and, on the other hand, that for $n < m$ they reduce to $\pm 10^{n-m}$; therefore the corresponding terms of our expression equal ± 1, and consequently the value of ratio (2) is an integer which may or may not be positive, but in any case is even or odd according to the parity of $m - 1$. Hence the sequence of ratios (2), since it is formed of integers of varying parity, cannot converge.

2. Lebesgue's Theorem on the Differentiation of a Monotonic Function. Sets of Measure Zero

We consider next the class of monotonic functions. We owe to Lebesgue the following theorem, one of the most striking and most important in real variable theory.

THEOREM. *Every monotonic function $f(x)$ possesses a finite derivative at every point x with the possible exception of the points x of a set of measure zero, or, as it is often phrased, almost everywhere.*

Before defining the expressions used, let us add that Lebesgue established his theorem using *the additional hypothesis of the continuity of $f(x)$*. He did this in 1904 in the first edition of his book on integration [*], and it appeared at the end of the last chapter as the final result of the entire theory. However, neither the idea of integral nor that of measure appear in the statement of the theorem. In fact, the idea of a set of measure zero does not depend essentially on the general theory of measure, and the main properties of these sets can be established in a few words.

According to Lebesgue, *a set of measure zero* is a set of values x which can be covered by a finite number or by a denumerable sequence of intervals whose total length (i. e., the sum of the individual lengths) is arbitrarily small. It follows immediately from this definition that every subset of such a set is also of measure zero. The same is true for the union of a finite number or of a denumerable sequence of such sets; in fact, we have only to cover these sets respectively by a system of intervals whose total length does not exceed $\frac{\varepsilon}{2^n}$; then the total length of all these intervals covering the union of our sets will not exceed the quantity ε. In particular, every finite or denumerable set of values of x is of measure zero.

It will sometimes be advantageous to give this definition the following form. A set E is of measure zero if it can be covered by a sequence of intervals of finite total length in such a way that every point of E is an interior point of an infinite number of these intervals. The two definitions are equivalent. The second implies the first since, when all the points of E belong to an infinite number of intervals of finite total length, we can decrease this total length at will by suppressing a finite number of intervals. Conversely, if E is of measure zero according to the first definition, we have only to cover it successively by systems of intervals in such a way that the total length of the n-th system is less than $\frac{1}{2^n}$, and if necessary to enlarge the intervals to the left and to the right (for instance by doubling their lengths); the union of all these systems will then satisfy the requirements of the second definition.

The term "almost everywhere" (abbreviated a. e.) is used to state that the fact in question holds everywhere except at the points of a set of measure zero.

Before proving the fundamental theorem of Lebesgue, we shall show that in a certain sense it gives the best possible result and cannot be improved upon. In fact, given a set E of measure zero, we shall construct an increasing function which does not possess a finite derivative at the points of E (the value of the derivative of our function will be infinite at these points). To do this we have only to cover E by intervals in the sense of the second definition and to set $f(x)$ equal to the sum of the lengths of those intervals or segments of intervals which lie to the left of the point x; the function so defined obviously has the required property.

3. Proof of Lebesgue's Theorem

We shall prove that monotonic functions are differentiable almost everywhere without using the theory of integration. The first proofs which take such independence into account are due to FABER [**1**] and G. C. YOUNG and W. H. YOUNG [**1**].

For convenience we shall assume at first that the function is continuous and monotonic and we shall indicate only at the end the modifications (which are almost obvious) that must be made to remove the hypothesis of continuity.

The proof will be based on the following

LEMMA.[3] *Let $g(x)$ be a continuous function defined in the interval $a \leq x \leq b$, and let E be the set of points x interior to this interval and such that there exists a ξ lying to the right of x with $g(\xi) > g(x)$. Then the set E is either empty or an open set, i. e., it decomposes into a finite number or a denumerable infinity of open and disjoint intervals (a_k, b_k), and*

$$g(a_k) \leq g(b_k)$$

for all these intervals.

[3] F. RIESZ [**17**].

To prove this lemma we first observe that the set E is open, since if $\xi > x_0$ and $g(\xi) > g(x_0)$, then, in view of the continuity, the relations $\xi > x$, $g(\xi) > g(x)$ remain valid when x varies in the neighborhood of the point x_0. This being true, let (a_k, b_k) be any one of the open intervals into which E decomposes; the point b_k will not belong to this set. Let x be a point between a_k and b_k; we shall prove that $g(x) \leq g(b_k)$; the inequality to be proved will follow by letting x tend to a_k. To prove the inequality for x, let x_1 be the largest number between x and b_k for which $g(x_1) \geq g(x)$; we have to show that x_1 coincides with b_k. If this were not true, the points ξ_1 which correspond to x_1 by the hypothesis of the theorem would lie beyond b_k and, since b_k does not belong to the set E, we would have $g(x_1) < g(\xi_1) \leq g(b_k) < g(x_1)$, which yields a contradiction.

The reader can readily verify that we have exactly $g(a_k) = g(b_k)$ except possibly when $a_k = a$. However, this fact is not important in the following application.

The lemma established, let $f(x)$ be a continuous and monotonic function for $a \leq x \leq b$; to fix our ideas, we shall assume it nondecreasing. To examine the differentiability of $f(x)$, we shall compare its derived numbers. The *upper and lower right-derived numbers* are respectively the limit superior and limit inferior of the ratio $\frac{1}{h}(f(x+h) - f(x))$ for $h > 0$, $h \to 0$, and are denoted by Λ_r and λ_r. The left derived numbers, Λ_l and λ_l, are defined in an analogous manner. Infinite values are admissible. A finite derivative exists at every point x where the four derived numbers have the same finite value.

To prove Lebesgue's theorem, we have only to prove that

$$1^0\ \Lambda_r < \infty; \qquad 2^0\ \Lambda_r \leq \lambda_l$$

almost everywhere. In fact, applying 2^0 to the function $-f(-x)$, it follows that we also have

$$\Lambda_l \leq \lambda_r$$

almost everywhere, and combining this with 1^0 and 2^0 we obtain

$$\Lambda_r \leq \lambda_l \leq \Lambda_l \leq \lambda_r \leq \Lambda_r < \infty;$$

hence the equality signs must hold, which was to be proved.

To verify assertion 1^0, that the set E_∞ of points x for which $\Lambda_r = \infty$ is of measure zero, we observe that this set is contained in the set E_C for which $\Lambda_r > C$, where C denotes a quantity chosen as large as we wish. But the relation $\Lambda_r > C$ implies the existence of a $\xi > x$, such that

$$\frac{f(\xi) - f(x)}{\xi - x} > C,$$

that is to say that $g(\xi) > g(x)$, where we have set $g(x) = f(x) - Cx$. Hence the set E_C is embedded in the intervals (a_k, b_k) of our lemma, and according to this lemma we have

$$f(b_k) - Cb_k \geq f(a_k) - Ca_k,$$

that is to say that

$$C(b_k - a_k) \leq f(b_k) - f(a_k).$$

This yields, by addition,

$$C \sum (b_k - a_k) \leq \sum [f(b_k) - f(a_k)] \leq f(b) - f(a),$$

which shows that, for C sufficiently large, the total length of the intervals (a_k, b_k) will be as small as we wish. That is, the set E_∞ is of measure zero.

The second statement is verified by analogous reasoning which is repeated alternately under two different forms. Let $c < C$ be two positive quantities. Let us form first of all the function $g(x) = f(-x) + cx$ and let Σ_1 be the system of corresponding intervals given by our lemma or, rather, that of their reflections through the origin; then, for reasons similar to those just used, Σ_1 will contain all the x for which $\lambda_l < c$. Let, moreover, Σ_2 be the system formed from the intervals (a_{kl}, b_{kl}) which correspond to the function $g(x) = f(x) - Cx$, but considered separately in the interior of each interval (a_k, b_k). Then for these intervals we shall have

$$f(b_k) - f(a_k) \leq c(b_k - a_k), \quad C(b_{kl} - a_{kl}) \leq f(b_{kl}) - f(a_{kl})$$

and it follows that

$$C\Sigma_2 \leq V_2 \leq V_1 \leq c\Sigma_1,$$

that is,

$$\Sigma_2 \leq \frac{c}{C} \Sigma_1,$$

where we have denoted by Σ_1, Σ_2 the total length of the two systems of intervals and by V_1, V_2 the sums of the corresponding variations of the function $f(x)$.

Repeating the two methods alternately, we shall obtain a sequence $\Sigma_1, \Sigma_2, \ldots$ of systems of intervals, each imbedded in the preceding and, in general, we shall have

$$\Sigma_{2n} \leq \frac{c}{C} \Sigma_{2n-1}.$$

It follows that

$$\Sigma_{2n} \leq \left(\frac{c}{C}\right)^n \Sigma_1 \to 0.$$

But the points x for which we have $\Lambda_r > C$ and $\lambda_l < c$ simultaneously are obviously contained in all the systems Σ_n; that is, they form a set E_{cC}

of measure zero. Finally, each point x such that $\Lambda_r > \lambda_l$ belongs to such a set, and we can even assume that c and C are rational numbers, because between two different real numbers we can always insert two rational numbers. That is, if we form the sets E_{cC} for all the rational couples, their union E^* will contain all the x for which $\Lambda_r > \lambda_l$. But there is only a denumerable infinity of rational couples; hence the set E^* is the union of a denumerable infinity of sets of measure zero and consequently E^*, and all the more the set considered, which is included in it, will themselves be of measure zero.

Thus the theorem is proved in the case where the monotonic function $f(x)$ is *continuous*. To extend it to the case of *discontinuous* functions, we observe that our lemma remains valid, after some modifications that are almost obvious, for discontinuous functions. In fact, for our purpose it suffices to consider the case where the limits $g(x - 0)$ and $g(x + 0)$ exist, which is obviously the case for monotonic functions $f(x)$, hence also for $f(x) - Cx$ and $f(-x) + cx$. For $a < x < b$, denote by $G(x)$ the greatest of the values $g(x - 0)$, $g(x)$, $g(x + 0)$, while at the points $x = a$ and $x = b$ set $G(a) = g(a + 0)$ and $G(b) = g(b - 0)$. The points x, if there are any, which are interior to (a, b) and for which there exists a $\xi > x$ with $g(\xi) > G(x)$, form an open set, and for the intervals (a_k, b_k) of which this set is composed we shall have $g(a_k + 0) \leq G(b_k)$.

The modifications that have to be made in the previous reasoning in order to prove the lemma in this extended form and to apply it to the case of discontinuous monotonic functions are so obvious that we shall omit the details. The only point which we shall insist upon is that the introduction of the function $G(x)$ does not disturb in any respect the points of continuity, and as to the points of discontinuity, since they form a denumerable[4] set and therefore a set of measure zero, we shall be able in any case to add them to the present set or to exclude them, according to our requirements.

4. Functions of Bounded Variation

We shall extend our result to a larger class of functions, namely the class of functions of bounded variation. This class plays a fundamental role in several branches of analysis, including the theory of Fourier series, the rectification

[4] It suffices to show that if k is an arbitrary positive integer, then the points x for which

$$|g(x + 0) - g(x - 0)| > \frac{1}{k}$$

are at most finite in number. In fact, if there were an infinite number of them we could select a monotonic convergent subsequence, either increasing or decreasing. Denoting the limit by ξ, at least one of the limits $g(\xi - 0)$, $g(\xi + 0)$ can not exist, contrary to the hypothesis.

of curves, and, of course, the theory of integration. We arrive at this class of functions (there are other approaches) by observing that on the one hand, if two functions $f_1(x)$ and $f_2(x)$ are differentiable almost everywhere then so is their difference $f(x) = f_1(x) - f_2(x)$ and that, on the other hand, when $f_1(x)$ and $f_2(x)$ are nondecreasing we obviously have

$$\sum_1^n |f(x_k) - f(x_{k-1})| \leq f_1(b) - f_1(a) + f_2(b) - f_2(a)$$

for every decomposition of the interval (a, b) into partial intervals (x_{k-1}, x_k) $(k = 1, 2, \ldots, n;\ x_0 = a, x_n = b)$.

Functions $f(x)$, continuous or not, for which the sum

$$\Sigma_{ab} = \sum_1^n |f(x_k) - f(x_{k-1})| \tag{3}$$

considered above does not surpass a finite bound, independent of the particular choice of the decomposition, are called *functions of bounded variation.* The least upper bound is called *the total variation of* $f(x)$ *in the interval* (a, b); we shall denote it by $T(a, b)$.

The total variation is an "additive interval function." That is, if c is a point between a and b, the function $f(x)$ is of bounded variation in (a, b) if and only if it is of bounded variation in (a, c) and in (c, b), and then

$$T(a, b) = T(a, c) + T(c, b).$$

To establish this property we have only to observe that since the sums Σ_{ab} can not decrease when we insert a new point of decomposition, it suffices to consider the decompositions of (a, b) which arise from a decomposition of (a, c) and from a decomposition of (c, b); then $\Sigma_{ab} = \Sigma_{ac} + \Sigma_{cb}$ and the proposition is verified by taking the least upper bounds.

We have just seen that the difference of two nondecreasing functions is of bounded variation. The converse is due to CAMILLE JORDAN, namely, the

THEOREM. *Every function of bounded variation is the difference of two nondecreasing functions.*

The proof of this fact is very simple: we have only to introduce the function $T(x) = T(a, x)$, the total variation of $f(x)$ calculated for the interval (a, x), which we propose to call, by analogy with indefinite integrals, the indefinite total variation of $f(x)$; $T(x)$ as well as $T(x) - f(x)$ are then nondecreasing and furnish the required decomposition

$$f(x) = T(x) - [T(x) - f(x)].$$

This is evident for $T(x)$; in fact, if $x < \xi$, we have

$$T(a, \xi) = T(a, x) + T(x, \xi);$$

hence

$$T(\xi) - T(x) = T(x, \xi) \geq 0.$$

To show that

$$T(x) - f(x) \leq T(\xi) - f(\xi),$$

or, equivalently, that

$$f(\xi) - f(x) \leq T(\xi) - T(x) = T(x, \xi),$$

we have only to observe that $|f(\xi) - f(x)|$ is a particular sum of the type $\Sigma_{x\xi}$ (where there are no interior points of decomposition), and that consequently

$$|f(\xi) - f(x)| \leq T(x, \xi).$$

The last inequality suggests still a second decomposition of $f(x)$ into monotonic functions:

$$f(x) = P(x) - N(x)$$

where

$$P(x) = \tfrac{1}{2}[T(x) + f(x)] \text{ and } N(x) = \tfrac{1}{2}[T(x) - f(x)].$$

$P(x)$ and $N(x)$ are, up to additive constants, what we call respectively *the positive and negative variations of* $f(x)$ in the interval (a, x); we shall also call them *the indefinite positive and negative variations of* $f(x)$.

Finally, since differentiation of a difference is carried out term by term, and since moreover the union of the two sets of exceptional points, each of which is of measure zero, is itself of measure zero, we can state our theorem in its final form:

LEBESGUE'S THEOREM. *Every function of bounded variation possesses a finite derivative almost everywhere.*

SOME IMMEDIATE CONSEQUENCES OF LEBESGUE'S THEOREM

5. Fubini's Theorem on the Differentiation of Series with Monotonic Terms

In what follows we shall derive some more or less immediate consequences of the fundamental theorem which we have just established. We begin with Fubini's theorem on the differentiation of series with monotonic terms.

FUBINI'S THEOREM.[5] *Let*

$$f_1(x) + f_2(x) + \dots = s(x) \tag{4}$$

[5] FUBINI [**2**]; see further TONELLI [**1**], RAJCHMAN–SAKS [**1**].

be a convergent series all of whose terms are monotonic functions of the same type, defined on the interval $a \leq x \leq b$. Then

$$f_1'(x) + f_2'(x) + \ldots = s'(x), \tag{5}$$

except perhaps on a set of measure zero; that is, term by term differentiation is possible almost everywhere.

Without loss of generality we can assume that $f_n(a) = 0$. For definiteness, we also assume that the f_n are nondecreasing. This done, we set

$$s_n(x) = f_1(x) + f_2(x) \ldots + f_n(x); \; s_n(x) \to s(x).$$

Except for a set E_0 of measure zero, namely the union of the denumerable infinity of sets of exceptional points, all these functions possess finite derivatives. Since we obviously have

$$s_n'(x) \leq s_{n+1}'(x) \leq s'(x),$$

the series in the first member of (5) has meaning and is convergent in the complement of E_0. Furthermore, since the s_n' increase, it will suffice to verify relation (5), i. e., the relation $s'(x) - s_n'(x) \to 0$, for a suitably chosen subsequence $\{s_{n_k}(x)\}$. We shall make this choice in such a way that $s(b) - s_{n_k}(b)$ decreases rapidly enough for the series formed of these differences and, with it, that formed of the differences

$$s(x) - s_{n_k}(x) \leq s(b) - s_{n_k}(b)$$

to be itself convergent; for example, we could chose the n_k so that

$$s(b) - s_{n_k}(b) < 2^{-k}.$$

Then, since the series formed from the $s - s_{n_k}$ is of the same type as (4), the series which results from it by term-by-term differentiation will converge almost everywhere and, a fortiori, we shall have

$$s'(x) - s_{n_k}'(x) \to 0$$

almost everywhere, which was to be proved.

6. Density Points of Linear Sets

The theorem we have just established has as an immediate application a theorem on the *density* of linear sets. To give this application we must first state what we mean by density points of a linear set. The definition is based on the notion of *exterior measure.*

The exterior measure $m_e(E)$ of a set E of points x is defined to be the greatest lower bound of the total lengths of all systems of intervals which contain E. It follows immediately from the definition that $m_e(E_1 \cup E_2) = m_e(E_1) + m_e(E_2)$

whenever E_1 and E_2 are contained in two disjoint intervals, a fact which we can also express by saying that the exterior measure of the part of a set E contained in a variable interval is an additive interval function.

We shall say that the point x, whether or not in the set E, is a *density point* of the set if

$$\frac{m_e(E; x-h, x+k)}{h+k} \to 1 \qquad (0 < h, k \to 0),$$

the expression in the numerator indicating the exterior measure of that part of E which is contained between $x - h$ and $x + k$. With this notation, the theorem in question can be stated in the following way:

THEOREM.[6] *Almost all points (i. e., all except those of a set of measure zero) of an arbitrary linear set are density points of that set.*

Let us state the theorem in an analytic form by introducing a function $f(x)$ on the interval (a, b) in which the set E lies, which is equal to the exterior measure of that part of E which is contained between a and x. The theorem asserts that $f'(x) = 1$ at almost every point of E. To prove the theorem, we have only to consider open sets (systems of intervals) $\Sigma_1, \Sigma_2, \ldots$ to which E is interior and whose total lengths tend rapidly to $m_e(E)$. Denote by $f_n(x)$ the function analogous to $f(x)$ which is formed with Σ_n instead of E; $f_n(x)$ is the total length of intervals or of their segments which belong to Σ_n and lie to the left of x. We have only to apply Fubini's theorem to the series composed of the differences $f_n(x) - f(x)$; it will follow that $f_n'(x) - f'(x)$ tends to zero almost everywhere in (a, b) and hence in E, and since $f_n'(x) = 1$ on E for all n, the theorem is proved.

It is proper to note that LEBESGUE stated the density theorem only for so-called measurable sets which we shall discuss later; otherwise, the two statements differ only in appearance; in reality, they are corollaries of one another.

7. Saltus Functions

Here is another corollary of Fubini's theorem.

Let $\{x_n\}$ be a finite or denumerable set in the interval (a, b) and consider the series

$$\sum f_n(x)$$

where

$$f_n(x) = 0 \text{ for } x < x_n,\ f_n(x_n) = u_n,\ f_n(x) = u_n + v_n \text{ for } x > x_n.$$

Assume that the series

$$\sum u_n,\ \sum v_n$$

[6] LEBESGUE [*] (1st Edition, pp. 123–124, and 2nd Edition, pp 185–187).

are absolutely convergent. In this case, the sum $s(x)$ of the series considered exists and is a so-called *saltus function*; it can be written in the explicit form

$$s(x) = \sum_{x_n \leqq x} u_n + \sum_{x_n < x} v_n. \tag{6}$$

In the case where $u_n \geq 0$ and $v_n \geq 0$ the functions $f_n(x)$ are increasing and $f_n'(x) = 0$ except at the point $x = x_n$; hence, by virtue of Fubini's theorem, we also have $s'(x) = 0$ *almost everywhere.*

The same holds even in the general case where the quantities u_n, v_n can be real numbers of arbitrary sign. The proof reduces to showing that *every saltus function $s(x)$ is of bounded variation and that its indefinite total variation $T(x)$ is a saltus function also, namely the function corresponding to the same points x_n and to the quantities $|u_n|$, $|v_n|$ instead of u_n, v_n. The indefinite positive and negative variations are then generated respectively by the positive and negative parts of u_n, v_n.*

To see this, it suffices to show that the total variation $T = T(b)$ is equal to $U = \Sigma(|u_n| + |v_n|)$; the assertion for $T(x)$ follows from this if we consider the partial interval (a, x) instead of (a, b).

Given $\varepsilon > 0$, let us choose the integer M so large that the sum of the first M terms of the series defining U comes within ε of U. We consider a decomposition of the interval (a, b) such that every closed subinterval $\alpha \leq x \leq \beta$ contains one and only one of the points $x_1, x_2, \ldots, x_M$, and this at its right or left extremity. For a subinterval of the type $\alpha \leq x \leq x_r$, where $r \leq M$, we have

$$|s(x_r) - s(\alpha)| = |u_r + \sum_{\alpha < x_n < x_r} u_n + \sum_{\alpha \leqq x_n < x_r} v_n| \geq |u_r| - (\sum_{\alpha < x_n < x_r} |u_n| + \sum_{\alpha \leqq x_n < x_r} |v_n|),$$

and all the u_n and v_n which appear in this formula except u_r have indices $> M$. The situation is analogous for the subintervals of the type $x_r \leq x \leq \beta$. It follows from this that the sum $\Sigma\,|s(\beta) - s(\alpha)|$, extended to all the subintervals of the decomposition considered, is $\geq \sum_1^M (|u_n| + |v_n|) - \varepsilon \geq$ $\geq U - 2\varepsilon$. Since, on the other hand, it is clear by the representation (6) of the function $s(x)$ that the analogous sum corresponding to an arbitrary decomposition of (a, b) can not surpass the quantity U, we conclude that the total variation T equals U, which completes the proof of the above propositions.

Being of bounded variation, the function $s(x)$ possesses limits from the left and right at every point. *We shall show that at the given points x_n, $s(x)$ has jumps from the left and right respectively equal to u_n and v_n, and that it is continuous at other points.*

To accomplish this, let us again make use of decompositions of the particular type considered above; we shall have for an interval of the type (a, x_r):

$$|s(x_r) - s(a) - u_r| < \varepsilon;$$

letting a tend to x_r, it follows that

$$|s(x_r) - s(x_r - 0) - u_r| \leq \varepsilon$$

for $r = 1, 2, \ldots, M$. But M was limited by ε only from below, hence this result is valid for $r = 1, 2, \ldots$. Since ε was arbitrary, we finally obtain that

$$s(x_r) - s(x_r - 0) = u_r \qquad (r = 1, 2, \ldots).$$

Analogous reasoning yields the relation

$$s(x_r + 0) - s(x_r) = v_r \qquad (r = 1, 2, \ldots).$$

The same reasoning can also be applied to a point x which is different from all the given points x_n; we have only to add it to the sequence $\{x_n\}$ and take the corresponding quantities u and v equal to 0, which does not change anything in the definition of $s(x)$. It follows that $s(x) - s(x - 0) = 0$ and $s(x + 0) - s(x) = 0$, i. e., x is a point of continuity.

The reason for interest in saltus functions is that *every function of bounded variation $f(x)$ can be decomposed into the sum of a continuous function of bounded variation $g(x)$ and a saltus function $s(x)$*; these are called the *continuous part* and the *saltus part of* $f(x)$. In fact, we have only to define $s(x)$ as the saltus function whose points of discontinuity and whose corresponding jumps are equal to those of $f(x)$; $g(x) = f(x) - s(x)$ is then everywhere continuous and, being the difference of two functions of bounded variation, is also of bounded variation.

We see immediately that *in the case where $f(x)$ is monotonic, its continuous part $g(x)$ and its saltus part $s(x)$ are also monotonic and of the same sense.*

8. Arbitrary Functions of Bounded Variation

We have just seen that for a saltus function $s(x)$, we have $s'(x) = 0$ almost everywhere. Since the indefinite total variation $T(x)$ of $s(x)$ is also a saltus function, we have $T'(x) = 0$ almost everywhere.

The equality of these two derivatives generalizes to the case of an arbitrary function $f(x)$ of bounded variation and its indefinite total variation $T(x)$.

THEOREM. $T'(x) = |f'(x)|$ *almost everywhere.*

We begin the proof by choosing a sequence $\{\Delta_n\}$ of decompositions of the interval (a, b) such that the sum (3) corresponding to the decomposition Δ_n is within 2^{-n} of the total variation $T = T(b)$. To the decomposition Δ_n we make correspond the following function $f_n(x)$. In each of the segments $x_{k-1} \leq x \leq x_k$ of the decomposition, let $f_n(x)$ equal

$$f(x) + \text{constant} \quad \text{or} \quad -f(x) + \text{constant},$$

according as

$$f(x_k) - f(x_{k-1}) \geqq 0 \text{ or } < 0;$$

the constants must be determined so that $f_n(a) = 0$ and that the values taken at the points x_k agree.

Then we have

$$f_n(x_k) - f_n(x_{k-1}) = |f(x_k) - f(x_{k-1})|$$

and consequently

$$T(b) - f_n(b) = T(b) - \sum_k [f_n(x_k) - f_n(x_{k-1})] \leq 2^{-n}.$$

On the other hand the function $T(x) - f_n(x)$ is *increasing*, or, what amounts to the same thing,

$$T(\xi) - T(x) \geq f_n(\xi) - f_n(x) \text{ for } x < \xi.$$

This follows immediately from the inequality

$$T(\xi) - T(x) \geq |f(\xi) - f(x)|$$

when x and ξ belong to the same segment (x_{k-1}, x_k). From here we pass to the case

$$x < x_k < \ldots < x_p < \xi$$

by adding the inequalities with respect to the segments

$$(x, x_k), (x_k, x_{k+1}), \ldots, (x_p, \xi).$$

Since the series

$$\sum [T(x) - f_n(x)]$$

is majorized by the convergent series $\Sigma 2^{-n}$, it is also convergent. By Fubini's theorem, the derived series converges almost everywhere, and consequently

$$T'(x) - f_n'(x) \to 0$$

almost everywhere. But we obviously have

$$f_n'(x) = \pm f'(x);$$

in as much as $T'(x) \geq 0$, since $T(x)$ is increasing, this proves that $T'(x) = |f'(x)|$ almost everywhere.

The above reasoning also permits us to establish relations between the *discontinuities* of $f(x)$ and those of its indefinite total variation $T(x)$. Namely, we shall prove the

THEOREM. *$f(x)$ and $T(x)$ have the same points of continuity and discontinuity and their jumps are equal except for sign. That is, at every point x we have*

$$T(x) - T(x-0) = |f(x) - f(x-0)|, \quad T(x+0) - T(x) = |f(x+0) - f(x)|.$$

In fact,

$$|T(\xi) - T(x) - f_n(\xi) + f_n(x)| \leq |T(\xi) - f_n(\xi)| + |T(x) - f_n(x)| \leq 2 \cdot 2^{-n}$$

for $x < \xi$; letting ξ tend to x, it follows that

$$|T(x+0) - T(x) - f_n(x+0) + f_n(x)| \leq 2^{1-n}.$$

Hence,

$$f_n(x+0) - f_n(x) \to T(x+0) - T(x) \text{ when } n \to \infty.$$

But the jumps of f_n are obviously equal to those of f except for sign, hence

$$|f(x+0) - f(x)| = T(x+0) - T(x).$$

The assertions about limits on the left are verified in the same way.

9. The Denjoy-Young-Saks Theorem on the Derived Numbers of Arbitrary Functions

Although we shall have no need for it in the sequel, it will be of interest to discuss here a very general theorem concerning the differentiation, or more precisely *the behavior of the four derived numbers, of an arbitrary function.*

The theorem is due to DENJOY and to Mrs. YOUNG,[7] who established it, independently of one another, for the case of continuous functions; then, Mrs. YOUNG extended it to measurable functions;[8] finally, SAKS showed that the theorem holds for arbitrary functions.[9] As we would expect in view of the great generality of the final statement of the theorem, the proof due to SAKS is of extreme simplicity.

We shall follow DENJOY in saying that two derived numbers are *associated* if they are taken on the same side, as for example λ_l and Λ_l, and *opposed* if they correspond to different sides and indices, as for example λ_l and Λ_r. The theorem in question can be stated as follows:

[7] DENJOY [**1**] (in particular pp. 174–195); G. C. YOUNG [**1**].
[8] G. C. YOUNG [**2**].
[9] SAKS [**1**]; see further HANSON [**1**], BLUMBERG [**1**].

THEOREM. *Except for a set of measure zero, only the following cases can occur: Two associated derivatives are either equal and finite or unequal with at least one infinite; two opposed derivatives are either finite and equal or infinite and unequal, with the one of higher index equal to* ∞ *and the other equal to* $-\infty$.

The first rule, that on associated derivatives, is an obvious consequence of the second, and we can restrict ourselves to the latter.

Before going into details, we call the reader's attention to the extreme simplicity of the rule which is obtained when we cease to distinguish between right and left and consider only the two *neutral* derived numbers, the lower and upper, defined for example as the limit inferior and limit superior of the ratio

$$\frac{f(x+h)-f(x-k)}{h+k} \qquad (h, k \geq 0, 0 < h+k \to 0).$$

Then, *except for a set of measure zero, only two extreme possibilities arise: either the two limits are infinite and of opposite sign, or the function possesses a finite derivative.*

In the particular case of a monotonic function, where infinities of opposite sign cannot occur, only one possibility remains — the existence of a finite derivative.

Let us sketch the proof. It will be sufficient to show that at almost all points where the derived number λ_l is not minus infinity, this derivative and its opposite Λ_r are equal and finite, for the general rule follows from this by replacing $f(x)$ successively by $-f(x)$, $f(-x)$ and $-f(-x)$. For definiteness, we shall assume that $f(x)$ is defined on the interval (a, b), which could, of course, be replaced by an arbitrary set. Let E be the set of points x for which λ_l differs from $-\infty$; this set can be considered as the union of a denumerable infinity of sets $E_{n,r}$, where $n = 0, 1, 2, \ldots$ and r runs through the rational numbers contained in the interval (a, b); the set $E_{n,r}$ consists of those points $x > r$ at which

$$\frac{f(x)-f(\xi)}{x-\xi} > -n$$

for all ξ lying between r and x. Since the sum of a denumerable infinity of sets of measure zero is itself of measure zero, it will suffice to prove that the rule in question is valid almost everywhere in each $E_{n,r}$; since, furthermore, the general case reduces to the case $n = 0, r = 0$ by replacing $f(x)$ by $f(x-r)+ nx$, we need only consider the set $E_0 = E_{0,0}$. Let us exclude those points of this set which are not density points (it would also suffice to exclude only those which are not density points with respect to the closure $\bar{E}_0$ of E_0) and also those where $f(x)$ does not possess a finite derivative with respect to the set E_0 (i.e., calculated so that we approach x without leaving the set E_0).

We have suppressed only a set of measure zero; in fact, in view of the definition of the set E_0, the function $f(x)$, when considered separately on E_0, is monotonic and consequently possesses almost everywhere on E_0 a finite derivative with respect to this set; this fact is an obvious corollary of Lebesgue's theorem.

Let us consider the points x of E_0 which remain. Consider the increment ratio

$$\frac{f(x') - f(x)}{x' - x}.$$

Let x' tend to x without leaving the set E_0. According to our hypothesis the ratio tends to a limit, which we denote by $f'_{E_0}(x)$. If x' does not belong to the set E_0, but is sufficiently close to x, then in view of the density hypothesis x' can be replaced by a $\xi > x'$ belonging to E_0 with the property that the difference $\xi - x'$ is infinitely small with respect to $x' - x$. Since by the definition of the set $E_0 = E_{0,0}$ we have $f(\xi) \geq f(x')$, the numerator of the ratio considered does not decrease when we replace x' by ξ; as for the denominator, it is not appreciably altered. Keeping in mind that the latter can be positive or negative, we see immediately that

$$\lambda_l \geq f'_{E_0}(x) \geq \Lambda_r.$$

On the other hand, from the definition of the quantity $f'_{E_0}(x)$, we see that it represents one of the limits, left as well as right, of the same increment ratio by which we defined the quantities λ_l and Λ_r, so that

$$\lambda_l \leq f'_{E_0}(x) \text{ and } \Lambda_r \geq f'_{E_0}(x).$$

Consequently only the equality sign can hold, and the theorem is proved.

INTERVAL FUNCTIONS

10. Preliminaries

We shall encounter important applications by considering other, not necessarily additive, interval functions. We define an interval function $f(I) = f(\alpha, \beta)$ to be a rule which assigns a definite quantity to the intervals $I = (\alpha, \beta)$ belonging to a certain family. We permit the interval function to be multivalent, as for example the function

$$f(\alpha, \beta) = (\beta - \alpha)f(\xi),$$

where ξ denotes a value which lies between α and β but is otherwise arbitrary, and $f(x)$ is an ordinary function. The function $f(\alpha, \beta)$ arises in the theory of integration or, more precisely, in the theory of what is called the Riemann integral. The upper and lower envelopes of this function, namely the interval functions

$$(7) \qquad (\beta - \alpha) \sup_{\alpha \leqq x \leqq \beta} f(x) \text{ and } (\beta - \alpha) \inf_{\alpha \leqq x \leqq \beta} f(x),$$

which are monovalent, serve, as is known, to define the upper and lower integrals, also called the Darboux integrals. The absolute variation

$$|f(\beta) - f(\alpha)|$$

of a function $f(x)$ occurs, as we have seen above, in the notion of a function of bounded variation. The length of the chord which joins the points with abscissas $x = \alpha$ and $x = \beta$ of a curve $y = f(x)$ or, more generally, that of the chord of a curve

$$x = x(t), \; y = y(t), \; z = z(t)$$

corresponding to the values α and β of the parameter t, is basic to the concepts of rectifiable curve and arc length. The multivalent function

$$(8) \qquad (g(\beta) - g(\alpha))f(\xi)$$

composed of two ordinary functions of which one, $g(x)$, is of bounded variation and the other, $f(x)$, is continuous, is used to define the Stieltjes integral.[10] We shall see that all these examples belong, under suitable hypotheses, to the class of *integrable interval functions*. Finally, let us mention the interval function

$$(9) \qquad \frac{(f(\beta) - f(\alpha))^2}{g(\beta) - g(\alpha)}$$

on which the Hellinger integral,[11] used by Hellinger in his study of the theory of quadratic forms with an infinite number of variables, is based.

The definition of an *integrable interval function and of its integral* is very simple; we use an immediate generalization of the Riemann or of the Darboux integrals. We divide the interval (a, b) into subintervals, form the sum of values which correspond to these intervals, and examine whether these sums tend to a finite limit when the subdivision is varied so that the length of the subintervals tends uniformly to zero, or, in other words, whether there corresponds to every $\varepsilon > 0$ a $\delta = \delta(\varepsilon) > 0$ such that for every decomposition of (a, b) into segments of length less than δ, the corresponding sums approach within ε of a definite limit. When this is the case the *interval function* is said to be *integrable* and the limit of the sum is called its integral and denoted by

$$\int_a^b f(\alpha, \beta).$$

[10] Stieltjes integrals will be analyzed in detail in Chapt. III.

[11] See HELLINGER [**1**] (in particular pp. 25—51) and [**2**] (in particular pp. 234—240).

To show the generality of this notion, we call to the reader's attention the obvious fact that all additive interval functions or, what amounts to the same thing, all functions $f(\alpha, \beta)$ of the form $f(\beta) - f(\alpha)$, are integrable no matter how singular the function $f(x)$.

Before examining the particular examples which we have just enumerated, we shall establish two theorems of a general nature concerning relations which exist between the integration and the differentiation of interval functions.

By the *derivative of an interval function* $f(\alpha, \beta)$ at the point x we understand the limit, when it exists, of the ratio

$$\frac{f(\alpha, \beta)}{\beta - \alpha},$$

as the interval (α, β) contracts to the point x. The derived numbers are defined in an analogous manner. When, in particular, $f(\alpha, \beta)$ is additive, these quantities are none other than those which correspond, in the ordinary sense, to the function $F(x) = f(a, x)$.

The problem of the differentiation of interval functions has been studied by several authors from a very general point of view.[12] The first theorem we shall prove contains only the essence of the principal results obtained.

11. First Fundamental Theorem

This is the following

THEOREM. *Let $f(\alpha, \beta)$ be a non-negative interval function which is integrable in (a, b), and assume further that the value of the integral is zero. Then $f(\alpha, \beta)$ is differentiable with derivative equal to zero almost everywhere in the interval (a, b).*

The proof of the theorem is almost immediate. Let $\delta_1, \delta_2, \ldots$ be positive quantities chosen so that the sum of the values of $f(\alpha, \beta)$ which corresponds to a decomposition of the entire interval (a, b) into segments whose lengths equal at most δ_n, does not surpass the n-th term of a predetermined convergent series, for example the series $\Sigma 2^{-n}$. This done, consider the functions $F_n(x)$ defined as follows. The function $F_n(x)$ is equal to the least upper bound of the sums of values of $f(\alpha, \beta)$ which correspond to decompositions of the interval (a, x) into segments whose lengths do not surpass δ_n. These functions $F_n(x)$ are obviously nondecreasing functions which form a convergent series, so that by Fubini's theorem, $F_n'(x)$ will tend to zero almost everywhere. Moreover, since

$$f(\alpha, \beta) \leq F_n(\beta) - F_n(\alpha) \qquad (\beta - \alpha \leq \delta_n),$$

[12] See, in particular, BURKILL [**1**]—[**3**], R. C. YOUNG [**1**], SAKS [*] (pp. 102—107) and [$\substack{*\\*}$] (pp. 165—169).

the derivatives of the F_n will be, wherever they exist, greater than the derived numbers of the interval function f. That is, these derived numbers are zero almost everywhere, which was to be proved.

12. Second Fundamental Theorem

Before discussing the second fundamental theorem, we note one fact which has not been mentioned previously. This is that *the integrability of* $f(\alpha, \beta)$ *on* (a, b) *implies its integrability on every subinterval* (c, d).

To prove this remark, consider an $\varepsilon > 0$ and the $\delta = \delta(\varepsilon)$ which corresponds to it when dealing with the entire interval (a, b) (Sec. 10). We consider two decompositions of (c, d) into segments of length less than δ and a decomposition of (a, c) and one of (d, b) of the same type; these decompositions define two decompositions of the entire interval (a, b). Let us form the difference of the two sums — sums which are within ε of the integral of $f(\alpha, \beta)$ on (a, b). The absolute value of this difference does not surpass 2ε, and the terms corresponding to the segments (a, c) and (d, b) cancel each other; consequently, the difference of the two sums corresponding to (c, d) is at most equal to 2ε; according to the Cauchy convergence criterion, this assures the existence of the limit, that is, of the integral on the segment (c, d).

Moreover, the convergence of the sums to the integrals is uniform with respect to all the segments. Finally, the integral on the subinterval (c, d) is evidently an *additive* interval function and is expressible in the form $F(d) - F(c)$ by means of the indefinite integral $F(x)$.

Now let us consider, for given $\varepsilon > 0$, a decomposition of the interval (a, b) into segments $I_1, I_2, \ldots, I_n$ whose lengths do not surpass the quantity $\delta = \delta(\varepsilon)$; for this decomposition, as well as for every other which arises from it by inserting new division points, the difference between the corresponding sums and the integral on (a, b) will be at most ε. In particular, keeping a part of the intervals I_k unaltered, dividing the others indefinitely and taking the limit, we shall arrive at a sum with terms of mixed type; some will be of the type $f(\alpha_k, \beta_k)$, the others of the type $F(\beta_k) - F(\alpha_k)$. Then, interchanging the role of the two types and forming the difference of the two sums, each of which approaches within ε of the integral of $f(\alpha, \beta)$ on (a, b), we arrive at the inequality

$$\sum_1^n \pm [f(\alpha_k, \beta_k) - (F(\beta_k) - F(\alpha_k))] \leq 2\varepsilon,$$

and since we still have the signs $\pm$ at our disposal in each term, it follows that

$$\sum_1^n |f(\alpha_k, \beta_k) - F(\beta_k) + F(\alpha_k)| \leq 2\varepsilon.$$

That is, the interval function

$$g(I) = g(\alpha, \beta) = |f(\alpha, \beta) - F(\beta) + F(\alpha)|,$$

evidently non-negative, is integrable and has integral zero; hence, we can apply our first fundamental theorem to it and this assures us that $g(I)$ possesses a derivative equal to zero almost everywhere. From this, we deduce the

THEOREM. *The integrable interval function $f(I)$ and its indefinite integral $F(x)$ possess the same derived numbers almost everywhere; in particular, almost everywhere that one of the two possess a finite derivative the other will, and conversely.*

13. The Darboux Integrals and the Riemann Integral

Let us return to the two interval functions (7) and form their integrals in the sense we have just introduced; then we obtain the lower and upper integrals of the function $f(x)$, also called Darboux integrals. The first of these integrals is in general less than the second; when they coincide we say that $f(x)$ is Riemann-integrable and we call the common value of the Darboux integrals the Riemann integral. That is, the condition for the integrability of $f(x)$ in the interval (a, b) in the Riemann sense is precisely that the non-negative interval function

$$(\beta - \alpha) [\sup_{\alpha \leqq x \leqq \beta} f(x) - \inf_{\alpha \leqq x \leqq \beta} f(x)] = (\beta - \alpha)\, \omega(f; \alpha, \beta), \tag{10}$$

in which $\omega(f; \alpha, \beta)$ denotes *the oscillation of $f(x)$ in the interval* (α, β), have integral zero. The derivative of this interval function, $\omega(x)$, *the oscillation of $f(x)$ at the point x*, exists and is zero at every point of continuity of $f(x)$, and conversely the relation $\omega(x) = 0$ at the point x assures the continuity of $f(x)$ at the point x.

Combining this fact with the first fundamental theorem, we arrive at *a necessary condition that $f(x)$ be integrable in the Riemann sense*, namely, that *the function $f(x)$ be continuous almost everywhere.*

Conversely, *when $f(x)$ is assumed bounded the same condition is also sufficient.* Generally, this fact as well as the necessity are proved without recourse to the theory of interval functions; perhaps it will be of interest to include a proof here which follows our present sequence of ideas. To do this we observe first that since $f(x)$ and hence $\omega(f; \alpha, \beta)$ are bounded by hypothesis, the indefinite integral of (10), or, more precisely, the function

$$\Omega(x) = \int_a^x (\beta - \alpha)\omega(f; \alpha, \beta)$$

has a bounded increment ratio, that is, it satisfies the Lipschitz condition

$$\Omega(\beta) - \Omega(\alpha) \leq C(\beta - \alpha) \qquad (\alpha < \beta); \tag{11}$$

moreover, it is nondecreasing. Finally, according to the second fundamental theorem, it possesses the same derivative as the interval function (10) almost everywhere, that is, under the hypothesis made, $\Omega'(x) = 0$ almost everywhere. Hence, the proof reduces to showing that *a nondecreasing function $\Omega(x)$ which satisfies condition* (11) *and has derivative zero almost everywhere is necessarily constant*; that is, the image of the interval (a, b) under the transformation $y = \Omega(x)$ reduces to a single point.

We consider the set E of points x for which $\Omega'(x)$ either does not exist or does not become zero, and the image of E by $y = \Omega(x)$, which we denote by $\Omega(E)$. E being of measure zero, we can enclose it in a system of intervals of arbitrarily small total length, say $< \varepsilon$; the image of these intervals, which by virtue of (11) will have a total length $< C\varepsilon$, will contain the set $\Omega(E)$. Consequently this last set is also of measure zero.

The same will be true of the set $\Omega(e)$, the image of the complementary set $e = (a, b) - E$. In fact, since $\Omega'(x) = 0$ on e, we can attach to each point x of e points $\xi > x$ for which $\Omega(\xi) - \Omega(x) < \varepsilon(\xi - x)$, where $\varepsilon > 0$ is fixed arbitrarily small. That is, for ε fixed, e is included in the set formed with respect to the function $g(x) = \varepsilon x - \Omega(x)$, which appears in the lemma of Section 3; denote the set by e_ε. According to this lemma, the open set e_ε consists of a system of intervals (a_k, b_k) for which $g(a_k) \leq g(b_k)$, or equivalently $\Omega(b_k) - \Omega(a_k) \leq \varepsilon(b_k - a_k)$; it follows that the total length of the intervals $(\Omega(a_k), \Omega(b_k))$ which make up the set $\Omega(e_\varepsilon)$ does not surpass $\varepsilon(b - a)$, and that consequently the set $\Omega(e)$, since it is contained in each set $\Omega(e_\varepsilon)$, has measure zero, which is what was to be proved.

Finally, since the interval $(\Omega(a), \Omega(b))$ is entirely covered by two sets of measure zero, it must itself be of measure zero.

However, the above reasoning will be complete only after we show that a set of measure zero cannot exhaust the entire interval. We have not had to use this fact up till now, but without it all our results of the type "almost everywhere" would be merely a play on words. To prove it, assume to the contrary that the interval (a, b) is of measure zero; then for any $\varepsilon > 0$, the closed interval $[a, b]$ can be covered by a sequence of intervals of total length $< \varepsilon$, and extending these intervals to the right and left (for example, by doubling them) we arrive at a sequence of intervals with the property that every point x of $[a, b]$ is interior to at least one of these intervals. According to the well-known theorem of BOREL, our sequence can be replaced by a finite number of its elements which still cover the interval $[a, b]$; from this it follows immediately that $2\varepsilon > b - a$, contrary to the hypothesis.

14. Darboux's Theorem

A second gap remains to be filled. At the beginning of the preceding section we based our argument, without proving it, on the existence of the lower and upper integrals of a bounded, but otherwise arbitrary, function and from that we deduced the integrability of the interval function (10). Now the existence theorem in question, the DARBOUX theorem, is proved in classical analysis books — among others, in the first chapter of CAMILLE JORDAN's text. However, we prefer to present it in the form of a general principle to which we shall have recourse several times in the sequel.

Assume that *the interval function* $f(\alpha, \beta)$ *is nondecreasing under decomposition*; this means that for $\alpha < \beta < \gamma$ one has

$$f(\alpha, \gamma) \leq f(\alpha, \beta) + f(\beta, \gamma)$$

or, equivalently, that the sums which serve to define the integral do not decrease when we insert a new division point. Alternately, we could have assumed $f(\alpha, \beta)$ to be nonincreasing in the analogous sense; in fact, this reduces to considering $-f(\alpha, \beta)$ instead of $f(\alpha, \beta)$.

It is necessary for us to make a second hypothesis, namely, *that the function* $f(\alpha, \beta)$ *be continuous everywhere.* By this we mean that for every point x, $f(\alpha, \beta)$ becomes infinitely small when the interval (α, β) contracts to the point x; i.e., for each x we can assign to every $\varepsilon > 0$ a $\delta = \delta(\varepsilon) > 0$ such that the hypothesis $\alpha \leq x \leq \beta$, $\beta - \alpha < \delta$ implies that $|f(\alpha, \beta)| < \varepsilon$.

Finally, we assume that *the sums* $\Sigma f(\alpha_{k-1}, \alpha_k)$ corresponding to all the possible decompositions of the interval (a, b) *possess a finite upper bound* L.

We shall prove the

THEOREM. *Under the three hypotheses made above, the function* $f(\alpha, \beta)$ *is integrable and its integral equals the least upper bound* L *of the sums* $\Sigma f(\alpha_{k-1}, \alpha_k)$.

To prove this theorem, consider a decomposition Δ_ε for which the corresponding sum Σ_ε surpasses the quantity $L - \frac{\varepsilon}{2}$; let ν be the number of the points of division which constitute Δ_ε. Choose δ such that for $\beta - \alpha < \delta$ we have $|f(\alpha, \beta)| < \frac{\varepsilon}{6\nu}$ about all these division points, and consider a decomposition Δ of (a, b) into segments of length less that δ; let the sum Σ correspond to Δ. Superposing the two decompositions Δ_ε and Δ, we obtain a decomposition Δ' and a sum Σ'. But since the interval function is nondecreasing under decomposition, it follows that $\Sigma' \geq \Sigma_\varepsilon > L - \frac{\varepsilon}{2}$. On the other hand, we can go from Δ to Δ' in ν steps by successively inserting each of the ν points of division which make up Δ_ε; at each step we increase the corresponding sum by at most $\frac{\varepsilon}{2\nu}$; hence $\Sigma' \leq \Sigma + \frac{\varepsilon}{2}$, and combining the two inequalities it follows that $\Sigma \geq L - \varepsilon$. Hence Σ approaches within ε of L, which was to be proved.

The lower integral of a bounded function $f(x)$ is obviously of the type considered, while the upper integral as well as the difference of the two integrals [the integral of the interval function (10)] are of the opposite type.

15. Functions of Bounded Variation and Rectification of Curves

Two other important concepts which when considered as integrals of interval functions fall into the type we have just dealt with are the total variation of functions of bounded variation and the length of rectifiable curves. Although the first is only a particular case of the second, it will be instructive to begin by considering it separately.

The total variation of a continuous function $f(x)$ of bounded variation, formed for an interval (a, b), is simply the integral over (a, b) of the interval function

(12) $$f(\alpha, \beta) = |f(\beta) - f(\alpha)|,$$

which is reflected in the generally adopted notation

$$T(x) = \int_a^x |df(x)|.$$

This integral is obviously of the type considered and if we apply our second fundamental theorem to it, it follows immediately that $T(x)$, the indefinite total variation of $f(x)$, has derivative $T'(x)$ equal almost everywhere, except for sign, to that of $f(x)$, that is, that $T'(x) = |f'(x)|$, as we have already seen above (Sec. 8), even for discontinuous functions.

We shall return to the case of discontinuous functions. First of all let us discuss *rectifiable curves*. Let the curve Γ be given by the parametric equations $x = x(t)$, $y = y(t)$, $z = z(t)$ $(a \leq t \leq b)$, where $x(t)$, $y(t)$ and $z(t)$ are continuous functions (we use 3-space rather than n-space in order to fix the ideas involved). Consider the inscribed polygonal lines, that is, the lines formed by decomposing the interval (a, b) into segments (t_{k-1}, t_k), where $k = 1, 2, \ldots, n$ and $t_0 = a$, $t_n = b$, and drawing the chords $P_{k-1}P_k$, where P_k denotes the point of the curve which corresponds to the value $t = t_k$. The curve Γ is said to be rectifiable if the length of the polygonal line $P_0 \ldots P_n$ does not surpass a certain finite bound, independent of the manner of decomposing the interval (a, b); the smallest of these bounds defines the length of the curve. Since the chords $P_{k-1}P_k$ are minorized by $|x(t_k) - x(t_{k-1})|$ and by the two other analogous quantities, and majorized by the sum of all three, we see immediately that a necessary and sufficient condition for the curve to be rectifiable is that the continuous functions $x(t)$, $y(t)$, $z(t)$ be of bounded variation. On the other hand, our general theorems assure us that this condition is equivalent to the condition that the interval function $f(\alpha, \beta)$, which is defined to be equal to the length of the chord joining those points $P(\alpha)$ and $P(\beta)$ which correspond to the values α and β of the parameter t, should be integrable, and that the length of Γ is simply the integral of this function taken from a to b. Moreover, denoting by $s(t)$ the length of the arc of our curve measured from a to t, our second fundamental theorem implies the most general extension of the classical formula

$$(s'(t))^2 = (x'(t))^2 + (y'(t))^2 + (z'(t))^2, \tag{13}$$

an extension given successively by LEBESGUE[13] and TONELLI[14] and valid, according to the latter, *for every rectifiable curve and almost everywhere with respect to the parameter t, which is arbitrary*. Choosing in particular the length s as parameter, we obtain

$$(x'(s))^2 + (y'(s))^2 + (z'(s))^2 = 1, \tag{14}$$

a relation which assures *the existence of a definite tangent almost everywhere with respect to s.*

The same results can be arrived at without recourse to the second fundamental theorem by applying the first theorem to the interval function $f(\alpha, \beta)$ which is defined to be the difference of the lengths of the arc and the chord joining the points $P(\alpha)$ and $P(\beta)$. This is obviously a non-negative function, non-increasing under decomposition and with integral 0. Relation (14) can then be written, using obvious notation, in the form

$$\frac{\overline{PR}}{\widehat{PQR}} \to 1$$

almost everywhere.

Let us return to the case of a single continuous function of bounded variation, say $x(t)$. It corresponds to the "curve" $x = x(t)$, $y = 0$, $z = 0$, which is obviously rectifiable, and the indefinite total variation

$$T(t) = \int_a^t |dx(t)|$$

coincides with the function $s(t)$. Hence relation (13) immediately yields that

$$T'(t) = |x'(t)|$$

almost everywhere, that is, that this fact, which was proved independently a

[13] LEBESGUE [*] (1st Edition, pp. 59–63, 125–129).
[14] TONELLI [**2**], LEBESGUE [*] (2nd Edition, pp. 198–201).

moment ago, is only a particular case of the generalization of the classical formula (13) which we have just discussed. This also suggests one of the methods of passing to discontinuous functions. For definiteness, let us first assume that $x(t)$ has only one discontinuity, say at $t = c$ $(a < c < b)$, and consider again the "curve" $x = x(t), y = 0, z = 0$, but now (in order that it not be discontinuous) completed by two segments, oriented so that the first goes from $x(c - 0)$ to $x(c)$ and the second from $x(c)$ to $x(c + 0)$. In the general case the same procedure must be repeated a finite number or a denumerable infinity of times in order to suppress all the points of discontinuity. Applying our results to the curve completed in this way, it follows easily that the relation $T'(t) = |x'(t)|$ almost everywhere remains valid for discontinuous functions of bounded variation.

The relation between functions of bounded variation and rectifiable curves is also made more precise, at least for plane curves $x = x(t)$, $y = y(t)$, if this plane is regarded as the plane of complex numbers $\zeta = x + iy$. In fact, the length of the curve is nothing else than the *total variation of the complex-valued function* $\zeta(t) = x(t) + iy(t)$, defined, just as in the real case, by the least upper bound of the sums

$$\sum |\zeta(t_k) - \zeta(t_{k-1})|.$$

CHAPTER II

THE LEBESGUE INTEGRAL

DEFINITION AND FUNDAMENTAL PROPERTIES

16. The Integral for Step Functions. Two Lemmas

During the second half of the last century, after CAUCHY and RIEMANN, numerous definitions of integral for bounded as well as for unbounded functions were successively proposed. But it was only at the beginning of the present century, in 1902, that HENRI LEBESGUE introduced, in his dissertation [**1**], a notion of integral that was to change the aspect of a great number of problems depending on integration. The reasons for such a change and, with them, the usefulness and beauty of the LEBESGUE theory, will be seen in the course of the following chapters; there is no point in speaking of them in advance.

In LEBESGUE's dissertation and in his lectures on integration given at the Collège de France, which followed it, the route leading to his results is still quite arduous; years passed by before his contemporaries became accustomed to the new methods. During these years mathematicians strove to obtain the results by easier methods; the approach was to replace the original definition by others which, as DE LA VALLÉE POUSSIN says, would permit the new theory to be included in the classical framework of mathematics where possible. In what follows we are going to discuss the theory starting from one of these definitions which is related to the idea of functional operator, whereas the original definition is related to the notion of measure. Afterward we shall give still other definitions, starting with that of Lebesgue; we shall compare them with one another and with the definition of which we shall speak now.

We start with the class of stepfunctions defined in a finite or infinite interval (a, b), that is, functions having a constant value c_k in each of a finite number of subintervals i_k of finite length $|i_k|$ and vanishing outside these intervals; as to the endpoints of these intervals, we can assign values to the functions there arbitrarily; in fact, in the following discussion we shall be permitted to omit sets of measure zero which cause us difficulty. We suppose the integral defined for these functions, as usual, by the sum

$$\sum c_k |i_k|,$$

and we shall extend this definition to functions of a much more general type by passage to the limit.

We note that we could choose the initial class to be the class of continuous functions or of functions integrable in the Riemann sense, vanishing, in the case where (a, b) is infinite, outside a finite interval which can vary with the function. We would arrive at the same final class with the same notion of integral with no substantial change in the details. We start with stepfunctions only to avoid assuming anything of the theory of integration. To distinguish them from other functions, we agree to denote them by Greek letters.

Our considerations will be based on two very simple lemmas.

LEMMA **A**. *For every sequence $\{\varphi_n(x)\}$ of step functions which decreases to 0 almost everywhere, the sequence of values of their integrals also tends to zero.*

LEMMA **B**. *If for an increasing sequence of stepfunctions $\{\varphi_n(x)\}$ the values of their integrals have a common bound, then the sequence $\{\varphi_n(x)\}$ tends almost everywhere to a finite limit.*

Proof of Lemma **A**. For arbitrary fixed $\varepsilon > 0$, we begin by enclosing the exceptional set E_0 of points of discontinuity of the functions φ_n in the interior of a finite number or of a denumerable sequence Σ_0 of intervals of total length less than ε. For the remaining points, $\varphi_n(x) \to 0$; therefore, to each of these points, say x_0, we can assign an n such that $\varphi_n(x_0) < \varepsilon$, and obviously $\varphi_n(x) < \varepsilon$ on the segment enclosing x_0 on which φ_n is constant. When we vary x_0, the set of these segments forms a second system of intervals, say Σ_1, to each of which is attached an index n. The two systems Σ_0 and Σ_1, taken together, cover the closed interval $[a_1, b_1]$ on whose complement $\varphi_1(x) = 0$, so we can apply Borel's theorem according to which a finite number of these segments suffice to cover the entire interval $[a_1, b_1]$. But those segments selected from Σ_0 have a total length less than ε, so their contribution to the integrals of $\varphi_n(x)$, for arbitrary n, does not exceed $M\varepsilon$ where M denotes a common bound of all the φ_n or, for definiteness, the maximum value of $\varphi_1(x)$. On the other hand, on each of the segments selected from Σ_1 one of the functions $\varphi_n(x)$ (and therefore all which follow) remains less than ε and consequently, denoting by N the largest of the indices n which appears, the function $\varphi_N(x)$ and all which follow are less than ε on each of these segments; therefore the contribution of these segments to the integral of either $\varphi_N(x)$ or any of the following functions does not exceed the bound $(b_1 - a_1)\varepsilon$. It follows that these integrals taken over the entire interval (a, b) remain less than $\varepsilon(M + b_1 - a_1)$ and become infinitely small with ε; this proves lemma **A**.

Proof of Lemma **B**. Assume given a nondecreasing sequence of stepfunctions $\{\varphi_n(x)\}$ and suppose their integrals over (a, b) remain less than some bound A. Without loss of generality we can also assume that the functions

φ_n are positive, otherwise we would have simply to consider the sequence $\{\varphi_n - \varphi_1\}$.

This assumed, the set E_0 of points x for which $\varphi_n(x)$ diverges is obviously enclosed in the set E_ε of points for which $\varphi_n(x) > A/\varepsilon$ from some point on. But the set E_ε is only a system of segments, the union of segments or systems of segments $\Sigma_{\varepsilon,n}$ for which $\varphi_n(x) > A/\varepsilon$, and the total length of each $\Sigma_{\varepsilon,n}$ multiplied by A/ε must be less than the integral of $\varphi_n(x)$ and hence less than A. Consequently these total lengths do not not exceed ε, and since

$$\Sigma_{\varepsilon,n} \subset \Sigma_{\varepsilon,n+1},$$

the same is true for the total length of the system E_ε. This means that for $\varepsilon > 0$, but otherwise arbitrary, we have enclosed the set E_0 in segments whose total length does not exceed ε, and consequently E_0 is of measure zero.

17. The Integral for Summable Functions

Denote the class of stepfunctions by C_0. Our two lemmas established, we can now extend the notion of integral to the *class C_1 of functions which are limits almost everywhere of the sequences $\{\varphi_n\}$ referred to in lemma* **B**.

In fact, according to that lemma, these limits $f(x) = \lim \varphi_n(x)$ exist almost everywhere, and moreover, according to the hypothesis, the integrals of $\varphi_n(x)$ have a common bound, and since they are increasing, converge to a finite limit. This suggests that we take this limit as the value of the integral of $f(x)$. As a formula,

$$\int_a^b f(x)dx = \lim_{n\to\infty} \int_a^b \varphi_n(x)dx.$$

But to justify this convention it must be shown that the limit in the second member does not depend on the particular choice of the functions φ_n, that is, it does not change when we replace the sequence $\{\varphi_n\}$ by another sequence $\{\psi_n\}$ of the same type, converging almost everywhere to the same function $f(x)$. More generally, we shall show that if the sequence $\{\psi_n\}$ increases almost everywhere to a limit function $g(x) \geq f(x)$, then we also have

$$J_2 = \lim_{n\to\infty} \int_a^b \psi_n(x)dx \geq \lim_{n\to\infty} \int_a^b \varphi_n(x)dx = J_1.$$

The case where $f(x) = g(x)$ almost everywhere is obviously implied by this more general fact; indeed, we have only to write our hypothesis in the form $g \geq f$ and $f \geq g$.

To verify our assertion, consider one of the functions φ, say $\varphi_m(x)$. Then by varying n, the positive part of $\varphi_m(x) - \psi_n(x)$ decreases to 0 almost everywhere and it follows by Lemma **A** that the integral of this positive part also tends to 0; therefore the integral of the difference itself, majorized by its positive part, tends to a negative limit or zero. This means that

$$\int_a^b \varphi_m(x)dx - J_2 \leq 0, \quad \int_a^b \varphi_m(x)dx \leq J_2,$$

and finally, letting m tend to infinity, it follows that $J_1 \leq J_2$, which was to be proved.

This done, we proceed to a more extensive new *class* C_2 by forming *the differences of functions belonging to the class* C_1. The integral of the difference $f_1 - f_2$ is defined by the formula

$$\int_a^b [f_1(x) - f_2(x)]dx = \int_a^b f_1(x)dx - \int_a^b f_2(x)dx.$$

To justify this convention it is necessary to show that if $f_1 - f_2 = g_1 - g_2$, the analogous equality is true for the respective integrals. But this statement can also be written in the form

$$\int_a^b f_1(x)dx + \int_a^b g_2(x)dx = \int_a^b f_2(x)dx + \int_a^b g_1(x)dx,$$

and on the other hand the hypothesis made is equivalent to $f_1 + g_2 = f_2 + g_1$. In this manner our assertion reduces to the additivity of the integral for functions belonging to the class C_1; but this additivity follows directly from the definition. Hence, our definition is legitimate.

For similar reasons the class C_2 is linear, that is, if $h_1(x)$ and $h_2(x)$ belong to C_2 then $c_1h_1(x) + c_2h_2(x)$ also belongs to C_2. Further, the integral is an additive interval function: if $a < b < c$ and $h(x)$ is integrable on (a, b) and (b, c), it is also integrable on (a, c) and conversely; the last integral is computed by adding the other two. This latter fact follows immediately from the same fact, which furthermore is obvious, concerning the class C_0.

The integral which we have just introduced possesses the important property that if $h(x)$ is integrable, or *summable* as Lebesgue called it, the same is true for its *modulus* $|h(x)|$ and its *positive* and *negative parts* $h^+(x)$ and $h^-(x)$. In fact, let us write $h = f_1 - f_2$, where f_1 and f_2 belong to the class C_1, and observe that their upper and lower envelopes, $\sup(f_1, f_2)$ and $\inf(f_1, f_2)$, are contained in the same class, which follows immediately from the analogous fact valid for the class C_0. But we have

$$|h| = \sup(f_1, f_2) - \inf(f_1, f_2),$$

$$h^+ = \sup(f_1, f_2) - f_2 = f_1 - \inf(f_1, f_2),$$

$$h^- = \sup(f_1, f_2) - f_1 = f_2 - \inf(f_1, f_2);$$

hence h, h^+ and h^- belong to the class C_2.

Another immediate consequence of our definition of the integral is that *for every summable function* $h(x)$ *there exists a sequence of step functions* $\varphi_n(x)$ *such that* $\varphi_n(x) \to f(x)$ *almost everywhere and*

$$\int_a^b |h(x) - \varphi_n(x)|\, dx \to 0 \qquad \textit{when } n \to \infty.$$

In fact, $h(x)$ is the difference of two functions, $f_1(x)$ and $f_2(x)$, belonging to the class C_1, and there exist increasing sequences of step functions $\varphi_{1n}(x)$, $\varphi_{2n}(x)$ tending almost everywhere to respectively $f_1(x)$ and $f(_2x)$. Setting $\varphi_n(x) = \varphi_{1n}(x) - \varphi_{2n}(x)$, we shall have

$$\int_a^b |h(x) - \varphi_n(x)|\, dx \leq \int_a^b [f_1(x) - \varphi_{1n}(x)]dx + \int_a^b [f_2(x) - \varphi_{2n}(x)]dx \to 0.$$

We want to establish the position of the functions *integrable in the Riemann sense* among those summable on the finite interval (a, b). Let us consider one of these Riemann-integrable functions $f(x)$ and form the sequence $\{\varphi_n(x)\}$, where φ_n is defined by decomposing the interval (a, b) into 2^n equal parts and taking for the constant value of φ_n on each segment the greatest lower bound of $f(x)$ on that segment. Since $f(x)$ is continuous almost everywhere, the increasing sequence $\{\varphi_n\}$ converges almost everywhere to $f(x)$; integrated term by term, it yields on the one hand the integral of $f(x)$ taken in the sense which we have just introduced, and on the other hand its lower integral in the Darboux sense, hence its integral in the Riemann sense also. That is to say, *the integrals coincide*; moreover $f(x)$ belongs to our class C_1, and for the same reasons the same is true of $-f(x)$. Conversely, when $f(x)$ as well as $-f(x)$ belong to the class C_1, the function $f(x)$ coincides almost everywhere with a function which is integrable in the Riemann sense.

We leave it up to the reader to construct functions of class C_2 which do not belong to the class C_1; by varying the examples one sees clearly that the notion of integral we have just considered is much more general than that of Riemann.

18. Term-by-Term Integration of an Increasing Sequence (Beppo Levi's Theorem)

Now let us pass to one of the most remarkable and surprizing facts of our theory. One would be tempted to apply the same procedure to the class C_1 or to C_2, hoping thus to define the integral for a larger class. We shall see that our procedure does not allow us to leave C_2 or, in other words, that our class is closed with respect to our procedure, and that moreover the sequences in question can be integrated term by term.

Let us begin by considering the increasing sequence $\{f_n\}$ whose elements are taken from C_1. We assume that

$$\int_a^b f_n(x)dx \leqq A$$

for all n. For each n let $\{\varphi_{nk}\}$ be an increasing sequence of step functions which converges almost everywhere to f_n. Set

$$\varphi_n = \sup_{i \leqq n} \{\varphi_{in}\}.$$

The step functions φ_n obviously form an increasing sequence and since

$$\varphi_{in} \leqq f_i \leqq f_n \quad \text{for} \quad i \leqq n,$$

we also have

$$\varphi_n \leqq f_n, \quad \text{hence} \quad \int_a^b \varphi_n \leqq \int_a^b f_n \leqq A;$$

consequently the sequence $\{\varphi_n\}$ converges almost everywhere to a limit $f(x)$. Since on the other hand

$$\varphi_k \geqq \varphi_{nk} \quad \text{for} \quad n \leqq k,$$

we also have, for $k \to \infty$,

$$f \geqq f_n.$$

Hence the $f_n(x)$, caught between the $\varphi_n(x)$ and their limit $f(x)$, have the same limit, and this limit obviously belongs to the class C_1. Moreover, since the integrals of the functions φ_n, f_n and f follow respectively in the same order, and since

$$\int_a^b \varphi_n(x)dx \to \int_a^b f(x)dx,$$

it follows that we also have

$$\int_a^b f_n(x)dx \to \int_a^b f(x)dx.$$

Let us now consider, more generally, an increasing sequence $\{h_n(x)\}$ of functions belonging to the class C_2, or, what amounts to the same thing, let us consider the series

$$h_1(x) + \Sigma k_n(x),$$

where we have set

$$k_n(x) = h_{n+1}(x) - h_n(x).$$

It thus is a question of the convergence and the integration of the series $\Sigma k_n(x)$, formed from non-negative functions belonging to the class C_2. But the hypothesis

$$\int_a^b h_n(x)dx \leq A$$

implies that

$$(1) \qquad \sum_1^\infty \int_a^b k_n(x)dx \leq A + \int_a^b |h_1(x)|\, dx = B;$$

hence the series in the first member is convergent. Set $k_n(x) = f_n(x) - g_n(x)$, where f_n and g_n are assumed to belong to the class C_1 and to be non-negative and where, furthermore,

$$\int_a^b g_n(x)dx < \frac{1}{2^n}.$$

In order to so choose f_n and g_n, we have only to start from a decomposition $k_n = f_n - g_n$ for which our inequality is not necessarily verified and to consider an increasing sequence of stepfunctions $\{\psi_k\}$ converging almost everywhere to g_n; we shall select from it a term of sufficiently high index so that

$$\int_a^b g_n(x)dx - \int_a^b \psi_{k_0}(x)dx < \frac{1}{2^n};$$

we then have only to replace f_n and g_n by respectively $f_n - \psi_{k_0}$ and $g_n - \psi_{k_0}$, and our last condition will be fulfilled. But this last condition assures the convergence of the series

$$\sum_1^\infty \int_a^b g_n(x)dx$$

and, by virtue of (1), that of

$$\sum_1^\infty \int_a^b f_n(x)dx$$

also. Hence the general case reduces to the particular case which we have just considered; in fact, we have only to apply the result obtained to the sequences formed from the partial sums of the series Σf_n and Σg_n; these partial sums obviously belong to the class C_1 since their individual terms do.

Thus, we have proved

BEPPO LEVI'S THEOREM.[1] *Every increasing sequence $\{h_n(x)\}$ of summable functions whose integrals on the interval (a, b) have a common bound converges almost everywhere to a summable function, and integration can be carried out term by term, that is, the order of taking the limit and integrating can be reversed.*

[1] B. LEVI [**1**].

A second form of the same theorem — in reality it is in this form that we have just proved it — concerns the integration of series with non-negative terms. But we shall state an immediate generalization:

Every series

$$\sum k_n(x)$$

of summable functions for which

$$\sum \int_a^b |k_n(x)|\, dx$$

converges, converges itself almost everywhere to a summable function, and the series can be integrated term by term.

The proof is obvious; it is carried out, for example, by considering separately the two series formed of respectively the positive and negative parts of the $k_n(x)$.

Here is a corollary of the last theorem:

The integral of $|k(x)|$ cannot be zero unless the function $k(x)$ is itself zero almost everywhere.

To obtain this corollary we have only to set $k_n(x) = k(x)$, $n = 1, 2, \ldots$.

Another corollary of our theorem which follows from it in an obvious way is:

When a sequence of summable functions is monotonic and tends to a summable function, it is then permissible to integrate term by term.

19. Term-by-Term Integration of a Majorized Sequence (Lebesgue's Theorem)

We could try to enlarge still further the notion of integral by using certain sequences or series of a more general type than those we have just considered. We could try, for instance, to assume nothing at all about the sequence other than that it converges almost everywhere. But the examples $f_n(x) = nx^n$, $g_n(x) = n^2x^n$ $(0 \leq x \leq 1)$, which converge to 0 except for $x = 1$ and for which the integral converges to 1 in the first example and increases indefinitely in the second, show that in order to assure term-by-term integrability it is necessary in any case to take certain precautions. One such precaution — with important applications — consists in assuming that $|f_n(x)| \leq g(x)$, where the majorant function $g(x)$ is assumed summable. We shall show that under this condition term-by-term integration of the relation $f_n \to f$ almost everywhere is justified, and that moreover our procedure does not take us outside the class C_2.

Let us note first of all that the upper envelope

$$g_1(x) = \sup\,(f_1(x),\ f_2(x),\ \ldots)$$

is summable: this is true for $\sup (f_1(x), f_2(x)) = [f_1(x) - f_2(x)]^+ + f_2(x)$, which is the sum of two summable functions, it follows by induction that the same is true for the envelopes $\sup (f_1, f_2, \ldots, f_m)$, and finally, since the integral of these functions remains less than those of g, i.e., below a bound independent of m, and since, moreover, these functions form an increasing sequence, their limit $g_1(x)$ is also summable. The same is true, for the same reasons, of the functions

$$g_n(x) = \sup (f_n(x), f_{n+1}(x), \ldots).$$

The g_n form a decreasing sequence which converges almost everywhere to the same limit function f as the f_n.

Operating in the same manner with the lower envelopes (or, what amounts to the same thing, with the upper envelopes of the $-f_n$), we arrive at the increasing sequence formed by the envelopes

$$h_n(x) = \inf (f_n(x), f_{n+1}(x), \ldots),$$

which also converges almost everywhere to the function $f(x)$.

Consequently $f(x)$ is summable and

$$\int_a^b g_n(x)dx \to \int_a^b f(x)dx, \quad \int_a^b h_n(x)dx \to \int_a^b f(x)dx.$$

Finally, since

$$h_n(x) \leq f_n(x) \leq g_n(x)$$

and therefore

$$\int_a^b h_n(x)dx \leq \int_a^b f_n(x)dx \leq \int_a^b g_n(x)dx,$$

it follows that

$$\int_a^b f_n(x)dx \to \int_a^b (fx)dx,$$

which was to be proved.

Let us state our result:

LEBESGUE'S THEOREM. *If the functions $f_n(x)$, assumed summable in the interval (a, b), converge almost everywhere to a function $f(x)$ and if furthermore there exists a summable function $g(x)$ such that*

$$|f_n(x)| \leq g(x)$$

for all n, then the function $f(x)$ is also summable and

$$\int_a^b f_n(x)dx \to \int_a^b f(x)dx.$$

Of course, LEBESGUE[2] proved this theorem starting with his definition, which we still have to show is equivalent to ours. The Lebesgue theorem and that of Beppo Levi are essentially equivalent; in other presentations of the theory they are deduced in the reverse order.

It is appropriate to mention a particular case of the theorem we have just proved, the case which is generally quoted and which corresponds to the hypothesis that $g(x)$ is a constant, that is, that the functions $f_n(x)$ *all remain bounded*. But this "little" theorem of Lebesgue is valid only in the case of a finite interval (a, b), since a constant different from 0 is not summable on an infinite interval.

It is also fitting to note several still more specialized cases of the theorem which are predecessors of it, belonging to the classical theory; they are the theorems of C. ARZELÀ[3] and of W. F. OSGOOD[4]. The first has to do with bounded sequences of integrable functions in the Riemann sense, the second with those of continuous functions. Compared with the Lebesgue theorem, of which they are corollaries, the essential difference is that in these particular theorems integrability or the continuity of the limit functions is not a consequence of the other hypotheses, but must be postulated separately. Moreover the theorem of Arzelà, of the year 1885, remained almost unobserved until it was rediscovered independently, in 1897, by Osgood, who stated it only for continuous functions.

20. Theorems Affirming the Integrability of a Limit Function

Here is another corollary of the Lebesgue theorem, but one which affirms only the integrability of the limit function and says nothing about term-by-term integration.

THEOREM. *When the functions $f_n(x)$ are summable in the interval (a, b) and converge almost everywhere to a function $f(x)$ such that*

$$|f(x)| \leq g(x),$$

where $g(x)$ is a summable function, then the function $f(x)$ is also summable.

In fact, we have only to apply the Lebesgue theorem to the sequence

$$\inf\,[g(x), \sup\,(f_n(x), -g(x))] \to f(x),$$

or, in other words, we have only to truncate the $f_n(x)$ above and below by respectively $g(x)$ and $-g(x)$, that is, to replace the $f_n(x)$, everywhere where

[2] LEBESGUE [**2**] (in particular p. 375).
[3] ARZELÀ [**1**] and [**3**] (in particular pp. 723–724).
[4] OSGOOD [**1**] (in particular pp. 183–189).

$|f_n(x)| > g(x)$, by $\pm g(x)$ according to its sign; by the hypothesis we made this does not modify the limit $f(x)$.

Here is an important corollary of this theorem, affirming the summability of *composite* functions.

Let $g(u_1, u_2, \ldots, u_r)$ be a continuous function in the space $(u_1, u_2, \ldots, u_r)$ and let $f_1(x), f_2(x), \ldots, f_r(x)$ be summable functions in (a, b). Furthermore, let us assume that there exists a function $h(x)$, summable in (a, b), such that we have

$$|g(f_1(x), f_2(x), \ldots, f_r(x))| \leq h(x).$$

Under these hypotheses, the function

$$G(x) = g(f_1(x), f_2(x), \ldots, f_r(x))$$

is summable in (a, b).

To prove it, we attach to each $f_k(x)$ a sequence $\{\varphi_{kn}(x)\}$ of step functions converging almost everywhere to $f_k(x)$. The existence of sequences of this type results immediately from the definition of summable functions. Then, by virtue of the continuity of the function g, we have almost everywhere

$$G_n(x) = g(\varphi_{1n}(x), \varphi_{2n}(x), \ldots, \varphi_{rn}(x)) \to G(x). \tag{2}$$

But, at least if the interval (a, b) is finite, the functions $G_n(x)$ are step functions, hence summable; and since by hypothesis we have $|G(x)| \leq h(x)$ with $h(x)$ summable, the preceding theorem applies to the sequence (2), and it follows that the function $G(x)$ is summable. The same reasoning applies also to the case of an infinite interval (a, b); we have only to modify the functions $G_n(x)$ by setting $G_n(x) = 0$ outside the interval $(-n, n)$.

Along the same lines is the theorem due to FATOU [**1**], which is important in applications. If affirms, as do the preceding theorems, the summability of the limit function, while with respect to term-by-term integration it gives only an approximation rather than a precise value. It is generally cited under the name of

FATOU'S LEMMA. *If the functions $f_n(x)$, non-negative and summable in (a, b), tend almost everywhere to a function $f(x)$, and if furthermore the sequence of values*

$$\int_a^b f_n(x)dx$$

is bounded, then the function $f(x)$ is summable and

$$\int_a^b f(x)dx \leq \liminf \int_a^b f_n(x)dx.$$

The proof of this lemma is nothing more, so to say, than the second half of the proof of Lebesgue's theorem. In fact, consider the functions

$$h_n(x) = \inf (f_n(x), f_{n+1}(x), \ldots),$$

forming an increasing sequence which converges almost everywhere to f. Since $h_n \leq f_{n+k}$, we also have

$$\int_a^b h_n(x)dx \leq \int_a^b f_{n+k}(x)dx \qquad (k = 0, 1, 2, \ldots)$$

and for $k \to \infty$, it follows that

$$\int_a^b h_n(x)dx \leq \liminf \int_a^b f_{n+k}(x)dx = \liminf \int_a^b f_k(x)dx.$$

Finally, making use of B. Levi's theorem, we deduce the summability of $f(x) = \lim h_n(x)$ and the relation

$$\int_a^b f(x)dx = \lim \int_a^b h_n(x)dx \leq \liminf \int_a^b f_k(x)dx$$

which was to be proved.

Fatou's lemma can also be stated in the following equivalent form.

If the functions $f_n(x)$, non-negative and summable in (a, b), tend almost everywhere to a function $f(x)$, and if furthermore

$$\int_a^b f_n(x)dx \leq A,$$

then the function $f(x)$ is summable and

$$\int_a^b f(x)dx \leq A.$$

21. The Schwarz, Hölder and Minkowski Inequalities

Let us return to the theorem on composite functions and point out its simplest particular cases, corresponding to particular choices of the function $g(u_1, u_2, \ldots, u_r)$. We have already encountered the case where $g(u_1, u_2) = c_1u_1 + c_2u_2$, that is, the integration of $c_1f_1 + c_2f_2$; we have also encountered $g(u) = |u|$, $g(u) = u^+$ and $g(u) = u^-$, and $g(u_1, u_2, \ldots, u_r) = \sup (u_1, u_2, \ldots, u_r)$. In all these cases there is no need for a supplementary hypothesis, since the role of the majorant $h(x)$ which appears in the theorem is played by respectively $|f|$, $C(|f_1| + |f_2|)$ and $|f_1| + |f_2| + \ldots + |f_r|$.

But this is not true for $g(u) = u^2$ or $g(u_1, u_2) = u_1u_2$, that is, for the integration of the square or the product, which are of fundamental importance in applications. The summability of f^2 must be assured either by postulating it expressly or by the existence of a summable majorant $g \geq f^2$. The summability of the product can be guaranteed by diverse hypotheses, of which two are of particular importance.

Firstly, when one of the two factors, say f_1, is bounded, $|f_1| \leq C$, and the other, f_2, is summable, then the product $f_1 f_2$ is summable and

$$\left|\int_a^b f_1(x) f_2(x) dx\right| \leq C \int_a^b |f_2(x)|\, dx.$$

This follows immediately from the fact that the summable function $C|f_2|$ is a majorant of the product $|f_1 f_2|$.

Secondly, when f_1 and f_2 are summable as well as their squares, then $f_1^2 + f_2^2$ is a majorant of $|f_1 f_2|$ and consequently the latter is summable. More precisely, we have

$$2|f_1 f_2| \leq \lambda f_1^2 + \frac{1}{\lambda} f_2^2,$$

therefore

$$2\left|\int_a^b f_1(x) f_2(x) dx\right| \leq \lambda \int_a^b f_1^2 (x) dx + \frac{1}{\lambda} \int_a^b f_2^2 x(dx),$$

where $\lambda > 0$ but is otherwise arbitrary; in particular, we can choose λ so that the two terms of the second member become equal, and then by direct calculation obtain the inequality

$$\left[\int_a^b f_1(x) f_2(x) dx\right]^2 \leq \int_a^b f_1^2(x) dx \int_a^b f_2^2(x) dx,$$

called *Schwarz's*[5] *inequality* — an extension to integrals of the well-known *Cauchy*[6] *inequality*

$$\left[\sum a_k b_k\right]^2 \leq \sum a_k^2 \sum b_k^2.$$

We should remark that in the particular case where f_1, f_2, or both are zero almost everywhere, just as in the case where all the a_k or all the b_k are zero, our calculation does not apply immediately, but the inequalities are obvious, or more precisely, the equality sign holds. In the general case, our calculation makes it clear immediately that the equality sign is valid only if f_1 and f_2 differ by at most a numerical factor (almost everywhere).

The following is a more general inequality, known by the name *Hölder's*[7] *inequality*. Consider two functions $f_1(x)$, $f_2(x)$, and assume that f_1, f_2, and their powers $|f_1|^p$ and $|f_2|^q$ are all summable, where

$$p > 1,\ q > 1 \text{ and } \frac{1}{p} + \frac{1}{q} = 1,$$

[5] After H. A. Schwarz; but it had already been stated (for classical integrals) by Bunyakovsky, see Hardy–Littlewood–Pólya [*] (p. 132–133).

[6] See Hardy–Littlewood–Pólya [*] (p. 16).

[7] O. Hölder proved the analogous inequality for series; the extension to integrals is due to F. Riesz; cf. Hardy–Littlewood–Pólya [*] (pp. 21–26, 146–150).

that is, $q = \frac{p}{p-1}$. The inequality in question can be written

$$(3) \qquad \left|\int_a^b f_1(x)f_2(x)dx\right| \leq \left[\int_a^b |f_1(x)|^p\,dx\right]^{\frac{1}{p}} \left[\int_a^b |f_2(x)|^q\,dx\right]^{\frac{1}{q}},$$

where the first member always has meaning when the second has meaning.

Instead of imitating the calculation we just made above, we prefer for reasons of clarity to modify the nomenclature, at least for the moment, by replacing $|f_1|^p$ by f, $|f_2|^q$ by g, and $\frac{1}{p}$ and $\frac{1}{q}$ by respectively α and $\beta = 1 - \alpha$. Without loss of generality, we can also assume that the functions f_1, f_2 are non-negative, hence that $f_1 = f^\alpha$, $f_2 = g^\beta$. Assume further, for the moment, that

$$(4) \qquad \int_a^b f(x)dx = \int_a^b g(x)dx = 1.$$

Having done this, we start with the inequality

$$t^\alpha \leq \alpha t + \beta = \alpha t + 1 - \alpha,$$

known from the elements of differential calculus (and expressing among other things that the tangent to the concave curve $y = x^\alpha$ at the point $x = 1$ lies above the curve), which can also be written, by replacing t by $\frac{t}{v}$, as

$$t^\alpha v^\beta \leq \alpha t + \beta v.$$

Inserting $f(x)$ for t and $g(x)$ for v in the last inequality and integrating, it follows that

$$\int_a^b f^\alpha(x)g^\beta(x)dx \leq \alpha + \beta = 1.$$

Finally, in order to dispense with hypothesis (4) we have only to "normalize" f and g, that is, to divide them by their respective integrals, which yields, after having multiplied by the respective powers of these integrals,

$$\int_a^b f^\alpha(x)g^\beta(x)dx \leq \left(\int_a^b f(x)dx\right)^\alpha \left(\int_a^b g(x)dx\right)^\beta,$$

from which, returning to the original nomenclature, we deduce inequality (3) immediately. However, it must be admitted that we have silently passed over the possibility that one or the other of the functions f and g or both are zero almost everywhere. But in this case the product in question is zero almost everywhere and our inequality becomes obvious. In the general case, it follows immediately from our calculation that the equality sign is valid only if $f(x) = \lambda g(x)$ almost everywhere. As to inequality (3), we deduce for it that the equality sign is valid only if $|f_1|^p$ and $|f_2|^q$ differ by at most a numerical factor almost everywhere and in addition the product $f_1(x)f_2(x)$ is almost everywhere of constant sign.

From Hölder's inequality there follows immediately another, that of *Minkowski*[8], which we shall make use of in the sequel. Let $f_1(x)$ and $f_2(x)$ be two non-negative summable functions whose p-th powers, f_1^p and f_2^p, are also summable ($p > 1$); then according to Hölder's inequality,

$$\int_a^b (f_1 + f_2)^p dx = \int_a^b f_1(f_1 + f_2)^{p-1} dx + \int_a^b f_2(f_1 + f_2)^{p-1} dx \leq$$

$$\leq [\int_a^b f_1^p dx]^{\frac{1}{p}} [\int_a^b (f_1 + f_2)^p dx]^{\frac{p-1}{p}} + [\int_a^b f_2^p dx]^{\frac{1}{p}} [\int_a^b (f_1 + f_2)^p dx]^{\frac{p-1}{p}}$$

and consequently, dividing by the last factor,

$$[\int_a^b (f_1 + f_2)^p dx]^{\frac{1}{p}} \leq [\int_a^b f_1^p dx]^{\frac{1}{p}} + [\int_a^b f_2^p dx]^{\frac{1}{p}};$$

a fortiori, the inequality remains valid for functions of arbitrary sign:

$$[\int_a^b |f_1 + f_2|^p dx]^{\frac{1}{p}} \leq [\int_a^b |f_1|^p dx]^{\frac{1}{p}} + [\int_a^b |f_2|^p dx]^{\frac{1}{p}}.$$

To complete the calculation, it is necessary to add that the existence of the integral of $(f_1 + f_2)^p$ which we have employed is obtained starting from the theorem on composite functions, Section 20, and from the relation

$$(f_1 + f_2)^p \leq 2^p \sup (f_1^p, f_2^p).$$

The question of the validity of the equality sign in the Minkowski inequality is again easy to discuss; we leave this discussion to the reader.

Finally, we observe that we shall have to deal in the sequel with functions taking on *complex* values. The extension of our results to these functions is evident. Another extension which we shall sometimes need occurs when *the interval of integration is infinite*. For definiteness, let us suppose that we have a function $f(x)$ defined on the interval $(-\infty, \infty)$ and we wish to integrate $|f|^2$. Then, even if $f(x) \geq 0$, this integral can exist without the integral of $f(x)$ existing, and instead of assuming f and f^2 summable, we can assume f^2 summable and f itself "measurable"; the meaning of this last expression will be made precise in the sequel.

22. Measurable Sets and Measurable Functions

We have introduced the integral for a certain class of functions defined in (a, b), the so-called *summable* functions. These functions are the limits almost

[8] MINKOWSKI proved the analogous inequality for finite sums; the extension to integrals is due to F. RIESZ; cf. HARDY–LITTLEWOOD–PÓLYA [*] (pp. 30–32, 146–150).

everywhere of sequences of step functions, but the converse is not true: there obviously are sequences of step functions which converge almost everywhere but whose limits are not summable. Let us agree to call every function which is *the limit almost everywhere of a sequence of step functions* a *measurable function.*

Summable functions are therefore measurable; the constant function $f(x) = c$ is measurable even on an infinite interval (a, b). By the first theorem of Section 20, every measurable function which is majorized in absolute value by a summable function is also summable. More generally, by truncating a measurable function $f(x)$ above and below by the summable functions $g(x)$ and $h(x)$ such that $g(x) < h(x)$, *i. e.*, by forming the function which is equal to $f(x)$ for points where $g(x) \leq f(x) \leq h(x)$, to $g(x)$ where $f(x) \leq g(x)$, and to $h(x)$ where $h(x) \leq f(x)$, we always obtain a summable function. If in the particular case of a finite interval we choose constants for $g(x)$ and $h(x)$, the function $f(x; c, d)$ which we obtain — the result of truncating $f(x)$ below at c and above at d — is summable.

It follows immediately from the definition that *the absolute value of a measurable function, and the sum, difference, product, and lower and upper envelopes of two measurable functions are likewise measurable.* The same holds for the *inverse* of a measurable function provided that $f(x) \neq 0$ almost everywhere. In fact, if $f(x)$ is the limit almost everywhere of the sequence $\varphi_n(x)$ of step functions, then its inverse $1/f(x)$ will be the limit almost everywhere of the sequence $\psi_n(x)$ of step functions which are defined as follows:

$$\psi_n(x) = 0 \text{ where } \varphi_n(x) = 0, \quad \psi_n(x) = 1/\varphi_n(x) \text{ where } \varphi_n(x) \neq 0.$$

But there are other operations which do not take us out of the class of measurable functions, for example, passage to the limit: the limit of a sequence of measurable functions which converges almost everywhere is itself a measurable function. To prove this, choose a strictly positive summable function $h(x)$; in the case of a finite interval (a, b) we can choose $h(x) = 1$, in the case of an infinite interval we can take, for example, the function which is equal to $1/n^2$ in the interval $n - 1 < x \leq n$, $(n = 0, \pm 1, \pm 2, \ldots)$. This done, if the functions $f_n(x)$ are measurable and tend almost everywhere to the function $f(x)$, then

$$g_n(x) = \frac{h(x) f_n(x)}{h(x) + |f_n(x)|} \to \frac{h(x) f(x)}{h(x) + |f(x)|} = g(x),$$

the functions $g_n(x)$ are also measurable, and

$$|g_n(x)| < h(x), \quad |g(x)| < h(x).$$

Consequently the functions $g_n(x)$ are summable, and by Lebesgue's theorem their limit $g(x)$ is also summable. By virtue of the relation

$$f(x) = \frac{h(x)g(x)}{h(x) - |g(x)|}$$

$f(x)$ is also measurable, q.e.d.

After having defined measurable function, a *measurable set* is defined to be a set whose characteristic function is measurable, where the characteristic function $e(x)$ of a set e is defined to be equal to 1 or to 0 according as x belongs to e or not. For a measurable set e, the measure $m(e)$ will be defined to be the value of the integral of the characteristic function $e(x)$ if the latter is summable, and to be ∞ if $e(x)$ is measurable without being summable.

It is obvious that every interval is a measurable set and that its measure equals the length of the interval (finite or infinite). It is also clear that the integral defining the measure $m(e)$ can be taken over an arbitrary interval (a, b) enclosing e.

A set of measure zero obviously has 0 for its measure. The converse is also true; it is contained in the corollary to B. Levi's theorem, Section 18.

The difference $e_1 - e_2$ of two measurable sets e_1, e_2 ($e_2 \subset e_1$), as well as the intersection $\bigcap e_k$ and the union $\bigcup e_k$ of a finite number or of a denumerable infinity of measurable sets are also measurable. All this follows immediately from what we have just said about operations which do not take us out of the class of measurable functions, if we note that the sets

$$\text{a)}\quad e_1 - e_2, \qquad \text{b)}\quad \bigcap_{k=1}^{n} e_k, \qquad \text{c)}\quad \bigcap_{k=1}^{\infty} e_k \qquad \text{d)}\quad \bigcup_{k=1}^{n} e_k, \qquad \text{e)}\quad \bigcup_{k=1}^{\infty} e_k$$

have, respectively, the characteristic functions

$$\text{a)}\quad e_1(x) - e_2(x), \quad \text{b)}\quad \prod_{k=1}^{n} e_k(x) = \inf_{1 \leqq k \leqq n} \{e_k(x)\}, \quad \text{c)}\quad \prod_{k=1}^{\infty} e_k(x) = \lim_{n \to \infty} \prod_{k=1}^{n} e_k(x),$$

$$\text{d)}\quad 1 - \prod_{k=1}^{n} (1 - e_k(x)) = \sup_{1 \leqq k \leqq n} \{e_k(x)\}, \quad \text{e)}\quad \lim_{n \to \infty} [\sup_{1 \leqq k \leqq n} \{e_k(x)\}].$$

The measure $m(e)$ is a "denumerably additive" function of the measurable set e, which means that if we form the union of a finite number or of a denumerable infinity of mutually disjoint (i. e., without common points) measurable sets $e_1, e_2, \ldots$, we have

$$m(e_1 \cup e_2 \cup \ldots) = m(e_1) + m(e_2) + \ldots.$$

In fact, since the set $e = \bigcup e_k$ then has $e(x) = \Sigma\, e_k(x)$ for characteristic function, and since $e(x)$ is obviously greater than all the terms and all the partial sums of the series appearing in the second member, it follows from the convergence theorems of Lebesgue and B. Levi (Sections 18—19) that the function $e(x)$ is summable if and only if all the terms of the series

$\Sigma e_k(x)$ are summable functions and the series formed of the integrals converges, and it follows that in this case it is permissible to integrate term by term. This proves our proposition.

The fact we have just proved is obviously equivalent to the following: *If $e_1, e_2, \ldots$ is an increasing sequence of measurable sets, we have*

$$m(\bigcup_{k=1}^{\infty} e_k) = \lim_{k\to\infty} m(e_k).$$

In an analogous manner, *if $e_1, e_2, \ldots$ is a decreasing sequence of measurable sets of finite measure, we have*

$$m(\bigcap_{k=1}^{\infty} e_k) = \lim_{k\to\infty} m(e_k).$$

The last statement follows immediately from Lebesgue's theorem if we apply it to the sequence of corresponding characteristic functions $\{e_k(x)\}$ which are summable and tend decreasingly to the characteristic function of the set product.

There exists a relation between sets and measurable functions which plays a very deep role in the theory of LEBESGUE. The relation is that for $f(x)$ measurable and for c arbitrary, the sets where

$$f(x) \leq c,\ f(x) < c,\ f(x) \geq c,\ f(x) > c$$

are measurable and that, conversely, if this is the case, even if it is assumed for only one of these inequalities, the function $f(x)$ is measurable. It will suffice to prove the first part of this assertion, without insisting, for the moment, on the converse. To achieve this we have only to consider the quotient

$$\frac{f(x; c+h, \infty) - f(x; c, \infty)}{h},$$

where $f(x; c, \infty)$ denotes the function $f(x)$ truncated below by c; since the two terms of the numerator are measurable functions, the same is true for the entire expression and for its limit for $0 < h \to 0$. But obviously this limit becomes zero everywhere where $(fx) > c$ and it is equal to 1 elsewhere; hence it is the characteristic function of the set for which $f(x) \leq c$; it follows that this set is measurable. The same limit, formed with negative h's, yields the characteristic function of the second set, and the two other sets follow from these by replacing f by $-f$ and c by $-c$.

INDEFINITE INTEGRALS
ABSOLUTELY CONTINUOUS FUNCTIONS

23. The Total Variation and the Derivative of the Indefinite Integral

Let $f(x)$ be a summable function; its indefinite integral (written as usual up to an additive constant)

$$F(x) = \int_a^x f(t)dt$$

is of bounded variation. This is obvious if $f(x)$ is non-negative, for then $F(x)$ is nondecreasing; the general case follows from this by decomposing $f(x)$ into positive and negative parts.

What is the total variation T of $F(x)$ over (a, b)?

Consider, for this purpose, a stepfunction $\varepsilon(x)$ taking on constant values ε_k ($|\varepsilon_k| \leq 1$) in the successive intervals (x_{k-1}, x_k); we have

$$\int_a^b \varepsilon(x)f(x)dx = \sum_k \varepsilon_k \int_{x_{k-1}}^{x_k} f(x)dx = \sum_x \varepsilon_k[F(x_k) - F(x_{k-1})].$$

The last member does not exceed

$$\sum_k |F(x_k) - F(x_{k-1})|$$

and therefore it does not exceed T, but it approaches indefinitely close to T if we let the function $\varepsilon(x)$ vary so that the maximum length of the intervals (x_{k-1}, x_k) tends to zero and that ε_k is always equal to $+1$, 0, or -1 according as $F(x_k) - F(x_{k-1})$ is > 0, $= 0$, or < 0. Hence we have

$$T = \sup_{|\varepsilon(x)| \leq 1} \int_a^b \varepsilon(x)f(x)dx \leq \int_a^b |f(x)|\, dx.$$

Now it is actually the equality sign which is valid here. In fact, let $\{\varphi_n(x)\}$ be a sequence of step functions tending almost everywhere to $f(x)$. Let $\varepsilon_n(x)$ equal the function $n\varphi_n(x)$, truncated above by 1 and below by -1. We have, for n approaching infinity, $\lim \varepsilon_n(x) = 1$ almost everywhere that $f(x) > 0$, and $\lim \varepsilon_n(x) = -1$ almost everywhere that $f(x) < 0$, hence almost everywhere

$$\lim \varepsilon_n(x)f(x) = |f(x)|.$$

Since on the other hand

$$|\varepsilon_n(x)f(x)| \leq |f(x)|,$$

we have

$$\lim \int_a^b \varepsilon_n(x)f(x)dx = \int_a^b |f(x)|\, dx.$$

Summarizing, we have the

THEOREM. *The indefinite integral $F(x)$ of the summable function $f(x)$ is of bounded variation; its total variation in the interval (a, b) is equal to*

$$T = \int_a^b |f(x)|\, dx.$$

Being of bounded variation, the function $F(x)$ possesses a derivative $F'(x)$ almost everywhere. The classical case in which $f(x)$ is continuous suggests the question of whether the relation $F'(x) = f(x)$ also exists in the general case.

It obviously suffices to examine the case in which $f(x)$ is of class C_1, that is, the limit of an increasing sequence of step functions $\varphi_n(x)$. But the affirmative answer for the φ_n is obvious, and in order to arrive at the same answer for $f(x)$, we have only to note that the integral $F(x)$ is furnished by the series

$$\Phi_1(x) + \sum_1^\infty (\Phi_{k+1}(x) - \Phi_k(x)),$$

formed from the integrals of the functions $\varphi_n(x)$, and then to apply the Fubini theorem for term-by-term differentiation.

Summarizing, we have the

THEOREM. *Every summable function is equal almost everywhere to the derivative of its indefinite integral.*

24. Example of a Monotonic Continuous Function Whose Derivative is Zero Almost Everywhere

The question arises: to what extent does the fact we have just established have a converse? Is is true that every function $F(x)$ is the integral of its derivative? Or, to content ourselves with a less profound question, is it true that it is determined by its derivative up to an additive constant? Now we are not thinking here of the classical theorems; in the order of the ideas which we are following, we require only that the derivative $F'(x)$ exist almost everywhere.

Let us begin by looking at the second question. Formulated in this general form, we shall give a negative answer to it by constructing in the interval $[0, 1]$ *a monotonic continuous function $F(x)$ whose derivative is zero almost everywhere and which, despite this fact, is not constant on any segment.*

To do this, take $0 < t < 1$ and define the sequence $\{F_n(x)\}$ by recursion in the following manner: let $F_0(x) = x$; let the functions $F_n(x)$ be continuous and, moreover, linear in the segments bounded by two consecutive points $k2^{-n}$, $(k + 1)2^{-n}$; let $F_{n+1}(x) = F_n(x)$ at these points, whereas at the midpoint of these segments, that is, at the new points of division, let

$$F_{n+1}\left(\frac{\alpha+\beta}{2}\right)=\frac{1-t}{2}F_n(\alpha)+\frac{1+t}{2}F_n(\beta),$$

where α and β denote the extremities of the respective segment. These functions $F_n(x)$ are obviously increasing. Since, moreover,

$$0 \leq F_n(x) \leq F_{n+1}(x) \leq 1,$$

the sequence $\{F_n(x)\}$ converges to a limit $F(x)$ which is nondecreasing. We shall show that $F(x)$ is continuous, strictly increasing, and that

$$F'(x)=0$$

almost everywhere.

Let x be an arbitrary point in $[0, 1]$ and consider the sequence of nested intervals (α_n, β_n) of the type

$$\alpha_n = k2^{-n},\ \beta_n=(k+1)2^{-n}$$

about x; we see immediately that

$$F_{n+1}(\beta_{n+1})-F_{n+1}(\alpha_{n+1})=\frac{1\pm t}{2}[F_n(\beta_n)-F_n(\alpha_n)].$$

Since $F_\nu(\alpha_p)=F(\alpha_p)$ and the same is true for β_p, we also have

$$F(\beta_{n+1})-F(\alpha_{n+1})=\frac{1\pm t}{2}[F(\beta_n)-F(\alpha_n)],$$

hence

$$F(\beta_n)-F(\alpha_n)=\prod_1^n \frac{1+\varepsilon_k t}{2} \quad (\varepsilon_k=\pm 1).$$

This shows that

$$F(\beta_n)-F(\alpha_n)>0,$$

and that

$$F(\beta_n)-F(\alpha_n)\leq\left(\frac{1+t}{2}\right)^n \to 0 \text{ for } n\to\infty,$$

hence F is strictly increasing and continuous. Furthermore, the derivative $F'(x)$ is equal, when it exists, to the limit of

$$\frac{F(\beta_n)-F(\alpha_n)}{\beta_n-\alpha_n}=\prod_1^n(1+\varepsilon_k t) \qquad (n\to\infty);$$

but the value of the infinite product $\Pi(1+\varepsilon_k t)$ obviously can only be 0, infinity, or indeterminate. Hence $F'(x)=0$ everywhere where $F(x)$ possesses a finite derivative, that is, *almost everywhere.*

25. Absolutely Continuous Functions. Canonical Decomposition of Monotonic Functions

We see by the above example that a function, even a function of bounded variation, is not necessarily an indefinite integral, not even when we add the assumption that it be continuous. Hence the class of continuous functions of bounded variation is larger than that of indefinite integrals. By what characteristic properties can we distinguish the latter? Obviously it suffices to consider the case where the interval of integration (a, b) is finite.

We shall see that *a necessary and sufficient condition that $F(x)$ be an indefinite integral is that it be absolutely continuous.* "Absolute continuity" means that not only $F(\beta) - F(\alpha)$ becomes infinitely small with $\beta - \alpha$, but that also the sum

$$\sum [F(\beta_k) - F(\alpha_k)]$$

becomes infinitely small with

$$\sum (\beta_k - \alpha_k),$$

where the (α_k, β_k) denote a finite or infinite system of non-overlapping intervals. Moreover, we can limit ourselves to finite systems, since the case of an infinite system reduces to this immediately. Note that, in particular, all functions satisfying the Lipschitz condition:

$$|F(\beta) - F(\alpha)| \leq C(\beta - \alpha),$$

are absolutely continuous.

We begin with the proof of the necessity of our condition. We shall consider a summable function $f(x)$, $a < x < b$, and its indefinite integral $F(x)$. If $f(x)$ is bounded, $|f(x)| \leq C$, the function $F(x)$ obviously satisfies the Lipschitz condition with the same constant C. When $f(x)$ is not bounded, we make use of the fact that for all $\varepsilon > 0$ we can decompose $f(x)$ into the sum of a bounded summable function $g(x)$ and a function $h(x)$ such that

$$\int_a^b |h(x)|\, dx < \varepsilon/2;$$

in fact, denoting by $f_n(x)$ the function which results from $f(x)$ by truncating it above and below to the levels $+n$ and $-n$, the decomposition

$$f(x) = f_n(x) + [f(x) - f_n(x)]$$

is of the required type when n is sufficiently large. Now when $|g(x)| \leq C$, we shall have for every non-overlapping system of intervals (α_k, β_k) whose total length is less than $\varepsilon/2C$:

$$|\sum\int_{a_k}^{\beta_k} f(x)dx| \leq \sum\int_{a_k}^{\beta_k}|g(x)|\,dx + \sum\int_{a_k}^{\beta_k}|h(x)|\,dx \leq$$

$$\leq \sum C(\beta_k - a_k) + \sum\int_{a_k}^{\beta_k}|h(x)|\,dx \leq C\frac{\varepsilon}{2C} + \frac{\varepsilon}{2} = \varepsilon,$$

hence $F(x)$ is absolutely continuous.

Conversely, assume that our condition is fulfilled, or equivalently, assume that for every $\varepsilon > 0$ there exists a δ such that the condition

$$\sum (\beta_k - a_k) < \delta,$$

where the intervals (a_k, β_k) do not overlap, implies that

$$\sum |F(\beta_k) - F(a_k)| < \varepsilon.$$

The two hypotheses are indeed equivalent; if the first were fulfilled while the second was false, we should have for some $\varepsilon > 0$ a sequence of systems of intervals for which

$$\sum (\beta_k - a_k) \to 0 \text{ and } \sum |F(\beta_k) - F(a_k)| \geq \varepsilon,$$

or, decomposing these systems of intervals into two parts according to the sign of $F(\beta_k) - F(a_k)$ and retaining only the suitable one of the two parts, we should have

$$\sum (\beta_k - a_k) \to 0 \text{ and } |\sum [F(\beta_k) - F(a_k)]| \geq \varepsilon/2,$$

in contradiction to the hypothesis made. On the other hand, the second hypothesis obviously implies the first.

This being the case, let us show first of all that our condition implies that $F(x)$ is of bounded variation. In fact, choose an $\varepsilon > 0$ and a corresponding δ, decompose the interval (a, b) into segments (x_{k-1}, x_k) of length $< \delta/2$ but otherwise arbitrary, and consider the expression

$$\Sigma = \sum |F(x_k) - F(x_{k-1})|. \tag{5}$$

Obviously, we could partition this sum into

$$m \leq \frac{2(b-a)}{\delta} + 1$$

groups such that for each group the total length of the respective segments (x_{k-1}, x_k) is less than δ and that consequently the contribution of each group to the sum (5) is less than ε. Therefore

$$\Sigma < \varepsilon\left(\frac{2(b-a)}{\delta} + 1\right)$$

for all decomposition of the type considered. It follows that $F(x)$ is of bounded variation.

It also follows immediately from the second variant of our condition that, simultaneously with $F(x)$, the indefinite total variation $T(x)$ of $F(x)$ is also absolutely continuous, and the same is true of the positive and negative variations. Consequently, without restricting the generality, we can limit ourselves to the case in which $F(x)$ is nondecreasing.

Finally, here is a last reduction. The ratio

$$\frac{F(x+h)-F(x)}{h}$$

is non-negative and tends, for almost all x, to $F'(x)$. On the other hand, its integral over the interval (α, β) is equal to

$$\frac{1}{h}\int_{\beta}^{\beta+h} F(x)dx - \frac{1}{h}\int_{\alpha}^{\alpha+h} F(x)dx,$$

hence, since $F(x)$ is continuous, this integral tends, for $h \to 0$, to $F(\beta)-F(\alpha)$. It follows from Fatou's lemma that $F'(x)$ is summable and that

$$\int_{\alpha}^{\beta} F'(x)dx \leq F(\beta) - F(\alpha).$$

Denoting by $G(x)$ the indefinite integral of $F'(x)$, the difference $F(x) - G(x)$ will therefore be nondecreasing. Since furthermore the functions $F(x)$ and $G(x)$ are absolutely continuous, the first by hypothesis and the second — an indefinite integral — by what we have just seen, it will be the same for their difference $F(x) - G(x)$.

Thus our problem has been reduced to the case where $F(x)$ is *nondecreasing, absolutely continuous, and* $F'(x) = 0$ *almost everywhere*. What we must show is that *under these hypotheses* $F(x)$ *is constant*. We have already carried out such a proof in Sec. 13, but with the stricter hypothesis of a Lipschitz condition instead of absolute continuity. The use we made of the Lipschitz condition was to show that under the transformation $y = F(x)$ the image $F(E)$ of a set E of measure zero is also of measure zero. But this fact is true even under the condition that $F(x)$ is absolutely continuous. In fact, letting δ denote the quantity which corresponds, under the hypothesis of absolute continuity, to a given $\varepsilon > 0$, we can cover E by a system of non-overlapping intervals of total length $< \delta$; the images of these intervals will be of total length $< \varepsilon$ and will cover the set $F(E)$. The proof given in Sec. 13 therefore extends to the more general case that we are considering. Consequently we have arrived at the theorem stated at the beginning of this section.

THEOREM.[9] *In order that the function $F(x)$ be an indefinite integral, it is necessary and sufficient that it be absolutely continuous.*

Let us point out further a decomposition of monotonic functions or, more generally, of functions of bounded variation — a decomposition which follows immediately from the preceding considerations. Let $F(x)$ be a nondecreasing continuous function which is otherwise arbitrary. By the Fatou lemma we show, exactly as above, that the function $F'(x)$ is summable and setting

$$G(x) = \int_a^x F'(t)dt,$$

that the function $H(x) = F(x) - G(x)$ is nondecreasing. Since $F'(x) \geq 0$, $G(x)$ is also nondecreasing, and we have $H'(x) = 0$ almost everywhere. Hence we have the

THEOREM. *Every monotonic continuous function $F(x)$ can be decomposed into the sum of two monotonic continuous functions, $G(x)$ and $H(x)$, of which $G(x)$ is absolutely continuous and $H(x)$ is "singular," that is, $H'(x) = 0$ almost everywhere.*

In the case of a monotonic function of general type we can add a third term, namely a saltus function. These facts extend immediately to all functions of bounded variation.

Let us return once more to the definition of absolute continuity in order to point out its relationship with the notion of a *set of zero variation* with respect to a function of bounded variation.

Let us agree to call the total variation of a continuous function of bounded variation $F(x)$ over a non-overlapping system of intervals the sum of the total variations of $F(x)$ formed on these intervals. A set e will be said to be of zero variation with respect to $F(x)$ if we can enclose it in a system of intervals on which the total variation of $F(x)$ becomes infinitely small. In case $F(x)$ is monotonic, this definition is obviously equivalent to requiring that the image of the set e on the y-axis given by the function $y = F(x)$ be of measure zero. A more profound analysis shows that the same is true in the general case, but we shall not insist upon this point nor upon the idea of total variation on other sets and we shall content ourselves with saying only that in these problems we can make use (for example) of an interesting theorem of Banach [**1**] which affirms that the total variation of $F(x)$ is equal to

$$\int_{-\infty}^{\infty} N(t)dt,$$

where $N(t)$ denotes the number (finite or infinite) of solutions x of the equation $t = F(x)$.

[9] Lebesgue [*] (2nd ed., p. 183 and p. 188).

What we do wish to point out here is that *in the definition of absolute continuity the hypothesis that the total variation become small on systems of intervals of sufficiently small total length can be replaced by the hypothesis that the total variation is zero on all sets of measure zero,* that is, that the sets of measure zero be also of zero variation with respect to $F(x)$.

Now it is clear that the first hypothesis implies the second; it remains for us to prove the converse. It obviously suffices to consider the case in which $F(x)$ is nondecreasing. Assume that the first hypothesis is not fulfilled; then for some $\varepsilon > 0$ there exists a system of intervals

$$\Sigma_1, \Sigma_2, \ldots, \Sigma_n, \ldots$$

of total length less than respectively

$$\frac{1}{2}, \frac{1}{4}, \ldots, \frac{1}{2^n}, \ldots$$

whose images under $y = F(x)$ each have a total length greater than ε. Set

$$\Sigma^{(n)} = \Sigma_n \cup \Sigma_{n+1} \cup \ldots;$$

the total length of $\Sigma^{(n)}$ is then less than

$$2^{-n} + 2^{-n-1} + \ldots = 2^{-n+1}.$$

Consequently the common part e_0 of the $\Sigma^{(n)}$, which decrease to this common part, is of measure zero. Repeating the procedure, but this time with the images of the Σ_n, all the total lengths which appear will be greater than ε, and consequently the image of e_0 will not be of measure zero, which contradicts the second hypothesis.

26. Integration by Parts and Integration by Substitution

The classical rules extend, with some precaution in their statements, to our notion of integral. First of all, let us consider integration by parts.

THEOREM. *If the functions $f(x)$ and $g(x)$ are assumed summable in the interval (a, b), and $F(x)$ and $G(x)$ denote their indefinite integrals, then the functions $F(x)g(x)$ and $G(x)f(x)$ are also summable and we have*

$$\int_a^b F(x)g(x)dx + \int_a^b G(x)f(x)dx = F(b)G(b) - F(a)G(a). \tag{6}$$

The theorem is evident in case $f(x)$ and $g(x)$ are constants; one can then pass immediately to the case of two step functions. To treat the general case, let us observe that we can assume without loss of generality that f and g are positive and belong to the class C_1; in fact, we have only to set $f = f_1 - f_2$ and $g = g_1 - g_2$, where f_1, f_2, g_1, g_2 are of class C_1 and are positive [functions of class C_1 are bounded below, consequently if f_1 and f_2 were not positive we could make them so by the addition of (the same) suitable constant to each; similarly for g_1 and g_2]; finally, we can combine the four formulas (6) corresponding to f_1, g_1; f_1, g_2; f_2, g_1; f_2, g_2 into a single formula.

To the assumption that f and g are positive and of class C_1, and consequently the limits almost everywhere of two increasing sequences of positive step functions φ_n and ψ_n, we add, momentarily, the assumption that the integrals $F(x)$, $G(x)$, $\Phi_n(x)$, and $\Psi_n(x)$ satisfy

$$F(a) = G(a) = \Phi_n(a) = \Psi_n(a) = 0.$$

If we then write formula (6) with φ_n, ψ_n, Φ_n, and Ψ_n instead of f, g, F, and G, we obtain a particular case which is already verified. Letting n approach infinity, the problem reduces to term-by-term integration of the increasing sequences $\{\Phi_n\psi_n\}$ and $\{\varphi_n\Psi_n\}$. Finally, the hypothesis that $F(a) = G(a) = 0$ does not restrict the generality; indeed, replacing $F(x)$, for example, by $F(x) + C$ has no effect other than the addition to both members of the same quantity $C(G(b) - G(a))$.

Another means of verifying formula (6) is to imitate the classical proof, based on the differentiation of products. We have $F'(x) = f(x)$ and $G'(x) = g(x)$ almost everywhere, therefore also

$$[F(x)G(x)]' = F'(x)g(x) + f(x)G'(x).$$

$F(x)$ and $G(x)$ are, according to Sec. 25, absolutely continuous, and the same must be true for their product, as follows immediately from the fact that

$$|F(\beta)G(\beta) - F(\alpha)G(\alpha)| = |[F(\beta) - F(\alpha)]G(\beta) + F(\alpha)[G(\beta) - G(\alpha)]| \leq$$
$$\leq M(|F(\beta) - F(\alpha)| + |G(\beta) - G(\alpha)|),$$

where M denotes a bound of $|F(x)|$ and of $|G(x)|$ in (a, b). Therefore $F(x)G(x)$ is the indefinite integral of its derivative, which proves (6).

Let us move on to integration by substitution.

THEOREM. *If the function $x(t)$ is nondecreasing and absolutely continuous in the interval $\alpha \leq t \leq \beta$, and $f(x)$ denotes a function summable in the interval $x(\alpha) = a \leq x \leq b = x(\beta)$, then the function $f(x(t))x'(t)$ is summable in (α, β) and we have*

(6a) $$\int_a^b f(x)dx = \int_\alpha^\beta f(x(t))x'(t)dt.$$

The proposition is immediately verified for step functions $f(x)$.

Let $f(x)$ be a function belonging to the class C_1, so that there exists a nondecreasing sequence of step functions $\varphi_n(x)$ tending to $f(x)$ everywhere except perhaps on a set E of measure zero. Since $x'(t) \geq 0$, the sequence of functions

(6b) $$\varphi_n(x(t))x'(t)$$

is also nondecreasing (we set aside the set of points t, of measure zero, where $x'(t)$ does not exist). This sequence obviously tends to

$$f(x(t))x'(t) \tag{6c}$$

at every point t where $x'(t) = 0$ and also at every point t for which the value of the function $x(t)$ does not belong to the exceptional set E. Hence the only exceptional set which remains is the set e of points t for which $x(t)$ is in the set E and the derivative $x'(t)$ exists and is positive. We shall show that this set e is also of measure zero.

To do this, let us cover the set E by a sequence of intervals $i_1, i_2, \ldots$ of finite total length such that every point of E is interior to an infinite number of these intervals, which is possible since E is of measure zero (see Sec. 2). Denote by $\psi_n(x)$ the sum of the characteristic functions of the intervals $i_1, i_2, \ldots, i_n$; we thus obtain a nondecreasing sequence of step functions tending to ∞ at every point of E. The sequence of functions $\psi_n(x(t))x'(t)$ is also nondecreasing and tends to ∞ at every point of the set e. Since the sequence of integrals

$$\int_\alpha^\beta \psi_n(x(t))x'(t)dt = \int_a^b \psi_n(x)dx$$

is bounded by the finite total length of the intervals considered, the set e is necessarily of measure zero.

Hence the sequence (6b) tends, increasingly, to the function (6c) almost everywhere in t. Since, furthermore,

$$\int_\alpha^\beta \varphi_n(x(t))x'(t)dt = \int_a^b \varphi_n(x)dx \to \int_a^b f(x)dx,$$

it follows, by virtue of Beppo Levi's theorem, that the limit function is also summable and that we have formula (6a).

With the theorem established for functions $f(x)$ of class C_1, we pass to functions of class C_2 by forming differences.

27. The Integral as a Set Function

Now let us consider the integral as a set function, that is, the integral taken over a measurable set e. This notion, which is an immediate generalization of measure, is defined by integrating the function which equals $f(x)$ on the set e and equals 0 elsewhere, or, what amounts to the same thing, by integrating the product $f(x)e(x)$, where the integration can be performed over an arbitrary interval containing the set e. The value of the integral is called the integral of $f(x)$ on e and is written

$$F(e) = \int_e f(x)dx.$$

Its existence is guaranteed by the fact that one of the factors of $f(x)e(x)$ is summable and the other measurable and bounded. Furthermore, in general the product $f(x)e(x)$ can be summable without $f(x)$ being summable over its entire domain, or even without the set e being measurable; in fact, for the latter it suffices that the subset where $f(x) \neq 0$ be measurable. It is obvious that when $f(x)$ is summable on e it is also summable on each measurable subset of e.

Since the measure of a set e is simply the integral of the constant 1 over this set, we are led to ask if the results obtained concerning measure are valid for the notion of integral which we have just considered. The answer will be affirmative, as is easy to see; of the results obtainable, we state the following.

When $e_1, e_2, \ldots$ are disjoint measurable sets, and $f(x)$ is summable on the set $e_1 \cup e_2 \ldots$, we have

$$F(e_1 \cup e_2 \cup \ldots) = F(e_1) + F(e_2) + \ldots,$$

that is, $F(e)$ is a denumerably additive set function.

The *absolute continuity* of the indefinite integral is presented here under the following more general form.

When the measurable set e varies in such a way that its measure tends to 0, the same holds for $F(e)$; in other words, to every $\varepsilon > 0$ there corresponds a $\delta > 0$ such that $m(e) < \delta$ implies $|F(e)| < \varepsilon$.

The proof can be carried out, just as in the case where e consists of non-overlapping intervals, by a decomposition of the function $f(x)$ into the sum of a bounded summable function $g(x)$ and a function $h(x)$ such that the integral of $|h(x)|$ is $< \varepsilon$.

THE SPACE L^2 AND ITS LINEAR FUNCTIONALS. L^p SPACES

28. The Space L^2; Convergence in the Mean; the Riesz-Fischer Theorem

Let us consider the measurable functions $f(x)$ defined on a set e of finite or infinite measure. In view of later applications, we shall permit the functions to take on complex values. We shall assume that the squares $|f|^2$ are summable; by the Schwarz inequality, $f(x)$ will then be summable on the subsets of finite measure. Furthermore, the same inequality implies that the product of any two of the functions considered is summable.

Let us agree to set

$$(f, g) = \int_e f(x)\overline{g(x)}dx, \quad \|f\| = \sqrt{(f, f)} = [\int_e |f(x)|^2 dx]^{\frac{1}{2}};$$

we shall call the first of these quantities the *scalar product* of f and g and the second the *norm* of f; the latter always has a positive value except when $f(x) = 0$ almost everywhere. From the Schwarz inequality we have

$$|(f, g)| \leq \|f\| \, \|g\|;$$

the Minkowski inequality shows that when f and g belong to the class considered so does $f + g$, and

$$\|f + g\| \leq \|f\| + \|g\|.$$

Writing $f_1 - f_2$ for f and $f_2 - f_3$ for g, and noting that along with f_1, f_2, and f_3 their differences also belong to the class considered, we obtain the "triangle" inequality

$$\|f_1 - f_3\| \leq \|f_1 - f_2\| + \|f_2 - f_3\|.$$

This inequality is the analogue of the inequality for the sides of a triangle, if we define the "*distance*" from f to g to be $\|f - g\|$; doing this, it is natural *not to distinguish between two functions whose "distance" is* 0, that is, which coincide almost everywhere; we agree to observe this convention in all that follows.

It is also obvious that whenever f and g belong to our class so do λf and λg and that

$$(\lambda f, g) = \lambda(f, g); \quad (f, \lambda g) = \bar{\lambda}(f, g); \quad \|\lambda f\| = |\lambda| \, \|f\|;$$

let us mention further that

$$\overline{(g, f)} = (f, g).$$

All this puts into evidence the analogy that our class, generally called *the (complex) functional space* L^2, shows with ordinary (complex) vector spaces.[10] The essential difference consists in the fact that the space L^2 is not of finite dimension, that is, that its elements are not all linear combinations of a given finite number of elements. It is this difference which gives rise to several important notions which are devoid of interest for spaces of finite dimension.

The first of these notions is that of *convergence in the mean*, also called *strong convergence*. We say that the sequence $\{f_n\}$ converges in the mean to f when $\|f - f_n\| \to 0$. (Later, when we deal with the abstract theory of Hilbert space, we shall simply write $f_n \to f$ to denote the same relation; at the moment, since we deal with ordinary functions, this nomenclature could cause some confusion.)

The most useful theorem concerning this notion is the following, the analogue of the classical criterion of Cauchy.

[10] The *real* space L^2 is obtained if one considers only the real-valued functions. Then $(f, g) = (g, f)$.

RIESZ-FISCHER THEOREM.[11] *If a sequence* $\{f_n\}$ *is given, then in order that there exist an element* f *toward which it converges in the mean it is necessary and sufficient that* $\|f_m - f_n\| \to 0$ *for* $m, n \to \infty$.

The condition is necessary; in fact, it follows from the triangle inequality that

$$\|f_n - f_m\| \leq \|f - f_n\| + \|f - f_m\| \to 0.$$

To see that it is also sufficient, let us choose the indices $m_1 < m_2 < \ldots$ in such a way that for $n > m_k$ we have

$$\|f_n - f_{m_k}\| < 2^{-k}.$$

Then in particular $\|f_{m_{k+1}} - f_{m_k}\| < 2^{-k}$, and since from the Schwarz inequality with $g(x)$ set equal to 1 we have

$$\int_{e'} |f_{m_{k+1}}(x) - f_{m_k}(x)|\, dx \leq \sqrt{m(e')}\|f_{m_{k+1}} - f_{m_k}\| \leq \sqrt{m(e')}2^{-k}$$

for every subset e' of e of finite measure $m(e')$, the series

$$\sum_1^\infty \int_{e'} |f_{m_{k+1}}(x) - f_{m_k}(x)|\, dx$$

is convergent; by Beppo Levi's theorem, Sec. 18, this implies the absolute convergence, almost everywhere, of the series

$$\sum_1^\infty (f_{m_{k+1}}(x) - f_{m_k}(x)),$$

hence, all the more, the convergence of $f_{m_k}(x)$ to a limit $f(x)$ almost everywhere. Since, furthermore, we see from the inequality

$$\|f_{m_k}\| \leq \|f_{m_1}\| + \|f_{m_k} - f_{m_1}\| \leq \|f_{m_1}\| + \tfrac{1}{2}$$

that the norms $\|f_{m_k}\|$, or equivalently the integrals of the $|f_{m_k}|^2$ remain bounded, the Fatou lemma, Sec. 20, assures us that $f(x)$ belongs to the space L^2. Moreover, applying the same lemma to the sequence $\{f_{m_k} - f_n\}$ with n fixed, $n > m_r$, and noting that when $k > r$,

$$\|f_{m_k} - f_n\| \leq \|f_{m_k} - f_{m_r}\| + \|f_n - f_{m_r}\| < 2^{-r+1},$$

it follows that

$$\|f - f_n\| \leq 2^{-r+1}.$$

Finally, for $n \to \infty$ we can also let r increase indefinitely, which gives

$$\|f - f_n\| \to 0,$$

which was to be proved. Further, let us note that the limit f is determined

[11] F. RIESZ [**1**], [**2**]; E. FISCHER [**1**].

by the sequence $\{f_n\}$ up to a set of measure zero; in fact, if f^* were also a limit in the mean of the same sequence, we would have

$$\|f - f^*\| \leq \|f - f_n\| + \|f_n - f^*\| \to 0,$$

that is, $\|f - f^*\| = 0$.

29. Weak Convergence

One of the consequences of the mean convergence of f_n to f is the relation

$$(f_n, g) \to (f, g), \tag{7}$$

valid for every element g of the space L^2. This follows immediately from the evaluation

$$|(f, g) - (f_n, g)| = |(f - f_n, g)| \leq \|f - f_n\| \, \|g\| \to 0.$$

But the converse is not true; relation (7), assumed valid for all the g's, does not imply mean convergence; this is shown by the example $f_n(x) = \sin nx$, $0 \leq x \leq \pi$, borrowed from the theory of Fourier series.

We express relation (7), assumed valid for all the g's, or also the equivalent relation $(g, f_n) \to (g, f)$, by saying that the sequence f_n *converges weakly to* f. In the example cited the sequence $\{\sin nx\}$ converges weakly to 0, but it does not converge in the mean to any limit, as can be seen by calculating, for $m \neq n$,

$$\|\sin nx - \sin mx\|^2 = \pi.$$

One of the essential differences between convergence in the mean and weak convergence is that the first implies that

$$\|f_n\| \to \|f\|, \tag{8}$$

whereas this relation does not follow from the second. For mean convergence, relation (8) is an immediate consequence of the inequalities

$$\|f_n\| \leq \|f_n - f\| + \|f\|, \quad \|f\| \leq \|f - f_n\| + \|f_n\|.$$

As for weak convergence, it suffices to recall the above example — the sequence $\{\sin nx\}$ — for which $\|f_n\| = \sqrt{\dfrac{\pi}{2}}$, but $\|f\| = 0$.

However, if we add to weak convergence the hypothesis that relation (8) is valid we do get convergence in the mean; in fact, it follows that

$$\|f - f_n\|^2 = \|f\|^2 - (f, f_n) - (f_n, f) + \|f_n\|^2 \to \|f\|^2 - (f, f) - (f, f) + \|f\|^2 = 0.$$

We add that in the case of weak convergence there is a sort of semi-continuity which remains valid, namely the inequality

$$\|f\| \leq \liminf \|f_n\|,$$

or, equivalently: the hypothesis $\|f_n\| \leq C$ implies that $\|f\| \leq C$. This results from the evaluation

$$C\|f\| \geq |(f_n, f)| \to (f, f) = \|f\|^2$$

after dividing by $\|f\|$.

30. Linear Functionals

The true nature of the concept of weak convergence will be made clearer by introducing the notion of linear functional. In fact, the scalar product (g, f), for $f(x)$ fixed and $g(x)$ varying in the space L^2, can be considered as a linear function, or rather, in conformity with generally adopted language, a *linear functional* of the variable function g.

In general, a *linear operation* or *linear functional* is an operation which assigns to every element g of L^2 a numerical value Ag and which is

1° *additive*: $A(g_1 + g_2) = Ag_1 + Ag_2$,

2° *homogeneous*: $A(cg) = cAg$, where c is an arbitrary numerical factor,

3° *bounded*: there exists a quantity M such that for all g

$$|Ag| \leq M\|g\|.$$

The smallest of these bounds M is denoted by M_A or by $\|A\|$ and is called the *norm* of the linear functional A.

We see immediately that the functional $Ag = (g, f)$ satisfies these conditions with $M_A = \|f\|$. But what is still more important is the converse, which follows.

THEOREM.[12] *Every linear functional Ag in the space L^2 can be put in the form*

$$Ag = (g, f),$$

where the generating function $f(x)$ is uniquely determined by the functional A.

The proof of this theorem is based on the following formula which is valid for two arbitrary elements of L^2:

$$\|u + v\|^2 + \|u - v\|^2 = 2\|u\|^2 + 2\|v\|^2. \tag{9}$$

This formula can be verified by an obvious calculation; its geometric interpretation is simply that in an arbitrary parallelogram the sum of the squares of the diagonals equals the sum of the squares of the sides.

To prove our theorem, consider the functional Ag; since the upper bound of $|Ag|$ on the "unit sphere" $\|g\| = 1$ is equal to M_A, there is a sequence $\{g_n\}$ such that

$$\|g_n\| = 1, \quad |Ag_n| \to M_A.$$

[12] This theorem was discovered independently by M. Fréchet and F. Riesz and published in 1907 in the same number of volume 144 of the *Comptes Rendus*. The following proof is due to F. Riesz [**15**].

We can even assume that the values Ag_n are real and non-negative and that consequently $Ag_n \to M_A$, by multiplying the g_n by suitable numerical factors of modulus 1. This done, let us set $u = g_m$, $v = g_n$ in (9), and note that

$$Ag_m + Ag_n = A(g_m + g_n) \leq M_A \|g_m + g_n\|;$$

then

$$\|g_m - g_n\|^2 = 4 - \|g_m + g_n\|^2 \leq 4 - \frac{1}{M_A^2}[Ag_m + Ag_n]^2 \to 4 - \frac{1}{M_A^2} 4M_A^2 = 0.$$

Consequently it follows from the Riesz-Fischer theorem that the sequence $\{g_n\}$ converges in the mean to an element g^*. Moreover, $\|g^*\| = 1$, and since

$$|Ag^* - Ag_n| = |A(g^* - g_n)| \leq M_A \|g^* - g_n\| \to 0,$$

that is, $Ag_n \to Ag^*$, we shall have precisely $Ag^* = M_A$. Hence the functional Ag attains its least upper bound M_A on the unit sphere—at the element $g = g^*$.

Let us show that for all g

$$Ag = (g, f), \tag{10}$$

with $f = M_A g^*$. Our assertion is obvious for $g = g^*$. Moreover, g^* is orthogonal to all the elements g for which $Ag = 0$; this means that for these elements

$$(g, g^*) = (g^*, g) = 0.$$

This follows from the relation

$$M_A^2 = (Ag^*)^2 - (A(g^* - \lambda g))^2 \leq M_A^2 \|g^* - \lambda g\|^2 =$$
$$= M_A^2 (1 - \lambda(g, g^*) - \bar{\lambda}(g^*, g) + \lambda\bar{\lambda}(g, g)),$$

whence

$$-\bar{\lambda}(g^*, g) - \lambda(g, g^*) + \lambda\bar{\lambda}(g, g) \geq 0;$$

in fact, setting

$$\lambda = \frac{(g^*, g)}{(g, g)} \text{ and } \bar{\lambda} = \frac{(g, g^*)}{(g, g)}$$

we obtain

$$-|(g, g^*)|^2 \geq 0;$$

hence $(g, g^*) = 0$ and thus $(g, f) = M_A (g, g^*) = 0$. Therefore formula (10) is valid, not only for $g = g^*$, but also for all the g_0 which make the functional Ag zero, and since any g can be put in the form $g_0 + \mu g^*$ by putting

$$\mu = \frac{Ag}{M_A} = \frac{Ag}{Ag^*},$$

formula (10) is valid for all the g's, which was to be proved.

31. Sequences of Linear Functionals; a Theorem of Osgood

Returning to weak convergence, it is clear that the weak convergence of f_n to f, when translated into the language of functionals, is expressed by saying that the sequence of functionals $A_n g = (g, f_n)$ converges everywhere in L^2 to the functional $Ag = (g, f)$. Considered from this point of view, the important fact that we are going to prove will be closely related to the basic principles of the theory of functions of real variables. We begin by recalling

OSGOOD'S THEOREM.[13] *Given a sequence $\{f_n(t)\}$ of continuous functions defined in the interval (a, b) with the property that for each t interior to (a, b) the sequence $\{f_n(t)\}$ is bounded (although not necessarily uniformly with respect to t), there exists a subinterval of (a, b) in which the sequence is uniformly bounded.*

We prove this theorem by contradiction, by assuming it is not valid for the sequence considered. If the sequence is not uniformly bounded in (a, b), there exists an n_1 and a point t_1 interior to (a, b) for which $|f_{n_1}(t_1)| > 1$, therefore by continuity we also have $|f_{n_1}(t)| > 1$ in some interval (a_1, b_1) surrounding t_1. For the same reason, there will be an $n_2 > n_1$ and an interval (a_2, b_2) interior to (a_1, b_1) such that in (a_2, b_2), $|f_{n_2}(t)| > 2$. Continuing, we obtain a sequence $n_1 < n_2 < \ldots$ and a sequence of intervals (a_k, b_k), each embedded in the preceding, such that in (a_k, b_k), $|f_{n_k}(t)| > k$. We can obviously assume in addition that $a_k < a_{k+1}$, $b_k > b_{k+1}$, and $b_k - a_k \to 0$. Then these intervals contract to a common point t_0, and for this point $|f_{n_k}(t_0)| > k$ for each k, contrary to the hypothesis.

We have mentioned this theorem and its proof without having need for it, since by means of slight modifications in the reasoning we shall arrive at some important results concerning sequences of linear functionals and also sequences of certain analogous operations of a more general type. We shall be concerned with these sequences not only in the space L^2, or more generally for the analogous spaces L^p about which we shall speak shortly, but also for a large category of functional and abstract spaces. But to fix the ideas involved we shall content ourselves for the moment with discussing convergent sequences of linear functionals in L^2 space. Here is the theorem we propose to establish:

THEOREM.[14] *The sequence $\{A_n\}$ of linear functionals cannot converge or even remain bounded for each element g of the space L^2 unless it is uniformly bounded in the unit sphere, that is, unless the sequence of norms M_{A_n} is bounded.*

The first part of the proof is only a repetition of the preceding reasoning, in which we replace the f_n by the A_n, the points t by elements g of L^2, the interval (a, b) by an arbitrary sphere of this space — for example the unit

[13] OSGOOD [1] (pp. 159–164).

[14] BANACH–STEINHAUS [1]; BANACH [*] (p. 80). Also, see LEBESGUE [3] (p. 70).

sphere — and the intervals (a_k, b_k) by spheres in L^2 with the g_{n_k} as centers and with radii $\varrho_k < \frac{1}{k}$ which are chosen sufficiently small that each sphere contains the succeeding spheres in its interior. Since for $k < l$

$$\|g_{n_l} - g_{n_k}\| < \frac{1}{k},$$

the sequence $\{g_{n_k}\}$ converges in the mean to an element g_0, interior to all the spheres, for which

$$|A_{n_k} g_0| > k \to \infty,$$

contrary to the hypothesis. It follows that there exists a sphere, say with center g^* and radius ϱ (i. e. the set of all elements $g^* + \varrho g$ with $\|g\| \leq 1$), in which the sequence A_n is uniformly bounded; therefore for an appropriate constant C,

$$|A_n(g^* + \varrho g)| \leq C_1$$

for all g lying in the unit sphere. From this we calculate, for the unit sphere,

$$|A_n g| = \frac{1}{\varrho} |A_n(g^* + \varrho g) - A_n g^*| \leq \frac{1}{\varrho} |A_n(g^* + \varrho g)| + \frac{1}{\varrho} |A_n g^*| \leq$$

$$\leq \frac{2}{\varrho} C_1 = C_2.$$

The theorem is thus proved.

In the language of weak convergence, our theorem is expressed by saying that *every weakly convergent sequence* $\{f_n\}$ *is necessarily bounded,*

$$\|f_n\| \leq C.$$

32. Separability of L^2. The Theorem of Choice

The last theorem admits a sort of converse, the analogue of the Bolzano-Weierstrass theorem.

THEOREM OF CHOICE. *From every bounded sequence* $\{f_n\}$ *of the space* L^2, $\|f_n\| \leq C$, *we can extract a weakly convergent subsequence.*

To prove this, we usually start from the fact that the space L^2 is *separable*, that is, that there exists a denumerable set of elements $\{g_n\}$ such that every element f of L^2 can be approximated in the mean arbitrarily closely by these g_n or their linear combinations $l = c_1 g_1 + \ldots + c_m g_m$. Since the various possibilities with regard to the set e of integration are obviously included in the case $e = (-\infty, \infty)$, and since, furthermore, when one approximates the real and imaginary parts of a function to within ε, the function itself is approximated to within 2ε, it will suffice to consider functions defined on the entire x-axis and belonging to the "real" space L^2. Let $f(x)$ be a function of this type and let $f_n(x)$ be the function obtained from $f(x)$ by setting it equal to zero outside the interval $(-n, n)$ and inside truncating it above and below by respectively n and $-n$. Then $f - f_n \to 0$ almost everywhere and $(f - f_n)^2 \leq f^2$, so we may integrate term by term. In this way we obtain $\|f - f_n\| \to 0$, that is, f can be approximated, in the sense considered, by bounded functions which are zero outside a finite interval (a, b). Now every function of the latter type is the limit almost everywhere of a sequence $\{\varphi_k\}$ of step functions which are defined and uniformly bounded in (a, b); if the step functions were not uniformly bounded we should have only to truncate them by the two bounds of the function to be approximated. Consequently, according to the "little" theorem of Lebesgue, the sequence $\{\varphi_k\}$ converges in the mean to its limit. Finally, a step function can obviously be approximated by step functions of a particular type, namely those whose discontinuities are at the rational points. One can use the triangle inequality to pass from one class to the other, by the same reasoning that one would use in ordinary space. We conclude that the element f can be approximated arbitrarily closely in the mean by step functions of the last type, or equivalently, by linear combinations of the characteristic functions of those intervals with rational endpoints. But these functions form a denumerable set; since the element f of L^2 was arbitrary, this proves our assertion that the space L^2 is separable.

With this established, one proves the theorem of choice as follows. The hypothesis $\|f_n\| \leq C$ implies that $|(g, f_n)| \leq C\,\|g\|$, that is, that for every given element g, the scalar products (g, f_n) form a bounded numerical sequence. Applying the Bolzano-Weierstrass theorem and the now classical diagonal process, we conclude that there exists a subsequence $\{f_{n_k}\}$ such that the sequences of scalar products (g, f_{n_k}) are convergent for the elements g of the denumerable set considered and therefore also for their linear combinations l. Using the language of functionals and setting

$$A_k g = (g, f_{n_k}),$$

this tells us that the sequence $\{A_k\}$ of linear functionals, with $M_{A_k} \leq C$, converges for a subset $\{l\}$ of the space L^2 which is everywhere dense in the sense of approximation in the mean. But the relation

$$|A_n g - A_m g| \leq |A_n g - A_n l| + |A_n l - A_m l| + |A_m l - A_m g| \leq$$
$$\leq 2C\|g - l\| + |A_n l - A_m l|$$

guarantees the convergence of the sequences $\{A_n g\}$ for all the g; in fact, for fixed g the first term of the last member can be made as small as we wish by a suitable choice of l, and then the second term will approach zero as $m, n \to \infty$. Finally, defining A by

$$\lim A_n g = Ag,$$

we shall obtain a functional A in the whole space L^2, which is obviously linear and whose generating function f is the weak limit of the sequence $\{f_{n_k}\}$. This completes the proof of the theorem of choice.

33. Orthonormal Systems

A finite or infinite system of elements of the space L^2 is said to be an *orthogonal and normalized* system, or briefly, an *orthonormal* system, if

$$(\varphi_m, \varphi_n) = 0 \text{ for } m \neq n \text{ and } (\varphi_n, \varphi_n) = \|\varphi_n\|^2 = 1.$$

We shall first consider a finite orthonormal system with elements $\varphi_1, \varphi_2, \ldots, \varphi_N$, and an arbitrary given element f of L^2. Let us try to approximate f, in the mean, by a linear combination of the φ_k, in the best possible way; that is, let us try to find the minimum of the distance

$$\|f - \sum_{k=1}^{N} c_k\varphi_k\|$$

by varying the coefficients c_k. We have, identically,

$$\|f - \sum_{k=1}^{N} c_k\varphi_k\|^2 = (f - \sum_{k=1}^{N} c_k\varphi_k, f - \sum_{j=1}^{N} c_j\varphi_j) =$$

$$= (f, f) - \sum_{j=1}^{N} \bar{c}_j(f, \varphi_j) - \sum_{k=1}^{N} c_k(\varphi_k, f) + \sum_{k=1}^{N} c_k\bar{c}_k =$$

$$= \|f\|^2 - \sum_{k=1}^{N} |(f, \varphi_k)|^2 + \sum_{k=1}^{N} |(f, \varphi_k) - c_k|^2.$$

This makes it clear that the minimum will be attained when

$$c_k = (f, \varphi_k) \qquad (k = 1, 2, \ldots, N).$$

Writing these values in our identity, we obtain the relation

$$\|f - \sum_{k=1}^{N} (f, \varphi_k)\varphi_k\|^2 = \|f\|^2 - \sum_{k=1}^{N} |(f, \varphi_k)|^2,$$

called *Bessel's identity*. Since the first member is obviously non-negative, it follows that

$$\sum_{k=1}^{N} |(f, \varphi_k)|^2 \leq \|f\|^2,$$

which is called *Bessel's inequality*.

From the inequality

$$\|f - \sum_{k=1}^{N} (f, \varphi_k)\varphi_k\| \leq \|f - \sum_{k=1}^{N} c_k\varphi_k\|,$$

which expresses the solution of the minimum problem we have just considered, we see in particular that if the element f can be approximated arbitrarily closely by linear combinations of the φ_k, then necessarily it is itself a linear combination of the φ_k, namely,

$$f = \sum_{k=1}^{n} (f, \varphi_k)\varphi_k;$$

in other words: the set of linear combinations $\sum_{k=1}^{N} c_k\varphi_k$ is *closed* with respect to convergence in the mean. It is, furthermore, impossible that this set should coincide with the entire space L^2, since then every system of $N + 1$ elements of L^2 would be linearly dependent, which obviously is not the case.

A set of elements of L^2 will be called *complete* (in L^2), if it determines the entire space, that is, if every element of the space can be approximated arbitrarily closely, in the mean, by the elements of this set, or by their linear combinations.

We just saw that a finite orthonormal system can not be complete in L^2. To prove the existence of a *denumerably infinite* complete orthonormal system, that is, of a *complete orthonormal sequence*, we can use the following *orthogonalization procedure* due to E. SCHMIDT [**1**].

Let there be given in L^2 a sequence (finite or infinite, complete or not) of elements $f_n \neq 0$ each of which is linearly independent of the preceding. We shall *orthogonalize* it—that is, replace it by an orthonormal sequence $\{\varphi_n\}$—in such a way that each φ_n is a linear combination of the f_m of index $m \leq n$, and conversely each f_n is a linear combination of the φ_m of index $m \leq n$:

(10a) $\varphi_n = c_{n1}f_1 + c_{n2}f_2 + \ldots + c_{nn}f_n, \quad f_n = \gamma_{n1}\varphi_1 + \gamma_{n2}\varphi_2 + \ldots + \gamma_{nn}\varphi_n.$

We begin by defining $\varphi_1 = f_1/\|f_1\|$; hence $c_{11}^{-1} = \gamma_{11} = \|f_1\|$.

To obtain φ_2, we subtract from f_2 a numerical multiple of φ_1 so that the difference

$$h_2 = f_2 - \gamma\varphi_1$$

is orthogonal to φ_1:

$$(h_2, \varphi_1) = (f_2, \varphi_1) - \gamma(\varphi_1, \varphi_1) = 0;$$

we shall have to choose $\gamma = (f_2, \varphi_1)$. Since f_2 does not depend linearly on f_1 or, equivalently, on φ_1, we necessarily have $h_2 \neq 0$ and we can set

$$\varphi_2 = h_2/\|h_2\|.$$

Then φ_2 is a linear combination of f_2 and φ_1, hence of f_2 and f_1, and conversely f_2 is a linear combination of h_2 and φ_1, or of φ_2 and φ_1.

In general, when we have already defined $\varphi_1, \varphi_2, \ldots, \varphi_{r-1}$ in such a way that we have relations (10a) for $n = 1, 2, \ldots, r - 1$, we construct φ_r as follows. We determine the constants λ_i in such a way that

$$h_r = f_r - \lambda_1\varphi_1 - \lambda_2\varphi_2 - \ldots - \lambda_{r-1}\varphi_{r-1}$$

is orthogonal to $\varphi_1, \varphi_2, \ldots, \varphi_{r-1}$, which gives $\lambda_i = (f_r, \varphi_i)$, $i = 1, 2, \ldots, r-1$. We necessarily have $h_r \neq 0$, because if this were not so f_r would be a linear combination of the $\varphi_1, \ldots, \varphi_{r-1}$, hence also of $f_1, \ldots, f_{r-1}$, which contradicts the hypothesis. Therefore we can set

$$\varphi_r = h_r/\|h_r\|.$$

Then the system $\varphi_1, \varphi_2, \ldots, \varphi_{r-1}$ is orthonormal; φ_r is a linear combination of f_r and of the φ_i of index $i \leq r-1$, hence a linear combination of the f_i, $i \leq r$; conversely, f_r is a linear combination of the h_r and the φ_i of index $i \leq r-1$, hence also a linear combination of the φ_i, $i \leq r$. We have therefore constructed φ_r so that the required relations (10a) hold also for $n = r$, and this completes our induction proof.

Now let us assume that the sequence $\{f_n\}$ we begin with is *complete*; such a sequence exists since the space L^2 is separable: we can take, for example, the characteristic functions—arranged in a sequence in some way—of all the intervals (r, R) with rational endpoints which are contained in the fundamental interval (a, b). We suppress from this sequence those terms—if there are any—which are linear combinations of the preceding; then we orthogonalize. The result is an orthonormal sequence $\{\varphi_n\}$ which obviously is *complete*.

A well-known example of a complete orthonormal system in the space $L^2(0, 1)$, taken from the theory of Fourier series, is formed by the functions

$$e^{2\pi i n x} \qquad (n = 0, \pm 1, \pm 2, \ldots).$$

We have already remarked that no *finite* orthonormal system can be complete in L^2. On the other hand, there is no nondenumerably infinite orthonormal system in L^2. This follows immediately from the general fact that *every subset of a separable metric space is itself separable.*[15] For a non-

[15] We restrict ourselves to the case of the space L^2. Let $\{f_n\}$ be a sequence everywhere dense in L^2 and let E be an arbitrary set of elements of L^2. For each pair of positive integers m, n, we choose one of the elements f of E, if there is one, such that

$$(*) \qquad \|f - f_n\| < \frac{1}{m}.$$

Denoting the element chosen by g_{mn}, the set of all these elements is obviously denumerable; we shall show that it is also everywhere dense in E. Let g be an arbitrary element of E and let $\varepsilon > 0$. First, we fix m so that $\frac{1}{m} < \frac{\varepsilon}{2}$. Since the set $\{f_n\}$ is everywhere dense, there exists an n for which $\|g - f_n\| < \frac{1}{m}$. For the integers m and n so chosen, the inequality (*) is satisfied by an element f of E (namely, by $f = g$); consequently, g_{mn} is defined and we have

$$\|g - g_{mn}\| \leq \|g - f_n\| + \|g_{mn} - f_n\| < \frac{1}{m} + \frac{1}{m} < \varepsilon,$$

which proves our assertion.

denumerable orthonormal system can not be separable since, due to orthogonality, no element of the system can be approximated arbitrarily closely by linear combinations of the other elements of the system.

We can summarize by saying that L^2 is a space whose *dimension* is *denumerably infinite.*

Consider now an infinite orthonormal sequence $\{\varphi_k\}$ and an arbitrary element f of L^2. Applying Bessel's inequality to the finite system $\varphi_1, \varphi_2, \ldots, \varphi_N$, then letting N tend to infinity, we obtain

$$\sum_{k=1}^{\infty} |(f, \varphi_k)|^2 \leq \|f\|^2;$$

therefore Bessel's inequality remains valid in this case. Since the series in the first member is convergent, we have

$$\|\sum_{k=m}^{n} (f, \varphi_k)\varphi_k\|^2 = \sum_{k=m}^{n} |(f, \varphi_k)|^2 \to 0 \quad \text{when } m, n \to \infty.$$

Hence, by the Riesz-Fischer theorem, the series

$$\sum_{k=1}^{\infty} (f, \varphi_k)\varphi_k \tag{10b}$$

converges in the mean to some element g of L^2. Since mean convergence implies weak convergence, we have

$$(g, \varphi_i) = \lim_{n\to\infty} \sum_{k=1}^{n} (f, \varphi_k)\,(\varphi_k, \varphi_i) = (f, \varphi_i),$$

hence

$$(f - g, \varphi_i) = 0 \qquad (i = 1, 2, \ldots),$$

that is, $f - g$ is orthogonal to all the elements of the given orthonormal sequence, and therefore to their linear combinations as well.

When the orthonormal sequence is *complete*, these linear combinations are everywhere dense in L^2, and consequently $f - g$ is orthogonal to every element of L^2; in particular it is orthogonal to itself, and consequently

$$\|f - g\|^2 = (f - g, f - g) = 0, \quad f - g = 0, \quad f = g.$$

Hence we have the

THEOREM. *Given a complete orthonormal sequence in L^2, every element f of the space L^2 admits a development, convergent in the mean:*

$$f = \sum_{1}^{\infty} (f, \varphi_k)\varphi_k.$$

It follows that for a *complete* orthonormal sequence the first member of Bessel's identity tends to 0 when $n \to \infty$, therefore Bessel's inequality can in this case be replaced by the following equation, called Parseval's formula:

$$\sum_1^\infty |(f, \varphi_k)|^2 = \|f\|^2.$$

The more general formula

$$\sum_1^\infty (f_1, \varphi_k)\,\overline{(f_2, \varphi_k)} = (f_1, f_2),$$

valid for two arbitrary elements f_1, f_2 of L^2, is also called Parseval's formula. The second formula follows from the first by applying the latter to $f_1 + f_2$, $f_1 - f_2$, $f_1 + if_2$, and $f_1 - if_2$ in place of f, and making use of the very useful identity

$$4(f_1, f_2) = \|f_1 + f_2\|^2 - \|f_1 - f_2\|^2 + i\|f_1 + if_2\|^2 - i\|f_1 - if_2\|^2.$$

In the proof of the convergence of the series (10b) we used only the fact that the squares of the moduli of its coefficients form a convergent series. Therefore, more generally, we have the

THEOREM. *Given an orthonormal sequence $\{\varphi_n\}$ in L^2 (complete or not) and an arbitrary sequence $\{a_n\}$ of numbers such that the series*

$$\sum_1^\infty |a_n|^2$$

converges, the series

$$\sum_1^\infty a_n\varphi_n$$

will converge in the mean to an element g of L^2 and we shall have

$$(g, \varphi_n) = a_n \qquad (n = 1, 2, \ldots).$$

It is in this form that the Riesz-Fischer theorem was originally stated by F. RIESZ.

34. Subspaces of L^2. The Decomposition Theorem

A subset E of the space L^2 is called a *linear set* (or a *linear manifold*) if it contains all the linear combinations of its elements. When the set E is also *closed* with respect to convergence in the mean it is called a *subspace* of L^2.

A linear set of finite dimension, that is, one which consists of the linear combinations

$$c_1f_1 + c_2f_2 + \ldots + c_Nf_N$$

of N fixed elements $f_1, f_2, \ldots, f_N$, is *always closed,* hence always a subspace of L^2. We proved this assertion in the preceding section for the case where these fixed elements form an orthonormal system; the general case can obviously be reduced to it by orthogonalization.

We have the following theorem, which is of great importance because of its applications.

DECOMPOSITION THEOREM. *If E is a subspace of L^2, every element h of L^2 can be uniquely decomposed into the sum of an element f belonging to E and an element g which is orthogonal to E (that is, orthogonal to all the elements of E).*

The *uniqueness* of the decomposition is immediate; in fact, if $h = f + g$ and $h = f' + g'$ are two decompositions of the type required, the difference $f - f'$ must belong to E; on the other hand, since it is equal to $g' - g$, it is orthogonal to E; in particular it is orthogonal to itself, which implies that

$$\|f - f'\|^2 = (f - f', f - f') = 0,$$

that is, $f = f'$ and $g = g'$ almost everywhere.

We can prove the *existence* of the decomposition by using an orthonormal sequence $\{\varphi_n\}$ of elements of E which is complete in the subspace E in the same sense that we have defined for the entire space L^2. We obtain a sequence of this type by orthogonalizing a sequence which is everywhere dense in E. Then, obviously, we have only to set

$$f = \sum_n (h, \varphi_n)\varphi_n \text{ and } g = h - f.$$

Here is another proof of the existence of the decomposition which is more direct and *does not make use of the fact that the space L^2 is separable.*[16]

Let h be fixed and let f be a variable element from the subspace E. Consider the "distances" $\|h - f\|$; let d be their greatest lower bound and let $\{f_n\}$ be an extremal sequence selected from E, i.e. a sequence such that

$$\|h - f_n\| \to d.$$

In equation (9), Sec. 30, put $h - f_m$ in place of u and $h - f_n$ in place of v, and therefore $h - \frac{f_m + f_n}{2}$ in place of $\frac{1}{2}(u + v)$. Since f_m and f_n belong to E, $\frac{f_m + f_n}{2}$ also belongs to E and consequently

$$\left\| h - \frac{f_m + f_n}{2} \right\| \geq d.$$

From this we obtain

$$\|f_m - f_n\|^2 \leq 2\|h - f_m\|^2 + 2\|h - f_n\|^2 - 4d^2 \to 2d^2 + 2d^2 - 4d^2 = 0;$$

hence, by virtue of the Riesz-Fischer theorem, the sequence $\{f_n\}$ converges

[16] See F. Riesz [**15**]; this is an adaptation of a previous argument by B. Levi in his research on the Dirichlet principle, see Levi [**2**] (§ 7).

in the mean to an element f^* of E for which $\|h - f^*\| = d$. Setting $g = h - f^*$, we assert that g is orthogonal to every element f of E. This is evident for $f = 0$; for other f, we have only to refer to the inequality

$$\|h - f^*\|^2 \leq \|h - f^* - \lambda f\|^2 = \|h - f^*\|^2 - \bar{\lambda}(h - f^*, f) - \lambda(f, h - f) + \lambda\bar{\lambda}(f, f)$$

and set

$$\lambda = \frac{(h - f^*, f)}{(f, f)}, \quad \bar{\lambda} = \frac{(f, h - f^*)}{(f, f)},$$

which yields

$$\frac{|(h - f^*, f)|^2}{\|f\|^2} \leq 0;$$

therefore $(h - f^*, f) = 0$, which was to be proved.

Let us add that this proof of the existence of an extremal element remains valid, without any modification, for *every convex, closed set* E, that is, for every set E which contains $c_1 f_1 + c_2 f_2$ ($c_1, c_2 \geq 0$, $c_1 + c_2 = 1$) with f_1 and f_2, as well as the limits of the mean convergent sequences whose elements belong to E.

Here is a corollary to our theorem.

Given a set E of elements f in L^2 — or more generally in a subspace F of L^2 — whose linear combinations, in the sense of convergence in the mean, are not everywhere dense in F, there are non-zero elements g in F which are orthogonal to all the elements of E.

Here is another variant of the theorem.

Given in L^2 a set E of elements f whose linear combinations are not everywhere dense in L^2 in the sense of mean convergence, the elements orthogonal to all the elements of E form a subspace and those orthogonal to these latter form a second subspace which, on the other hand, is only the set of the above-mentioned linear combinations and of their limits, or, equivalently, it is the smallest subspace containing E; it is also called the subspace determined or subtended by E. Furthermore, each element f of L^2 decomposes in one and only one way into $f = g + h$, where g and h belong respectively to the two "complementary" orthogonal subspaces.

35. Another Proof of the Theorem of Choice. Extension of Functionals

Here are two interesting consequences of this theorem. The first is a proof of the theorem of choice, which differs from that given in Sec. 32 in that it does not make use of the separability of the space L^2.

Let $\{f_n\}$ be a bounded sequence of elements of L^2. By the diagonal process, we establish the existence of a subsequence $\{f_{n_k}\}$ for which the sequence of scalar products (f_m, f_{n_k}) is convergent for any fixed m. The sequence (f, f_{n_k}) is then convergent for every element f of L^2. This is verified first for linear combinations of the f_m and for the limits of these combinations just as in Sec. 32. If these elements exhaust L^2, our assertion is proved. Otherwise, we shall use the decomposition $f = g + h$, where g belongs to the subspace determined by the elements f_m, and h is orthogonal to these elements and therefore in particular to the elements f_{n_k}. It follows that the sequence

$$(f, f_{n_k}) = (g, f_{n_k}) + (h, f_{n_k}) = (g, f_{n_k})$$

is also convergent.

The second consequence of the decomposition theorem which we have in mind concerns the problem of the extension of linear functionals.

THEOREM. *Let A be a functional defined only for the elements f of the set E and assume that there exists a bound M such that, for every linear combination formed from elements of E,*

$$\left|\sum_1^n c_k A f_k\right| \leq M \left\|\sum_1^n c_k f_k\right\|.$$

Then the functional can be extended to a linear functional A operating on the entire space L^2 and such that $M_A \leq M$.

In fact, to arrive at this result we define A, first for linear combinations by

$$A\left(\sum_1^n c_k f_k\right) = \sum_1^n c_k A f_k,$$

then for their limits by an obvious passage to the limit, and finally by setting A equal to zero for the elements orthogonal to E; then if we use the above-mentioned decomposition $f = g + h$ and set $Af = Ag + Ah$, A will be defined for all f.

36. The Space L^p and its Linear Functionals

Let us discuss briefly the spaces L^p $(p \geq 1)$, which are made up of the measurable functions f, defined in a measurable set e, for which $|f|^p$ is summable. Here the norm $\|f\|$ is defined by

$$\|f\| = \left[\int_e |f(x)|^p dx\right]^{\frac{1}{p}};$$

we have $\|f\| = 0$ if and only if $f(x) = 0$ almost everywhere in e; we agree not to distinguish between two functions equal almost everywhere, just as in the case of L^2.

Minkowski's inequality, or equivalently the triangle inequality, is valid for this norm, as we have seen, whenever $p > 1$, and obviously also for $p = 1$ and for the limiting case $p = \infty$. As for this last case, we understand by L^∞ the space of measurable functions which are bounded or are equal almost everywhere to bounded functions, and we define the norm $\|f\|$ in L_∞ to be the "true maximum" of $|f(x)|$, that is, the smallest value of M for which $|f(x)| \leq M$ almost everywhere. The reason for considering this space as a limiting case of the L^p spaces is that on sets of finite measure, as is easily seen, the norms of f corresponding to different exponents p tend, as $p \to \infty$, to the bound M considered.

To arrive at the idea of scalar product, it is necessary to consider, besides L^p, a second space L^q, where $\frac{1}{p} + \frac{1}{q} = 1$ and therefore $q = \frac{p}{p-1}$, $p = \frac{q}{q-1}$; in the limiting cases $p = 1$ and $p = \infty$ we have $q = \infty$ and $q = 1$. The scalar product can be defined in a way analogous to that in L^2; however in the general case one of the two factors must belong to L^p, the other to L^q.

To simplify the writing we shall consider only the real space L^p of real-valued functions—the modifications necessary for the case of complex-valued functions can be made without difficulty.

Mean convergence is defined for $p \geq 1$ just as for $p = 2$, where of course we now denote by $\|f - f_n\|$ and $\|f_n - f_m\|$ the respective norms formed with the exponent p. The convergence criterion which corresponds here to the Riesz-Fischer theorem — briefly, *the Cauchy criterion* — is established just as in L^2, and even more easily in the particular case $p = 1$. As to the limiting case $p = \infty$, it is essentially only a question there of uniform convergence, that is, of a classical criterion. We should add that the definitions also apply to positive exponents $p < 1$, but an extension in this direction brings little of interest.[17]

We shall define *weak convergence* in the space L^p, $1 \leq p < \infty$, by saying that the sequence $\{f_n\}$ in L^p converges weakly to f in L^p if for every g of the "conjugate" space L^q,

$$(f_n, g) = \int_e f_n(x)g(x)dx \to (f, g).$$

This can also be expressed in terms of linear functionals, just as in the case $p = 2$. In fact the product (f, g), with f fixed in L^p and g variable in L^q, defines in the latter a linear functional with properties analogous to those it possesses for $p = 2$. This is evident; what is less immediate is the converse, namely that, at least in the case $1 \leq q < \infty$, every linear functional in L^q (the function-

[17] For $p < 1$, there are no linear functionals in L^p except the linear functional which is identically zero; see DAY [1].

al is of course supposed real in the case we are considering) can be written in the form (f, g), where the generating function $f(x)$ belongs to the space L^p.

Before taking up this problem we shall establish a necessary and sufficient condition that a given function $F(x)$, defined in a finite or infinite interval (a, b), should be the integral of a function $f(x)$ belonging to L^p.

LEMMA.[18] *A necessary and sufficient condition that the function $F(x)$ be the integral of an element $f(x)$ of the space L^p $(1 < p < \infty)$, is that the sums*

$$\sum_{1}^{m} \frac{|F(x_k) - F(x_{k-1})|^p}{(x_k - x_{k-1})^{p-1}}, \tag{10}$$

formed for every finite system of points

$$x_0 < x_1 < \ldots < x_{m-1} < x_m$$

lying in (a, b), have a finite least upper bound. The least upper bound is simply the integral of $|f(x)|^p$ on (a, b).

First, let us note that if the lemma is valid for every finite interval it is also valid for every infinite interval. Hence it suffices to prove it for finite intervals (a, b).

The necessity follows immediately from Hölder's inequality applied to $f(x)$ and 1,

$$|F(x_k) - F(x_{k-1})|^p \leq [\int_{x_{k-1}}^{x_k} |f(x)|dx]^p \leq (x_k - x_{k-1})^{p-1} \int_{x_{k-1}}^{x_k} |f(x)|^p dx. \tag{11}$$

Now assume that the sums (10) are bounded; let B^p be their least upper bound. The sum analogous to (10), formed for a system of non-overlapping intervals (α_k, β_k) (not necessarily covering the interval (a, b)) remains, a fortiori, below B^p; from this, applying Hölder's inequality for finite sums,[19] we arrive at the inequality

$$\sum |F(\beta_k) - F(\alpha_k)| \leq$$

$$\leq \left[\sum \frac{|F(\beta_k) - F(\alpha_k)|^p}{(\beta_k - \alpha_k)^{p-1}}\right]^{\frac{1}{p}} [\sum (\beta_k - \alpha_k)]^{\frac{p-1}{p}} \leq B[\sum(\beta_k - \alpha_k)]^{\frac{p-1}{p}}$$

which proves that the function $F(x)$ is absolutely continuous and that, consequently, it possesses almost everywhere a derivative $F'(x)$ of which it is the indefinite integral. Now this derivative is the limit of a sequence of step functions $f_n(x)$ which are formed with the aid of a sequence of decompositions of the interval (a, b); one way to obtain such a sequence is to divide (a, b)

[18] F. RIESZ [**6**] (§ 5).

[19] $|\Sigma a_k b_k| \leq \{\Sigma |a_k|^p\}^{\frac{1}{p}} \{\Sigma |b_k|^q\}^{\frac{1}{q}} \left(p > 1, \quad q = \frac{p}{p-1}\right).$

successively into 2^n equal segments and to set $f_n(x)$ equal, in each segment, to the increment ratio

$$\frac{F(\beta) - F(\alpha)}{\beta - \alpha}$$

corresponding to the segment. It is clear that the sum (10) which corresponds to the decomposition considered represents precisely the integral of $|f_n|^p$ over (a, b). It follows from Fatou's lemma that $|F'(x)|^p$ is summable and that its integral does not exceed the bound B^p. On the other hand, in view of (11) the sums (10), and therefore also B^p, do not exceed this integral; consequently the integral equals B^p, which was to be proved.

Let us return to the problem of the representation of linear functionals in the form of an integral. Without loss of generality, we can restrict ourselves to the case where the measurable set e in question is simply a finite or infinite interval (a, b).[20] Let Ag be a linear functional in L^q space $(1 \leq q < \infty)$ and set $F(x)$ equal to Ag_x, where $g_x(\xi)$ is the characteristic function of the interval (a, x).

First consider the case where $1 < q < \infty$. We shall show that $F(x)$ satisfies our lemma with $p = \dfrac{q}{q-1}$ and that consequently it is the indefinite integral of a function $f(x)$ belonging to L^p. In fact, consider the step function $\varphi(\xi)$ which on the segments (x_{k-1}, x_k) assumes respectively the constant values

$$\frac{|F(x_k) - F(x_{k-1})|^{p-1} \operatorname{sgn} (F(x_k) - F(x_{k-1}))}{(x_k - x_{k-1})^{p-1}}.$$[21]

Then

$$A\varphi = \sum_1^m \frac{|F(x_k) - F(x_{k-1})|^p}{(x_k - x_{k-1})^{p-1}};$$

on the other hand, since $q(p-1) = q\left(\dfrac{q}{q-1} - 1\right) = p$,

$$A\varphi \leq M_A\|\varphi\| = M_A \left[\int_a^b |\varphi(\xi)|^q d\xi\right]^{\frac{1}{q}} = M_A\left[\sum_1^m \frac{|F(x_k) - F(x_{k-1})|^p}{(x_k - x_{k-1})^{p-1}}\right]^{\frac{1}{q}},$$

[20] In fact, if e is a measurable subset of an interval (a, b), every linear functional of the space L^q of functions defined in e can be considered as a linear functional of the space L^q of functions defined in (a, b)—a functional which depends only on the values which these functions take on in e. If f is the generating function of this functional, we have, therefore,

$$\int_a^b g(x)f(x)dx = 0$$

for every function $g(x)$ belonging to L^q which takes on the value zero in e; this implies that $f(x)$ is zero in the complementary set.

[21] $\operatorname{sgn} a = 1, 0, -1$, according as $a > 0$, $a = 0$, $a < 0$.

hence

$$\sum_1^m \frac{|F(x_k) - F(x_{k-1})|^p}{(x_k - x_{k-1})^{p-1}} \leq M_A^p.$$

Consequently $F(x)$ is the indefinite integral of a function $f(x)$ belonging to the space L^p and such that

$$\int_a^b |f(x)|^p dx \leq M_A^p.$$

Finally, the relation

$$Ag = \int_a^b g(x)F'(x)dx = (g, f)$$

is obvious for step functions g and since the set of step functions is everywhere dense in L^q, the relation holds in the entire space. Applying Hölder's inequality, we establish that

$$|Ag| = |\int_a^b g(x)f(x)dx| \leq [\int_a^b |g(x)|^q dx]^{\frac{1}{q}} [\int_a^b |f(x)|^p dx]^{\frac{1}{p}},$$

hence that

$$M_A^p \leq \int_a^b |f(x)|^p dx;$$

combining this with the inequality obtained above, it follows that the equality sign holds.

Now let us consider the simple case $p = 1$. Then we have for every pair of points x_1, x_2 in (a, b):

$$|F(x_2) - F(x_1)| = |A(g_{x_2} - g_{x_1})| \leq M_A \int_a^b |g_{x_2}(\xi) - g_{x_1}(\xi)|d\xi = M_A|x_2 - x_1|$$

hence $F(x)$ satisfies the Lipschitz condition with the constant M_A. It follows from this (Sec. 25) that $F(x)$ is the indefinite integral of a function $f(x)$ such that $|f(x)| \leq M_A$. It can be shown just as in the preceding case that

$$Ag = \int_a^b g(x)f(x)dx = (g, f).$$

It follows that

$$|Ag| \leq [\text{true max } |f(x)|] \int_a^b |g(x)|dx,$$

hence

$$M_A \leq \text{true max } |f(x)|;$$

since we also have $|f(x)| \leq M_A$, it follows that M_A is equal to the true maximum of $|f(x)|$.

Thus, interchanging the letters p and q, f and g, we have the

THEOREM.[22] *Every linear functional Af defined in the space L^p $(1 \leq p < \infty)$ gives rise to a generating function $g(x)$ in the space L^q which it determines almost everywhere, and by means of which it can be written in the form*

$$Af = (f, g);$$

the norm of the functional equals the norm of the generating function:

$$M_A = [\int_e |g(x)|^q dx]^{\frac{1}{q}} \text{ for } p > 1,\ q < \infty,$$

$$M_A = \text{true max } |g(x)| \text{ for } p = 1,\ q = \infty.$$

We add that, as we shall see later (Sec. 87), this theorem does not hold for $p = \infty$; that is, there exist linear functionals in L^∞ which are not generated by functions belonging to L^1.

By use of the theorem we have just proved, the theorem of choice as well as its proof given in Sec. 32 extend almost word for word to L^p spaces.

37. A Theorem on Mean Convergence

Here is another theorem which, as we have seen, is established by a very easy calculation when $p = 2$, but which requires a more profound examination in the general case.

THEOREM.[23] *If the sequence $\{f_n\}$ of elements of the space L^p $(1 < p < \infty)$ converges weakly to f, and if, furthermore,*

$$\|f_n\| \to \|f\|,$$

then the sequence $\{f_n\}$ also converges in the mean to f.

First, assume $p \geq 2$; then we have, for every real value of z,

$$(12) \qquad |1 + z|^p \geq 1 + pz + c|z|^p,$$

where c is some positive constant independent of z. In fact, consider the ratio

$$(13) \qquad \frac{|1 + z|^p - 1 - pz}{|z|^p};$$

the second derivative of the expression appearing in the numerator, which equals

$$p(p - 1)|1 + z|^{p-2},$$

is positive and becomes zero only for $z = -1$; since, furthermore, the expression itself becomes zero for $z = 0$ together with its first derivative, it will

[22] F. Riesz [**6**] (§ 11).

[23] Radon [**1**] (p. 1363); also, see F. Riesz [**12**].

be strictly positive for $z \neq 0$. For $z = 0$, the ratio (13) becomes infinite or equal to 1 according as $p > 2$ or $p = 2$; hence our ratio will be strictly positive for every finite value of z; since, finally, for $|z| = \infty$ it tends to a limit equal to 1, it follows that it possesses a strictly positive minimum c, which was to be proved.

In the inequality we have just established, replace z by

$$\frac{f_n(x) - f(x)}{f(x)};$$

multiplying by $|f|^p$ and integrating, we find that

$$\int_e |f_n|^p dx \geq \int_e |f|^p dx + p \int_e |f|^{p-2} f(f_n - f)\,dx + c \int_e |f_n - f|^p dx,$$

and since, according to the weak convergence hypothesis, the second of the integrals appearing in the second member tends to 0 with $1/n$, while, according to the second hypothesis made, the first of these integrals is the limit of the integral in the first member, it follows that

$$\int_e |f_n - f|^p dx \to 0; \tag{14}$$

this is the theorem that was to be proved.

When $1 < p < 2$ we consider, instead of ratio (13), the function given by this same expression for $|z| \geq 1$, and by

$$\frac{|1 + z|^p - 1 - pz}{z^2}$$

for $|z| \leq 1$. For $z = 0$ this function has the limit $\frac{1}{2}p(p - 1)$, and since, just as in the first case, it is strictly positive for every other finite or infinite value of z, it possesses a positive minimum. If we now apply the reasoning of the first case, we do not obtain immediately relation (14), which is to be proved, but only the following relation:

$$\int_{e_n} |f_n - f|^p dx + \int_{e-e_n} (f_n - f)^2 |f|^{p-2} dx \to 0,$$

where we have denoted by e_n the set of values x for which

$$|f_n(x) - f(x)| \geq |f(x)|.$$

Hence these two integrals tend to zero. In order to deduce relation (14) from this, we must show that the relation

$$\int_{e-e_n} (f_n - f)^2 |f|^{p-2} dx \to 0$$

implies the relation

$$\int_{e-e_n} |f_n - f|^p dx \to 0.$$

To do this, we have only to apply the Schwarz inequality, keeping in mind the inequality $|f_n(x) - f(x)| < |f(x)|$ which is valid by hypothesis for x in the set $e - e_n$. In fact, we have

$$\int_{e-e_n} |f_n - f|^p dx \leq \int_{e-e_n} |f|^{p-1}|f_n - f|\, dx \leq [\int_{e-e_n} |f|^p dx]^{\frac{1}{2}} [\int_{e-e_n} (f_n - f)^2 |f|^{p-2} dx]^{\frac{1}{2}} \leq$$
$$\leq [\int_e |f|^p dx]^{\frac{1}{2}} [\int_{e-e_n} (f_n - f)^2 |f|^{p-2} dx]^{\frac{1}{2}} \to 0.$$

Thus the theorem is also proved for the case $p < 2$.

38. A Theorem of Banach and Saks

In this same order of ideas we mention the following

THEOREM. *Given in L^p a sequence $\{f_n\}$ which converges weakly to an element f, we can select a subsequence $\{f_{n_k}\}$ such that the arithmetic means*

$$\frac{f_{n_1} + f_{n_2} + \ldots + f_{n_k}}{k}$$

converge in the mean to f.

This theorem is due to the two Polish geometers S. BANACH and S. SAKS, whose work and, in particular, the importance of whose research in the topics treated in this book are widely acknowledged. Their proof is similar to the one we have just presented, but is more laborious in its details; by referring the reader to the memoir cited below,[24] we content ourselves with the consideration of the case $p = 2$.

Replacing f_n by $f_n - f$, we can assume $f = 0$. We shall successively choose the n_k in the following manner. Beginning for definiteness with $n_1 = 1$, let n_2 be an index or, if we wish, the first index such that $|(f_1, f_n)| \leq 1$; this choice is possible since $(f_1, f_n) \to 0$. In general, after having chosen $f_{n_1}, f_{n_2}, \ldots, f_{n_k}$, we choose $n_{k+1} > n_k$ so that

$$|(f_{n_1}, f_{n_{k+1}})| \leq \frac{1}{k}, \ldots, |(f_{n_k}, f_{n_{k+1}})| \leq \frac{1}{k},$$

which is possible since $(f_{n_i}, f_n) \to 0$ ($i = 1, 2, \ldots, k$; $n \to \infty$). Since, furthermore, the norms $\|f_n\|$ form a bounded sequence, say $\|f_n\| \leq B$, it follows that

$$\left\| \frac{f_{n_1} + f_{n_2} + \ldots + f_{n_k}}{k} \right\|^2 \leq \frac{kB^2 + 2\cdot 1 + 4\cdot\frac{1}{2} + \ldots + 2(k-1)\dfrac{1}{k-1}}{k^2} <$$
$$< \frac{B^2 + 2}{k} \to 0,$$

[24] BANACH–SAKS [1]. We remark that the same theorem has been established for the functional space C which we shall consider in Chapt. III and even for every "Banach" space, which we shall define in Chapt. V; see MAZUR [1].

which was to be proved. Instead of writing these formulas, we could have said briefly that we successively choose the functions f_{n_k} so that each is almost orthogonal to its predecessors.

Finally, here is an important corollary to this theorem which follows from it immediately.

THEOREM. *If a linear set, or more generally a convex set, in L^p is closed in the sense of convergence in the mean, it is also closed in the sense of weak convergence.*

FUNCTIONS OF SEVERAL VARIABLES

39. Definitions. Principle of Transition

We shall consider only functions of two variables; this will suffice to make the general case clear.

In order to construct a theory of integration, we shall need only to imitate the definitions and the methods which we used for functions of a single variable, starting, however, with step functions of two variables, that is, with functions constant over rectangles (or, if we wish, over squares) whose sides are parallel to the axes. The majority of the paragraphs will transfer to this case without modification, with the rectangles or squares playing the role of the intervals; of course, "set of measure zero" and "almost everywhere" will have to be defined in terms of rectangles and their areas instead of intervals and their lengths. One of the new problems to arise will be that of successive integrations; furthermore, we shall have to pay special attention to some of the details and difficulties concerning the reciprocal relationship of integration and differentiation.

Between the integral of functions of two variables thus defined and the integral of functions of a single variable there is not only an analogy, but a close connection which permits us to transfer most of the theorems proved in the case of a single variable to the case of two variables.

Let us begin by observing that in the theory which we have just developed we could have started from a special class of step functions, namely those whose discontinuities occur at the triadic fractions $m3^{-n}$, or equivalently, those which are linear combinations of the characteristic functions of intervals of the type $(m3^{-n}, (m+1)3^{-n})$. Indeed, all step functions are limits of increasing sequences of functions of this type. We obtain analogous functions in the (x, y) plane by considering the characteristic functions of the squares

$$m3^{-n} \leq x < (m+1)3^{-n}, \quad p3^{-n} \leq y < (p+1)3^{-n}$$

and the linear combinations of these functions. For definiteness we have taken the squares to be half-open; in the case of a single variable this corresponds to the half-open intervals $m3^{-n} \leqq x < (m+1)3^{-n}$. However, the points which we thus keep from counting several times form a set of measure zero, so this convention is not essential.

For step functions of this type the integral is defined in the same way as in the case of a single variable, but of course using areas instead of lengths; this suggests establishing a one-to-one correspondences between the two types of functions — a correspondence leaving the integrals invariant. This is done in an obvious way by establishing first a one-to-one correspondence between the intervals and the squares in such a way that the measure is preserved. We begin with $n = 0$, that is, by decomposing the x-axis into intervals $(m, m + 1)$ of unit length and setting up a fixed, but arbitrary, one-to-one correspondence between these intervals and the meshes of the square grill of unit mesh. We continue by setting up successive one-to-one correspondences between the squares of width 3^{-n} and the intervals of length 9^{-n}, always taking care to pair the squares and the intervals in such a way that each square and interval which correspond are contained in a square and an interval which were paired in the preceding subdivision. This procedure will then introduce in an obvious way a correspondence between the two classes of step functions which we have just considered — functions of respectively one and two variables. The correspondence clearly preserves the values of integrals, and this property immediately extends to the limit functions. Also, we shall establish an "almost" one-to-one correspondence between the points of the x-axis and the points of the (x, y) plane by pairing those points which belong to corresponding nests of intervals and squares. The points for which this correspondence is not uniquely determined have finite triadic abscissas or ordinates and thus form a set of measure zero. Furthermore, if we consider in particular the characteristic functions of sets which correspond, we see that these sets are either simultaneously measurable or not measurable, and that when they are measurable the two measures are equal. Moreover, we also see immediately that two sets which correspond always have the same exterior measure.

We observe that a particular correspondence which satisfies our requirements, and to which we could have restricted ourselves, can be defined by writing the coordinates x, y in the triadic system:

$$x = \ldots a_{-2}a_{-1}a_0 . a_1 a_2 \ldots; \quad y = \ldots b_{-2}b_{-1}b_0 . b_1 b_2 \ldots$$

and letting the point on the x-axis

$$x = \ldots a_{-2}b_{-2}a_{-1}b_{-1}a_0 b_0 . a_1 b_1 a_2 b_2 \ldots$$

correspond to the point (x, y). We could also have used the diadic system, the decimal system, or other analogous systems.[25] We could even have

[25] The particular choice of the triadic system was made with certain later applications in mind.

made the correspondence continuous by arranging it to make neighboring squares correspond to neighboring intervals, as is done in the construction of Peano's curve and analogous curves.

For most of the notions introduced and results established in the case of a single variable, our principle of transition yields immediately the corresponding notions and results in the case of two variables, and the same principle, slightly modified, is applicable to functions of several or even of an infinite number of variables.[26] It is superfluous to formulate all that follows in an obvious fashion from it; rather we shall pass on to the new problems which arise, such as that of the computation of the double integral by successive integrations, and to other problems for which our principle furnishes only a partial result; the most important of these are problems concerning differentiation.

40. Successive Integrations. Fubini's Theorem

With regard to the calculation of double integrals by successive integrations, let us begin by establishing the relation which exists between the sets of measure zero of two sorts, linear and planar.

A plane set E of measure zero can be enclosed, by definition, in a finite number or a denumerable infinity of squares which belong to our square grills and whose total area is arbitrarily small. Or equivalently, just as in the linear case, E can be enclosed in a sequence of squares whose areas form a series with finite sum and which has the property that each point of E belongs to an infinite number of squares. Since the areas of these squares are simply the integrals of their characteristic functions $e_n(x, y)$, these integrals form a convergent series, and therefore Beppo Levi's theorem, or our Lemma B, Sec. 16, is applicable. Since the calculation of the double integral by successive integrations is clearly justified for step functions, the series of the integrals of the $e_n(x, y)$ with respect to y is convergent for almost all x. That is, for almost all x the points of E with abscissa x are contained, each an infinite number of times, in intervals which belong to the vertical line with abscissa x and whose lengths form a convergent series. Consequently, these points constitute a set of linear measure zero. Obviously, the same conclusion remains when we interchange the roles of x and y.

As for the general problem, it is clear that it will suffice to consider functions $f(x, y)$ defined in the rectangle

$$R = [a < x < b,\ c < y < d]$$

and belonging to the class C_1, that is, functions for which there exists a non-

[26] LEBESGUE [**2**] (p. 367); RIESZ [**6**] (p. 497); JESSEN [**1**].

decreasing sequence of step functions $\varphi_n(x, y)$ tending everywhere in R, except perhaps in a set E of (planar) measure zero, to $f(x, y)$, and for which

$$\iint_R \varphi_n(x, y)dx\,dy = \int_a^b \left(\int_c^d \varphi_n(x, y)dy\right)dx \to \iint_R f(x, y)dx\,dy. \tag{15}$$

By virtue of B. Levi's theorem, the sequence of functions

$$\Phi_n(x) = \int_c^d \varphi_n(x, y)dy$$

will then be convergent for almost all x. Let us fix a value $x = \xi$ for which the sequence $\{\Phi_n(\xi)\}$ converges and such that the points of E with abscissa ξ constitute a set of linear measure zero. We shall then have

$$\varphi_n(\xi, y) \to f(\xi, y)$$

almost everywhere with respect to y; therefore, by B. Levi's theorem, $f(\xi, y)$ is a summable function of y and we shall have

$$\int_c^d \varphi_n(\xi, y)dy \to \int_c^d f(\xi, y)dy. \tag{16}$$

Now the possible values of ξ fill the interval (a, b) except for a set of measure zero, therefore

$$\int_c^d f(x, y)dy \tag{17}$$

has a meaning for almost all x, and it follows from (15) and (16), again by B. Levi's theorem, that (17) is a summable function of x, and that

$$\int_a^b \left(\int_c^d f(x, y)dy\right)dx = \iint_R f(x, y)dx\,dy.$$

This result is expressed by the following theorem.

FUBINI'S THEOREM.[27] *The integral of a summable function of two variables can be calculated by successive integrations.*

The same result holds, for the same reasons, for summable functions of more than two variables; we can integrate successively, either separately with respect to each variable or by groups, exactly as is done in classical analysis.

41. The Derivative Over a Net of a Non-negative, Additive Rectangle Function. Parallel Displacement of the Net

For problems concerning differentiation it is primarily a question of the differentiation of additive functions of a rectangle — or *rectangle functions*.

[27] FUBINI [**2**].

Recall that in the case of a single variable, a nondecreasing function $f(x)$ can be considered as defining a non-negative, additive interval function $f(\alpha, \beta) = f(\beta) - f(\alpha)$, or if we wish, a distribution of non-negative masses, and conversely, a distribution of non-negative masses in an interval (a, b) gives rise to a nondecreasing function $f(x) = f(a, x)$. Other, slightly different, conventions would consist of setting $f(\alpha, \beta) = f(\beta - 0) - f(\alpha - 0)$ or $f(\alpha, \beta) = f(\beta + 0) - f(\alpha + 0)$. Obviously, for continuous functions and in particular for indefinite integrals these conventions are not important, and even in the general case it is easy to see that these distinctions have no essential influence on the derivatives, except at most at points of discontinuity; but these points form a set which is at most denumerable and, a fortiori, of measure zero.

In the case of two variables we start directly from *rectangle functions* $f(R)$, where R denotes a variable rectangle whose sides remain parallel to the axes. We can obviously attach a function of x, y to f by choosing for R the rectangle which has as opposite vertices the origin (or some other fixed point) and the point (x, y). But this is of little interest to us and intervenes only when we wish to express the results in terms of the mixed second derivative of a function $f(x, y)$. Let us assume that the function $f(R)$ is additive in an obvious sense, and begin by considering the question of the existence of its derivative. Because of the analogy with monotonic functions and also in view of the most important applications, let us assume first of all that $f(R)$ is *non-negative*; we shall proceed from here to a more general class — the analogue of the class of functions of bounded variation, or rather of the interval functions which they determine — by exactly the same steps as in the case of a single variable.

We shall work with the net of our successive square grills, and in terms of this net we shall define in an obvious manner the quantities which, until stated otherwise, will play the role of the derived numbers and of the derivative of $f(R)$: the upper and lower derived numbers, with respect to our net, at the point $P = (x, y)$ are simply the upper and lower limits of the sequence of ratios

$$\frac{f(R)}{|R|},$$

where we have denoted by $|R|$ the area of the rectangle and where we let R run through the sequence of those meshes of our square grills which contain the point P; when these two quantities, $\overline{D}f(P)$ and $\underline{D}f(P)$, coincide, their common value $Df(P)$ will be the derivative of $f(P)$ at the point P.

This done, our principle of transition assures us immediately that *the derivative $Df(P)$ of an additive, non-negative rectangle function exists almost everywhere.* However, as we have just remarked, we are dealing here with a *derivative of a very particular type.* Our goal is to show that in reality this derivative is *independent* of the net, at least almost everywhere.

The first step will be to compare the derivative $Df(P)$ with the derivative $D'f(P)$ which is entirely analogous to it but is calculated after making *a parallel displacement of the coordinate system.* We shall see that

$$D'f(P) = Df(P)$$

almost everywhere.

To do this, we have only to repeat, mutatis mutandis, the reasoning, or rather a part of the reasoning, we used to establish the existence of the derivative of a monotonic function. Let $c < C$ be two positive quantities and consider the set E_{cC} of points for which

$$Df(P) < c \text{ and } D'f(P) > C.$$

Then, if P runs through the square grills of the first net, there will be, for each of these P, a first mesh containing P with the property that

$$\frac{f(R)}{|R|} < c.$$

Let Σ_1 be the system of all these squares. Let Σ_2 be the system of the first meshes of the second net which contain the P's and which are contained, in their turn, in the meshes of which Σ_1 is formed, and for which

$$\frac{f(R)}{|R|} > C.$$

Then we shall obviously have

$$C|\Sigma_2| \leq V_2 \leq V_1 \leq c|\Sigma_1|,$$

where we have denoted by $|\Sigma_i|$ the total area of Σ_i, and by V_i the sum of the respective values of $f(R)$.

Repeating the two procedures alternately, we shall obtain a sequence $\Sigma_1, \Sigma_2, \ldots$ of systems of squares, each embedded in the preceding, and we shall have in general

$$|\Sigma_{2n}| \leq \frac{c}{C} |\Sigma_{2n-1}|,$$

from which, recalling that the Σ_n decrease,

$$|\Sigma_{2n+1}| \leq |\Sigma_{2n}| \leq \left(\frac{c}{C}\right)^n |\Sigma_1| \to 0;$$

consequently E_{cC} is of measure zero. Finally, we consider all the rational values of c and C for which $c < C$, and form the union of the denumerable infinity of sets E_{cC} which results; from this we arrive at the result that the set of points for which $Df(P) < D'f(P)$ is of measure zero. The same holds for $D'f(P) < Df(P)$. That is, $D'f(P) = Df(P)$ almost everywhere; q.e.d.

42. Rectangle Functions of Bounded Variation. Conjugate Nets

The result we have just obtained extends immediately to additive rectangle functions $f(R)$ of arbitrary sign but *of bounded variation,* that is, such that, for every rectangle R, the sums

$$\sum |f(R_k)| \tag{18}$$

corresponding to the decompositions of R into rectangles R_k have a finite upper limit which we shall denote by

$$Tf(R)$$

and which we shall call the indefinite total variation of $f(R)$. In fact, we have the decomposition of $f(R)$ into the difference of two non-negative additive rectangle functions:

$$f(R) = Tf(R) - [Tf(R) - f(R)],$$

just as in the case of a single variable (Sec. 4).

Other results concerning functions of bounded variation of a single variable have analogues for rectangle functions of bounded variation. We consider one for which we shall have immediate use.

The equation

$$DTf(P) = |Df(P)| \tag{19}$$

is valid almost everywhere.

This is the analogue of the fact established in Sec. 8 for functions of bounded variation of a single variable:

$$T'(x) = |f'(x)| \quad \text{almost everywhere.} \tag{19a}$$

It is even a consequence of this fact, by our transition principle, at least if the rectangle function $f(R)$ is continuous in an obvious sense, which is the case, in particular, if it is the indefinite integral of a summable function; we then have only to observe that for a rectangle R belonging to our net we can restrict ourselves in the definition of $Tf(R)$ to decompositions of R into rectangles R_k which also belong to the net. In the case where $f(R)$ is not continuous this reduction is not possible in general, and then we must prove (19) directly, by adapting, for example, the proof of (19a) in Sec. 8. The adaptation is almost obvious, including the proof of Fubini's theorem which, in its present form, is stated as follows: *Every convergent series whose terms are non-negative, additive rectangle functions can be differentiated term by term almost everywhere with respect to the given net.*

It is on relation (19) that we shall base the following step, which will consist of adapting to our needs an idea which LEBESGUE used with great success for Fourier series. We could have presented this step in any of the sections on functions of a single variable.

Let us apply relation (19) to the function

$$g(R;\lambda) = f(R) - \lambda|R|$$

instead of to $f(R)$, where λ denotes a real parameter, and to the function $Tg(R;\lambda)$ which corresponds to it. We shall show that *our relation remains valid almost everywhere for every value of* λ.

This would be evident if it were a question simply of a denumerable infinity of values of λ, since the union of a denumerable infinity of sets of exceptional points, each of measure zero, is itself a set of measure zero. Therefore we begin by letting λ run through only a denumerable everywhere dense set, or for definiteness, all the rational numbers; then the relation

$$DTg(P;\lambda) = |Dg(P;\lambda)| \tag{20}$$

will be satisfied almost everywhere for all the rational numbers $\lambda = r$. Now for the other values of λ we shall be able to write

$$g(R;\lambda) = g(R;r) + (r-\lambda)\,|R|,$$

and it follows immediately that

$$|Tg(R;\lambda) - Tg(R;r)| \leq |r-\lambda|\,|R|;$$

then, letting r tend to λ, we conclude from this that relation (20) is satisfied almost everywhere for every value of λ. Finally, setting in particular, for a point P_0, $\lambda = Df(P_0)$, we shall have $Dg(P_0;\lambda) = 0$, hence also $DTg(P_0;\lambda) = 0$. Therefore we have the

THEOREM. *Given an additive rectangle function $f(R)$ of bounded variation, the indefinite total variation of the rectangle function*

$$f(R) - Df(P_0)\,|R|$$

possesses, for almost all points P_0, a derivative which takes on the value zero at the point P_0.

In order to see more clearly what the situation is here, let us consider the particular — but perhaps the most important — case in which the rectangle function $f(R)$ is the integral of a summable function $h(P) = h(x,y)$. Let us recall that, as proved in Sec. 23, the total variation of an indefinite integral is equal to the integral of the modulus of the function in question, a fact which, together with its proof, also can be extended to functions of several variables. Then what our result affirms is that *for almost every point P_0 the integral of the function $|h(P) - h(P_0)|$, considered as a rectangle function, possesses a derivative which takes on the value zero at the point P_0.*

Our result generalizes this fact to rectangle functions which are not necessarily integrals.

It is now possible for us to complete the proof of the fact that the derivative $Df(P)$ of the additive rectangle function $f(R)$ of bounded variation does not depend on the net. We already know that it is invariant with respect to a parallel displacement of the net (Sec. 41).

The final step which remains to be made was invented by DE LA VALLÉE POUSSIN, under the name of *method of conjugate nets.*[28]

Let us displace our net of square grills by $\frac{1}{2}$ in the x direction, or in the y direction, or even in both directions simultaneously; we obtain three other nets, called the conjugates of the first. The advantage of operating simultaneously with these four nets instead of with a single one consists in the following fact. Consider a square with sides parallel to the axes, of width a, and suppose that

$$\frac{1}{2\cdot 3^{n+1}} \leq a \leq \frac{1}{2\cdot 3^{n}};$$

this square is then certainly contained in one of the meshes of the four square grills of width 3^{-n}, and its area is not less than $\frac{1}{36}$ of that of the mesh. It follows that the quotient of $T(f(R) - \lambda|R|)$ by $|R|$, formed for the above-named square of width a, is majorized by 36 times the same quotient formed for a mesh of one of the four nets, of width not exceeding 6a; then, setting $\lambda = Df(P_0)$ and letting $a \to 0$ in the reasoning just above, we see that the first quotient tends to 0 just as the second. This implies, a fortiori, the following result:

THEOREM. *The derivative of $f(R)$ exists at the point P_0 and equals $Df(P_0)$ even if, instead of approaching P_0 by particular meshes belonging to our square nets, we use squares which are parallel to the axes but otherwise arbitrary to contract to the point P_0.*

Moreover, we can go further and replace these squares by *rectangles,* always parallel to the axes, requiring only that *the ratio of the adjacent sides does not exceed a given bound q*; in fact to see this we have only to replace 36 by $36q$.

43. Additive Set Functions. Sets Measurable (B)

Finally let us observe, without going into details, that we can pass from additive rectangle functions $f(R)$ to *additive set functions $f(e)$* in the same way that we pass from the area of rectangles to the measure of sets. Now it is almost evident that after the introduction of these set functions we

[28] DE LA VALLÉE POUSSIN [*] (pp. 68–70).

can substitute them, in what precedes, for the function $f(R)$ and that among other things the quotient of $f(e)$ by the measure of the set e tends to $Df(P_0)$ almost everywhere, whenever we let e run through a sequence of sets $\{e_n\}$ such that the ratio of the measure of e_n to that of the smallest circumscribed square (or circle) remains above a bound $d > 0$.

With this point established the relations between integration and differentiation—differentiation in the sense which we have just introduced—will be easy to explain, since their study does not differ substantially from that of the case of a single variable. Instead of summarizing or formulating theorems, it will suffice for us to say that the role of intervals contracting to a point, but otherwise arbitrary, is played by rectangles which are not permitted to become infinitely narrow. But this difference disappears when we pass to the differentiation of set functions; in fact, we have to take here the same precautions for linear sets as for planar sets.

It is appropriate to observe further that we could have considered set functions instead of rectangle functions right from the beginning. For example, we would have been able to begin with additive functions of an open set, a concept which appears to us to be more appropriate for the idea of distribution of mass. In general, authors do not make use of this idea; perhaps they are afraid of the inconvenience which the decomposition of an open set into open sets presents. One can remove this inconvenience by defining additivity by the relation

$$f(e_1 \cup e_2) + f(e_1 \cap e_2) = f(e_1) + f(e_2).$$

Another point of departure is to suppose that the set functions are defined for all Borel measurable sets. The concept of these sets is due to EMILE BOREL who, shortly before LEBESGUE, introduced a concept of measure by a constructive means, by passing successively to more and more general sets, but without attaining in the end the generality of the notion of LEBESGUE.[29] But this does not prevent the notion of Borel measure from sufficing for the majority of problems which depend on the Lebesgue integral and its generalizations; it sometimes even has the advantage that in certain lines of reasoning we know, a priori, that all the sets which arise are Borel-measurable.

According to one of the definitions, the class of Borel-measurable sets is the smallest class containing respectively all the intervals or rectangles, and closed with respect to the formation of the union or the intersection of a finite number or a denumerable infinity of sets from the class.

We also speak, in an analogous sense, of *Borel-measurable functions*. This class of functions can be characterized as the smallest class which contains the continuous functions, which is closed with respect to the formation of the sum, the difference, or the product of a finite number of functions, and

[29] Cf. BOREL [*] (pp. 46—50), or, also, LEBESGUE [**5**].

which contains the limit of every sequence of functions from the class which converges *everywhere*. Or, equivalently, this class is formed from the continuous functions and all other functions which can be obtained from them by iterated passages to the limit.[30] In particular, step functions are Borel-measurable. It is clear that every Borel measurable function is also measurable in the sense of Sec. 23.

The converse is not true. But in any case, we have the

THEOREM. *To every measurable function $f(P)$ there corresponds a Borel measurable function $g(P)$ which differs from $f(P)$ only on a set of measure zero.*

For definiteness in the exposition, we take the case of functions of a single variable. We shall first show that every set E of measure zero is contained in a Borel measurable set E^* which is also of measure zero. In fact, E can be covered by a set J_n, the sum of a finite number or a denumerable infinity of intervals of total length $< \frac{1}{n}$, for each $n = 1, 2, \ldots$; but all these J_n, and therefore their intersection

$$E^* = \bigcap_n J_n$$

are Borel-measurable and since $m(J_n) < \frac{1}{n}$, we have $m(E^*) < \frac{1}{n}$, hence $m(E^*) = 0$.

With this established, let $f(P)$ be the measurable function in question; by definition there exists a sequence $\{f_n(P)\}$ of summable functions which converge almost everywhere to $f(P)$. Each $f_n(P)$ is the limit almost everywhere of a sequence $\{\varphi_{nk}(P)\}$ of step functions. Let E be the sum of all the exceptional sets which arise; since E is a set of measure zero, it is contained in a Borel measurable set E^* of measure zero. Since the characteristic function $e(P)$ of the complement of E^*, as well as all the step functions, are Borel-measurable, their products and their limits are also Borel-measurable; in particular

$$f_n(P)e(P) = \lim_{k\to\infty} \varphi_{nk}(P)e(P)$$

and

$$f(P)e(P) = \lim_{n\to\infty} f_n(P)e(P)$$

will be Borel-measurable. But obviously the last function differs from $f(P)$ only on the set of measure zero E^*.

[30] The classification of functions according to the number (finite or transfinite) of passages to the limit is due to R. BAIRE [**1**]. For more details, see DE LA VALLÉE POUSSIN [*] (Chapt. II).

OTHER DEFINITIONS OF THE LEBESGUE INTEGRAL

44. Sets Measurable (*L*)

In their attempts to simplify the Lebesgue theory, mathematicians have formulated various definitions of the Lebesgue integral, among them the definition which we have discussed. Before treating some of the others, let us compare this one with the original definition of Lebesgue. The latter consists of three stages, namely the definition of the measure of sets, the definition of measurable functions, and finally the definition of summable functions and their integrals. In order to distinguish them, we shall write measure (L), measurable (L), and integral (L). So as not to complicate the language and notation, we consider the case of functions *of a single variable* and a *finite* interval (a, b).

With regard to measure, we have already defined (Sec. 6) the exterior measure of a set E, which we denoted by $m_e(E)$, as the greatest lower bound of the total lengths of systems of intervals containing E. Note that it is immaterial whether these intervals are taken open or closed; in fact in the latter case, in order to replace them by open intervals, we have only to prolong them in both directions by an arbitrarily small fractional part of their length, which will not change the greatest lower bound of their total lengths.

The set E is a part of the finite interval (a, b); let $CE = (a, b) - E$ be its complement. Then we have, in general,

$$m_eE + m_eCE \geqq b - a. \tag{21}$$

In fact, enclose E and CE in systems of intervals whose total lengths are less than respectively $m_eE + \frac{\varepsilon}{2}$ and $m_eCE + \frac{\varepsilon}{2}$; as we have just seen we can assume that the intervals in question are open. Furthermore, without loss of generality we can assume that the interval (a, b) is closed. The two systems of intervals cover the intervals (a, b), and consequently, by virtue of the well-known Borel theorem, we can choose a finite number of these intervals with the same property. The total length of these intervals is obviously less than $m_eE + m_eCE + \varepsilon$. We have, a fortiori, $b - a \leqq m_eE + m_eCE + \varepsilon$, and inequality (21) follows when $\varepsilon \to 0$.

We agree to say that *the set E is measurable* (L), *and has measure* (L) *equal to $mE = m_eE$, whenever the equality sign holds in* (21). It is clear this definition does not depend on the particular choice of the interval (a, b).

We shall show that the measure (L) is identical with that which we defined before as equal to the integral, according to our first definition, of the characteristic function $e(x)$ of the set E. We shall use the following lemma:

LEMMA. *Whenever the function $f(x)$, defined in (a, b), can be enclosed in (a, b) between two summable functions $g(x)$ and $h(x)$ in such a way that*

$$g(x) \leq f(x) \leq h(x); \quad \int_a^b [h(x) - g(x)]dx \leq \varepsilon$$

for all positive ε, then $f(x)$ itself is summable.

We successively set $\varepsilon = 2^{-n}$ and let g_n and h_n be the corresponding functions. Then since the series

$$\sum_1^\infty \int_a^b [h_n(x) - g_n(x)]dx \leq \sum_1^\infty 2^{-n}$$

is convergent, the series

$$\sum_1^\infty [h_n(x) - g_n(x)]$$

will be convergent almost everywhere, and consequently we shall have $h_n - g_n \to 0$, that is, $g_n \to f$, $h_n \to f$. In view of the inequality $g_1 \leq f \leq h_1$, our assertion follows from this and the first theorem in Sec. 20.

Now assume that the set E is measurable (L), and consider the two systems of intervals, say Σ_1 and Σ_2, of which we were just speaking; let $|\Sigma_1|$ and $|\Sigma_2|$ be their total lengths, and let $\sigma_1(x)$ and $\sigma_2(x)$ be the sums of the characteristic functions of the intervals of which the two systems are composed; the existence almost everywhere of these sums is assured by our lemma B, Sec. 16. In addition, the functions $\sigma_1(x)$ and $\sigma_2(x)$ are summable and their integrals are equal respectively to $|\Sigma_1|$ and $|\Sigma_2|$. Furthermore, since $\sigma_1(x) \geq e(x)$, $\sigma_2(x) \geq 1 - e(x)$ and

$$\int_a^b [\sigma_1(x) - (1 - \sigma_2(x))]dx = |\Sigma_1| + |\Sigma_2| - (b-a) \leq mE + mCE + \varepsilon - (b-a) = \varepsilon,$$

we have only to set $g(x) = 1 - \sigma_2(x)$, $h(x) = \sigma_1(x)$, and $f(x) = e(x)$ in order to conclude from our lemma that the function $e(x)$ is summable. Finally, since its integral is contained between those of $g(x)$ and $h(x)$, that is, between

$$b - a - |\Sigma_2| \geq b - a - mCE - \frac{\varepsilon}{2} = mE - \frac{\varepsilon}{2}$$

and

$$|\Sigma_1| \leq mE + \frac{\varepsilon}{2},$$

for every $\varepsilon > 0$, we have precisely

$$\int_a^b e(x)dx = mE.$$

Conversely, suppose that the characteristic function $e(x)$ is summable;

we shall show that under this hypothesis the set E is measurable (L). In order to arrive at this conclusion, let us consider a sequence $\{\varphi_n(x)\}$ of step functions which converges to $e(x)$. We can assume that the functions φ_n are the characteristic functions of certain sets Σ_n, where each Σ_n is formed from a finite number of intervals; in fact, if this were not the case we would have only to replace the values of the $\varphi_n(x)$ by 1 wherever $\varphi_n(x) > \frac{1}{2}$ and by 0 elsewhere, and the sequence thus modified would still converge to $e(x)$. The function $e(x)$ will also be the limit of a monotonic sequence formed from the functions

$$g_n = \sup(\varphi_n, \varphi_{n+1}, \ldots);$$

g_n is the characteristic function of the set $\Sigma^{(n)}$ which is obtained by forming the union of the sets $\Sigma_n, \Sigma_{n+1}, \ldots$. Now $\Sigma^{(n)}$ is composed of a finite number or of a denumerable infinity of intervals, and we can assume these to be non-overlapping by suppressing in each Σ_k, if necessary, the points contained in the sets Σ_i of lower index. The total length of these intervals is equal to the integral of the respective characteristic function g_n, and this total length tends to the integral of the function $e(x)$ as $n \to \infty$. Furthermore, each system $\Sigma^{(n)}$ covers all of the set E with the possible exception of a subset of measure zero. It follows that

$$m_e E \leq \int_a^b e(x)dx.$$

An analogous calculation yields

$$m_e CE \leq \int_a^b [1 - e(x)]dx = b - a - \int_a^b e(x)dx.$$

These two inequalities, compared with (21), give precisely

$$m_e E + m_e CE = b - a,$$

that is, the measurability (L) of the set E. Therefore the two definitions of measure are equivalent and in the future we shall not have to distinguish between measure as we have defined it and measure (L).

45. Functions Measurable (*L*) and the Integral (*L*)

The second stage in the definition of LEBESGUE is marked by the concept of *measurable function.*

The function $f(x)$, defined in the finite interval (a, b), is said to be measurable (L) in (a, b) when, for $A < B$ but otherwise arbitrary, the set of points for which

$$A \leq f(x) < B$$

is measurable.

From what we know of the sum and the product of measurable sets, we deduce immediately that the hypothesis stated is equivalent to the following: for each value of A, the set where $f(x) \leq A$ is measurable. The same is true of the analogous hypothesis formed with one or another of the inequalities $f > A$, $f \geq A$, $f < A$.

We have also seen, in Sec. 23, that the functions which are measurable in our sense satisfy these hypotheses. What remains to be shown is that, conversely, every function $f(x)$ measurable (L) is also measurable according to the former definition, or, equivalently, that it is either summable or the limit of a sequence of summable functions.

Let

$$\ldots, l_{-2}, l_{-1}, l_0, l_1, l_2, \ldots$$

be an increasing sequence of numbers running from $-\infty$ to ∞ with the property that the differences $l_n - l_{n-1}$ remain below a bound δ. Let us denote by E_n the set of values for which $l_{n-1} \leq f(x) < l_n$, and set $g(x) = l_{n-1}$ on E_n when $|n| < N$, $g(x) = 0$ elsewhere. The set E_n is measurable by hypothesis, and the function $g(x)$ is obviously summable. Assume further that $\delta \leq 2^{-N}$. Letting N increase indefinitely, $g(x)$ runs through a sequence $\{g_N(x)\}$ which tends to $f(x)$.

Therefore the equivalence of the two notions is also established in the second stage.

In order to pass to the *integral* (L), let $f(x)$ be a measurable function which we assume at first to be non-negative, let

$$l_0, l_1, l_2, \ldots$$

be a sequence with $l_0 = 0$ and $l_n - l_{n-1} \leq \delta$, and assume that the series

$$\sum_1^\infty l_n mE_n \tag{22}$$

is convergent. When this is true for one sequence $\{l_n\}$, it is also true for analogous sequences. In fact, let $\{\bar{l}\}$ be a second sequence, with $\bar{l}_n - \bar{l}_{n-1} \leq \bar{\delta}$, and let $\bar{E}_n$ be the corresponding sets. Denote by $e_n(x)$ and $\bar{e}_n(x)$ the characteristic functions of the sets E_n and $\bar{E}_n$. Then we have

$$\sum_1^\infty \bar{l}_{n-1}\bar{e}_n(x) \leq f(x) \leq \sum_1^\infty l_n e_n(x);$$

this last series converges almost everywhere and it represents a summable function, as follows immediately from the convergence of series (22). It then follows from Lebesgue's theorem (Sec. 19) that the series on the left can be integrated term by term, in particular that the series $\Sigma \bar{l}_{n-1} m\bar{E}_n$ converges, and finally, since

$$(23)\quad \sum_{1}^{\infty} \bar{l}_n m\bar{E}_n = \sum_{1}^{\infty} \bar{l}_{n-1} m\bar{E}_n + \sum_{1}^{\infty} (\bar{l}_n - \bar{l}_{n-1}) m\bar{E}_n \leqq \sum_{1}^{\infty} \bar{l}_{n-1} m\bar{E}_n + \delta(b-a),$$

it follows that the series on the left is convergent, which was to be proved.

From the above argument we see that

$$\sum_{1}^{\infty} \bar{l}_{n-1} m\bar{E}_n \leq \sum_{1}^{\infty} l_n mE_n,$$

that is, that the "lower sums" of the type considered do not exceed any of the "upper sums." Keeping in mind inequality (23), it follows that these two categories of sums, which are arbitrarily close to one another, are separated by a uniquely determined quantity, and *it is by this quantity that the integral* (*L*) *of the function* $f(x)$ *is defined.*

As for functions of variable sign, the definition of the integral (*L*) is extended to them by writing them as the difference of two non-negative and summable (*L*) functions, in particular by using their positive and negative parts. Hence the function $f(x)$ is summable (*L*) under the condition that it be measurable and that one of the series of the type

$$\sum_{-\infty}^{\infty} l_n mE_n \qquad (l_n - l_{n-1} \leq \delta)$$

be absolutely convergent. In particular, every measurable and bounded function is summable (*L*).

It also follows immediately from this reasoning that every function summable (*L*) is also summable according to our definition and that the values of the two integrals coincide. What we must still show is that summability according to our definition implies summability (*L*). Obviously, we can restrict ourselves to the case of a non-negative function $f(x)$.

We already know that every function which is summable by our definition is measurable, hence also measurable (*L*). This being the case, consider again the sequence $\{l_n\}$ and the sets E_n with the corresponding functions $e_n(x)$. Since

$$f(x) \leq \sum_{1}^{\infty} l_n e_n(x) \leq f(x) + \delta,$$

and since the functions f and e_n are summable, the series $\Sigma l_n e_n$ can be integrated term by term. We deduce from this, in particular, the convergence of the integrated series, that is, of the series $\Sigma l_n\, mE_n$. Therefore the function $f(x)$ is summable (*L*), which was to be proved.

46. Other Definitions. Egoroff's Theorem

As we have said, there are various other definitions of integral besides the one which we have used which are equivalent to that of Lebesgue. To mention several, let us begin by pointing out one formulated by W. H. YOUNG,[31] to which basically ours is closely related. It too is based on the integration of monotonic sequences, but it can also be formulated very simply, at least for bounded functions, by modifying the definition of the Riemann integral given by Darboux. Let us define *the lower-upper integral* of a function $f(x)$ by the greatest lower bound of the lower integrals of the functions $h(x)$ which are lower semi-continuous and greater than $f(x)$, and its *upper-lower integral* in the inverse manner; when these two integrals coincide we take their common value to be the *integral* of the function $f(x)$. Finally, in order to pass to the integration of unbounded functions, we have only to express them as the limits of the bounded functions which are obtained by truncating them.

A second definition which it is appropriate to mention and which was pointed out by one of the authors[32] also has its simplest form when applied to bounded functions. It is based on the fact, which is easy to prove, that every bounded sequence of step functions which converges almost everywhere can be integrated term by term, that is, that the sequence of integrals is convergent. This suggests defining the integral of the limit function $f(x)$ by means of the limit of the integrals; of course we have to show that two sequences which converge almost everywhere to the same limit give the same value for the integral, or equivalently that in the case $f(x) \equiv 0$ the limit in question is equal to 0. All this follows immediately from the facts already established concerning the integration of monotonic sequences, but we can arrive at the same facts from a different order of ideas.

The first of these is centered about

EGOROFF'S THEOREM.[33] *Every convergent sequence of measurable functions which are defined in a measurable set E contained in a finite interval (a, b) can be made uniformly convergent by removing from E a subset of arbitrarily small measure.*

It is easy to establish this theorem using the facts concerning measure,[34] but perhaps it is of interest to see how it is formulated and proved independently[35] of these facts. Let us prove the theorem first of all in the following special form where the question of measure does not arise:

Every convergent sequence of continuous functions $f_n(x)$ which are defined in a bounded and closed set E can be made uniformly convergent by removing from E the points of suitably chosen open intervals whose total length is arbitrarily small.

[31] W. H. YOUNG [**1**] (in particular, pp. 25–35).
[32] F. RIESZ [**10**].
[33] EGOROFF [**1**].
[34] SAKS [*] (p. 18).
[35] RIESZ [**11**].

Denote by $E_{\nu,k}$ the set of points x for which

$$|f_m(x) - f_n(x)| \leq \frac{1}{k} \tag{24}$$

whenever $m, n \geq \nu$. The sets $E_{\nu,k}$ are closed; assume for the moment that the same is true of their complements $E - E_{\nu,k}$. For k fixed and ν increasing indefinitely, these complements form a decreasing sequence which must converge to the empty set; in fact, if x_0 belonged to all the sets there would be arbitrarily large m and n with

$$|f_m(x_0) - f_n(x_0)| > \frac{1}{k},$$

and the sequence $\{f_n(x_0)\}$ would diverge. But a decreasing sequence of closed sets can not converge to the empty set unless the sets themselves are eventually empty. That is, for some $\nu = \nu(k)$ the set $E - E_{\nu,k}$ is empty, and consequently $E_{\nu,k} = E$, so the integral (24) is valid without exception, and since k is arbitrary the uniform convergence is verified.

However, to arrive at this conclusion it was necessary to assume that the sets $E - E_{\nu,k}$ are closed, which is not true in general. But if we can succeed in showing that this hypothesis is satisfied after removing from E the points contained in a system of intervals of arbitrarily small total length, our proposition will be justified. To do this, let (a, b) be an open interval containing the set E and let ε be a given positive quantity, arbitrarily small. The set $(a, b) - E_{\nu,k}$, being open, is the union of a denumerable infinity of non-overlapping (but possibly touching) closed intervals of finite total length; we retain a finite number of them such that the remainder have a total length less than $2^{-\nu-k-1}\varepsilon$. Let $\Sigma_{\nu,k}$ be the system of intervals belonging to this remainder, each doubled by dilating it about its center. The sum of these doubled intervals, which we take to be open, is less than $2^{-\nu-k}\varepsilon$. Let $\bar{E}_{\nu,k}$ be the part of E contained in the intervals retained, finite in number; it is closed and has no point in common with $E_{\nu,k}$. Finally, form the set E^* by suppressing in E all the points which are contained in one of the systems of intervals $\Sigma_{\nu,k}$, where $\nu, k = 1, 2, \ldots$; they are contained in a system of intervals of total length less than $\sum_{\nu,k=1}^{\infty} 2^{-\nu-k}\varepsilon = \varepsilon$. But E^* is obviously closed and the same holds for its intersection $\bar{E}^*_{\nu,k}$ with $\bar{E}_{\nu,k}$. The set $E^*_{\nu,k} = E^* - \bar{E}^*_{\nu,k}$, being the intersection of E^* and $E_{\nu,k}$, is composed precisely of the points x of E^* for which inequality (24) is valid whenever $m, n \geq \nu$. But since the sets $\bar{E}^*_{\nu,k} = E^* - E^*_{\nu,k}$ are closed, as we have just seen, our hypothesis is satisfied for the set E^*, and our proposition is proved.

Before passing to the general Egoroff theorem, let us show that the special case which we have just proved has as a simple consequence

LUSIN'S THEOREM.[36] *Every measurable function $f(x)$ defined in a measurable set E can be made continuous by removing from E the points contained in suitably chosen open intervals whose total length is arbitrarily small.*

It obviously suffices to consider the case where E can be enclosed in a finite interval which we can assume closed, say in the interval $[a, b]$. We extend the function $f(x)$ to the entire interval by setting $f(x) = 0$ in the complementary set, thus obtaining a function which is measurable in $[a, b]$ and is consequently the limit almost everywhere of a sequence of step functions $\varphi_n(x)$. The points where the sequence does not converge to $f(x)$ and the points of discontinuity of the functions $\varphi_n(x)$ form a set of measure zero; they can therefore be covered by open intervals of total length less than $\varepsilon/2$. In the closed set which remains after having removed these intervals from $[a, b]$ all the functions $\varphi_n(x)$ are continuous and tend everywhere to $f(x)$. We can therefore remove from this closed set the points contained in a system of suitably chosen open intervals, of total length less than $\varepsilon/2$, so that in the set which remains the sequence $\varphi_n(x)$ tends *uniformly* to $f(x)$, and the proof is completed by applying the classical theorem which states that the limit of a uniformly convergent sequence of continuous functions is itself continuous.

With this fact established, Egoroff's theorem in its general form is proved by an argument analogous to that which we have just made. Let $\{f_n(x)\}$ be a sequence of measurable functions defined in the measurable set $E \subset [a, b]$ which converges almost everywhere. We extend the functions $f_n(x)$ to the entire interval by setting $f_n(x) = 0$ in the complementary set. Given an arbitrary positive quantity ε, we first enclose the set of measure zero of points where the sequence $\{f_n(x)\}$ does not converge with a sequence of intervals whose total length is less than $\varepsilon/4$. Then we remove for each n the points belonging to a system of intervals whose total length is less than $2^{-n-1}\varepsilon$ and which has the property that the function $f_n(x)$ is continuous in the set which remains. Finally, by the special case of the theorem already proved, we can remove the points belonging to intervals of total length less than $\varepsilon/4$ and have the convergence of the $f_n(x)$ uniform in the set which remains. Since the total length of the intervals removed in the course of this procedure is less than ε, the theorem is proved.

The Egoroff and Lusin theorems suggest that we reverse our approach and *define* measurable functions in an interval by the hypothesis that they can be made continuous by removing intervals of arbitrarily small total length, and furthermore that we define measurable sets as those possessing measurable characteristic functions. From this follows a definition of summable functions and their integrals which is equivalent to that of Lebesgue; we first integrate the functions which are continuous on closed sets and then

[36] LUSIN [**1**]

pass to the limit, using a procedure which is immediately obvious. This theory, the idea of which is due to E. BOREL, was developed by H. HAHN.[37]

It is appropriate here to say a few words about the notion of *convergence in measure*, beginning with a fact which is apparently less significant than that of Egoroff, but which suffices to replace it in most applications.[38] The fact in question is that *every sequence of measurable functions* $\{f_n(x)\}$ *which converges to* $f(x)$ *in a measurable set E has the property that, for every* $\delta > 0$, *the measure of the sets for which* $|f_n(x) - f(x)| \leq \delta$ *converges to* 0 *when* $n \to \infty$, *or as we say, the sequence* $\{f_n(x)\}$ *converges in measure.*

This theorem is easily proved within the framework of the general theory and it is also easy to deduce it from the Egoroff theorem or from the following theorem of choice, which has been applied by several authors to prove the Riesz-Fischer theorem: *From every sequence which is convergent in measure we can select a subsequence which converges in the ordinary sense almost everywhere.*

We shall not insist upon the proofs. In the order of ideas which we have just been following, it is more interesting to observe that *we can define "convergence in measure" without using the notion of "measure" by making the hypothesis that every subsequence of* $\{f_n(x)\}$ *contains sequences which converge almost everywhere to* $f(x)$. We leave it to the reader to convince himself of this on the basis of the preceding facts.[39]

47. Elementary Proof of the Theorems of Arzelà and Osgood

Let us return to our main subject. In our discussion of the various definitions of the Lebesgue integral we were concerned with demonstrating in as elementary a manner as possible that a bounded sequence of step functions $\{\varphi_n(x)\}$ which converges almost everywhere can be integrated term by term, that is, that the sequence of integrals is convergent. This problem is in reality the problem of finding an elementary proof of the theorems of ARZELÀ and OSGOOD, since the more general conditions of the latter do not substantially change the arguments involved. Now the importance of these theorems, and of their generalization by Lebesgue, leads us to give yet another, very

[37] BOREL [**1**] or [*] (pp. 248–250); HAHN [**2**].

[38] LEBESGUE [$\ast\atop\ast$] (p. 10).

[39] It is proper to note that we are led to this definition by an idea borrowed from the theory of abstract topological sets. In the space of measurable functions, let us first define the limit of a sequence by convergence almost everywhere. Furthermore, let us agree to call $f(x)$ a limit element of a sequence or set of functions if the latter contains a subsequence of which $f(x)$ is the limit. Then let us forget our definition of limit and try to reconstruct it by starting with limit elements and defining the limit $f(x)$ of a sequence $f_n(x)$ by requiring that $f(x)$ be a limit element of every subsequence of the $f_n(x)$. But this, instead of bringing us back to our original definition, obviously arrives at the definition of convergence in measure.

elementary, proof which possesses the great advantage of many elementary arguments, namely, a very wide applicability; in fact, our proof also applies, as we shall see later, to the integration of functions defined in abstract sets. We present it here for step functions,[40] but it can be immediately adapted to the general case.

We can assume without loss of generality that $\varphi_n(x) \geqq 0$ and that $\varphi_n(x) \to 0$ almost everywhere. In fact, if for such functions the integral converges to 0, we shall have in the general case

$$\int_a^b |\varphi_m(x) - \varphi_n(x)|\, dx \to 0 \qquad (m, n \to \infty),$$

for if not there would be $m_k, n_k \to \infty$ and $\psi_k = |\varphi_{m_k} - \varphi_{n_k}| \to 0$ almost everywhere without having $\int_a^b \psi_k(x) dx \to 0$.

Let us now recall that we have already proved in a very elementary manner, in Lemma **A**, Sec. 16, that $\int_a^b \varphi_n(x) dx \to 0$ whenever $\varphi_n(x) \to 0$ almost everywhere and the $\varphi_n(x)$ form a decreasing sequence. The problem,therefore, is to remove this last hypothesis. Let us also recall, by way of comparison, how we would have to procede in the theory of Lebesgue. There, we would consider the functions

$$f_n(x) = \sup (\varphi_n, \varphi_{n+1}, \ldots)$$

which are summable and which clearly not only converge almost everywhere to 0, as do the φ_n, but also form a decreasing sequence; consequently, according to Beppo Levi's theorem, $\int_a^b f_n(x) dx \to 0$. Since $0 \leq \varphi_n(x) \leq f_n(x)$, it follows that $\int_a^b \varphi_n(x) dx \to 0$. If we wish to imitate this argument without resorting to the Lebesgue integral, it will be necessary for us to replace the $f_n(x)$ with step functions which are less than, but sufficiently close to, the $f_n(x)$. To attain this goal, set

$$\varphi_m^n(x) = \sup (\varphi_m, \varphi_{m+1}, \ldots, \varphi_n)$$

and denote by I_m^n the integral of φ_m^n. For m fixed, the φ_m^n and I_m^n increase with n and the I_m^n converge to a limit J_m. Since the J_m are non-negative and form a decreasing sequence, they converge to a limit $J \geqq 0$; since, moreover,

$$\int_a^b \varphi_m(x) dx \leq I_m^n \leq J_m, \tag{25}$$

the relation $J = 0$, which we shall verify, will prove our proposition.

[40] Cf. RIESZ [**8**] and LANDAU [**1**].

Let $\varepsilon > 0$ be arbitrarily small and let $n = n_1$ be the first index for which $I_1^n > J_1 - \frac{\varepsilon}{2}$; in general, let $n = n_m$ be the first index after n_{m-1} for which $I_m^n > J_m - 2^{-m}\varepsilon$. Set

$$\psi_m(x) = \inf(\varphi_1^{n_1}(x), \varphi_1^{n_2}(x), \ldots, \varphi_m^{n_m}(x)).$$

Then we shall have

$$\int_a^b \psi_m(x)dx > J_m - (1 - 2^{-m})\varepsilon.$$

We shall verify this by induction; it is valid for $m = 1$. Assuming it valid for $m - 1$, that is, that

$$\int_a^b \psi_{m-1}(x)dx > J_{m-1} - (1 - 2^{-m+1})\varepsilon,$$

we observe that

$$\psi_m(x) = \inf(\psi_{m-1}(x), \varphi_m^{n_m}(x)) = \psi_{m-1}(x) + \varphi_m^{n_m}(x) - \sup(\psi_{m-1}(x), \varphi_m^{n}(x))$$

and that these last two functions, hence also their upper envelope, remain below the $\varphi_{m-1}^{n_m}(x)$. Consequently,

$$\psi_m(x) \geq \psi_{m-1}(x) + \varphi_m^{n_m}(x) - \varphi_{m-1}^{n_m}(x),$$

from which we obtain by integration

$$\int_a^b \psi_m(x)dx \geq \int_a^b \psi_{m-1}(x)dx + I_m^{n_m} - I_{m-1}^{n_m} \geq$$

$$\geq J_{m-1} - (1 - 2^{-m+1})\varepsilon + J_m - 2^{-m}\varepsilon - J_{m-1} = J_m - (1 - 2^{-m})\varepsilon,$$

as we have alleged. From this, a fortiori,

$$J_m \leq \int_a^b \psi_m(x)dx + \varepsilon,$$

and since the $\psi_m(x)$, which are step functions, decrease and converge almost everywhere to 0, their integral also converges to 0 by virtue of the lemma indicated; consequently, $\lim J_m \leq \varepsilon$, or, ε being arbitrary, $J_m \to 0$; hence finally, keeping (25) again in mind,

$$\int_a^b \varphi_m(x)dx \to 0,$$

which was to be proved.

48. The Lebesgue Integral Considered as the Inverse Operation of Differentiation

Finally, we discuss a definition of the Lebesgue integral based on differentiation, just as the classical integral was formerly defined in many textbooks on analysis. A similar definition, if only for bounded functions, was already formulated in the first edition of LEBESGUE's *Leçons sur l'intégration*, but without being followed up: "A bounded function $f(x)$ is said to be summable, if there exists a function $F(x)$ with bounded derived numbers such that $F(x)$ has $f(x)$ for derivative, except for a set of values of x of measure zero. The integral in (a, b) is then, by definition, $F(b) - F(a)$." We shall speak of a closely related idea which is a little more general, since it applies also to unbounded functions.

Before doing this, let us mention that DENJOY, KHINTCHINE, and PERRON[41] developed theories of integration of a very broad scope whose final goal is the investigation of primitive functions, that is, functions with given derivatives. The first two authors start with the Lebesgue integral, but PERRON defines the integral, or rather the lower and upper integrals, by considering functions whose derived numbers are respectively less than and greater than the given function. This definition, whose scope, as we have just said, is much broader than that of the Lebesgue integral, reduces to the latter when $f(x)$ is bounded and also when it is summable. But when we limit ourselves to these particular cases and when, moreover, we already possess, as we have from the first pages of this book, the theorem asserting the existence of the derivative of monotonic functions or functions of bounded variation, we are lead naturally to a very simple definition; this is the one which we are going to formulate.

It is particularly for non-negative functions $f(x)$ that our definition is immediate: *The function $f(x) \geq 0$ will be called summable in an interval (a, b) if there exist nondecreasing functions $F(x)$ whose derivatives equal $f(x)$ almost everywhere.* The integral in the interval (a, b) is defined to be the lower bound of the increments $F(b) - F(a)$ of all these functions. We prove without difficulty that there exists among the $F(x)$ an extremal function which is determined up to an additive constant; it is this which we can take as the indefinite integral, since it furnishes the lower bound in question, hence the integral, not only for the interval (a, b), but also for all the subintervals.

As to the integration of functions of variable sign, they reduce by direct decomposition to the case which we have just considered. Therefore we have the following complete *definition*:

The function $f(x)$ is summable in (a, b) if there exist functions of bounded variation $F(x)$ of which it is the derivative almost everywhere; the indefinite integral

[41] DENJOY [**2**], [**3**]; KHINTCHINE [**1**], [**2**]; PERRON [**1**].

of $f(x)$ is then that one among the $F(x)$ whose total variation is the smallest possible; it is determined up to an additive constant.

For the details we refer the reader to the paper cited below.[42] We point out only that these details, except for the theorem asserting the existence of the derivative of monotonic functions, are based on the following theorem:

Given in (a, b) an arbitrary function $g(x)$, let $\{G\}$ be the set, if such a set exists, of functions $G(x)$ which are nondecreasing and have the property that $G'(x) \geq g(x)$ almost everywhere. The set $\{G\}$ has an extremal element $G^*(x)$ such that $G^*(d) - G^*(c) \leq G(d) - G(c)$ for $a \leq c < d \leq b$ and for all the elements $G(x)$ of the set $\{G\}$, or in other words, such that all the differences $G(x) - G^*(x)$ are nondecreasing; $G^*(x)$ is uniquely determined up to an additive constant.

All we have just said concerning the various definitions of the Lebesgue integral can be extended, with more or less obvious modifications, to functions of several variables.[43]

[42] RIESZ [**18**].

[43] BOURBAKI [*] and STONE [**3**] have recently suggested a method of introducing the Lebesgue integral which is similar to the one we have followed in that it does not make use of measure theory, but which, in contrast, makes use of a notion of *upper integral* which is the analogue of the notion of exterior measure. First the upper integral is defined for lower semi-continuous functions $h \geq 0$ as the (finite or infinite) least upper bound of the elementary integrals of those continuous functions g with "compact support" (i.e., which are zero in the exterior of a finite interval) such that $0 \leq g \leq h$. For an arbitrary (finite or infinite) function $f \geq 0$, the upper integral $N(f)$ is then defined as the greatest lower bound of the upper integrals of the lower semi-continuous functions such that $h \geq f$; then $N(f_1 + f_2) \leq N(f_1) + N(f_2)$ and $N(cf) = cN(f)$ $(c > 0)$. For the characteristic function of a set e, the upper integral equals the exterior measure of e; for continuous functions $f \geq 0$ (with compact support), $N(f)$ coincides with the elementary integral of f. A function f is said to be integrable if there exists a sequence $\{g_n\}$ of continuous functions with compact support such that $N(|f - g_n|) \to 0$. The integral of g_n then tends to a limit which is independent of the particular choice of the sequence $\{g_n\}$; the integral of f is defined to be this limit. Cf. the definition of the integral given by W. H. YOUNG, Sec. 46.

CHAPTER III

THE STIELTJES INTEGRAL AND ITS GENERALIZATIONS

LINEAR FUNCTIONALS ON THE SPACE OF CONTINUOUS FUNCTIONS

49. The Stieltjes Integral

Given, on $a \leq x \leq b$, a continuous function $f(x)$ and a nondecreasing function $\alpha(x)$, we define the Stieltjes integral of f with respect to α, denoted by

$$\int_a^b f d\alpha = \int_a^b f(x) d\alpha(x), \tag{1}$$

to be the limit of the sums

$$\Sigma = \sum_1^n f(\xi_k) [\alpha(x_k) - \alpha(x_{k-1})] \tag{2}$$

$$(a = x_0 < x_1 < \dots < x_n = b;\ x_{k-1} \leq \xi_k \leq x_k)$$

when $\max(x_k, x_{k-1}) \to 0$. Or, what amounts to the same thing, we form the integral, in the sense previously stated, of the additive, multivalent interval function

$$f(c, d) = f(\xi) [\alpha(d) - \alpha(c)] \qquad (c \leq \xi \leq d):$$

we could also form its upper and lower integrals, the two coinciding according to a classical argument.

The function $\alpha(x)$ can be assumed to be of bounded variation instead of nondecreasing, the former case reducing to the latter by decomposing $\alpha(x)$ into its indefinite positive and negative variations. The generalization to continuous functions with complex values $f(x) = f_1(x) + if_2(x)$ and to complex functions of bounded variation $\alpha(x) = \alpha_1(x) + i\alpha_2(x)$ (cf. Sec. 15) is immediate; we have only to take the corresponding linear combination of the integrals

$$\int_a^b f_j(x) d\alpha_k(x) \qquad (j, k = 1, 2).$$

The Stieltjes integral and its generalizations are of great importance in many problems and concepts of analysis, mechanics, mathematical physics, and probability theory, for example in curvilinear and surface integrals and various moments and potentials. STIELTJES himself used them, in 1894, in the study of continued fractions and in what is called the problem of moments.[1] Fifteen years later, when there was little interest in the problem, one of the authors succeeded in attracting general interest to the question by establishing a close connection between this integral and an important class of linear functionals, those which are defined on the field of continuous functions on a finite interval.

50. Linear Functionals on the Space C

We denote the set of real continuous functions on the interval $a \leq x \leq b$ by C. An operation which assigns to each element f of C a real number Af is said to be a *linear functional* if it is

1) *additive*: $A(f_1 + f_2) = Af_1 + Af_2$,
2) *homogeneous*: $A(cf) = cAf$,
3) *bounded*: there exists a constant M such that

$$|Af| \leq M \cdot \max |f(x)|.$$

We shall denote the smallest of the possible bounds M by M_A or by $\|A\|$ and call it the *norm of the linear functional* A. If we agree further to write

$$\|f\| = \max |f(x)|$$

and to call it the *norm of the function* f, we arrive at a complete analogy with our previous linear functionals defined on the space L^p.

But there is an essential difference. Our functional is defined only for continuous functions and we do not know *a priori* if it can be extended to a larger class. Moreover, in general it cannot be put in the form of an integral, at least not in the sense previously adopted; as an example, for x_0 fixed, $Af = f(x_0)$ is one of our functionals and it can not be put into the form of an integral other than a Stieltjes integral with

$$\alpha(x) = 0 \text{ for } x < x_0, \quad \alpha(x) = 1 \text{ for } x > x_0,$$

and the value of $\alpha(x_0)$ arbitrary.

To consider the problem in all its generality, we observe first that integral (1), formed with $\alpha(x)$ fixed and of bounded variation, always defines a linear functional of the type considered. In fact, it follows immediately from formula (2) that

$$|\Sigma| \leq \max |f| \times \text{total variation of } \alpha,$$

and, taking the limit, we arrive at the result:

$$\left|\int_a^b f(x)d\alpha(x)\right| \leq \max |f| \times \text{total variation of } \alpha.$$

[1] STIELTJES [**1**] (in particular, pp. 68–76).

Hence, this functional is bounded; additivity and homogeneity are obvious.

We postpone till later the question of under what conditions integral (1) can be put into the form of the integral of an ordinary product as, for instance, the functionals in the space L^p, and for the time being we content ourselves with showing that every linear functional in the space C can be written as a Stieltjes integral.

The main idea in the proof consists in extending the functional A beyond C to a larger function field. To do this we have only to imitate, with some slight modifications, the procedure which has just led us from the integration of step functions to that of summable functions.

Let us consider a sequence of continuous functions $f_n(x)$ which is increasing and bounded and consequently tends to a bounded function $f(x)$. The sequence of values Af_n will tend to a finite limit. In fact, the partial sums of the series

$$|Af_2 - Af_1| + |Af_3 - Af_2| + \dots \tag{3}$$

are the values which correspond, by means of A, to the respective partial sums of the series

$$\pm [f_2(x) - f_1(x)] \pm [f_3(x) - f_2(x)] \pm \dots$$

where we assume the signs suitably chosen. But all these partial sums are in absolute value less than or equal to

$$[f_2(x) - f_1(x)] + [f_3(x) - f_2(x)] + \dots = f(x) - f_1(x),$$

and consequently they are less in absolute value than a constant B. Hence the partial sums of the series (3) are all less than BM_A. It follows that the series

$$Af_1 + (Af_2 - Af_1) + (Af_3 - Af_2) + \dots$$

converges absolutely, and a fortiori its partial sums, which equal Af_n, tend to a finite limit.

We agree to denote this limit by Af and to assign it to the function f — a function which is not necessarily continuous. To justify this convention, it is necessary to show that if two increasing sequences $\{f_n\}$ and $\{g_n\}$ of continuous functions have the same limit f, then the sequences $\{Af_n\}$ and $\{Ag_n\}$ also tend, in their turn, to the same limit. Without loss of generality we can assume that the two sequences of functions are increasing in the strict sense, $f_n < f_{n+1}$, $g_n < g_{n+1}$; in the contrary case, we would have only to use the sequences $\left\{f_n - \frac{1}{n}\right\}$, $\left\{g_n - \frac{1}{n}\right\}$. We choose an element f_m which we hold fixed and run through the second sequence; then, starting with a certain index, we shall have $f_m < g_n$. In fact, if this were not the case the points x such that $f_m(x) \geq g_1(x)$, $f_m(x) \geq g_2(x)$, ..., would constitute a nested sequence of closed, non-empty sets and there would exist at least one point x^* included in all the sets. We would have $f_m(x^*) \geq \lim g_n(x^*) = f(x^*)$, contrary to the hypothesis which guarantees that $f_m < f$. By the same reasoning, there exists for each g_m an infinity of larger functions f_n. Hence we can form an increasing sequence $f_{m_1} < g_{m_2} < f_{m_3} < g_{m_4} < \ldots$ tending to f, and the sequence of values which correspond to it under the functional A tends to a definite limit. Obviously this limit must coincide with that of the Af_n and at the same time with that of the Ag_n; hence these two limits are equal, which was to be proved.

Thus the functional A has been defined uniquely for every bounded function which is the limit of an increasing sequence of continuous functions. Let f and g be two functions of the type considered, then $f + g$ will be of the same type and $A(f + g) = Af + Ag$. On the other hand, the difference $f - g$ belongs, in general, neither to the type f nor to the opposite type $-f$. We agree to define

$$A(f - g) = Af - Ag.$$

To justify this convention, we have only to observe that, just as in the case of the Lebesgue integral, the relation $f - g = f_1 - g_1$ can also be written $f + g_1 = f_1 + g$; hence

$$Af + Ag_1 = A(f + g_1) = A(f_1 + g) = Af_1 + Ag,$$

which yields $Af - Ag = Af_1 - Ag_1$. Therefore the convention is justified and the operation A is uniquely defined for every function of the type $f - g$.

It is almost evident that the functional so extended remains additive and homogeneous; let us show that it also remains bounded. To be precise, we shall show that

$$|A(f - g)| \leq M_A \cdot \sup_{a \leqq x \leqq b} |f(x) - g(x)|.$$

For brevity we denote this least upper bound by μ. Let $\{f_n\}$, $\{g_n\}$ be two increasing sequences of continuous functions tending respectively to f and g. We form the auxiliary functions $h_n(x)$ by setting

$$h_n(x) = f_n(x) \quad \text{when} \quad |f_n(x) - g_n(x)| \leq \mu,$$

$$h_n(x) = g_n(x) + \mu \quad \text{when} \quad f_n(x) - g_n(x) > \mu,$$

$$h_n(x) = g_n(x) - \mu \quad \text{when} \quad f_n(x) - g_n(x) < -\mu.$$

We see immediately that the functions $h_n(x)$ are continuous and that they tend, increasingly, to f. Moreover,

$$\sup_{a \leqq x \leqq b} |h_n(x) - g_n(x)| \leq \mu.$$

Consequently we have

$$|A(f-g)| = |\lim (Ah_n - Ag_n)| = \lim |(Ah_n - g_n)| \leq M_A \mu,$$

which was to be proved.

For our immediate purposes it is unnecessary to examine the function field under consideration more closely. It will suffice to mention that in addition to the continuous functions it contains, among others, certain very simple discontinuous functions, which we shall point out, and moreover, that when $f_1, f_2, \ldots, f_n$ belong to the field their linear combinations likewise belong to the field. Let $f_{c,d}(x)$ be the characteristic function of the closed interval $c \leq x \leq d$.[2] This function is the limit of the decreasing sequence formed by the continuous functions $f_n(x)$ which are zero outside $\left(c - \frac{1}{n}, d + \frac{1}{n}\right)$, equal to 1 on (c, d), and linear on the two segments which remain. Hence the function $f_{c,d}(x)$ belongs, in the interval $a \leq x \leq b$, to the function field considered.

The introduction of the characteristic functions $f_{c,d}$ permits us, first of all, to define the function of bounded variation $\alpha(x)$ which appears in our assertion. We set $\alpha(a) = 0$ and

$$\alpha(x) = Af_{a,x}$$

for $a < x \leq b$. We shall show that this function $\alpha(x)$ is of *bounded variation* and that its total variation does not surpass the norm M_A. To do this, we decompose (a, b) in the usual manner and consider the expression

$$\sum_1^n |\alpha(x_k) - \alpha(x_{k-1})|.$$

This is the value which corresponds, under the functional A, to the function

$$f(x) = \varepsilon_1 f_{a,x_1}(x) + \sum_2^n \varepsilon_k[f_{a,x_k}(x) - f_{a,x_{k-1}}(x)],$$

where ε_k equals $-1, 0$, or 1 according to the sign of $\alpha(x_k) - \alpha(x_{k-1})$. This function, being a linear combination of the functions f_{a,x_k}, certainly belongs to the field under consideration; its absolute value is ≤ 1; consequently,

$$\sum_1^n |\alpha(x_k) - \alpha(x_{k-1})| = Af \leq M_A.$$

The upper limit of these sums, that is, the total variation of $\alpha(x)$, is then also $\leq M_A$, which was to be shown.

Now let $f(x)$ be a continuous function. Let us make the usual decompo-

[2] In particular, $f_{c,c}(x)$ is the characteristic function of the point c; hence, it equals 1 for $x = c$ and zero elsewhere.

sition of (a, b), denote by ξ_k one of the points of the interval $I_k = (x_{k-1}, x_k)$, and define the function $\varphi(x)$ as follows: it is constant and equal to $f(\xi_k)$ for $x_{k-1} < x \leq x_k$; $\varphi(a) = f(\xi_1)$. This function belongs to the field considered, and is written precisely as

$$\varphi(x) = f(\xi_1)f_{a,x_1}(x) + \sum_2^n f(\xi_k)\,[f_{a,x_k}(x) - f_{a,x_{k-1}}(x)].$$

Consequently,

$$A\varphi = f(\xi_1)\alpha(x_1) + \sum_2^n f(\xi_k)[\alpha(x_k) - \alpha(x_{k-1})];$$

or finally, observing that $\alpha(x_0) = \alpha(a) = 0$, it follows that

$$A\varphi = \sum_1^n f(\xi_k)\,[\alpha(x_k) - \alpha(x_{k-1})].$$

We note that the second member is expression (2), which serves to define integral (1). On the other hand, denoting by ω the maximum oscillation of f in the subdivision intervals, we have $|f(x) - \varphi(x)| \leq \omega$, hence

$$|Af - A\varphi| = |A(f - \varphi)| \leq \omega M_A.$$

For max $(x_k - x_{k-1}) \to 0$, $\omega \to 0$; hence $A\varphi \to Af$, that is to say, expression (2) converges to Af, so Af equals integral (1), which was to be proved. It also follows that the total variation of $\alpha(x)$ is a bound for the functional, hence

$$\text{total variation of } \alpha \geq M_A.$$

But we have seen above that the first member can not surpass the second; hence the equality sign holds.

Summarizing, we have the

THEOREM.[3] *The Stieltjes integral*

$$\int_a^b f(x)d\alpha(x)$$

formed with a fixed function of bounded variation $\alpha(x)$, defines a linear functional in the space C of continuous functions $f(x)$, and conversely, every linear functional can be written in this integral form.

This theorem remains valid for the space of continuous functions with *complex* values, with the same definition of the norm, $\|f\| = \max |f(x)|$. Of course, the function of bounded variation $\alpha(x)$ then can also have complex values, as well as the linear functional Af; furthermore, the property $A(cf) = cAf$ must also be required for complex coefficients c. In the proof one has to use, instead of the signs $\pm$, suitable complex factors of unit modulus.

[3] RIESZ [3], [7], [22].

51. Uniqueness of the Generating Function

An analysis which is a little more detailed shows that the functional A determines the function $\alpha(x)$, at its points of continuity, up to an additive constant. In fact, this question reduces immediately to the following: *Which functions of bounded variation $\alpha(x)$ have the property that*

$$\int_a^b f(x)d\alpha(x) = 0$$

for all continuous $f(x)$?

To answer this question, consider a point d of continuity of $\alpha(x)$, interior to the interval (a, b), choose for $f(x)$ the continuous function equal to 1 for $a \leq x \leq d$, zero for $d + \frac{1}{n} \leq x \leq b$, and linear between d and $d + \frac{1}{n}$. Then obviously our integral decomposes into three integrals, corresponding respectively to the segments $a \leq x \leq d$, $d \leq x \leq d + \frac{1}{n}$, $d + \frac{1}{n} \leq x \leq b$; the first of these integrals equals $\alpha(d) - \alpha(a)$, the last is zero, and the second is majorized by the total variation of $\alpha(x)$ on the second segment; this variation tends to 0 with $\frac{1}{n}$. Consequently, $\alpha(d) - \alpha(a) = 0$, that is to say $\alpha(x) = \alpha(a)$ at every point of continuity of $\alpha(x)$ interior to the interval (a, b). Again, setting $f(x) = 1$ everywhere, we see that $\alpha(b) = \alpha(a)$ even if b is not a point of continuity.

Summarizing, *in order that the integral be zero for every continuous function $f(x)$, it is necessary that $\alpha(x)$ be constant at its points of continuity, and that its constant value coincide with $\alpha(a)$ and $\alpha(b)$.*

Obviously this condition is also *sufficient*, since the points of discontinuity of $\alpha(x)$ form only a denumerable set and therefore the set of points of continuity is everywhere dense in (a, b), and of such a nature that the points of subdivision x_k which serve to define the integral can be chosen from this set.

Our results can obviously also be stated in the following form:

THEOREM. *A necessary and sufficient condition that the integral*

$$\int_a^b f(x)d\alpha(x)$$

be zero for every element $f(x)$ of the space C is that the function $\alpha(x)$ be constant on a set everywhere dense in (a, b) which includes the two endpoints a and b.

At points of discontinuity we generally agree to define $\alpha(x)$ by one of its limit values $\alpha(x \pm 0)$ (formed by running through points of continuity) or by the arithmetic mean of these two values; in all these cases, and more generally *under the condition that* $\alpha(x)$ *remain between* $\alpha(x-0)$ *and* $\alpha(x+0)$, *the total variation of* $\alpha(x)$ *equals the norm* M_A.

In fact, since this is true for the generating function constructed in the preceding section, it follows for an arbitrary function of the type considered by virtue of the uniqueness which we have just proved.

52. Extension of a Linear Functional

Let us proceed to a related problem whose analogue we have already studied in the space L^2. We assume that the functional A is given for only a subset E of the (real) space C. Can we extend it so as to define a linear functional A of norm $M_A \leq M$ in the entire space C?

An evident necessary condition is that for all linear combinations of elements of E,

$$|\sum_1^n c_k Af_k| \leq M \, \|\sum_1^n c_k f_k\|. \tag{4}$$

We shall see that this condition is also sufficient, with M_A being the smallest value of M for which it is fulfilled. To attain this goal we shall use an idea found in a 1912 note [**1**] of E. Helly, an idea which is also applicable to linear functionals on the spaces L^p and even, as was observed by Hahn and Banach,[4] on abstract spaces of a very general type.

To simplify the writing we set $M = 1$, which can be done without loss of generality.

We start by defining the functional A for all the elements g which can be represented as linear combinations of elements of E: if

$$g = \sum_1^n c_k f_k, \text{ set } Ag = \sum_1^n c_k Af_k.$$

Let us show that this definition is single-valued, that is to say that if

$$\sum_1^n c_k f_k = \sum_1^n c_k' f_k, \text{ then } \sum_1^n c_k Af_k = \sum_1^n c_k' Af_k.$$

Taking differences, the problem reduces to showing that

$$\sum_1^n c_k'' f = 0 \text{ implies } \sum_1^n c_k'' Af_k = 0 \qquad (c_k'' = c_k - c_k');$$

but this follows immediately from condition (4).

These linear combinations g form a linear manifold E', and the functional A, extended to E', is obviously additive, homogeneous, and by virtue of (4), also bounded by $M = 1$.

[4] Hahn [**3**] (in particular p. 217); Banach [**1**], (Chapt. IV, § 2); also see M. Riesz [**1**].

With this established, we proceed by successively adjoining new elements.

We denote by g_1 and g_2 two arbitrary elements of E' and by Ag_1 and Ag_2 the corresponding values of the extension of A to E'. Let f be an element of the space C which does not belong to E'. Since

$$Ag_1 - Ag_2 = A(g_1 - g_2) \leq \|g_1 - g_2\| \leq \|g_1 - f\| + \|g_2 - f\|,$$

it follows that

$$Ag_1 - \|g_1 - f\| \leq Ag_2 + \|g_2 - f\|.$$

So if we vary g in such a way that it runs through all the elements of E', the quantities

$$Ag - \|f - g\| \text{ and } Ag + \|f - g\|$$

form, for fixed f, two classes of real numbers, the first of which lies to the left of the second, and consequently there are one or more numbers included between the two classes. Set Af equal to one of these values; then, for every element g of E',

$$Ag - \|f - g\| \leq Af \leq Ag + \|f - g\|,$$

and replacing g by $-g$ (which is permissible since E' is a linear manifold), we have

$$|Af + Ag| \leq \|f + g\|. \tag{5}$$

Let us now extend the functional A to linear combinations of f and elements g of E' by defining

$$A(cf + g) = cAf + Ag.$$

We shall have

$$|A(cf + g)| \leq \|cf + g\|;$$

for $c = 0$ this is nothing other than condition (4), and for $c \neq 0$ it follows from (5) by replacing g by $\frac{1}{c}g$ there; in fact,

$$|A(cf + g)| = |cAf + Ag| = |c|\left|Af + A\frac{g}{c}\right| \leq |c|\left\|f + \frac{g}{c}\right\| = \|cf + g\|.$$

Therefore condition (4) is also fulfilled in the linear set E'' formed by linear combinations of f and elements g of E'. Moreover, the functional A when extended to E'' remains additive, homogeneous, and bounded by 1.

Consequently we have only to choose successively for f a sequence of functions, for example $1, x, x^2, \ldots$, whose linear combinations, among themselves or with elements of E' are everywhere dense in C in the sense of uniform convergence; the functional A will be defined successively for this everywhere dense set without its being necessary to change the bound $M = 1$; finally, we need only pass to the limit to extend A to the entire space. Therefore we have proved the

THEOREM. *A necessary and sufficient condition that the functional A, given in a set E of the space C, be extendable to the entire space C so as to define there a linear functional of norm $\leq M$ is that*

$$\left|\sum_1^n c_k A f_k\right| \leq M \left\|\sum_1^n c_k f_k\right\|$$

for every linear combination of elements of E.

This theorem also holds for the space of *complex* continuous functions. The necessity of condition (4), with complex coefficients c_k, is evident; we shall show how the proof of the sufficiency can be reduced to the real case.[5]

We first extend the functional A to the linear set E' formed by complex linear combinations of elements of E, just as was done in the real case. On E', A will be homogeneous even with respect to a complex numerical factor, therefore, in particular we shall have $A(ig) = iAg$ for all elements g of E'. Denote by $A_1 g$ the real part of Ag; inequality (4) holds for A_1 in place of A, if we consider only linear combinations of elements of E' with *real* coefficients c_k. Then we extend this real-valued functional A_1 to the entire space $C_{complex}$, by the same procedure as in the real case; A_1, extended in this manner, will be real-valued, additive, homogeneous with respect to *real* factors, and bounded by M. We shall show that

$$Bf = A_1 f - iA_1(if)$$

furnishes the desired extension of the functional A.

To accomplish this it is necessary to show that the functional B is *additive, homogeneous* with respect to *complex* factors, *bounded* by M, and finally, that it coincides on E' with the functional A. Additivity follows immediately from the additivity of A_1. A little more calculation is necessary to establish homogeneity with respect to a complex factor $c = a + ib$:

$$B(cf) = A_1(af + bif) - iA_1(aif - bf) = aA_1 f + bA_1(if) - iaA_1(if) + ibA_1 f =$$
$$= (a + ib)\,(A_1 f - iA_1(if)) = cBf;$$

in this calculation we have made use of the additivity of A_1 and of its homogeneity with respect to real factors. To show that B is bounded by M, we set

$$Bf = re^{it} \quad (r \geq 0)$$

for arbitrary fixed f; then we shall have

$$|Bf| = e^{-it}Bf = B(e^{-it}f) = A_1(e^{-it}f) \leq M\,\|e^{-it}f\| = M\|f\|$$

(the third equation is motivated by the fact that $B(e^{-it}f)$, being equal to r,

[5] Cf. Bohnenblust–Sobczyk [1]; Soukhomlinoff [1].

is real-valued). Finally, to show that $Bg = Ag$ for every element g of E', we note that $-A_1(ig)$ is by definition equal to the real part of $-A(ig) = -iAg$, hence equal to the imaginary part of Ag, and that consequently

$$Ag = A_1 g - iA_1(ig),$$

which completes the proof.

53. The Approximation Theorem. Moment Problems

Here is a corollary of the preceding theorem which merits notice. Choose for E an arbitrary set E_0 of the space C and adjoin an element f_0 which does not already belong to it. We try to define the functional A so that $Af_0 = 1$ and A becomes zero for elements of E_0. Since the linear combinations formed from E can be written, up to a constant factor, as $f_0 - g$, where g denotes linear combinations formed from E_0, condition (4) can be written as follows:

$$1 \leq M_A \|f_0 - g\|,$$

to be satisfied for all the g's. Denoting by d the greatest lower bound of $\|f_0 - g\|$, or as can be said, the "distance" from f_0 to the linear manifold generated by E_0, the condition is also expressed by $M_A \geq \frac{1}{d}$. Hence we have the

THEOREM.[6] *Given a set E_0 and an element f in the space C, a necessary and sufficient condition that it be possible to approximate uniformly to within d of f by linear combinations of elements of E_0 is that for every linear functional A which becomes zero in E_0 and for which $Af = 1$, one has $M_A \geq \frac{1}{d}$. In particular a necessary and sufficient condition that it be possible to uniformly approximate arbitrarily closely every element of C by these linear combinations is that there exist no linear functional except $Af = 0$ which becomes zero in E_0, or what amounts to the same thing, that there exist no function of bounded variation $\alpha(x)$ such that*

$$\int_a^b g(x)d\alpha(x) = 0$$

for every element of E_0, unless this is true for every element of C.

The theorem established in the preceding section also furnishes an answer to the following problem. A function $\alpha(x)$ of bounded variation in (a, b) is sought, certain of whose "moments"

$$\mu_n = \int_a^b f_n(x)d\alpha(x) \qquad (n = 0, 1, \ldots)$$

[6] RIESZ [**4**], [**4a**].

are prescribed; the $f_n(x)$ are given continuous functions. If in particular $f_n(x) = x^n$, these are moments of a distribution of masses or electric charges, as they are defined in physics.

Denoting by E the set of given $f_n(x)$, the problem is to find a bounded linear functional A, on the space C of all continuous functions in (a, b), which assumes prescribed values μ_n on E. In as much as the norm of A is equal to the total variation of the generator $\alpha(x)$, we have the

THEOREM. *A necessary and sufficient condition that the problem of moments under consideration admit a function $\alpha(x)$ of total variation $\leq M$ as solution is that*

$$|\sum_{k=0}^{n} c_k \mu_k| \leq M \max |\sum_{k=0}^{n} c_k f_k(x)|$$

for arbitrary integral n and arbitrary numbers c_k (real or complex, depending on whether the functions and moments are real or complex).

In the real case, this condition is satisfied, in particular, if one of the given functions, say $f_0(x)$, is equal to 1, and if the problem is of *positive* type, that is, if $\Sigma c_k \mu_k \geq 0$ whenever $\Sigma c_k f_k(x) \geq 0$. In fact, we then have, for an arbitrary real linear combination $h(x) = \Sigma c_k f_k(x)$:

$$\|h\| f_0(x) \pm h(x) \geq 0, \text{ hence also } \|h\| \mu_0 \pm \sum c_k \mu_k \geq 0,$$

and, consequently,

$$|\sum c_k \mu_k| \leq \mu_0 \|h\|.$$

Moreover, in this case the solution $\alpha(x)$ is *nondecreasing*, as follows from the equation

$$\mu_0 = \int_a^b f_0(x) d\alpha(x) = \int_a^b d\alpha(x),$$

and from the fact that the total variation of $\alpha(x)$ is $\leq \mu_0$.

We consider as an example the *problem of trigonometric moments*[7]

$$\mu_n = \int_0^{2\pi} e^{inx} d\alpha(x) \qquad (n = 0, \pm 1, \pm 2, \ldots),$$

where $\mu_{-n} = \bar{\mu}_n$, or rather the equivalent "real" problem

$$\varkappa_n = \tfrac{1}{2}(\mu_n + \mu_{-n}) = \int_0^{2\pi} \cos nx \, d\alpha(x) \qquad (n = 0, 1, 2, \ldots),$$

$$\sigma_n = \frac{1}{2i}(\mu_n - \mu_{-n}) = \int_0^{2\pi} \sin nx \, d\alpha(x) \qquad (n = 1, 2, \ldots).$$

[7] F. Riesz [**4a**]; Herglotz [**1**].

In order that the problem admit a nondecreasing solution $\alpha(x)$, it is obviously necessary that the sequence $\{\mu_n\}$ $(n = 0, \pm 1, \ldots)$ be *positive definite,* which means that

$$\sum_{m,n=-N}^{N} \mu_{n-m}\varrho_n\bar{\varrho}_m \geq 0, \tag{6}$$

for arbitrary integral N and arbitrary complex numbers ϱ_n; in fact, this double sum equals

$$\int_0^{2\pi} \sum_{m,n=-N}^{N} e^{i(n-m)x}\varrho_n\bar{\varrho}_m\, d\alpha(x) = \int_0^{2\pi} |\sum_{m=-N}^{N} e^{inx}\varrho_n|^2 d\alpha(x) \geq 0.$$

Condition (6) is also sufficient. To prove this we have only to show that (6) implies that the problem is of positive type, that is, that for every positive trigonometric polynomial

$$t(x) = \sum_{k=-n}^{n} c_k e^{ikx} \geq 0,$$

we also have

$$\sum_{k=-n}^{n} c_k \mu_k \geq 0.$$

To see this, we make use of a lemma due to L. FEJÉR and F. RIESZ which states that *every trigonometric polynomial $t(x) \geq 0$ can be represented as the square of the modulus of another trigonometric polynomial $q(x)$ whose coefficients are, in general, complex*:

$$t(x) = |q(x)|^2 = |\sum_j \varrho_j e^{ijx}|^2 = \sum_{j,k} \varrho_j\bar{\varrho}_k e^{i(j-k)x}.$$ [8]

In fact, this representation implies that

$$At = \sum_{j,k} \varrho_j\bar{\varrho}_k \mu_{j-k},$$

hence, by (6), $At \geq 0$.

On the other hand, it would have sufficed to consider only strictly positive polynomials $t(x)$, since the general case can be reduced to this case by adding to $t(x)$ the constant term $\varepsilon > 0$, which then is made to tend to 0. For a polynomial

$$t(x) = \sum_{k=-n}^{n} c_k e^{ikx} > 0$$

the lemma in question is proved as follows.

[8] Cf. FEJÉR [**1**]. Denoting the real and imaginary parts of $q(x)$ by $u(x)$ and $v(x)$, we have the representation

$$t(x) = u^2(x) + v^2(x)$$

of $t(x)$ as the sum of the squares of two *real* trigonometric polynomials.

We observe first that $c_k = \overline{c_{-k}}$ and that we can assume $c_n \neq 0$. Consider the polynomial

$$P(z) = c_{-n} + c_{-n+1}z + \ldots + c_n z^{2n};$$

it obviously satisfies the relation

$$P(z) = z^{2n} P\left(\frac{1}{\bar{z}}\right). \tag{7}$$

Moreover, since

$$e^{-inx}P(e^{ix}) = t(x),$$

$P(z)$ can have no zeros on the circle $|z| = 1$. Denote by $\alpha_1, \alpha_2, \ldots$ its zeros in the interior and by $\beta_1, \beta_2, \ldots$ those in the exterior of the unit circle. Let the multiplicities of these zeros by equal to respectively $r_1, r_2, \ldots$ and $s_1, s_2, \ldots$. Then we have the factorization

$$P(z) = c \prod_k (z - \alpha_k)^{r_k} z^S \prod_j \left(\frac{1}{z} - \frac{1}{\beta_j}\right)^{s_j}$$

where $S = \Sigma s_j$. Relation (7) shows that if β is a zero exterior to the unit circle, $\alpha = 1/\bar{\beta}$ is a zero interior to this circle, of the same multiplicity as β, and conversely. Hence we can enumerate the β so that we have $\beta_k = 1/\bar{\alpha}_k$; then $r_k = s_k$ and

$$S = \Sigma s_k = \Sigma r_k = \tfrac{1}{2}\Sigma(s_k + r_k) = n.$$

We shall therefore have

$$t(x) = e^{-inx}P(e^{ix}) = c \prod_k (e^{ix} - \alpha_k)^{r_k} \prod_k (e^{-ix} - \bar{\alpha}_k)^{r_k}.$$

The coefficient c is necessarily positive, and the trigonometric polynomial

$$q(x) = \sqrt{c} \prod_k (e^{ix} - \alpha_k)^{r_k}$$

satisfies the requirements of the theorem.

54. Integration by Parts. The Second Theorem of the Mean

Among the rules of calculation for the Stieltjes integral — rules which for the most part are analogous to those known in classical analysis and also to those valid for the Lebesgue integral, and are obtained in the same manner — there are some which merit being mentioned separately. Among these are the formula for integration by parts and certain of its consequences.

The formula in question is written

$$\int_a^b f(x)d\alpha(x) + \int_a^b \alpha(x)df(x) = [f(x)\alpha(x)]_a^b.$$

What is interesting is that it can be established without assuming anything about the nature of the functions f and α except the existence, according to the STIELTJES definition, of one of the two integrals, say, for definiteness, the convergence of the sums (2) to a limit. In fact, this follows immediately from the identity

$$\sum_{0}^{n} \alpha(\xi_i)\,(f(x_{i+1}) - f(x_i)) = [\alpha(x)f(x)]_a^b - \sum_{0}^{n+1} f(x_i)\,(\alpha(\xi_i) - \alpha(\xi_{i-1}))$$

where $\xi_{-1} = x_0 = a$ and $\xi_{n+1} = x_{n+1} = b$.

In particular, when $f(x)$ is continuous and $\alpha(x)$ is of bounded variation, our rule guarantees the existence of the integral of $\alpha(x)$ with respect to $\mathrm{f}(x)$. Let us now consider a more special case in which $\alpha(x)$ is *monotonic* or, for definiteness, *nondecreasing* and let us replace $f(x)$ (which we merely assume to be summable in the Lebesgue sense) by its indefinite integral

$$F(x) = \int_a^x f(\xi)d\xi.$$

Noting that, on the one hand, the value of the integral

$$\int_a^b F(x)\mathrm{d}\alpha(x)$$

lies between the two extremal values of $F(x)$ multiplied by $\alpha(b) - \alpha(a)$, and that consequently, due to the continuity of $F(x)$, it can be written in the form

$$F(\xi)\,(\alpha(b) - \alpha(a)) = (\alpha(b) - \alpha(a)) \int_a^{\xi} f(x)dx,$$

and that, on the other hand, as we shall see later (Sec. 58) under even more extensive assumptions,

$$\int_a^b \alpha(x)dF(x) = \int_a^b \alpha(x)f(x)dx,$$

we obtain

$$\int_a^b \alpha(x)f(x)dx = \alpha(b)\int_a^b f(x)dx - (\alpha(b) - \alpha(a))\int_a^{\xi} f(x)dx;$$

hence

$$\int_a^b \alpha(x)f(x)dx = \alpha(a)\int_a^{\xi} f(x)dx + \alpha(b)\int_{\xi}^{b} f(x)dx,$$

a formula which expresses the *second theorem of the mean.*

55. Sequences of Functionals

Here is another application in which, however, integration by parts plays only a secondary role. We consider a sequence $\{A_n\}$ of linear functionals, or what amounts to the same thing, a sequence $\{\alpha_n(x)\}$ of functions of bounded variation; we can assume that $\alpha_n(a) = 0$. Under what conditions does the sequence $\{A_n f\}$ converge to a linear functional Af with $\alpha(x)$ for generating function for every continuous function $f(x)$?

A necessary condition is established by repeating almost word for word the reasoning we used to answer the analogous question asked for the space L^2. It is that the norms M_{A_n} remain less than some finite bound.

Now it is obvious that this condition is also sufficient if we add to it that $A_n f \to Af$ for a complete set in C, that is to say for a set of continuous functions $f(x)$ possessing the property that every continuous function can be approximated uniformly by linear combinations of these functions. In fact, let f be an element of C and, fixing $\varepsilon > 0$, choose a linear combination $g = \sum_1^m c_k f_k$ such that $\| f - g\| < \varepsilon$. Since, by hypothesis,

$$A_n g = \sum_{k=1}^{m} c_k A_n f_k \to \sum_{k=1}^{m} c_k A f_k = Ag \quad \text{as } n \to \infty,$$

we shall have $|Ag - A_n g| < \varepsilon$ for large enough n. Then, denoting by B the common bound of the functionals A_n, we have

$$|Af - A_n f| \leq |A(f - g)| + |Ag - A_n g| + |A_n(g - f)| \leq M_A \varepsilon + \varepsilon + B\varepsilon,$$

which verifies our assertion.

In order to write the condition in a sufficiently simple final form which uses the generating functions $\alpha_n(x)$ and $\alpha(x)$, we must choose a suitable complete set. Such a set is the set formed by the constant 1 and by the positive parts of the functions $\xi - x$, where ξ takes on values in the interval (a, b), that is, the functions $f(x; \xi) = \xi - x$ for $x \leq \xi$ and 0 for $x \geq \xi$. In fact, every continuous function in (a, b) with linear traces (i.e., whose image is a polygonal line) is a linear combination of the chosen functions; on the other hand, every continuous function in (a, b) can be approximated uniformly by polygonal functions of this type. Therefore the additional condition can be written

$$\int_a^b d\alpha_n(x) \to \int_a^b d\alpha(x), \qquad \int_a^\xi (\xi - x) d\alpha_n(x) \to \int_a^\xi (\xi - x) d\alpha(x);$$

the first relation can also be written as $\alpha_n(b) \to \alpha(b)$ and the second, after integration by parts, becomes

$$\int_a^\xi \alpha_n(x) dx \to \int_a^\xi \alpha(x) dx.$$

We have thus arrived at the answer to the question.

THEOREM. *Let A_n and A be linear functionals operating on the space of continuous functions on the interval $a \leq x \leq b$, and let $\alpha_n(x)$ and $\alpha(x)$ be their generating functions, with $\alpha_n(a) = \alpha(a) = 0$. Under this assumption, a necessary* *and sufficient condition that for every continuous function $f(x)$ the sequence $\{A_n f\}$ converge to Af is that*

$$\int_a^\xi \alpha_n(x)dx \to \int_a^\xi \alpha(x)dx \qquad (a < \xi \leq b), \tag{8}$$

$$\alpha_n(b) \to \alpha(b),$$

and that the norms of the functionals A_n remain less than some finite bound.[9]

We add that, as can easily be seen, we can slightly weaken the first condition so that it is required only for the points ξ of a set everywhere dense in (a, b).

Another simplification is possible when the functionals A_n are positive, that is, when the functions $\alpha_n(x)$ are nondecreasing:

THEOREM.[10] *When the functions $\alpha_n(x)$ are nondecreasing, it is permissible to replace condition* (8) *in the preceding theorem by the condition that the functions $\alpha_n(x)$ themselves converge to the function $\alpha(x)$ at every point of continuity of $\alpha(x)$, or equivalently, at every point x of a set everywhere dense in (a, b).*

In fact, since the functions α_n are nondecreasing, we have

$$\alpha(x) = \lim \alpha_n(x) \leq \liminf \alpha_n(\xi) \leq \limsup \alpha_n(\xi) \leq \lim \alpha_n(x') = \alpha(x')$$

for every point of continuity ξ of $\alpha(x)$ and for $x < \xi < x'$, where x and x' belong to our everywhere dense set; letting x and x' approach ξ, it follows that $\alpha_n(\xi) \to \alpha(\xi)$. Moreover, the points for which convergence is not assured form a denumerable set, and since $\alpha_n(x) \leq \alpha_n(b) \to \alpha(b)$ — that is, the $\alpha_n(x)$ are uniformly bounded — term-by-term integration of the $\alpha_n(x)$ is permissible.

Conversely, hypothesis (8) and the fact that the functions $\alpha_n(x)$ are increasing imply that

$$\frac{1}{h}\int_{\xi-h}^{\xi} \alpha_n(x)dx \leq \alpha_n(\xi) \leq \frac{1}{k}\int_{\xi}^{\xi+k} \alpha_n(x)dx, \qquad (h, k > 0)$$

and taking the limit,

$$\frac{1}{h}\int_{\xi-h}^{\xi} \alpha(x)dx \leq \liminf \alpha_n(\xi) \leq \limsup \alpha_n(\xi) \leq \frac{1}{k}\int_{\xi}^{\xi+k} \alpha(x)dx.$$

But since at every point of continuity ξ of $\alpha(x)$ the two ratios converge to $\alpha(\xi)$ as $h, k \to 0$, the last inequality establishes the existence of the $\lim \alpha_n(\xi)$ and the relation

$$\alpha_n(\xi) \to \alpha(\xi),$$

which was to be proved.

[9] Riesz [**3**].

[10] Karamata [**1**].

GENERALIZATIONS OF THE STIELTJES INTEGRAL

56. The Riemann-Stieltjes and Lebesgue-Stieltjes Integrals

When we wish to extend the sense of the Stieltjes integral we can choose from several different methods.

First, we can ask whether, when $\alpha(x)$ is assumed of bounded variation, the definition of integral (1) as the limit of the sums (2) is applicable to functions $f(x)$ other than the continuous functions. If we restrict ourselves to the same considerations used in the discussion of the Riemann integral, we find that a necessary and sufficient condition on the function $f(x)$ is that its discontinuities form a set of zero variation with respect to $\alpha(x)$, that is, that this set can be enclosed in intervals in such a way that the sum of the total variations of $\alpha(x)$ formed on these intervals is arbitrarily small.

To go further, one tries to modify the definition in such a way that it applies to larger classes, for example by excluding points of discontinuity of $\alpha(x)$ from the points of subdivision or by restricting the choice of the ξ_k to the points of continuity of $f(x)$. To see the effect of this modification, let $\alpha(x)$ equal 0 for $x < x_0$ and 1 for $x > x_0$, while its value for $x = x_0$ remains at our disposal; moreover, let $f(x) = 0$ except for $x = x_0$, where $f(x) = 1$. Then when x_0 is not a point of subdivision, the sum Σ is 0 or 1 depending on the choice of the ξ_k; when x_0 is a point of the subdivision, Σ again depends on the value of $\alpha(x_0)$. But when we exclude x_0 from both the x_k and the ξ_k, we always have $\Sigma = 0$ and thus the value of integral (1) will be determined. Several modifications of the definition of the Stieltjes integral in this direction have been proposed by different authors; the reader would do well to refer to a note by HILDEBRANDT [**2**] where these definitions are analyzed and compared.

The second method is suggested by the developments of Section 50 and is aided by the relation between the Stieltjes integral and the functional Af which we have just studied. The idea is simply to replace the integral by the functional and to extend the latter beyond the field C of continuous functions. One advantage of this method is that it spares us the indecision caused by the discontinuities of $\alpha(x)$. In fact, let us consider again the above example. Certainly for continuous functions $f(x)$ the value of the functional Af is the same independent of the choice of $\alpha(x_0)$, namely $Af = f(x_0)$; as for the discontinuous function $f(x)$ which is under discussion, if we approximate it in an obvious way by a decreasing sequence of continuous functions, we likewise can only assign the value $f(x_0)$ to it. In general, at least as far as we have already gone, since the extension of the functional Af by the method of monotonic sequences is uniquely determined we have avoided the difficulties which the discontinuities of the function $\alpha(x)$ could have caused. To be precise, let $\alpha(x)$ and $\beta(x)$ be two functions of bounded variation which differ by at most a constant on a set which is everywhere dense in the interval (a, b) and includes the endpoints a, b; we shall then have (with obvious notation) $Af = Bf$, first in the field C and then for the extensions of the two functionals.

The extension of an operator A is accomplished by the method of monotonic sequences which we used in Section 50 for a special problem, and is modeled in its details on the theory of the Lebesgue integral as we have developed it; there are a few differences, but these cause no serious difficulty. The first, of no importance, is that for the Lebesgue integral we started with step functions and here we start with continuous functions. The second difference, which forced us in Section 50 to juggle the signs + and −, arises from the fact that we did not limit ourselves there to positive functionals, that is, functionals such that $Af \geqq 0$ when $f \geqq 0$.

It is easy to remove the latter inconvenience, since decomposing the generating function $\alpha(x)$ into the difference of two nondecreasing functions, say $\alpha(x) = \alpha_1(x) - \alpha_2(x)$, results in a decomposition of the functional A into the difference of two positive linear functionals, $A = A_1 - A_2$.

There is still another difference: for our special problem in Section 50 it was not necessary for us to carry the extension process beyond bounded functions. However, all that is necessary in order to include the integration of unbounded functions in the theory from the beginning, just as in the case of the Lebesgue integral, is to consider, instead of increasing and uniformly bounded sequences $\{f_n\}$, increasing sequences such that $A_1 f_n$ and $A_2 f_n$ remain bounded. Of course, the role of sets of measure zero will be played by the sets which have zero variation with respect to both $\alpha_1(x)$ and $\alpha_2(x)$; these sets can be characterized by the fact that an increasing sequence $\{f_n\}$ can increase to infinity in them while the sequences of $A_1 f_n$ and $A_2 f_n$ remain bounded. For definiteness we agree to choose from among the possible decompositions of $\alpha(x)$ the canonical decomposition into the difference of indefinite positive and negative variations, $\alpha(x) = p(x) - n(x)$, and the decomposition $A = P - N$ which corresponds to it. With this agreement made, the sets in question are also of zero variation with respect to $\alpha(x)$, or equivalently, with respect to the indefinite total variation $\tau(x)$ of $\alpha(x)$.

The only essentially new problems are those which concern the relations between several types of integrals, for example the relations between the Lebesgue integral and the Stieltjes integral or its extension — which is called the Lebesgue-Stieltjes integral — or between two integrals of the latter type formed with respect to two different functions $\alpha(x)$. We now proceed to these problems.

57. Reduction of the Lebesgue-Stieltjes Integral to That of Lebesgue

A final method for carrying out the extension in question is due to LEBESGUE;[11] it is suggested by the following idea. We assume, in order to avoid secondary difficulties at the beginning, that $\alpha(x)$ is continuous and increasing in the strict sense; it then admits an inverse function $x(\alpha)$ of the same type and, for $f(x)$ continuous, the relation

$$\int_a^b f(x)d\alpha(x) = \int_{\alpha(a)}^{\alpha(b)} f(x(\alpha))d\alpha \tag{9}$$

is immediate. In fact, by choosing the points x_k sufficiently close in (a, b) — indeed, so close that the points $\alpha_k = \alpha(x_k)$ are sufficiently close in the corresponding interval — and setting $\beta_k = \alpha(\xi_k)$, we can guarantee that the sums

$$\sum_1^n f(\xi_k)\,[\alpha(x_k) - \alpha(x_{k-1})], \qquad \sum_1^n f(x(\beta_k))\,[\alpha_k - \alpha_{k-1}], \tag{10}$$

which equal one another, will be as close as we please to the two integrals in (9), which proves our proposition.

When we relax the assumptions on $\alpha(x)$ so that, while still nondecreasing, it may have jumps or intervals where it is constant, we must redefine the inverse $x(\alpha)$. In the case of jumps, let us agree to consider the function $\alpha(x)$ as multivalent, taking on at the point x all the values between $\alpha(x - 0)$ and $\alpha(x + 0)$ [where $\alpha(a - 0) = \alpha(a)$, $\alpha(b + 0) = \alpha(b)$]. With this convention made, we shall define $x(\alpha)$ as the solution of $\alpha(x) = \alpha$; it will be multivalent at its discontinuities, namely at the values of α taken by $\alpha(x)$ along a segment. Now, in order to compare the two members of (9) we have only to consider the sums (10) and to identify them; in fact, if we choose the x_k sufficiently close and then insert more points so that the α_k are also sufficiently close, the two approximation sums will be as close to the respective integrals as we please.

Finally, when we pass from the Stieltjes integral to its extension, equality (9) extends to all functions $f(x)$ which are summable with respect to $\alpha(x)$, or for which $f(x(\alpha))$ is summable in the ordinary sense. It is this relation which gave Lebesgue the idea of *defining* the Stieltjes integral, that is the first member of (9), by the second.

When $\alpha(x)$ is of bounded variation without being monotonic, we can make use of the usual decomposition of $\alpha(x)$. But we can arrive at the same result by a single change of variable, even when the function $\alpha(x)$ is discontinuous, by using the indefinite total variation $\tau(x)$ of $\alpha(x)$ and its inverse $x(\tau)$. Let us consider the function $\alpha(x(\tau))$; we shall avoid the indecision caused

[11] LEBESGUE [**4**] or [*] (2nd Edition, pp. 252–313).

by the discontinuities of $\alpha(x)$ and $\tau(x)$ by agreeing to complete our function by linear traces joining the points of the image of the function which correspond to the values $\tau(x-0)$, $\tau(x)$ and $\tau(x)$, $\tau(x+0)$; these will evidently be lines forming an angle $\pm\frac{\pi}{4}$ with the τ-axis. The function $\alpha(x(\tau))$ so defined is not only continuous and of bounded variation, but it also satisfies the Lipschitz condition, since its increment ratios do not surpass the bound 1 in absolute value; consequently it is the integral of its derivative

$$\varepsilon(\tau) = [\alpha(x(\tau))]'$$

— a derivative which exists almost everywhere. All this follows from the easily verified fact that the indefinite total variation of $\alpha(x(\tau))$ is simply equal to τ up to an additive constant. Moreover, since a function of bounded variation and its indefinite total variation have the same derivative almost everywhere except for sign (see Section 8), we have $|\varepsilon(\tau)| = \tau' = 1$ almost everywhere.

We therefore have, for two arbitrary points x_1, x_2 and continuous $\alpha(x)$,

$$\alpha(x_2) - \alpha(x_1) = \alpha(x(\tau_2)) - \alpha(x(\tau_1)) = \int_{\tau_1}^{\tau_2} \varepsilon(\tau)d\tau,$$

where $\tau_1 = \tau(x_1)$, $\tau_2 = \tau(x_2)$. From this we obtain the equation

(11) $$\int_a^b f(x)d\alpha(x) = \int_A^B f(x(\tau))\varepsilon(\tau)d\tau \qquad (A = \tau(a),\ B = \tau(b))$$

first for step functions $f(x)$ whose jumps occur at points of continuity of $\alpha(x)$, then, by a uniform passage to the limit, for every continuous function $f(x)$. Since, on the other hand,

$$\int_a^b f(x)d\tau(x) = \int_A^B f(x(\tau))d\tau,$$

we shall have for the indefinite positive and negative variations of $\alpha(x)$, $p(x) = \frac{1}{2}[\tau(x) + \alpha(x)]$ and $n(x) = \frac{1}{2}[\tau(x) - \alpha(x)]$:

$$\int_a^b f(x)dp(x) = \int_A^B f(x(\tau))\,\frac{1+\varepsilon(\tau)}{2}\,d\tau = \int_{e^+} f(x(\tau))d\tau,$$

$$\int_a^b f(x)dn(x) = \int_A^B f(x(\tau))\,\frac{1-\varepsilon(\tau)}{2}\,d\tau = \int_{e^-} f(x(\tau))d\tau,$$

where e^+ and e^- denote respectively the sets on which $\varepsilon(\tau)$ equals $+1$ or -1.

The last three formulas extend, by our method of monotonic sequences, to all functions $f(x)$ which are summable with respect to $p(x)$ and $n(x)$, or equivalently, which are summable with respect to $\tau(x) = p(x) + n(x)$; we see that these are precisely the functions for which $f(x(\tau))$ is summable in (A, B)

in the ordinary sense. Taking the difference of the two last formulas, we again obtain (11).

We thus see that the function $f(x)$ is summable with respect to $\alpha(x)$ if and only if $f(x(\tau))$ is summable in the ordinary sense, and that when this is true we have relation (11). This verifies Lebesgue's observation that it is possible to define the first member of (11) by the second.[12]

58. Relations Between Two Lebesgue-Stieltjes Integrals

An argument analogous to that made in the preceding section leads us to the relation

$$\int_a^b f(x)d\alpha(x) = \int_{\beta(a)}^{\beta(b)} f(x(\beta)) \, [\alpha(x(\beta))]' d\beta, \tag{12}$$

interpreted in an analogous manner, where $\beta(x)$ *denotes a nondecreasing function with respect to which* $\alpha(x)$ *is assumed to be absolutely continuous*; this means that for every system of non-overlapping intervals (a_k, b_k) for which the sum

$$\sum (\beta(b_k) - \beta(a_k))$$

is sufficiently small, the sum

$$\sum |\alpha(b_k) - \alpha(a_k)|$$

becomes, in its turn, as small as we please.

A case of particular interest arises when $\beta(x) = x$ and $\alpha(x)$ *is absolutely continuous in the ordinary sense; then we have*

$$\int_a^b f(x)d\alpha(x) = \int_a^b f(x)\alpha'(x)dx.$$

[12] We remark that the Hellinger integrals, that is, the integrals of the interval functions

$$\frac{[f(\beta) - f(\alpha)]^2}{g(\beta) - g(\alpha)},$$

or more generally the integrals of the interval functions

$$\frac{|f(\beta) - f(\alpha)|^p}{|g(\beta) - g(\alpha)|^{p-1}},$$

formed with an increasing function $g(x)$, admit an analogous reduction to Lebesgue integrals. In the particular case where $g(x) = x$, this reduction follows from our considerations in Section 36: the Hellinger integral exists if $f(x)$ is the indefinite integral of a function $\varphi(x)$ belonging to L^p, and in this case only; it is then equal to $\int_a^b |\varphi|^p \, dx$. The extension to the case of an arbitrary increasing function $g(x)$ can be found in HAHN [**1**] for $p = 2$, and in RADON [**1**] (pp. 57–87) for $p > 1$.

We observe that this formula includes the formula for *integration by substitution* which was extended to Lebesgue integrals in Section 26. In fact, let $\alpha(x)$ be absolutely continuous and nondecreasing, and let $h(\alpha)$ be an arbitrary function which is defined and summable in the interval $A = \alpha(a) \leq \alpha \leq \alpha(b) = B$. Setting $f(x) = h(\alpha(x))$, we shall have $f(x(\alpha)) = h(\alpha)$, and using relation (9), we obtain

$$\int_A^B h(\alpha)d\alpha = \int_A^B f(x(\alpha))d\alpha = \int_a^b f(x)d\alpha(x) = \int_a^b f(x)\alpha'(x)dx = \int_a^b h(\alpha(x))\alpha'(x)dx.$$

Let us return to formula (12). The function $[\alpha(x(\beta))]'$, which is defined almost everywhere with respect to β, takes on the same value in the intervals where the function $x(\beta)$ is constant; in fact, in these intervals we have defined the function $\alpha(x(\beta))$ by linear traces. Consequently $[\alpha(x(\beta))]'$ can be considered as a function of $x(\beta)$,

$$[\alpha(x(\beta))]' = g(x(\beta)),$$

and by virtue of (9) and (12)

$$\int_a^b f(x)d\alpha(x) = \int_a^b f(x)g(x)d\beta(x). \tag{13}$$

Choosing $f(x)$ equal to the characteristic function of the interval (a, ξ), we obtain in particular the relation

$$\alpha(\xi) = \int_a^\xi g(x)d\beta(x). \tag{14}$$

Therefore *a function $\alpha(x)$ which is absolutely continuous with respect to a nondecreasing function $\beta(x)$ is the indefinite integral of a function $g(x)$, summable with respect to $\beta(x)$.* Conversely, *every indefinite integral of this type is absolutely continuous with respect to $\beta(x)$*, which can be seen exactly as in the case of the ordinary integral, Section 25.

We observe that if we already know that the function $\alpha(x)$ is of the form (14), we can arrive at relation (13) in a more direct manner, even in the case where $\beta(x)$ is of bounded variation but not monotonic; decomposing $\beta(x)$ into the difference of its indefinite positive and negative variations, and in addition decomposing $g(x)$ into the difference of its positive and negative parts, we reduce the proof of (13) to the case where $\beta(x)$ is nondecreasing and $g(x)$ is non-negative. Then this relation, which is evident for step functions $f(x)$, extends to functions $f(x)$ which are summable with respect to $\beta(x)$ by the reasoning we used in Section 26 in proving the rule for integration by substitution.

We give now a relation for *composite functions*. Let $g(x)$ be a function which is summable with respect to the nondecreasing function $\beta(x)$, $a \leq x \leq b$. Let $e_y(x)$ be the characteristic function of the set of points x for which $g(x) \leq y$, and set

$$\alpha(y) = \int_a^b e_y(x) d\beta(x);$$

this is evidently a nondecreasing function of y, defined on the entire y-axis. Let $f(y)$ be a function summable with respect to $\alpha(y)$, $-\infty < y < \infty$.

We shall show that, under these hypotheses, the composite function $f(g(x))$ is summable with respect to $\beta(x)$ and

$$\int_a^b f(g(x)) d\beta(x) = \int_{-\infty}^{\infty} f(y) d\alpha(y). \tag{15}$$

This relation is evident when $f(y)$ is the characteristic function of an interval $c < y \leq d$, because then $f(g(x)) = e_d(x) - e_c(x)$. Therefore it will also be valid for every step function $f(x)$. Finally, it can be extended to all functions which are summable with respect to $\alpha(y)$ by the reasoning to which we have already referred many times.

59. Functions of Several Variables. Direct Definition

There is little which remains to be said about the extensions of the results which we have just established to the case of several variables. The methods which we used for the analogous problem concerning the Lebesgue integral also apply to the Lebesgue-Stieltjes integral, and the reader himself will be able to formulate the results and try to establish them. Therefore we shall only give some indications of the changes which are not completely obvious, instead of discussing the details.

Let us speak first of linear functionals and their representations by Stieltjes integrals. When this theory was first developed, the case of several variables appeared to be much more difficult than that of one variable; the main reason, it seems, consisted of the fact that the domains of definition of the functions cannot in general be decomposed into a finite number of rectangles. But this difficulty was bound to disappear as soon as mathematicians began to think in terms of set functions instead of rectangle functions. [13]

For definiteness, let us assume that we are dealing with the field of continuous functions which are defined on a bounded and closed set E; without loss of generality, we can also assume E to be interior to the unit square $0 \leq x \leq 1$, $0 \leq y \leq 1$. Consider a linear functional A operating on our field, just as in the case of a single variable. Call A' the functional which is defined, for every continuous function $f(x)$ on the unit square, by the value of Af formed for the restriction of $f(x)$ to the set E. Having passed to the

[13] See in particular RADON [1], or STONE [*] (pp. 198—221), SAKS [*] (Chapter III), [⁂].

unit square, the method of Section 50 applies almost word for word, with the result that there exists a function $\alpha(x, y)$ of two variables which is of bounded variation in a sense which is easily made precise[14] and which has the property that

$$A'f = \int_0^1\int_0^1 f(x, y)d\alpha(x, y),$$

where the integral is the limit of the sums

$$\sum_{i,k=1}^{m,n} f(\xi_i, \eta_k)\,[\alpha(x_i, y_k) - \alpha(x_{i-1}, y_k) - \alpha(x_i, y_{k-1}) + \alpha(x_{i-1}, y_{k-1})]$$

$$(x_0 = y_0 = 0;\ x_m = y_n = 1;\ x_{i-1} \leq \xi_i \leq x_i;\ y_{k-1} \leq \eta_k \leq y_k).$$

The function $\alpha(x, y)$ is equal to the value of the functional A' — or more precisely to that of its extension — evaluated for the characteristic function of the closed rectangle bounded by the abscissas 0 and x, and the ordinates 0 and y. Or equivalently, it is equal to the value of the functional A — or its extension — evaluated for the characteristic function of that part of E included in the above-mentioned rectangle. We can also say that it represents the mass which is supported by this subset and corresponds to the functional A. But it is necessary to observe that the functional A — or rather its extension — assigns masses to many other sets, namely to all those for whose characteristic functions this functional is defined, in particular to all the closed subsets of the set E. In fact, let e be a closed set and let $f_n(x, y)$ be the function equal to the positive part of $1 - n{\cdot}d(x, y)$, where $d(x, y)$ denotes the distance from the point (x, y) to the set e. The functions $f_n(x, y)$ are continuous and tend, decreasingly, to the characteristic function $e(x, y)$ of the set e; therefore Ae is defined by the limit of Af_n.

Perhaps it is of interest to point out that the Stieltjes integral of a continuous function of two variables — and even that of a function of more than two variables — can also be defined without making precise decompositions of the domain of integration. Assume that A is of positive type, that is, that

$$\alpha(x, y) - \alpha(x', y) - \alpha(x, y') + \alpha(x', y') \geq 0 \text{ for } x \geq x',\ y \geq y';$$

the general case reduces to this particular type by analogy with the case of a single variable. Then the upper and lower integrals of the positive function

[14] Instead of the interval function $\varphi(x_1, x_2) = |\varphi(x_2) - \varphi(x_1)|$ which appears in the definition of the total variation of a function of a single variable $\varphi(x)$, here we must proceed from the rectangle function

$$\alpha(R) = |\,\alpha(x_2, y_2) - \alpha(x_2, y_1) - \alpha(x_1, y_2) + \alpha(x_1, y_1)\,|,$$

where R denotes the rectangle with vertices (x_1, y_1), (x_1, y_2), (x_2, y_1), (x_2, y_2).

$f(x, y)$ are defined — among other ways — in the following manner. We form the sum

$$\sum_1^n m_k A e_k(x, y)$$

where the $e_k(x, y)$ are the characteristic functions of the closed non-overlapping subsets e_k which may or may not exhaust the set E, and m_k denotes the minimum of $f(x, y)$ on the set e_k. Then the lower integral of $f(x, y)$ on E is the least upper bound of all the sums formed in this manner. On the other hand, if we require that the e_k, overlapping or not, exhaust the set E, and replace m_k by the maximum M_k, the greatest lower bound of the sums thus formed defines the upper integral. When $f(x, y)$ is continuous, and also for many other functions, the two integrals coincide and yield the integral to be calculated.

60. Definition by Means of the Principle of Transition

But we can also reduce the Stieltjes integral, and more generally the Stieltjes-Lebesgue integral, to the Lebesgue integral of a function of a single variable; this can be done whether we are dealing with functions of a single variable or of several variables. We use the same principle of transition as when, in the Lebesgue theory, we passed from functions of several variables to functions of a single variable, although with a certain precaution. To take a definite case, suppose that we are dealing with a layer of positive mass distributed on a set E. Consider the mass situated to the left of the abscissa x and let x vary; we obtain a nondecreasing function of x, which is necessarily continuous except at most for a denumerable infinity of values of x where it admits jumps. Let the same be true for the mass lying below the ordinate y. To avoid the inconvenience which could be caused by these jumps, consider a net formed by successive division of the (x, y) plane into squares with the mesh becoming infinitely small, arranged in such a way that the lines through the above abscissas and ordinates which correspond to jumps are never among the lines of subdivision. As an example of how to attain this result in the particular case when we wish to form successive square subdivisions of side length 2^{-n}, we determine first a point (x_0, y_0) such that the values $x_0 + m2^{-n}$ $(m = \ldots -1, 0, 1, \ldots; n = 1, 2, \ldots)$ are not among the abscissas of the jumps — let us call them $x_1, x_2, \ldots$ — and the values $y_0 + m2^{-n}$ do not coincide with any ordinate y_k of the jumps. Now the differences $x_k - m2^{-n}$ $(k = 1, 2, \ldots; m = \ldots, -1, 0, 1, \ldots; n = 1, 2, \ldots)$ form a denumerable set and therefore do not cover the x-axis; any value which remains can be chosen for x_0. The same holds for y_0. Having taken these precautions, the lines of the subdivision into squares will not support any mass and we shall not have to be concerned with whether or not we are using meshes completed by their contour. We can then determine a correspondence between the points of the set E and those of an interval which transforms the mass into a linear measure. This will be done as in Section 39; the difference that we operate with the dyadic system instead of the triadic system is of no consequence; before we had a special reason for using the latter, namely to be able to introduce conjugate nets. Another difference, this time an essential one, consists in the fact that it is not the area of a mesh, but rather the mass contained in it, which determines the length of the interval which is set in correspondence with the mesh — or, if one prefers, to the part of the set E which is contained in the mesh. We easily see that after the exclusion of two exceptional sets, of which one supports zero mass and the other has linear measure zero, this process yields a correspondence between the points of the set E and those of an interval whose length equals the mass supported by E, and that the subsets of E which are measurable with respect to $\alpha(x, y)$ will have as images sets whose linear measure equals the mass supported by the (corresponding) subsets of E. However, we should note that it is possible to have singular points (x_0, y_0) which support masses different from 0; that is, such that for the function $f(x, y)$ which equals 1 at (x_0, y_0) and takes on the value zero elsewhere, the value of the extension of the functional A does not become zero. Although in this case the segments which correspond to meshes which contract to the point (x_0, y_0) still form a decreasing and nested sequence, they do not contract to a single point, but rather to an entire segment whose length equals the mass supported by (x_0, y_0) and it is this segment which must be made to correspond to the point (x_0, y_0), just as in the case of a monotonic function $\alpha(x)$ which admits jumps.

Finally, having represented the set E on a linear interval in accordance with our principle of transition, we can immediately set up a correspondence between functions $f(x, y)$ and functions on the interval by setting $f(t) = f(x, y)$ when t corresponds to the point (x, y). We see immediately that since this correspondence is defined in terms of a representation which preserves the respective values of masses and measures, it in its turn preserves integrability and the values of integrals, thus permitting us to immediately extend the properties of the Lebesgue integral to Stieltjes-Lebesgue integrals of functions of several variables.

THE DANIELL INTEGRAL

61. Positive Linear Functionals

Our principle of transition has forced us to upset the structure of sets and to give up continuity, but this was done with very little loss. Is it possible to abandon completely the sets with geometric character and integrate functions defined in abstract sets?

This is what was done in 1915 by FRÉCHET [1], the founder of the topology of abstract spaces, in extending the method of Lebesgue to this general problem, and a little later by DANIELL [1], using the idea of the extension of linear functionals. Since then, the first mentioned order of ideas has been repeated and extended by several authors, mostly in papers and books on the theory of probability and related subjects.[15] We shall discuss only the second, which is closer to the method we made use of for the Lebesgue integral.

Let E be an arbitrary set or, as we say, an abstract set. Denoting the elements of E by x, let us keep fixed a class C_0 of real-valued functions $\varphi(x)$, which contains with $\varphi_1(x)$ and $\varphi_2(x)$ the combinations $c_1\varphi_1(x) + c_2\varphi_2(x)$, and also contains $|\varphi(x)|$ with $\varphi(x)$. Then the class C_0 will contain, with $\varphi_1(x)$ and $\varphi_2(x)$, their envelopes

$$\inf(\varphi_1, \varphi_2) = \tfrac{1}{2}(\varphi_1 + \varphi_2) - \tfrac{1}{2}|\varphi_1 - \varphi_2|; \ \sup(\varphi_1, \varphi_2) = \tfrac{1}{2}(\varphi_1 + \varphi_2) + \tfrac{1}{2}|\varphi_1 - \varphi_2|.$$

In particular, it will contain the positive and negative parts of its elements,

$$\varphi^+ = \sup(\varphi, 0) = \tfrac{1}{2}(\varphi + |\varphi|), \ \varphi^- = \sup(-\varphi, 0) = \tfrac{1}{2}(|\varphi| - \varphi) = (-\varphi)^+.$$

In order not to complicate matters, we assume further that C_0 contains the unit, the function $\varphi(x) \equiv 1$; when this is not the case, the role of unity can be played by other functions which are strictly positive on all of E or at least on the subsets in which we are interested.

With these assumptions made, we consider a functional $A\varphi$ which plays the role of integral of φ. It is assumed *additive and homogeneous*:

$$A(c_1\varphi_1 + c_2\varphi_2) = c_1 A\varphi_1 + c_2 A\varphi_2.$$

Until stated otherwise, we also assume it to be *positive*,

$$A\varphi \geq 0 \ \text{ for } \ \varphi(x) \geq 0;$$

we then have, more generally:

$$A\varphi \geq A\psi \ \text{ for } \ \varphi(x) \geq \psi(x).$$

We shall say, briefly, that A is a *positive linear functional.*

[15] See, among others, NIKODYM [1]; HAHN [*]; KOLMOGOROFF [*], [1]; SAKS [*], [⁑]; BANACH (Note II in SAKS [⁑]); CARATHÉODORY [1]; WECKEN [3].

In order to be able to reproduce our theory of the Lebesgue integral, giving the role of step functions to elements of the class C_0, we shall need certain facts which correspond to the two fundamental lemmas **A** and **B**, the first asserting that $A\varphi_n \to 0$ when $\varphi_n(x)$ decreases to 0 almost everywhere, the second that for increasing sequences φ_n the hypothesis that $A\varphi_n$ remain bounded implies that $\varphi_n(x)$ converges almost everywhere to a finite limit.

But we have not yet said what we mean by "almost everywhere" or by a set of measure zero. To do so, we have only to replace lemma **B** by a *definition* introducing sets of measure zero as those on which an increasing sequence $\{\varphi_n(x)\}$ can diverge without having the sequence $\{A\varphi_n\}$ cease being bounded. We leave it to the reader to give other equivalent forms to this hypothesis which are closer to the usual definition of sets of measure zero.

As for lemma **A**, we have to replace it by a *hypothesis*, at least partially; namely, we shall assume

HYPOTHESIS **A**. $A\varphi_n \to 0$ *when* $\varphi_n(x)$ *decreases to zero everywhere.*

We shall show that the term "everywhere" can be replaced by "almost everywhere." In fact, let us assume only that $\varphi_n(x) \to 0$ except on a set e_0 of measure zero and let $\{\psi_n(x)\}$ be an increasing sequence chosen so that it diverges on e_0 and that $A\psi_n$ remains bounded; replacing ψ_n by $\psi_n - \psi_1$ and if necessary multiplying by a constant, we can suppose that

$$\psi_n(x) \geq 0, \quad A\psi_n < \varepsilon \qquad (n = 1, 2, \ldots).$$

Then the functions $(\varphi_n - \psi_n)^+$ tend everywhere to 0 and

$$A(\varphi_n - \psi_n)^+ \to 0,$$

hence

$$A\varphi_n = A(\varphi_n - \psi_n) + A\psi_n \leq A(\varphi_n - \psi_n) + \varepsilon \leq A(\varphi_n - \psi_n)^+ + \varepsilon \to \varepsilon;$$

but since ε is arbitrarily small, it follows that $A\varphi_n \to 0$, which was to be proved.

Now that we have disposed of the two fundamental lemmas, the theory is developed parallel to that of the Lebesgue integral, and yields the analogues of the theorems in Sections 18 to 22 with the exception of the theorem on composite functions (Sec. 20), the proof of which was based on the fact that the composite functions which one obtains by replacing the variables of a continuous function (of one or several variables) by step functions φ of class C_0 belong to the same class C_0. In general, this fact will not be valid for the present class C_0, nevertheless the theorem itself remains in force; a proof can be based on the fact that every continuous function $g(u_1, u_2, \ldots, u_r)$ is the limit of a sequence of functions $\varphi_n(u_1, u_2, \ldots, u_r)$ each of which is obtained from linear functions $c_0 + c_1 u_1 + \ldots c_r u_r$ by the formation of upper and lower envelopes a finite number of times. When we make use of this theorem only to pass from f to f^2, we can also reason as follows:

$$f^2 = \sup (2\lambda|f| - \lambda^2),$$

where λ runs through the positive semi-axis; in fact,

$$f^2 - 2\lambda\,|f| + \lambda^2 = (|f| - \lambda)^2 \geq 0$$

and for each particular x the equality sign is valid when $\lambda = |f(x)|$. Therefore, it clearly suffices to let λ run through a sequence $\{\lambda_n\}$ which is everywhere dense, for example that of the rational numbers. Hence f^2 is the limit of the increasing sequence of envelopes

$$g_n = \sup_{m=1,2,\ldots,n} (2\lambda_m|f| - \lambda_m^2);$$

it follows immediately from the general theory that the g_n are summable, and that their limit f^2 is also summable whenever there exists a majorizing summable function $h(x) \geq f^2(x)$ — in particular whenever f is bounded — and that, in any case, f^2 is measurable.

This result suffices to extend to our general case the ideas and facts concerning the functional space L^2.[16] In particular, the theorem of Section 30, which asserts that every functional can be represented in the form (f, g) extends to the general case; we shall make use of this result later.

62. Functionals of Variable Sign

It is time to lift the restriction that the functional A be *positive*. Instead, we make the hypothesis that A — always assumed *additive* and *homogeneous* — admits a *positive majorant*, that is to say, a positive linear functional B such that $Af \leq Bf$ for every function $f(x) \geq 0$. These are the functionals A which we shall call *linear functionals* in the rest of this chapter.

A functional A of this type always has a smallest positive majorant, that is, a majorant P such that for every other majorant B of A one has $Pf \leq Bf$ whenever $f(x) \geq 0$. We construct P by forming, for every $f(x) \geq 0$, the upper limit of the values Af_1, where Af_1 varies subject to the condition $0 \leq f_1(x) \leq f(x)$. Denoting this bound by Pf, we shall have $Pf \geq A0 = 0$. If we form Pg and $P(f + g)$, where $g(x) \geq 0$, the relations

$$0 \leq f_1 \leq f, \quad 0 \leq g_1 \leq g, \quad Af_1 \geq Pf - \varepsilon, \quad Ag_1 \geq Pg - \varepsilon$$

yield

$$P(f + g) \geq A(f_1 + g_1) = Af_1 + Ag_1 \geq Pf + Pg - 2\varepsilon,$$

[16] We could make the corresponding extension for the spaces L^p in an analogous manner, but we shall not do so.

and since $\varepsilon > 0$ is arbitrary,

$$P(f+g) \geq Pf + Pg.$$

On the other hand, let $0 \leq h(x) \leq f(x) + g(x)$, where $h(x)$ is chosen so that $Ah \geq P(f+g) - \varepsilon$. Set

$$f_1 = \inf(f, h), \quad g_1 = h - f_1;$$

then

$$g_1 = h - \inf(f, h) = h - f - h + \sup(f, h) \leq h - f - h + f + g = g,$$

so

$$P(f+g) - \varepsilon \leq Ah = Af_1 + Ag_1 \leq Pf + Pg$$

and consequently, $\varepsilon > 0$ being arbitrary,

$$P(f+g) \leq Pf + Pg.$$

We have thus shown that for non-negative functions $P(f+g) = P(f) + P(g)$; for the same functions and for $c \geq 0$ we obviously have $P(cf) = cPf$. If for functions of variable sign we define P in the usual manner, by applying it to the positive and negative parts of the functions, these relations obviously remain valid, and consequently P is a positive linear functional. Finally, the relations $Af \leq Pf \leq Bf$ for $f(x) \geq 0$ are an immediate result of our construction.

Let us agree to call P the *positive part* of A. Since to each positive majorant B of A there corresponds a positive majorant of $-A$, namely $B - A$, and reciprocally, $N = P - A$ is the smallest positive majorant of $-A$ and can be called the *negative part* of A. Finally, we consider $2P - A$, the sum of these two parts, which also can be defined independently as the smallest common majorant of A and $-A$; we call it the *modulus* of A and denote it by $|A|$.

It is easy to see that in the case of ordinary Stieltjes integration with respect to a function of bounded variation $\alpha(x)$, Pf and Nf are the integrals of $f(x)$ with respect to the positive and negative indefinite variations of $\alpha(x)$.

Thus we have come back to positive functionals; but it still remains for us to see what becomes of our two fundamental lemmas, or more precisely, the definition of sets of measure zero and the hypothesis **A** which replaces lemma **A**.

The simplest way to define sets of measure zero is to replace A by $|A|$ in their definition. As for our hypothesis **A**, we retain it by showing that when it is verified for A it is also verified for P, and therefore also for $N = P - A$ and for $|A|$. But first it is necessary for us to admit that of what precedes, all which concerns the majorant P is only for the initial field C_0, despite the fact that we have neglected to denote the functions by Greek letters. In fact, we would be unable to extend the functional A before deciding on the generalizations of the two lemmas or of what replaces them. With this point made precise, the extension is made in an obvious way by extending the functionals P and N.

Continuing to denote the elements of the field C_0 by Latin letters, let $\{f_n(x)\}$ be a decreasing sequence which converges everywhere to 0. From hypothesis **A** we deduce that $Af_n \to 0$. We shall show, following DANIELL, that we then also have $Pf_n \to 0$.

Choose g_n such that $0 \leq g_n \leq f_n$ and

$$Ag_n \geq Pf_n - 2^{-n}\varepsilon, \tag{16}$$

and set

$$h_1 = g_1, \; h_n = \inf(g_1, g_2, \ldots, g_n) = \inf(h_{n-1}, g_n).$$

Then the h_n form a decreasing sequence which converges to 0, consequently $Ah_n \to 0$; we shall show that

$$Pf_n \leq Ah_n + (2^{-1} + 2^{-2} + \ldots + 2^{-n})\,\varepsilon < Ah_n + \varepsilon. \tag{17}$$

The proof is by induction; we assume that inequality (17) — shown by (16) to be valid for $n = 1$ — is true for a certain n and we shall pass from that to the case $n + 1$. Since

$$0 \leq h_n \leq f_n, \; 0 \leq g_{n+1} \leq f_{n+1} \leq f_n,$$

we have

$$0 \leq \sup(h_n, g_{n+1}) \leq f_n,$$

and therefore

$$A(\sup(h_n, g_{n+1})) \leq Pf_n.$$

It follows from this and the fact that

$$h_{n+1} = \inf(h_n, g_{n+1}) = h_n + g_{n+1} - \sup(h_n, g_{n+1}),$$

using (16) and (17), that

$$Ah_{n+1} = Ah_n + Ag_{n+1} - A(\sup(h_n, g_{n+1})) \geq Ah_n + Pf_{n+1} - 2^{-n-1}\varepsilon - Pf_n \geq$$
$$Pf_{n+1} - 2^{-n-1}\varepsilon - (2^{-1} + 2^{-2} + \ldots + 2^{-n})\varepsilon \geq Pf_{n+1} - \varepsilon,$$

that is to say that (17) is valid with n replaced by $n + 1$. Finally, since $Ah_n \to 0$, it follows from (17) that $\limsup Pf_n \leq \varepsilon$; since $\varepsilon > 0$ is arbitrary and $Pf_n \geq 0$, it follows that $Pf_n \to 0$, which was to be proved.

With this result established, the study of the functional A and of its extension proceeds ,as we have already said, in a manner analogous to the development of the Lebesgue integral or the Lebesgue-Stieltjes integral and gives rise to new problems only when the notion of derivative intervenes.

63. The Derivative of One Linear Functional With Respect to Another

For functions defined on an abstract set, without a system of coordinates, the idea of a derivative seems to escape our grasp. But let us recall the problems concerning integration where the derivative intervenes. The simplest was the inversion of integration: to find a function whose indefinite integral is given. Now even in our general case there is an indefinite integral; it no longer is a point function, but rather an additive set function $F(e)$ which is defined for measurable subsets e of E, that is, for sets e whose characteristic functions $e(x)$ are summable; it is equal to the integral of $f(x)$ on e, or in other words, to the integral of the product $e(x)\,f(x)$on the set E. The inverse problem consists in determining the conditions under which a set function $F(e)$ is an integral of this sort and to compute the function $f(x)$ of which it is the integral. As to the conditions, they are completely analogous to those given in the case of the Lebesgue integral; but to calculate $f(x)$ it will be necessary to use a different method due to the lack of a derivative. One of these methods consists in graphing, so to speak, the level lines or rather the level sets of the function $f(x)$, that is, to determine the sets on which the function is (almost everywhere, of course) greater than a given constant.[17] Another method consists in reducing the problem to a closely related problem which is easily solved; leaving the task of relating these two problems to the reader, we shall discuss only the second.[18]

This problem is connected with formula (13); essentially, it is a problem of changing the variable of integration. In the language of functionals, it is formulated as follows. *Given two positive linear functionals A and B, defined at first for the same initial field C_0 and extended by the method we have just discussed, can we determine a function $g(x)$ such that*

$$Af = B(gf) \tag{18}$$

for all functions $f(x)$ which are measurable with respect to A and B and for which one or the other member has a meaning?

First we consider the particular case where B is a *majorant* of A, that is, where $Af \leq Bf$ for every $f(x) \geq 0$. Note that it suffices to make this hypothesis for the initial field C_0, since it is clearly preserved under extension and moreover functions summable or measurable with respect to B will also be so with respect to A. We now denote by L_B^2 the space of functions f which together with their squares are summable with respect to B, and consider A to be a functional defined in L_B^2. Since, according to an obvious extension

[17] See RADON [1] (in particular pp. 1342–1351); DANIELL [3]; NIKODYM [1] (in particular pp. 167–179).

[18] Due to J. v. NEUMANN, see [8] (in particular pp. 127–130).

of the Schwarz inequality,

$$|Af|^2 \leq (A1)\,(Af^2) \leq (A1)\,(Bf^2),$$

A is bounded in the space L_B^2, and according to the equally obvious extension of the theorem proved in Section 30 for linear functionals in the space L_B^2, there exists a function g belonging to L_B^2 which satisfies relation (18) for every element f of L_B^2. Furthermore, $g(x) \geq 0$ almost everywhere; in fact, if this were not true we could set $f(x)$ equal to the negative part g^- of g, and formula (18) would yield

$$Ag^- = B(gg^-) = B(-(g^-)^2) = -B(g^-)^2 < 0,$$

contrary to the hypothesis that A is a positive functional. If we replace A by $B - A$, and therefore g by $1 - g$, it follows similarly that $1 - g \geq 0$, $g \leq 1$ almost everywhere. This function $g(x)$, which is uniquely determined up to a set of measure zero with respect to B, solves our problem, but at first only for f belonging to the space L_B^2. To include all functions $f(x)$ measurable with respect to B (and hence also measurable with respect to A), we note that it suffices to treat the case of non-negative functions, since the general case can be reduced to this by means of a decomposition. Therefore, let us consider a measurable function $f(x) \geq 0$ and set $f_n(x) = \inf(f(x), n)$; since $f_n(x)$ is bounded it belongs to L_B^2, hence

$$Af_n = B(gf_n).$$

But the f_n increase everywhere to f and the gf_n to gf; therefore it is permissible to take the limit, which yields (18).

Now let A and B be two positive linear functionals which have the same initial field C_0, but are otherwise arbitrary; we can apply our result to A and $C = A + B$ instead of to A and B. It follows that for all functions $f(x)$ which are measurable with respect to A and B, and hence also with respect to C, the equation

$$Af = C(hf) \tag{19}$$

and with it the equation

$$Bf = C((1-h)f) \tag{20}$$

can be satisfied in such a way that if one of the two members has meaning, so has the other. Moreover, $0 \leq h(x) \leq 1$. Let us replace f by $\dfrac{h}{1-h}f$ in equation (20); we obtain

$$B\left(\frac{h}{1-h}f\right) = C(hf);$$

together with equation (19), this yields

$$Af = B\left(\frac{h}{1-h}f\right).$$

Therefore our problem is solved by setting

$$g(x) = \frac{h(x)}{1 - h(x)}$$

However, for the time being our solution is only formal and far from being exact; this is due to the fact that the factor $g(x)$ becomes infinite wherever $h(x) = 1$ and that the set e_1 of these points could be involved in an essential way in the argument; it suffices to consider the extreme case where $B = 0$, hence $C = A$, and consequently $h(x) = 1$ almost everywhere with respect to C, or if we wish $h(x) \equiv 1$. Now let us replace $f(x)$ by $e_1(x)$, the characteristic function of the set e_1, in equations (19) and (20). Since $h(x)e_1(x) \equiv e_1(x)$, it follows that

$$Ae_1 = Ce_1, \quad Be_1 = 0,$$

that is to say, e_1 is of measure zero with respect to B. On the other hand, to be able to let $\frac{h}{1-h} f$ take the role of f, it is necessary that this function be measurable and in particular finite almost everywhere with resepct to C, and consequently, with respect to A, which implies that the set e_1 must be of measure zero with respect to A. It is only under this hypothesis that our reasoning becomes exact. The most direct method of ensuring the hypothesis is suggested to us by the particular case of our problem; the method has already been alluded to in Section 58, and consists in assuming that the functional A is absolutely continuous with respect to the functional B. We mean by this that the sets which are of measure zero with respect to B are also of measure zero with respect to A, or what obviously amounts to the same thing, that the relations $f(x) \geq 0$, $Bf = 0$ imply that $Af = 0$. In this hypothesis every function measurable with respect to B is also measurable with respect to A and to C. Hence we have the following:

THEOREM. *Let A and B be two positive linear functionals of the general type considered in Section* 61, *operating on the same initial field C_0 and satisfying hypothesis* **A**. *Furthermore, let A be absolutely continuous with respect to B, that is to say that every function $f(x) \geq 0$ which is summable with respect to B and satisfies $Bf = 0$ is also summable with respect to A and satisfies $Af = 0$. Then there exists a function $g(x) \geq 0$, measurable with respect to B, such that*

$$Af = B(gf)$$

for all functions $f(x)$ which are measurable with respect to A and B and for which one or the other of the two members has a meaning.

In particular, when Af is majorized by B, that is to say when $Af \leq Bf$ for every function $f(x) \geq 0$ of the initial field C_0, we have $0 \leq g(x) \leq 1$.

We note the following corollary:

Let B be a positive linear functional which satisfies hypothesis **A**. *Let A be a linear functional on the space L_B of functions which are summable with respect to B, that is, let Af be defined for every element of L_B, and moreover be additive, homogeneous and bounded in the following sense:*

$$|Af| \leq M \cdot B|f|, \tag{21}$$

with a numerical constant M. Then there exists a function $g(x)$ which is measurable with respect to B, bounded in absolute value by M, and such that

$$Af = B(gf) \tag{22}$$

for every element f of the space L_B.

To see this, we have only to apply the theorem to the positive and negative parts of the functional A; condition (21) assures us that these parts divided by M are majorized by B. For the particular case of the space L^1 of functions which are summable in the ordinary sense of Lebesgue, we have already established representation (22) in Section 36.

PART TWO

INTEGRAL EQUATIONS
LINEAR TRANSFORMATIONS

CHAPTER IV

INTEGRAL EQUATIONS

THE METHOD OF SUCCESSIVE APPROXIMATIONS

64. The Concept of an Integral Equation

The general theory of integral equations is the work of our century. However, the geometers of the nineteenth century sometimes encountered particular equations of this type. Probably the first example was the problem of the inversion of the integral

$$g(x) = \frac{1}{\sqrt{2\pi}} \int_{-\infty}^{\infty} e^{ixy} f(y)dy,$$

a very important problem for certain branches of mathematical physics, which was solved in 1811 by FOURIER in the form

$$f(y) = \frac{1}{\sqrt{2\pi}} \int_{-\infty}^{\infty} e^{-ixy} g(x)dx.$$

In 1823 ABEL was led by his research on a generalization of the "tautochrone" problem to the equation

$$g(x) = \int_a^x \frac{f(y)}{(x-y)^\alpha} dy \qquad (0 < \alpha < 1,\ g(a) = 0),$$

for which he found the solution

$$f(y) = \frac{\sin \pi\alpha}{\pi} \int_a^y \frac{g'(x)}{(y-x)^{1-\alpha}} dx.[1]$$

Abel's equation was the object of numerous later investigations. It belongs to the class of integral equations which are today called Volterra equations and which are of the type

$$\int_a^x H(x, y)f(y)dy = g(x), \tag{1}$$

[1] ABEL [1], [2].

where $H(x, y)$ and $g(x)$ are given functions and $f(x)$ is a function to be determined. When $H(x, y)$ has a continuous derivative $H_x(x, y)$, equation (1) can be transformed, by taking the derivative of each side, into the equation

$$f(x)H(x, x) + \int_a^x H_x(x, y)f(y)dy = g'(x),$$

or, when in addition $H(x, x) \neq 0$, into the equation

$$f(x) + \int_a^x K(x, y)f(y)dy = g_1(x), \tag{2}$$

where

$$K(x, y) = \frac{H_x(x, y)}{H(x, x)} \text{ and } g_1(x) = \frac{g'(x)}{H(x, x)}.$$

Linear differential equations with initial or limit conditions transform, in general, into integral equations. For example, such a transformation was already used by J. Liouville [**1**] in 1837 in his investigations of linear differential equations of the second order. Consider, for example, the equation

$$f''(x) + [\varrho^2 - \sigma(x)]f(x) = 0$$

with the initial conditions $f(a) = 1$, $f'(a) = 0$. Since the related equation

$$f''(x) + \varrho^2 f(x) = g(x)$$

has, as we easily see, the single solution

$$f(x) = \cos \varrho(x - a) + \frac{1}{\varrho} \int_a^x g(y) \sin \varrho(x - y)dy$$

which satisfies the given initial conditions, the solution of the original equation must satisfy the integral equation

$$f(x) - \frac{1}{\varrho} \int_a^x \sigma(y) \sin \varrho(x - y)f(y)dy = \cos \varrho(x - a),$$

which is of type (2).

Equations of types (1) and (2), which were studied for the first time by V. Volterra [**1**] and J. le Roux [**1**], are particular cases of integral equations of the types

$$\int_a^b K(x, y)f(y)dy = g(x), \tag{1'}$$

$$f(x) - \int_a^b K(x, y)f(y)dy = g(x), \tag{2'}$$

where the limits of integration are fixed quantities. Types (1) and (2) clearly correspond to functions $K(x, y)$ which are zero for $x < y$. The function $K(x, y)$ is called the *kernel* of the equation.

In what follows we shall concern ourselves with integral equations of the *second type*, that is, of type (2′). These are also called FREDHOLM integral equations, after the illustrious Swedish geometer who first (in 1900) gave a complete theory for them.[2]

The results which we shall obtain can be extended to much more general functional equations. Let us content ourselves for the moment with pointing out one of these generalizations, the most immediate: it is that the fundamental interval (a, b) can be replaced by an arbitrary domain of integration D, lying in a space of any number of dimensions, a domain which may or may not be bounded, and can be for example, a curve or a surface. We must understand by x and y two points of the domain D, and the integral is to be taken over D. It can also be replaced by a Stieltjes integral with respect to a distribution of positive mass, or by an arbitrary functional of the type considered in Sec. 61.

65. Bounded Kernels

In order to orient ourselves, we consider first an equation

$$f(x) - \int_a^b K(x, y)f(y)dy = g(x) \tag{3}$$

whose kernel $K(x, y)$ is bounded and continuous except perhaps for $y = x$. (We allow discontinuities on the straight line $y = x$ in order to include the case of Volterra kernels, which are continuous on the triangle $a \leq y \leq x \leq b$ and zero elsewhere.) The given function $g(x)$ is assumed summable.

Whatever the summable function $h(x)$, the integral

$$\int_a^b K(x, y)h(y)dy$$

has a meaning and is a continuous function of x which we denote by $Kh(x)$, or by Kh. Let us write our equation in the form

$$f = g + Kf$$

and try to solve it by the method of successive approximations, taking the function $f_0(x) = 0$ for the first approximation. We obtain the successively better approximations:

$$f_1 = g,\ f_2 = g + Kf_1 = g + Kg,\ f_3 = g + Kf_2 = g + Kg + K^2g, \ldots$$

and in general,

$$f_n = g + Kf_{n-1} = g + Kg + \ldots + K^{n-1}g,$$

[2] FREDHOLM [**1**], [**2**].

where the "iterates" $K^n g$ are defined by the recurrence relation:

$$K^1 g = Kg, \ K^n g = K(K^{n-1} g) \text{ for } n \geq 2.$$

We are thus led to the series of functions called the *Neumann series*,[3]

$$g + Kg + K^2 g + \ldots + K^n g + \ldots. \tag{4}$$

If this series converges uniformly, its sum f will be a solution of (3). In fact, since term-by-term integration is then permissible, we have

$$Kf = K(g + Kg + K^2 g + \ldots) = Kg + K(Kg) + K(K^2 g) + \ldots =$$
$$= Kg + K^2 g + K^3 g + \ldots = f - g.$$

Now in order to assure the uniform convergence of series (4), it suffices to assume that

$$M = \max |K(x, y)| < \frac{1}{b - a}, \tag{5}$$

since starting with its second term series (4) is then majorized by the convergent series

$$AM + AM^2(b - a) + AM^3(b - a)^2 + \ldots + AM^n(b - a)^{n-1} + \ldots,$$

where

$$A = \int_a^b |g(x)| dx.$$

Condition (5) is far from being necessary for the uniform convergence of the Neumann series. Here are two examples:

1) $K(x, y) = \alpha(x)\beta(y)$, where $\alpha(x)$ and $\beta(x)$ are two arbitrary continuous functions, subject only to the condition:

$$\int_a^b \alpha(x)\beta(x) dx = 0.$$

Then we have

$$Kg(x) = \int_a^b \alpha(x)\beta(y)g(y) dy = c\alpha(x),^4$$

$$K^2 g(x) = \int_a^b \alpha(x)\beta(y)c\alpha(y) dy = c\alpha(x) \int_a^b \beta(y)\alpha(y) dy = 0,$$

$$K^3 g = K(K^2 g) = 0, \qquad K^4 g = K(K^3 g) = 0, \ldots$$

[3] Carl Neumann was the first to give a rigorous proof of the convergence of the successive approximations, in his solution of the Dirichlet problem in potential theory; cf. C. Neumann [*]. An analogous development is found in the paper of Liouville which was referred to above.

[4] $c = \int_a^b \beta(y) g(y) dy.$

Hence, series (4) reduces to its first two terms, and equation (3) has the solution

$$f(x) = g(x) + Kg(x).$$

2) *Volterra kernels*, characterized by the condition:

$$K(x, y) = 0 \text{ for } a \leq x < y \leq b.$$

Keeping the notation

$$M = \max |K(x, y)|, \quad A = \int_a^b |g(x)| dx,$$

we have

$$|Kg(x)| = |\int_a^x K(x, y) g(y) dy| \leq AM,$$

$$|K^2 g(x)| = |\int_a^x K(x, y) [Kg(y)] dy| \leq \int_a^x M \cdot AM dy = AM^2(x - a),$$

$$|K^3 g(x)| = |\int_a^x K(x, y) [K^2 g(y)] dy| \leq \int_a^x M \cdot AM^2 (y - a) dy = AM^3 \frac{(x - a)^2}{2},$$

and in general,

$$|K^n g(x)| \leq AM^n \frac{(x - a)^{n-1}}{(n - 1)!} \qquad (n = 1, 2, \ldots).$$

Therefore, starting with its second term series (4) is majorized by the series

$$AM + AM^2(b - a) + AM^3 \frac{(b - a)^2}{2} + \ldots + AM^n \frac{(b - a)^{n-1}}{(n - 1)!} + \ldots,$$

which is convergent no matter how large M may be.

Hence for an arbitrary Volterra kernel the Neumann series is always uniformly convergent and furnishes the solution of the equation.

We add that condition (5) has the inconvenience that it has no meaning when the finite interval (a, b) is replaced by an infinite interval.

66. Square-Summable Kernels. Linear Transformations of the Space L^2

Condition (5) can be replaced, as was observed by E. Schmidt [**2**], by the less restrictive condition

$$\int_a^b \int_a^b |K(x, y)|^2 dx dy < 1. \tag{6}$$

We can even dispense with any continuity hypothesis; it suffices to assume that the kernel $K(x, y)$ is square summable, that is, that it belongs to the (complex) space L^2 corresponding to the domain of integration $a \leq x \leq b$, $a \leq y \leq b$. We shall denote this functional space by $\mathbf{L}^2$, preserving the notation L^2 for the space of functions of the single variable x defined on the interval $a \leq x \leq b$. Scalar products and norms will be denoted in L^2 by

$$(\, , \,) \quad \text{and} \quad \| \quad \|,$$

and in $\mathbf{L}^2$ by

$$\mathbf{(} \, , \, \mathbf{)} \quad \text{and} \quad \mathbf{|} \quad \mathbf{|}.$$

Moreover, if we thus generalize our hypotheses on the kernel, it will be necessary for us to make a certain restriction on the given function $g(x)$, namely that it be not only summable, but also square summable, that is, that it belong to the space L^2.

Let $K(x, y)$ be a function belonging to $\mathbf{L}^2$. It follows from Fubini's theorem on successive integrations (see Sec. 40) that the integral

$$\int_a^b |K(x, y)|^2 dy$$

exists for almost all x and that

$$\int_a^b [\int_a^b |K(x, y)|^2 dy] dx = \int_a^b \int_a^b |K(x, y)|^2 dx dy = \mathbf{|} K \mathbf{|}^2.$$

Therefore the function

$$k(x) = [\int_a^b |K(x, y)|^2 dy]^{\frac{1}{2}}$$

is an element of the space L^2 and $\|k\| = \mathbf{|} K \mathbf{|}$.

Let $h(x)$ be an arbitrary function belonging to L^2. The integral

$$\int_a^b K(x, y) h(y) dy$$

has a meaning for all the points x where $k(x)$ has a definite and finite value, and defines a function which also belongs to L^2 and which we shall denote by $Kh(x)$. In fact, by Schwarz's inequality,

$$\text{(7)} \quad \begin{aligned} &|\int_a^b K(x, y) h(y) dy|^2 \leq \int_a^b |K(x, y)|^2 dy \cdot \int_a^b |h(y)|^2 dy = k^2(x) \, \|h\|^2, \\ &\|Kh\|^2 = \int_a^b | \int_a^b K(x, y) h(y) dy|^2 dx \leq \int_a^b k^2(x) dx \cdot \|h\|^2 = \mathbf{|} K \mathbf{|}^2 \, \|h\|^2. \end{aligned}$$

Therefore, since the kernel $K(x, y)$ belongs to the space $\mathbf{L}^2$, it generates a *transformation*

$$h(x) \rightarrow Kh(x)$$

of the space L^2 into itself; we obviously have $K(h_1 + h_2) = Kh_1 + Kh_2$, $K(ch) = cKh$, and as we have just seen, $\|Kh\| \leq |K| \cdot \|h\|$.

In general, we shall say that a transformation T of the space L^2 into itself which transforms the element h into the element Th is *linear* if it is

1) *additive*: $T(h_1 + h_2) = Th_1 + Th_2$,
2) *homogeneous*: $T(ch) = cT(h)$,
3) *bounded*: there exists a constant M such that

$$\|Th\| \leq M\|h\|.$$

The smallest of these bounds M is called the *norm* of the linear transformation T and is denoted by $\|T\|$.

Every linear transformation T is also *continuous* in the sense that if the sequence $\{h_n\}$ of elements of L^2 converges in the mean to the element h, the sequence $\{Th_n\}$ converges in the mean to Th. In fact,

$$\|Th_n - Th\| = \|T(h_n - h)\| \leq \|T\| \, \|h_n - h\| \to 0.$$

Conversely, every transformation T which is additive, homogeneous and continuous is also bounded, and hence *linear*. In fact, in the contrary case there would be a sequence $\{h_n\}$ such that $\|Th_n\| > n\|h_n\|$; setting

$$g_n = \frac{1}{n\|h_n\|} h_n,$$

we should have

$$\|g_n\| = \frac{1}{n} \to 0 \quad \text{and} \quad \|Tg_n\| > 1,$$

which contradicts the continuity hypothesis.

Later, in Chapter VIII, we shall need to extend the notion of linear transformation to include *unbounded* transformations, but until then we shall keep the definition we have just given.

For linear transformations, addition and multiplication are defined in an obvious manner:

$$(cT)h = cTh, \quad (T_1 + T_2)h = T_1h + T_2h, \quad (T_1T_2)h = T_1(T_2h);$$

we see immediately that

$$\|cT\| = |c| \, \|T\|, \; \|T_1 + T_2\| \leq \|T_1\| + \|T_2\|, \; \|T_1T_2\| \leq \|T_1\| \, \|T_2\|.$$

In particular, we have for the iterated transformations $T^2 = TT$, $T^3 = T(T^2), \ldots$:

$$\|T^2\| \leq \|T\|^2, \; \|T^3\| \leq \|T\|^3, \ldots, \|T^n\| \leq \|T\|^n, \ldots.$$

The *convergence* of a sequence $\{T_n\}$ of linear transformations to a linear transformation T can be defined in several ways. Namely, we can require that one or another of the following conditions be satisfied:

a) for every f, $T_n f$ converges *weakly* to Tf,
b) for every f, $T_n f$ converges *strongly* (*in the mean*) to Tf,
c) $\|T_n - T\| \to 0$.

We speak of respectively a) *weak convergence,* b) *strong convergence* (or simply *convergence*), c) *convergence in the norm.*

Obviously, strong convergence implies weak convergence.

If the sequence $\{T_n\}$ tends to T in the *norm,* we have for every element f:

$$\|T_n f - Tf\| = \|(T_n - T)f\| \leq \|T_n - T\| \, \|f\| \to 0,$$

hence $T_n f$ tends strongly to Tf. Moreover, this convergence is uniform in f in every bounded set of the space L^2, that is, in every set of elements f whose norms $\|f\|$ are bounded by the same quantity C; that is why *convergence in the norm* is also called *uniform convergence.* Cauchy's criterion is valid for it: in order that the sequence $\{T_n\}$ be uniformly convergent, it is necessary and sufficient that $\|T_n - T_m\| \to 0$ when $m, n \to \infty$.

*

Having introduced these notions, which are fundamental in what follows, let us return to the linear transformation K of the space L^2 which is generated by the kernel $K(x, y)$ belonging to $\mathbf{L}^2$.

According to (7), we have

(8) $$\|K\| \leq |K|$$

where of course $|K|$ denotes the norm of the function $K(x, y)$ as an element of $\mathbf{L}^2$ and $\|K\|$ denotes the norm of the linear transformation K on the space L^2. Therefore hypothesis (6) implies that

(9) $$\|K\| < 1.$$

Given any function $g(x)$ belonging to the space L^2, the Neumann series

(10) $$g + Kg + K^2 g + \ldots$$

converges in the mean, since, by virtue of (9),

$$\|K^m g + K^{m+1} g + \ldots + K^n g\| \leq (\|K\|^m + \|K\|^{m+1} + \ldots + \|K\|^n) \, \|g\| \to 0$$

when $m, n \to \infty$. Let us denote the sum in the mean of the series by f; it is an element of L^2 and we obviously have

(11) $$\|f\| \leq (1 + \|K\| + \|K\|^2 + \ldots) \, \|g\| = (1 - \|K\|)^{-1} \, \|g\|.$$

In view of the continuity of the linear transformation K, it is permissible to calculate Kf by applying K to the series (10) term by term, which yields

$$Kf = Kg + K^2 g + K^3 g + \ldots = f - g;$$

hence f is a solution of the integral equation

$$f - Kf = g.$$

Furthermore, the series (10) also converges in the ordinary sense, at least almost everywhere, since we have, by virtue of the first of the inequalities (7),

$$|K^n g(x)| \leq k(x)\, \|K^{n-1} g\| \leq k(x)\, \|K\|^{n-1}\, \|g\|.$$

The sum of (10) in the ordinary sense is necessarily equal almost everywhere to the sum in the mean, $f(x)$.

Thus we have proved the

THEOREM. *If the kernel $K(x, y)$ satisfies condition* (6), *the integral equation*

$$f - Kf = g$$

has a solution f belonging to L^2 for every function g belonging to L^2; this solution is obtained as the sum in the mean, and even as the sum in the ordinary sense almost everywhere, of the Neumann series (10). *Furthermore, instead of* (6) *it suffices to assume that $K(x, y)$ belongs to $\mathbf{L}^2$ and that the linear transformation K satisfies condition* (9).

Under this condition the solution is even *unique*. In fact, if $f_1 - Kf_1 = g$ and $f_2 - Kf_2 = g$, we have $(f_1 - f_2) - K(f_1 - f_2) = 0$, hence

$$\|f_1 - f_2\| = \|K(f_1 - f_2)\| \leq \|K\|\, \|f_1 - f_2\|;$$

since $\|K\| < 1$, we necessarily have $\|f_1 - f_2\| = 0$; therefore $f_1(x) = f_2(x)$ almost everywhere.

67. Inverse Transformations. Regular and Singular Values

Denote by I the identity transformation of L^2, that is, the transformation which leaves each element of L^2 invariant, and by O the transformation which maps each element of L^2 into 0. Our integral equation can then be written in the form

$$(I - K)f = g.$$

Let us set up a correspondence — always assuming condition (9) — from the solution f to the given function g; we thus obtain a transformation $f = Rg$ of the space L^2 into itself which is obviously additive, homogeneous, and by virtue of (11), also bounded; hence it is a linear transformation. We have $(I - K)Rg = (I - K)f = g$ for every g; therefore

$$(I - K)R = I. \tag{12}$$

In particular, it follows that

$$(I - K)R(I - K)h = (I - K)h,$$

for every element h of L^2; but the equation $(I - K)\varphi = (I - K)h$ has the unique solution $\varphi = h$; therefore $R(I - K)h = h$, and consequently

$$R(I - K) = I. \tag{13}$$

We can summarize the two equations (12), (13) by saying that R is the *inverse* of $I - K$; this is written

$$R = (I - K)^{-1}.$$

In general, we say that the linear transformation T has an *inverse* when there exists a linear transformation S such that

$$TS = I \text{ and } ST = I;$$

S is called the inverse of T and is denoted by T^{-1}.

Later, in Chapter VIII, we shall generalize this notion, at the same time that we generalize the concept of linear transformation.

It is immediate that if T_1 and T_2 have inverses the same is true of their product, and that

$$(T_1T_2)^{-1} = T_2^{-1}T_1^{-1}.$$

The inverse, if it exists, is *uniquely* defined. One can say even more: if T has a "right inverse" S_1 and a "left inverse" S_2, these are necessarily equal. In fact, if

$$TS_1 = I \text{ and } S_2 T = I,$$

we have

$$S_1 = IS_1 = (S_2 T)S_1 = S_2(TS_1) = S_2 I = S_2.$$

The existence of a "right inverse" of T does not assure, in general, the existence of a "left inverse" and therefore the existence of an inverse T^{-1}. But if we know that T transforms *distinct* elements into *distinct* elements, that is, that

$$Tf_1 = Tf_2 \text{ implies } f_1 = f_2,$$

then the equation $TS = I$ implies the equation $ST = I$, for the same reasons that (12) implied (13).

We have seen that every linear transformation of the form $I - K$, where $\|K\| < 1$, has an inverse which can be constructed by means of the Neumann series (10). This can be expressed by the formula

$$(I - K)^{-1} = I + K + K^2 + \ldots + K^n + \ldots;$$

the series even converges in norm, since

$$\|(I - K)^{-1} - (I + K + \ldots + K^{n-1})\| \leq \sum_{m=n}^{\infty} \|K\|^m = \|K\|^n(I - \|K\|)^{-1} \to 0 (n \to \infty).$$

Of course, the inverse of $I - K$ can exist even if $\|K\| \geq 1$, of which we can easily convince ourselves by considering the examples given at the end of Sec. 65.

Let us introduce in our integral equation a complex parameter λ by writing

$$(I - \lambda K)f = g.$$

A value of λ will be said to be *regular* if $(I - \lambda K)^{-1}$ exists; we then set

$$(I - \lambda K)^{-1} = I + \lambda K_\lambda. \tag{14}$$

This equation uniquely determines the linear transformation K_λ for every regular value of λ except for $\lambda = 0$; we define K_0 by setting $K_0 = K$; this is equivalent, as we shall see, to defining K_0 by continuity. K_λ is called the *resolvent transformation*. The non-regular values of λ are called *singular*.

Applying our results to λK instead of to K, we see that *every value of λ such that $|\lambda| < \dfrac{1}{\|K\|}$ is regular, and that for these values we have*

$$(I - \lambda K)^{-1} = I + \lambda K + \lambda^2 K^2 + \lambda^3 K^3 + \ldots \tag{15}$$

in the sense of convergence in the norm.

By its definition the transformation K_λ also satisfies the equations

$$(I + \lambda K_\lambda)\,(I - \lambda K) = (I - \lambda K)\,(I + \lambda K_\lambda) = I$$

which are equivalent for $\lambda \neq 0$ to the equations

$$\lambda K_\lambda K = \lambda K K_\lambda = K_\lambda - K; \tag{16}$$

obviously these latter equations also hold for $\lambda = 0$.

Let λ and μ be two arbitrary regular values. From equations (16) and from the same equations with λ replaced by μ, we see that

$$(\lambda K_\lambda K)\mu K_\mu = (K_\lambda - K)\mu K_\mu, \;\; \lambda K_\lambda(\mu K K_\mu) = \lambda K_\lambda(K_\mu - K).$$

Since the first members are equal, it follows that

$$\mu K_\lambda K_\mu - \mu K K_\mu = \lambda K_\lambda K_u - \lambda K_\lambda K,$$

hence by (16),

$$(\lambda - \mu)K_\lambda K_\mu = \lambda K_\lambda K - \mu K K_\mu = (K_\lambda - K) - (K_\mu - K) = K_\lambda - K_\mu,$$

and

$$K_\lambda K_\mu = \frac{K_\lambda - K_\mu}{\lambda - \mu}. \tag{17}$$

The second member of (17) is symmetric in λ and μ, from which it follows that

$$K_\lambda K_\mu = K_\mu K_\lambda,$$

that is, the transformations K_λ and K_μ are *permutable*.

We shall show the set of regular values is an *open* set. More precisely, *if μ is a regular value then all the values λ such that*

$$|\lambda - \mu| < \frac{1}{\|K_\mu\|},$$

are also regular and we have, always in the sense of convergence in the norm,

$$K_\lambda = K_\mu + (\lambda - \mu)K_\mu^2 + (\lambda - \mu)^2 K_\mu^3 + \ldots .$$

To see this observe that, according to the hypothesis, $\lambda - \mu$ is a regular value with respect to the transformation K_μ, and that the corresponding transformation, which we can denote by $(K_\mu)_{\lambda-\mu}$, is represented by the series in question. Hence the problem reduces to showing that $(K_\mu)_{\lambda-\mu} = K_\lambda$, that is, that

$$\lambda(K_\mu)_{\lambda-\mu}K = \lambda K(K_\mu)_{\lambda-\mu} = (K_\mu)_{\lambda-\mu} - K.$$

Since we obviously have $(K_\mu)_{-\mu} = K$, our assertion follows from (17) by applying it to K_μ instead of to K, and to the values $\lambda - \mu$ and $-\mu$ instead of to λ and μ.

In particular, it follows from the development obtained that the norm of K_λ is a continuous function of λ at every regular point μ. In fact,

$$| \|K_\lambda\| - \|K_\mu\| | \leq \|K_\lambda - K_\mu\| \leq |\lambda - \mu| \, \|K_\mu\|^2 + |\lambda - \mu|^2 \, \|K_\mu\|^3 + \ldots =$$

$$= |\lambda - \mu| \frac{\|K_\mu\|^2}{1 - |\lambda - \mu| \, \|K_\mu\|} \to 0 \text{ for } \lambda \to \mu.$$

Conversely, let $\{\lambda_n\}$ be a sequence of regular values tending to λ^*, and assume that the norms $\|K_{\lambda_n}\|$ remain less than a number M. We shall show that λ^* is then a regular value also. In fact, it follows from (17) that

$$\|K_{\lambda_n} - K_{\lambda_m}\| \leq |\lambda_n - \lambda_m| \, \|K_{\lambda_n}\| \, \|K_{\lambda_m}\| \leq |\lambda_n - \lambda_m| M^2 \to 0$$

for $m, n \to \infty$. Therefore the sequence of transformations K_{λ_n} satisfies Cauchy's convergence condition with respect to convergence in the norm, hence it converges in the norm to a linear transformation K^*. Relations (16), written for $\lambda = \lambda_n$, become in the limiting case:

$$\lambda^* K^* K = \lambda^* K K^* = K^* - K,$$

which implies that λ^* is a regular value and that $K_{\lambda^*} = K^*$.

These facts can be summarized as follows:

THEOREM. *The resolvent transformation K_λ is a regular analytic function of λ at every point μ which is a regular value for K. It cannot be continued analytically beyond this set: when we approach a singular value through regular values, the norm of K_λ becomes infinite.*

This justifies the nomenclature regular and singular values.

We could continue by discussing singular points and the corresponding principal parts of the analytic function K_λ, applying to it the analogue of the calculus of residues, which would lead to the reduction of the case of general kernels to the simple case of kernels of finite rank.[5] We content ourselves here with pointing this out; we shall not return to it until Chapter XI.

68. Iterated Kernels. Resolvent Kernels

We showed in Sec. 66 that every function $K(x, y)$ belonging to the space $\boldsymbol{L}^2$ generates a linear transformation K of the space L^2 whose norm $\|K\|$ does not exceed the norm of $K(x, y)$ as an element of $\boldsymbol{L}^2$, $\|K\| \leq |K|$. But afterwards we operated with the transformations themselves, without making use of the kernels which generated them. This has the advantage that most of the results obtained extend immediately to more general functional equations, in which the linear transformation K is no longer of integral type.

The problem arises of examining how operations with transformations can be interpreted in terms of operations with the kernels themselves.

It is evident that if the transformations F and G are generated by the kernels $F(x, y)$ and $G(x, y)$ belonging to $\boldsymbol{L}^2$, *the transformations cF and $F + G$ are generated by the kernels $cF(x, y)$ and $F(x, y) + G(x, y)$*. Let us show that *the transformation $H = FG$ is generated by the kernel*

$$H(x, y) = \int_a^b F(x, z)G(z, y)dz,$$

which also belongs to $\boldsymbol{L}^2$, and that furthermore we have

$$|H| \leq |F||G|.$$

In fact, $F(x, z)$ and $G(z, y)$ are measurable, square summable functions with respect to z for all x and y with the possible exception of two sets, respectively e_F and e'_G, of (linear) measure zero. Consequently $H(x, y)$ is defined for all the points (x, y) for which neither coordinate is exceptional, that is, for almost all points of the square $a \leq x \leq b$, $a \leq y \leq b$. For x and y not belonging to the sets e_F and e'_G respectively, we have, according to Schwarz's inequality:

$$|H(x, y)|^2 \leq \int_a^b |F(x, z)|^2 dz \cdot \int_a^b |G(z, y)|^2 dz;$$

integrating over the square $a \leq x \leq b$, $a \leq y \leq b$ and taking square roots, we obtain $|H| \leq |F||G|$. For an arbitrary element h of L^2, Hh and $F(Gh)$ are defined by the integrals

$$\int_a^b \Big[\int_a^b F(x, z)G(z, y)dz\Big]h(y)dy, \quad \int_a^b F(x, z)\Big[\int_a^b G(z, y)h(y)dy\Big]dz;$$

[5] F. Riesz [*] (pp. 113–121).

we shall show that these integrals exist and equal one another for almost all x, in particular for the x not belonging to e_F. By virtue of Fubini's theorem, this reduces to showing that for such a point $x = x_0$ the function $F(x_0, z)\,G(z,y)h(y)$ is summable in the square $a \leq z \leq b$, $a \leq y \leq b$. This follows from the fact that the functions $G(x, y)$ and $F(x_0, x)h(y)$ belong to $\boldsymbol{L}^2$.

Finally, it follows immediately from inequality (8), Sec. 66, that the *convergence in the mean of the kernels implies the convergence in the norm of the corresponding transformations.* In fact, if $F_n(x, y) \to F(x, y)$ in the mean, we have

$$\|F_n - F\| \leq |F_n - F| \to 0.$$

It follows from what has been said that if the transformation K is generated by the kernel $K(x, y)$, the transformations $K^2, K^3, \ldots$ will be generated by the kernels $K^{(2)}(x, y)$, $K^{(3)}(x, y)$, ... which are called *iterated kernels*; they are defined by the recursion formula:

$$K^{(n)}(x, y) = \int_a^b K(x, z)K^{(n-1)}(z, y)dz \qquad (n = 2, 3, \ldots),$$

where $K^1(x, y)$ is given by

$$K^{(1)}(x, y) = K(x, y).$$

Furthermore, we shall have

$$|K^{(2)}| \leq |K||K| = |K|^2, |K^{(3)}| \leq |K|\,|K^{(2)}| \leq |K|^3, \ldots, |K^{(n)}| \leq |K|^n, \ldots.$$

Consider the series

$$K(x, y) + \lambda K^{(2)}(x, y) + \lambda^2 K^{(3)}(x, y) + \cdots \tag{18}$$

where λ is a complex parameter. It is majorized in the norm $|\ |$ by the numerical series

$$|K| + |\lambda|\,|K|^2 + |\lambda|^2\,|K|^3 + \cdots.$$

From this it follows that for every value of the parameter λ such that

$$|\lambda| < \frac{1}{|K|}, \tag{19}$$

series (18) converges in the mean; its sum in the mean, $K_\lambda(x, y)$, is thus an element of $\boldsymbol{L}^2$. The transformation K_λ which generates it is identical with the one we defined in the preceding section. In fact, since the convergence in the mean of series (18) implies the convergence in the norm of the corresponding transformations, we have

$$K_\lambda = K + \lambda K^2 + \lambda^2 K^3 + \cdots;$$

our assertion follows by comparing this development with formulas (14) and (15).

Moreover, under condition (19) series (18) converges not only in the mean, but also in the ordinary sense, at least almost everywhere, and necessarily to the same limit $K_\lambda(x, y)$. This results from the inequality, valid for $n > 2$,

$$|K^{(n)}(s, t)| = |\int_a^b\int_a^b K(s, x)K^{(n-2)}(x, y)K(y, t)dxdy| = |(K^{(n-2)}, H_{st})| \leq \|K\|^{n-2}\|H_{st}\|$$

where $H_{st}(x, y)$ denotes the function $\overline{K(s, x)K(y, t)}$ which belongs to $\boldsymbol{L}^2$ for almost all pairs of values s, t.

Let us summarize: *The values of* λ *which satisfy inequality* (19) *are all regular with respect to the transformation* K; *the resolvent transformation* K_λ *is also of integral type, that is, generated by a kernel* $K_\lambda(x, y)$ *belonging to* $\boldsymbol{L}^2$; *this kernel is the sum in the mean and — almost everywhere — also the sum in the ordinary sense of the series* (18).

Condition (19) is sufficient but not necessary for the convergence in the mean of series (18). In fact, for the examples given at the end of Sec. 65, series (18) converges (in the mean and even uniformly) for any value of λ. This is obvious for the kernel $K(x, y) = \alpha(x)\beta(y)$ with

$$\int_a^b \alpha(x)\beta(x)dx = 0,$$

all of whose iterates are zero. For a kernel $K(x, y)$ of the *Volterra type*, $|K(x, y)| \leq M$, all the iterates are of the same type, and we have

$$|K^{(n)}(x, y)| \leq M^n \frac{(x-y)^{n-1}}{(n-1)!} \qquad \text{for } a \leq y \leq x \leq b,$$

which is obvious for $n = 1$ and can be verified by induction for general n. Therefore series (18) is majorized by the convergent series whose general term is

$$|\lambda|^n M^n \frac{(b-a)^{n-1}}{(n-1)!}.$$

In these examples, all complex values λ are therefore regular with respect to the transformation K, and the resolvent transformation K_λ is generated by the kernel $K_\lambda(x, y)$, which is the sum of the series (18).

But for certain transformations K it is possible that a value of λ be regular, that is, that the resolvent transformation K_λ exist, without having the series (18) converge, and the question then arises whether or not the transformation K_λ is still of integral type. If we multiply the two members of relation (14) on the right by the transformation K and then use relation (16), we obtain for K_λ the explicit expression

$$K_\lambda = (I - \lambda K)^{-1}K.$$

Now in Sec. 69 we shall prove the following lemma:

LEMMA. *Every transformation which can be represented as the product TK of two linear transformations, of which K is of integral type, is also of integral type.*

Consequently *for every regular value λ the resolvent transformation K_λ is generated by a kernel $K_\lambda(x, y)$, called the resolvent kernel.*

The proof of the lemma will be based on an approximation theorem which will also play an essential rôle in the sequel.

It is appropriate to note that there exist linear transformations in the space L^2 which are not of integral type, for example, the identity transformation I. In fact, if I were generated by a kernel $I(x, y)$, we should have

$$\int_a^b\int_a^b I(x, y)f(y)\overline{g(x)}dydx = (If, g) = (f, g),$$

for any f and g in L^2; let us choose in particular f and g equal to the characteristic functions of two arbitrary nonoverlapping intervals; then $(f, g) = 0$, and the first member is the integral of $I(x, y)$ over a rectangle in the x, y plane whose sides are parallel to the axes and which is not intersected by the line $y = x$. Therefore the integral of $I(x, y)$ is equal to 0 over every rectangle of this type, and consequently $I(x, y) = 0$ almost everywhere, which is impossible.

69. Approximation of an Arbitrary Kernel by means of Kernels of Finite Rank

A particularly simple class of *linear transformations* of the space L^2 is formed of those which are said to be of *finite rank*; these are the transformations which can be represented in the form

$$Kf = \sum_{i=1}^{r} (f, \psi_i)\varphi_i$$

where $\varphi_1, \ldots, \varphi_r$ and $\psi_1, \ldots, \psi_r$ are given elements of the space L^2. Every transformation of finite rank is of integral type with the kernel

$$K(x, y) = \sum_{i=1}^{r} \varphi_i(x)\overline{\psi_i(y)};$$

the kernel is also said to be of *finite rank*; it obviously belongs to the space $\mathbf{L}^2$.

We can pass from the study of kernels of finite rank to that of more general kernels by making use of the following theorem:

THEOREM. *Every kernel $K(x, y)$ belonging to $\mathbf{L}^2$ can be approximated as closely as we wish by means of a kernel of finite rank, where the approximation is in the sense of the metric of $\mathbf{L}^2$, that is, in the mean.*

We know, in fact, that the double Fourier series of the square-summable function $K(x, y)$ converges in the mean to this function; therefore we have only to take a partial sum with sufficiently high index.

If one prefers not to use the theory of Fourier series, he can reason as follows:

Let $K_N(x, y) = K(x, y)$ when $|K(x, y)| \leq N$ and let $K_N(x, y) = 0$ elsewhere. Letting N tend to infinity, the function

$$|K(x, y) - K_N(x, y)|^2$$

tends to 0 while remaining majorized by the summable function $|K(x, y)|^2$. By virtue of Lebesgue's theorem, the integral

$$\int_a^b\int_a^b |K(x, y) - K_N(x, y)|^2 dxdy$$

will also tend to 0. Given a quantity $\varepsilon > 0$, we can therefore choose N so that $\| K - K_N \| < \frac{\varepsilon}{2}$. The function $K_N(x, y)$, being summable, is the limit almost everywhere of a sequence $\{\varphi_{Nn}(x, y)\}$ of step functions which can be chosen so that $|\varphi_{Nn}(x, y)| \leq N$. The sequence of functions

$$|K_N(x, y) - \varphi_{Nn}(x, y)|^2 \qquad (n = 1, 2, \ldots)$$

is bounded (by $4N^2$) and tends to 0; hence, the sequence of integrals also tends to 0. Therefore for large enough n we shall have $\| K_N - \varphi_{Nn} \| < \frac{\varepsilon}{2}$. It follows from Minkowski's inequality that

$$\| K - \varphi_{Nn} \| \leq \| K - K_N \| + \| K_N - \varphi_{Nn} \| < \varepsilon,$$

and the proof is completed by noting that every step function of two variables x, y can be put in the form

$$\sum_{i=1}^{r} \varphi_i(x)\overline{\psi_i(y)}.$$

This reasoning also applies to the case where the fundamental domain

$$a \leq x \leq b, \quad a \leq y \leq b$$

is infinite, that is, where $a = -\infty$ or $b = \infty$, except that we shall then have to also put $K_N(x, y) = 0$ at the points which are *exterior* to the finite square

$$-N \leq x \leq N, \quad -N \leq y \leq N,$$

and likewise choose the functions $\varphi_{Nn}(x, y)$ so that they are zero exterior to this finite square.

*

With the approximation theorem thus established, here are two of its applications.

First we shall prove the lemma stated in the preceding section concerning linear transformations of type TK where K is generated by a kernel $K(x, y)$ belonging to $\boldsymbol{L}^2$. We choose a sequence of kernels of finite rank

$$K_n(x, y) = \sum_{i=1}^{r_n} \varphi_{ni}(x)\overline{\psi_{ni}(y)}$$

converging in the mean to $K(x, y)$. Let $\chi_{ni} = T\varphi_{ni}$ and set

$$H_n(x, y) = \sum_{i=1}^{r_n} \chi_{ni}(x)\overline{\psi_{ni}(y)}.$$

For every fixed value of y, $K_n(x, y)$ and $H_n(x, y)$ are elements of L^2, and we have

$$H_n(x, y_0) = TK_n(x, y_0),$$

from which

$$\int_a^b |H_n(x, y) - H_m(x, y)|^2 dx \leq \|T\|^2 \int_a^b |K_n(x, y) - K_m(x, y)|^2 dx,$$

and, integrating with respect to y,

$$\mathbf{|} H_n(x, y) - H_m(x, y) \mathbf{|}^2 \leq \|T\|^2 \mathbf{|} K_n(x, y) - K_m(x, y) \mathbf{|}^2.$$

The mean convergence of the sequence $\{K_n(x, y)\}$ therefore implies, in view of the Riesz-Fischer theorem, the mean convergence of the sequence $\{H_n(x, y)\}$; let us denote the limit of the latter by $H(x, y)$. Denoting the corresponding linear transformations by the same letters, we have

$$\|H_n - H\| \leq \mathbf{|} H_n - H \mathbf{|} \to 0,$$

and

$$\|TK_n - TK\| = \|T(K_n - K)\| \leq \|T\| \, \|K_n - K\| \leq \|T\| \, \mathbf{|} K_n - K \mathbf{|} \to 0.$$

Now for every element f of L^2 we have

$$H_n f = \sum_i (f, \psi_{ni})\chi_{ni} = T \sum_i (f, \psi_{ni})\varphi_{ni} = TK_n f;$$

hence

$$H_n = TK_n,$$

and consequently

$$H = TK;$$

therefore the transformation TK is generated by the kernel $H(x, y)$, q.e.d.

As a second application of our approximation theorem we shall show that *distinct kernels always generate distinct transformations*; of course, two kernels which differ only on a (plane) set of measure zero are considered identical.

It suffices to show that the transformation K which is generated by the kernel $K(x, y)$ cannot be equal to the transformation O (defined to be zero for all the elements of L^2) unless $K(x, y)$ is zero almost everywhere.

Suppose, therefore, that $K = O$; then for two arbitrary functions $f(x)$ and $g(x)$ belonging to L^2, and for the function $F(x, y) = g(x)\overline{f(y)}$ belonging to $\boldsymbol{L}^2$, we have

$$(K, F) = \int_a^b\int_a^b K(x, y)\overline{F(x, y)}dxdy = \int_a^b\left[\int_a^b K(x, y)f(y)dy\right]\overline{g(x)}dx = (Kf, g) = 0.$$

It follows that $K(x, y)$ is orthogonal to all the kernels of finite rank, and since these are everywhere dense in $\boldsymbol{L}^2$, $K(x, y)$ is orthogonal to itself: $\|K\|^2 = (K, K) = 0$; therefore $K(x, y) = 0$ almost everywhere, which was to be proved.

THE FREDHOLM ALTERNATIVE

70. Integral Equations With Kernels of Finite Rank

In the case where the kernel is of finite rank,

$$K(x, y) = \sum_{i=1}^{r} \varphi_i(x)\overline{\psi_i(y)},$$

the study of the integral equation

$$f - Kf = g$$

reduces to the study of a system of linear algebraic equations. Moreover, we can assume that the functions φ_i are linearly independent;[6] also the functions ψ_i. If this were not so it could be achieved by selecting new functions; the kernel would then have an analogous form, but with fewer terms.

The integral equation is now of the form

$$f - \sum_{i=1}^{r} (f, \psi_i)\varphi_i = g, \tag{20}$$

and every solution f must be of the form

$$f = g + \sum_{j=1}^{r} \xi_j \varphi_j, \tag{21}$$

with numerical coefficients ξ_j which are to be determined. Let us substitute for f in (20) its expression (21); there results a homogeneous linear relation among

[6] That is, a linear combination with numerical coefficients, $\sum_{i=1}^{r} c_i\varphi_i(x)$, can be zero almost everywhere only if $c_1 = c_2 = \ldots = c_r = 0$.

the φ_j. Since these functions are linearly independent, all the coefficients in this relation must be equal to 0. Thus we obtain the equations

(22) $$\xi_j - \sum_{i=1}^{r} c_{ij}\xi_i = \eta_j \qquad (j = 1, 2, \ldots, r)$$

where

$$c_{ij} = (\varphi_i, \psi_j), \quad \eta_j = (g, \psi_j).$$

It is obvious that, conversely, every solution $\{\xi_j\}$ of the system of equations (22) furnishes, by means of formula (21), a solution f of (20).

When the determinant

$$d = |\delta_{ij} - c_{ij}| \quad (\delta_{ij} = 0 \text{ for } i \neq j,\ \delta_{ii} = 1)$$

of system (22) is not zero, we have the unique solution

$$\xi_i = \frac{1}{d} \sum_{j=1}^{r} d\binom{i}{j}\eta_j, \qquad (i = 1, 2, \ldots, r)$$

where we have denoted by $d\binom{i}{j}$ the minor associated with the element of indices (i, j). Equation (20) then has for unique solution

$$f = g + \frac{1}{d}\sum_{i=1}^{r}\sum_{j=1}^{r} d\binom{i}{j}(g, \psi_j)\varphi_i = (I + K_1)g,$$

the transformation K_1 being generated by the kernel

(23) $$K_1(x, y) = \frac{1}{d}\sum_{i=1}^{r}\sum_{j=1}^{r} d\binom{i}{j}\varphi_i(x)\overline{\psi_i(y)}.$$

In particular, the *homogeneous* equations

(24) $$f - \sum_{i=1}^{r}(f, \psi_i)\varphi_i = 0,$$

(25) $$\xi_j - \sum_{i=1}^{r} c_{ij}\xi_i = 0 \qquad (j = 1, 2, \ldots, r)$$

then have the unique solutions $f = 0$ and $\xi_j = 0$ $(j = 1, 2, \ldots, r)$.

On the other hand, when $d = 0$ (25), and consequently also (24), have non-zero solutions, and the number of linearly independent solutions is equal to the "nullity" of the matrix $(\delta_{ij} - c_{ij})$, that is, to the difference between its order r and its rank.

Consider also the integral equation

(26) $$f' - K^*f' = g'$$

whose kernel $K^*(x, y)$ is the "adjoint" of the kernel $K(x, y)$ according to the relation

$$K^*(x, y) = \overline{K(y, x)},$$

that is,

$$K^*(x, y) = \sum_{i=1}^{r} \psi_i(x)\overline{\varphi_i(y)}.$$

Let c_{ij}^*, d^*, $d^*\begin{pmatrix} i \\ j \end{pmatrix}$ be the quantities corresponding to the kernel $K^*(x, y)$. We have

$$c_{ij}^* = (\psi_i, \varphi_j) = \overline{(\varphi_j, \psi_i)} = \overline{c_{ji}},$$

from which it follows that $d^* = \bar{d}$, $d^*\begin{pmatrix} i \\ j \end{pmatrix} = \overline{d\begin{pmatrix} j \\ i \end{pmatrix}}$, and that the two matrices $(\delta_{ij} - c_{ij}^*)$ and $(\delta_{ij} - c_{ij})$ have the same nullity.

Therefore when $d \neq 0$ we have $d^* \neq 0$, and equation (26) has a unique solution f' for any g', whereas when $d = d^* = 0$, equation (24) and the equation

$$f' - \sum_{i=1}^{r} (f', \varphi_i)\psi_i = 0 \tag{27}$$

have the same number, ν, of linearly independent solutions, $\nu \geq 1$.

For certain g, equation (20) has a solution even when $d = 0$, namely for functions g which are orthogonal to all the solutions f' of equation (27). In fact, setting

$$f = g + \sum \xi_j \varphi_j, \quad f' = \sum \xi'_j \psi_j, \quad \eta_j = (g, \psi_j),$$

we pass to the equivalent systems of algebraic equations

$$\left.\begin{aligned} \xi_j - \sum_{i=1}^{r} c_{ij}\xi_i &= \eta_j \\ \xi'_j - \sum_{i=1}^{r} \overline{c_{ji}}\xi'_i &= 0 \end{aligned}\right\} \qquad (j = 1, 2, \ldots, r),$$

and observe that the orthogonality of the functions g and f' is equivalent to that of the vectors $\eta = \{\eta_j\}$ and $\xi' = \{\xi'_j\}$, since

$$(g, f') = (g, \sum \xi'_j \psi_j) = \sum \overline{\xi'_j}\,(g, \psi_j) = \sum \overline{\xi'_j}\eta_j = (\eta, \xi').$$

Therefore the problem reduces to the known theorem from algebra that the system of equations

$$\sum_{i=1}^{r} a_{ji}x_i = y_j \qquad (j = 1, 2, \ldots, r) \tag{28}$$

has a solution $x = \{x_i\}$ if and only if the vector $y = \{y_i\}$ is orthogonal to all the vectors $x' = \{x_i'\}$ — the solutions of the homogeneous adjoint system:

$$\sum_{i=1}^{r} \overline{a_{ij}} x_i' = 0 \qquad (j = 1, 2, \ldots, r).^{7} \tag{29}$$

Hence for integral equations with kernel of *finite rank*, we have the following theorem.

THEOREM. *Either the integral equations*

$$\text{(I)} \quad f - Kf = g, \qquad \text{(I*)} \quad f' - K^*f' = g'$$

with kernels

$$K(x, y), \qquad K^*(x, y) = \overline{K(y, x)},$$

have unique solutions f, f', *whatever be the given functions* g, g', *and in particular have the unique solutions* $f = 0$, $f' = 0$ *when* $g = 0$, $g' = 0$, *or the homogeneous equations*

$$\text{(H)} \quad \varphi - K\varphi = 0, \qquad \text{(H*)} \quad \varphi' - K^*\varphi' = 0$$

also have non-zero solutions and the number ν *of linearly independent solutions is finite and the same for the two homogeneous equations.*

In the second case, a necessary and sufficient condition that equations (I) *and* (I*) *have solutions is that* g *be orthogonal to all the solutions* φ' *of equation* (H*), *and that* g' *be orthogonal to all the solutions* φ *of equation* (H).

We shall see later that the alternative expressed by this theorem, also called *the Fredholm alternative,* holds even for equations whose kernels *are not of finite rank.*

Now consider the integral equation $f - \lambda Kf = g$ with the parameter λ. In the above calculations, let us replace $K(x, y)$ by $\lambda K(x, y)$; the quantities c_{ij} will be replaced by the quantities λc_{ij}, and the determinant d and its minors $d\begin{pmatrix} i \\ j \end{pmatrix}$ will be polynomials in λ which we shall denote by $d(\lambda)$ and $d\begin{pmatrix} i \\ j \end{pmatrix}; \lambda\Big)$. Since $d(0) = 1$, $d(\lambda)$ is not identically zero and consequently it has at most a finite number of zeros, say $\lambda_1, \lambda_2, \ldots, \lambda_s$. All the values of λ different from these zeros are *regular* and the resolvent kernel is equal to

$$K_\lambda(x, y) = \frac{1}{d(\lambda)} \sum_{i=1}^{r} \sum_{j=1}^{r} d\left(\begin{matrix} i \\ j \end{matrix}; \lambda\right) \varphi_i(x)\overline{\psi_j(y)}.$$

[7] This is a consequence of the identity

$$\sum_j \left(x_j' \overline{\sum_i a_{ji} x_i}\right) = \sum_i \left(\sum_j \overline{a_{ji}} x_j'\right)\bar{x}_i;$$

in fact, the identity implies that the vectors y which are expressible in the form (28) and the vectors x' which are solutions of (29) form two orthogonal and complementary subspaces of the r-dimensional (complex) vector space.

The values $\lambda_1, \lambda_2, \ldots, \lambda_s$ are *singular*, since for these values, the equation $f - \lambda Kf = g$ can not be solved for all g, and consequently the transformation $I - \lambda K$ does not possess an inverse.

Therefore *for an integral equation with kernel of finite rank there are at most a finite number of singular values; the resolvent kernel is a rational function of the parameter λ which has its poles at the singular values.*

71. Integral Equations With Kernels of General Type

If we wish to extend the results obtained in the preceding section to the case of equations with arbitrary kernel we can use the approximation theorem proved in Section 69. Several methods are available. One of these is to replace the integral equation with given kernel $K(x, y)$ by an infinite sequence of equations with kernels of finite rank $L_n(x, y)$ such that $\| K - L_n \| \to 0$, and to study the convergence of the solutions of these equations when $n \to \infty$.[8] Another way is to approximate $K(x, y)$ by a single kernel of finite rank $L(x, y)$ such that the difference $K(x, y) - L(x, y)$ is "small" enough that we can apply the method of successive approximations to it. It is this second way, due to E. SCHMIDT [2], that we shall follow.[9]

Let us consider, therefore, the integral equation

$$(I - \lambda K)f = g$$

whose kernel $K(x, y)$ belongs to $\boldsymbol{L}^2$. Given a positive quantity ω, approximate $K(x, y)$ by a kernel of finite rank

$$L(x, y) = \sum_{i=1}^{r} \varphi_i(x)\overline{\psi_i(y)}$$ [10]

so that the difference $H(x, y) = K(x, y) - L(x, y)$ satisfies the inequality

$$\| H \| \leq \frac{1}{\omega}.$$

According to Sec. 67, all the values of λ lying in the interior of the circle

$$C_\omega: \quad |\lambda| = \omega$$

are regular with respect to the transformation H generated by the kernel $H(x, y)$, and for these values

$$(I - H)^{-1} = I + \lambda H_\lambda = I + \lambda H + \lambda^2 H^2 + \lambda^3 H^3 + \ldots.$$

[8] GOURSAT [1] and [*] (pp. 386–390); LEBESGUE [6]; COURANT [1]; COURANT–HILBERT [*] (Chapt. III, 3).

[9] In the form developed by RADON [2].

[10] We assume that the φ_i, and the ψ_i, are linearly independent.

We have, then,

$$I-\lambda K = I-\lambda H-\lambda L = (I-\lambda H)\,[I-\lambda(I-\lambda H)^{-1}L] = (I-\lambda H)[I-\lambda L(\lambda)] \tag{30}$$

where the transformation

$$L(\lambda) = (I - \lambda H)^{-1}L$$

is of finite rank:

$$L(\lambda)f = \sum_{i=1}^{r} (f, \psi_i)\varphi_{\lambda,i}$$

with

$$\varphi_{\lambda,i} = (I - \lambda H)^{-1}\varphi_i = \varphi_i + \lambda H\varphi_i + \lambda^2 H^2 \varphi_i + \dots .$$ [11]

Form the quantities

$$c_{ij} = c_{ij}(\lambda), \quad d = d(\lambda), \quad d\begin{pmatrix} i \\ j \end{pmatrix} = d\begin{pmatrix} i \\ j \end{pmatrix}; \lambda$$

corresponding, in the sense of Sec. 70, to the transformation of finite rank $\lambda L(\lambda)$. Since

$$c_{ij}(\lambda) = \lambda(\varphi_{\lambda,i}, \psi_j) = \lambda(\varphi_i, \psi_j) + \lambda^2(H\varphi_i, \psi_j) + \lambda^3(H^2\varphi_i, \psi_j) + \dots,$$

these quantities are holomorphic functions of λ in the interior of the circle C_ω. Since the determinant $d(\lambda)$ is obviously equal to 1 for $\lambda = 0$, it does not vanish identically, and consequently its zeros cannot have any limit point in the interior of C_ω.

Now we know (Sec. 70) that the inverse transformation

$$[I - \lambda L(\lambda)]^{-1} = I + \lambda L_\lambda(\lambda)$$

exists if and only if $d(\lambda) \neq 0$. On the other hand, inasmuch as $(I - \lambda H)^{-1}$ exists for all λ in C_ω, it follows from relation (30) that $(I - \lambda K)^{-1}$ exists for a λ in C_ω if and only if $[I - \lambda L(\lambda)]^{-1}$ exists; in this case

$$I + \lambda K_\lambda = (I - \lambda K)^{-1} = [I - \lambda L(\lambda)]^{-1}\,(I - \lambda H)^{-1} = [I + \lambda L_\lambda(\lambda)]\,(I + \lambda H_\lambda),$$

hence

$$K_\lambda = L_\lambda(\lambda) + \lambda L_\lambda(\lambda)H_\lambda + H_\lambda.$$

According to the preceding section, the transformation $L_\lambda(\lambda)$ is of finite rank, namely

$$L_\lambda(\lambda)f = \frac{1}{d(\lambda)} \sum_{i=1}^{r} \sum_{j=1}^{r} d\begin{pmatrix} i \\ j \end{pmatrix}; \lambda\,(f, \psi_j)\varphi_{\lambda,i}.$$

[11] The functions $\varphi_{\lambda,i}$ $(i = 1, 2, \dots, r)$ are linearly independent simultaneously with the φ_i; in fact, $\Sigma\, c_i \varphi_{\lambda,i} = 0$ implies that

$$(I - \lambda H) \sum_i c_i \varphi_{\lambda,i} = \sum_i c_i \varphi_i = 0.$$

Replacing f by $H_\lambda f$, we see that the transformation $L_\lambda(\lambda)H_\lambda$ is also of finite rank:

$$L_\lambda(\lambda)H_\lambda f = \frac{1}{d(\lambda)} \sum_{i=1}^{r} \sum_{j=1}^{r} d\begin{pmatrix} i \\ \cdot ; \lambda \\ j \end{pmatrix} (f, H_\lambda^* \psi_j)\varphi_{\lambda,i};$$

we have denoted by H_λ^* the transformation generated by the adjoint kernel

$$H_\lambda^*(x, y) = \overline{H_\lambda(y, x)}.$$

Let us summarize:

All the interior points of the circle C_ω are regular values for the transformation K except the points where $d(\lambda)$ is zero, and these points do not have an accumulation point in the interior of C_ω. The resolvent transformation is given by the formula

$$K_\lambda = \frac{G(\lambda)}{d(\lambda)} + H_\lambda, \tag{31}$$

where $G(\lambda)$ denotes the transformation of finite rank

$$G(\lambda)f = \sum_{i=1}^{r} \sum_{j=1}^{r} d\begin{pmatrix} i \\ \cdot ; \lambda \\ j \end{pmatrix} [(f, \psi_j)\varphi_{\lambda,i} + \lambda(f, H_\lambda^* \psi_j)\varphi_{\lambda,i}].$$

Choosing ω sufficiently large, we arrive in this manner at the construction of the resolvent transformation K_λ for an arbitrary *regular* value of λ, and if we know K_λ we also know the solution of the equation $f - Kf = g$, whatever be the given function g; it is $f = g + \lambda K_\lambda g$.

The resolvent kernel $K_\lambda(x, y)$ is also obtained without difficulty; namely, we have

$$H_\lambda(x, y) = H(x, y) + \lambda H^{(2)}(x, y) + \lambda^2 H^{(3)}(x, y) + \ldots$$

(cf. Sec. 68), and $G(x, y; \lambda)$ is deduced immediately from the formula for the transformation of finite rank $G(\lambda)$.

It remains to study the case where λ is a *singular* value, which we shall do in the following sections. We know that these values cannot have an accumulation point in the interior of the circle C_ω. But ω could have been chosen arbitrarily large. From this follows the

THEOREM. *The singular values cannot have an accumulation point in the finite part of the complex plane; hence they form either a finite sequence (perhaps void), or an infinite sequence tending in modulus to ∞.*

72. Decomposition Corresponding to a Singular Value

Let λ_0 be a singular value of the transformation K. Let us consider the representation (31) of the resolvent transformation K_λ for the regular values in a circle C_ω about 0 containing λ_0. Denoting by ν the multiplicity of λ_0 as a zero of the function $d(\lambda)$, we can develop $\frac{1}{d(\lambda)}$ about λ_0 in a Laurent series whose first term contains $(\lambda - \lambda_0)^{-\nu}$. The entire series of $d\left(\begin{matrix} i \\ j \end{matrix}; \lambda\right)$, $\varphi_{\lambda,i}$, and H_λ are majorized in the ordinary sense, respectively in the norm, by convergent numerical series, uniformly in every circle interior to C_ω; from this it follows that we can rearrange them in powers of $\lambda - \lambda_0$ and multiply them in the Cauchy sense. We thus obtain the decomposition

$$K_\lambda = \frac{A(\lambda - \lambda_0)}{(\lambda - \lambda_0)^\nu} + B(\lambda - \lambda_0) \tag{32}$$

where the linear transformation $A(\lambda - \lambda_0)$ is a polynomial in $\lambda - \lambda_0$ of degree at most $\nu - 1$, and the linear transformation $B(\lambda - \lambda_0)$ is represented for small values of $|\lambda - \lambda_0|$ by an entire series in $\lambda - \lambda_0$, convergent in the norm:

$$A(\lambda - \lambda_0) = A_0 + (\lambda - \lambda_0)A_1 + \ldots + (\lambda - \lambda_0)^{\nu-1}A_{\nu-1},$$

$$B(\lambda - \lambda_0) = B_0 + (\lambda - \lambda_0)B_1 + \ldots + (\lambda - \lambda_0)^n B_n + \ldots;$$

furthermore, $A(\lambda - \lambda_0)$ and its coefficients A_k are of finite rank.

The singular value λ_0 is therefore a pole of the resolvent transformation, considered as a function of λ, and (32) is simply the decomposition into the "principal" and "regular" parts corresponding to this pole. Since the principal part

$$S(\lambda) = \frac{A(\lambda - \lambda_0)}{(\lambda - \lambda_0)^\nu}$$

has a polynomial for numerator, it is defined for all the complex values $\lambda \neq \lambda_0$. The regular part

$$R(\lambda) = B(\lambda - \lambda_0)$$

will then be determined by equation (32) for all values of λ for which K_λ is defined, that is, for all the regular values, and it will have, as an analytic function of λ, the same singularities as K_λ except for λ_0.

Let us substitute expression (32) of the resolvent transformation into equation (17) of Sec. 67:

$$(\lambda - \mu)K_\lambda K_\mu = K_\lambda - K_\mu.$$

Setting

$$u = \lambda - \lambda_0, \; v = \mu - \lambda_0,$$

we obtain

$$(u - v)\,[u^{-\nu}A(u) + B(u)]\,[v^{-\nu}A(v) + B(v)] = u^{-\nu}A(u) + B(u) - v^{-\nu}A(v) - B(v),$$

or, multiplying by $\dfrac{u^\nu v^\nu}{u - v}$,

(33) $$[A(u)+u^\nu B(u)]\,[A(v)+v^\nu B(v)] = \frac{v^\nu A(u)-u^\nu A(v)}{u-v} + \frac{B(u)-B(v)}{u-v}\,u^\nu v^\nu.$$

For $|u|$ and $|v|$ sufficiently small we replace $A(u)$, $A(v)$, $B(u)$, $B(v)$ by their expressions in polynomials of degree $\leq \nu - 1$ or in entire series; then we carry out the divisions by $u - v$, which is possible since

$$\frac{v^\nu u^\varkappa - u^\nu v^\varkappa}{v - u} = v^{\nu-1}u^\varkappa + v^{\nu-2}u^{\varkappa+1} + \ldots + v^\varkappa u^{\nu-1} \text{ for } 0 \leq \varkappa < \nu.$$

Comparing the terms containing u to a power $< \nu$, we obtain from (33) that

(34) $$A(u)\,[A(v) + v^\nu B(v)] = \frac{v^\nu A(u) - u^\nu A(v)}{u - v};$$

then, comparing in (34) the terms containing v to a power $< \nu$, we find that

(35) $$A(u)A(v) = \frac{v^\nu A(u) - u^\nu A(v)}{u - v}.$$

(34) and (35) imply that

(36) $$A(u)B(v) = 0.$$

Interchanging the roles of u and v, we obtain

(37) $$B(u)A(v) = 0.$$

Since $A(u)$ is a polynomial and $B(u)$ is an analytic function of u, relations (35)—(37), established first for small values of $|u|$ and $|v|$, remain valid for all values of u and v, respectively for the values for which $B(u)$ and $B(v)$ are defined. Finally, comparing (33) with relations (35)—(37), we obtain:

(38) $$B(u)B(v) = \frac{B(u) - B(v)}{u - v}.$$

Since $\lambda = 0$ is a regular value and $K_0 = K$, we have, by virtue of (32),

$$K = S + R,$$

where we have set $S = S(0)$, $R = R(0)$.

Let us write relations (35) and (38), first for $u = \lambda - \lambda_0$, $v = -\lambda_0$, then for $u = -\lambda_0$, $v = \lambda - \lambda_0$; we obtain

$$\lambda S(\lambda)S = \lambda SS(\lambda) = S(\lambda) - S,$$

$$\lambda R(\lambda)R = \lambda RR(\lambda) = R(\lambda) - R;$$

the equations in the first line are valid for all values of λ different from λ_0, and those in the second line for all values of λ which are regular with respect to the transformation K, and also for $\lambda = \lambda_0$. According to what was said in Sec. 67, these equations imply that

$$I + \lambda S(\lambda) = (I - \lambda S)^{-1} \text{ for } \lambda \neq \lambda_0,$$

$$I + \lambda R(\lambda) = (I - \lambda R)^{-1} \begin{cases} \text{for all values of } \lambda \text{ which are regular with respect} \\ \text{to the transformation } K, \text{ and for } \lambda = \lambda_0. \end{cases}$$

Setting $u = v = -\lambda_0$ in (36) and (37), we obtain

$$SR = RS = 0.$$

Let us observe, finally, that since the transformation S is of finite rank, it is generated by a kernel of finite rank $S(x, y)$, and R is generated by the kernel $R(x, y) = K(x, y) - S(x, y)$.

Thus we have proved the

DECOMPOSITION THEOREM. *To each singular value λ_0 of K corresponds a decomposition of K into the sum of two linear transformations S and R such that S is of finite rank, $K = S + R$, and*

1° $SR = RS = 0$,

2° *the value λ_0 is regular with respect to R and singular with respect to S,*

3° *all the values $\lambda \neq \lambda_0$ are regular with respect to S,*

4° *a value $\lambda \neq \lambda_0$ is regular with respect to R if and only if it is regular with respect to K.*

There is a corresponding decomposition of the kernel: $K(x, y) = S(x, y) + R(x, y)$, where $S(x, y)$ is of finite rank.

73. The Fredholm Alternative for General Kernels

This decomposition provides us with a means of proving the Fredholm alternative in the case of an arbitrary kernel $K(x, y)$ belonging to $\boldsymbol{L}^2$ by reducing the problem to the case of kernels of finite rank. We remark that in the proof we shall not make use of properties 3° and 4° of the decomposition.

Consider the integral equations

$$f - Kf = g \text{ and } f' - K^*f' = g'$$

whose kernels are

$$K(x, y) \text{ and } K^*(x, y) = \overline{K(y, x)}.$$

There are two cases according as the transformation $I - K$ does or does not possess an inverse.

First case. There exists a resolvent transformation K_1, generated by a resolvent kernel $K_1(x, y)$, such that

$$K_1K = KK_1 = K_1 - K \text{ (cf. Sec. 67, formula (16)).}$$

Let us write the corresponding relations for the kernels, interchange the variables x and y, take the complex conjugates, and then return to the transformations: we thus arrive at the relations

$$K^*K_1^* = K_1^* K^* = K_1^* - K^*.$$

(In this argument we have used the fact, proved in Sec. 69, that a kernel is uniquely determined by the transformation which generates it.) These relations, in their turn, imply that $(I - K^*)^{-1}$ exists and is equal to $I + K_1^*$.

Therefore in this case the two integral equations possess unique solutions

$$f = (I + K_1)g, \; f' = (I + K_1^*)g',$$

whatever be the given functions g and g' belonging to L^2.

Second case. $\lambda = 1$ is a singular value with respect to K. Let $K = S + R$ be the corresponding decomposition. Reasoning as above with the kernels, we conclude from the relations

$$K = S + R, \; SR = RS = 0$$

that

$$K^* = S^* + R^*, \; R^*S^* = S^*R^* = 0.$$

Consequently we have the relations:

a) $I - K = (I - S)(I - R)$, a*) $I - K^* = (I - R^*)(I - S^*)$,

b) $I - K = (I - R)(I - S)$, b*) $I - K^* = (I - S^*)(I - R^*)$.

Since $(I - R)^{-1}$ exists, $(I - R^*)^{-1}$ also exists (first case), and

c) $(I - K)(I - R)^{-1} = I - S$, c*) $(I - R^*)^{-1}(I - K^*) = I - S^*$,

d) $(I - R)^{-1}(I - K) = I - S$, d)* $(I - K^*)(I - R^*)^{-1} = I - S^*$.

We deduce from relations b) and d) that the homogeneous integral equations

$$(I - K)f = 0, \; (I - S)f = 0$$

possess the same solutions f. The same holds, by virtue of a*) and c*) for the equations

$$(I - K^*)\, f' = 0, \; (I - S^*)\, f' = 0.$$

But S is of finite rank and has 1 for a singular value; hence by the theorem of Sec. 70 the equations $(I - S)f = 0$ and $(I - S^*)f' = 0$ possess the same finite number $s \geq 1$ of linearly independent solutions.

As for the non-homogeneous equation $(I - K)f = g$, we deduce from a) and c) that it possesses a solution f if and only if the equation $(I - S)\varphi = g$ possesses a solution φ; these solutions f and φ are connected by the relations $\varphi = (I - R)f$, $f = (I - R)^{-1}\varphi$. But we know from Sec. 70 that the equation $(I - S)\varphi = g$ has a solution only if g is orthogonal to all the solutions f'

of the equation $(I - S^*)f' = 0$, or equivalently, of the equation $(I - K^*)f' = 0$. When this condition is satisfied, the general solution of the equation $(I - K)f = g$ will obviously be the sum of a particular solution of this equation and of the general solution of the homogeneous equation $(I - K)f = 0$.

This completes the proof of the fact that *the Fredholm alternative, proved in Sec. 70 for integral equations with kernels of finite rank, remains valid for integral equations with arbitrary kernel belonging to* $\mathbf{L}^2$.

FREDHOLM DETERMINANTS

74. The Method of Fredholm

We shall sketch the method by means of which IVAR FREDHOLM [**1**] [**2**] was the first to solve the integral equation

$$f(x) - \lambda \int_a^b K(x, y) f(y) dy = g(x), \tag{39}$$

under the hypothesis that the kernel $K(x, y)$ is a *bounded* and *continuous* function.

FREDHOLM started from the idea, already used by VOLTERRA, of replacing the integral in (39) by the sum

$$\sum_{j=1}^{n} K(x, \xi_j) f(\xi_j) h \quad \left(h = \frac{b-a}{n}, \ \xi_j = a + jh \right),$$

and considering (39) at the points $x = \xi_i$ $(i = 1, 2, \ldots, n)$. The integral equation thus gives rise to a system of linear algebraic equations with unknowns $f_j = f(\xi_j)$ $(j = 1, 2, \ldots, n)$:

$$f_i - \lambda h \sum_{j=1}^{n} K_{ij} f_j = g_i \qquad (i = 1, 2, \ldots, n), \tag{40}$$

where we have set $K_{ij} = K(\xi_i, \xi_j)$ and $g_i = g(\xi_i)$.

The determinant of this system can be developed in the following form:

$$d_\lambda = 1 - \lambda h \sum_i K_{ii} + \frac{\lambda^2 h^2}{2!} \sum_{i,j} \begin{vmatrix} K_{ii} & K_{ij} \\ K_{ji} & K_{jj} \end{vmatrix} - \ldots + \tag{41}$$
$$+ (-1)^n \frac{\lambda^n h^n}{n!} \sum_{i_1, i_2, \ldots, i_n} \begin{vmatrix} K_{i_1 i_1} & K_{i_1 i_2} & \ldots & K_{i_1 i_n} \\ K_{i_2 i_1} & K_{i_2 i_2} & \ldots & K_{i_2 i_n} \\ \vdots & & & \vdots \\ K_{i_n i_1} & K_{i_n i_2} & \ldots & K_{i_n i_n} \end{vmatrix}.$$

Since d_λ is a polynomial in the parameter λ which is not identically zero, system (40) has a unique solution for all values of λ with the exception of at most a finite number of values; furthermore, this solution $(f_1, f_2, \ldots, f_n)$ appears in the form of n fractions having for denominator the determinant d_λ and for numerators determinants constructed from the minors of d_λ and from the given g_i in a well-known manner.

Instead of investigating whether the solution of (40) tends, for $n \to \infty$, to a limit furnishing a solution of (39), a problem which was solved later, however — for continuous $K(x, y)$ — by HILBERT,[12] FREDHOLM used the classical formula only to deduce from it, by a purely formal passage to the limit, an expression which he then showed in a direct manner converges and furnishes a solution of (39), that is, he showed this without appeal to the formal origin of the expression.

To simplify the writing we introduce the following notation

$$K\begin{pmatrix} x_1 x_2 \ldots x_n \\ y_1 y_2 \ldots y_n \end{pmatrix} = \begin{vmatrix} K(x_1, y_1) & K(x_1, y_2) \ldots K(x_1, y_n) \\ K(x_2, y_1) & K(x_2, y_2) \ldots K(x_2, y_n) \\ \vdots & \vdots \qquad\qquad \vdots \\ K(x_n, y_1) & K(x_n, y_2) \ldots K(x_n, y_n) \end{vmatrix}$$

and remark that this determinant changes sign when we interchange the variables x_i and x_k or the variables y_i and y_k, and that it consequently remains unaltered when we carry out an even number of these transformations, in particular, when we permute the n couples $\begin{pmatrix} x_i \\ y_i \end{pmatrix}$ in an arbitrary manner.

This being the case, the coefficient of $\dfrac{\lambda^m}{m!}$ in the development (41) can be written in the abridged form

$$h^m \sum_{i_1, i_2, \ldots, i_m} K\begin{pmatrix} \xi_{i_1} \xi_{i_2} \ldots \xi_{i_m} \\ \xi_{i_1} \xi_{i_2} \ldots \xi_{i_m} \end{pmatrix}.$$

This sum has the formal limit

$$\int_a^b \int_a^b \ldots \int_a^b K\begin{pmatrix} \xi_1 \xi_2 \ldots \xi_m \\ \xi_1 \xi_2 \ldots \xi_m \end{pmatrix} d\xi_1 \, d\xi_2 \ldots d\xi_m.$$

The formal limit of the determinant d is therefore the sum of the entire series

$$\begin{aligned} 1 - \lambda \int_a^b K\begin{pmatrix} \xi \\ \xi \end{pmatrix} d\xi + \frac{\lambda^2}{2!} \int_a^b \int_a^b K\begin{pmatrix} \xi_1 \xi_2 \\ \xi_1 \xi_2 \end{pmatrix} d\xi_1 \, d\xi_2 - \ldots + \\ + (-1)^n \frac{\lambda^n}{n!} \int_a^b \int_a^b \ldots \int_a^b K\begin{pmatrix} \xi_1 \xi_2 \ldots \xi_n \\ \xi_1 \xi_2 \ldots \xi_n \end{pmatrix} d\xi_1 \, d\xi_2 \ldots d\xi_n + \ldots \end{aligned} \tag{42}$$

This series is convergent for all values of λ. To show this we make use of the fact that a determinant of order n, all of whose elements are in absolute

[12] HILBERT [*] (1. Note).

value less than M, is at most equal in absolute value to $M^n n^{n/2}$. This is a corollary of the Hadamard inequality which we shall prove a little later.

It follows that series (42) has the convergent majorant

$$1 + |\lambda|(b-a)M + \frac{|\lambda^2|}{2!}(b-a)^2 M^2 \cdot 2 + \ldots + \frac{|\lambda|^n}{n!}(b-a)^n M^n n^{\frac{n}{2}} + \ldots,$$

where M denotes a bound for $|K(x, y)|$. The sum δ_λ of the series (42) is therefore an entire function of λ and is not identically zero, since $\delta_0 = 1$. It is called the *Fredholm determinant*.

The numerator of the classical algebraic formula leads, by an analogous formal passage to the limit, to the series

$$K\begin{pmatrix} x \\ y \end{pmatrix} - \lambda \int_a^b K\begin{pmatrix} x\xi \\ y\xi \end{pmatrix} d\xi + \frac{\lambda^2}{2!}\int_a^b\int_a^b K\begin{pmatrix} x\xi_1\xi_2 \\ y\xi_1\xi_2 \end{pmatrix} d\xi_1 d\xi_2 - \ldots +$$

$$(43) \qquad + (-1)^n \frac{\lambda^n}{n!}\int_a^b \ldots \int_a^b K\begin{pmatrix} x\xi_1\ldots\xi_n \\ y\xi_1\ldots\xi_n \end{pmatrix} d\xi_1 \ldots d\xi_n + \ldots.$$

It converges for all values of λ for the same reason; its sum $\delta_\lambda\begin{pmatrix} x \\ y \end{pmatrix}$ is called the *minor* of the Fredholm determinant.

Developing it by the elements of the first row, we obtain:

$$K\begin{pmatrix} x\xi_1\ldots\xi_n \\ y\xi_1\ldots\xi_n \end{pmatrix} = K\begin{pmatrix} x \\ y \end{pmatrix} K\begin{pmatrix} \xi_1\ldots\xi_n \\ \xi_1\ldots\xi_n \end{pmatrix} - K\begin{pmatrix} x \\ \xi_1 \end{pmatrix} K\begin{pmatrix} \xi_1\xi_2\ldots\xi_n \\ y\xi_2\ldots\xi_n \end{pmatrix} +$$

$$+ K\begin{pmatrix} x \\ \xi_2 \end{pmatrix} K\begin{pmatrix} \xi_1\xi_2\xi_3\ldots\xi_n \\ y\xi_1\xi_3\ldots\xi_n \end{pmatrix} - \ldots + (-1)^n K\begin{pmatrix} x \\ \xi_n \end{pmatrix} K\begin{pmatrix} \xi_1\xi_2\ldots\xi_n \\ y\xi_1\ldots\xi_{n-1} \end{pmatrix} =$$

$$= K\begin{pmatrix} x \\ y \end{pmatrix} K\begin{pmatrix} \xi_1\ldots\xi_n \\ \xi_1\ldots\xi_n \end{pmatrix} - K\begin{pmatrix} x \\ \xi_1 \end{pmatrix} K\begin{pmatrix} \xi_1\xi_2\ldots\xi_n \\ y\xi_2\ldots\xi_n \end{pmatrix} -$$

$$- K\begin{pmatrix} x \\ \xi_2 \end{pmatrix} K\begin{pmatrix} \xi_2\xi_1\xi_3\ldots\xi_n \\ y\xi_1\xi_3\ldots\xi_n \end{pmatrix} - \ldots - K\begin{pmatrix} x \\ \xi_n \end{pmatrix} K\begin{pmatrix} \xi_n\xi_1\xi_2\ldots\xi_{n-1} \\ y\xi_1\xi_2\ldots\xi_{n-1} \end{pmatrix}.$$

Since the variable in a definite integral can be denoted by any other letter, it follows that

$$\int_a^b \ldots \int_a^b K\begin{pmatrix} x\xi_1\ldots\xi_n \\ y\xi_1\ldots\xi_n \end{pmatrix} d\xi_1\ldots d\xi_n = K\begin{pmatrix} x \\ y \end{pmatrix}\int_a^b \ldots \int_a^b K\begin{pmatrix} \xi_1\ldots\xi_n \\ \xi_1\ldots\xi_n \end{pmatrix} d\xi_1\ldots d\xi_n -$$

$$- n\int_a^b \ldots \int_a^b K\begin{pmatrix} x \\ z \end{pmatrix} K\begin{pmatrix} z\xi_1\ldots\xi_{n-1} \\ y\xi_1\ldots\xi_{n-1} \end{pmatrix} dz\, d\xi_1\ldots d\xi_{n-1}.$$

Comparing (42) with (43), we therefore see that

$$\delta_\lambda\begin{pmatrix} x \\ y \end{pmatrix} = K(x, y)\delta_\lambda + \lambda\int_a^b K(x, z)\delta_\lambda\begin{pmatrix} z \\ y \end{pmatrix} dz.$$

An analogous calculation, starting with the development of the determinants $K\begin{pmatrix} x\xi_1\ldots\xi_n \\ y\xi_2\ldots\xi_n \end{pmatrix}$ by the elements of the first column, furnishes the relation

$$\delta_\lambda\begin{pmatrix} x \\ y \end{pmatrix} = K(x, y)\delta_\lambda + \lambda\int_b^a \delta_\lambda\begin{pmatrix} x \\ z \end{pmatrix} K(z, y)dz.$$

These two relations express the fact that the function

$$\frac{\delta_\lambda\begin{pmatrix} x \\ y \end{pmatrix}}{\delta_\lambda}$$

is identical with the function $K_\lambda(x, y)$, the resolvent kernel associated with the kernel $K(x, y)$, for all values of λ for which $\delta_\lambda \neq 0$. These values are therefore all regular with respect to $K(x, y)$. On the other hand, the zeros of δ_λ are singular values. In fact, a zero λ_0 is a pole of $K_\lambda(x, y)$, because $\delta_\lambda\begin{pmatrix} x \\ y \end{pmatrix}$ is not divisible by the same power of $(\lambda - \lambda_0)$; this is an immediate consequence of the relation

(44) $$-\frac{d\delta_\lambda}{d\lambda} = \int_b^b \delta_\lambda\begin{pmatrix} x \\ x \end{pmatrix} dx$$

which, in its turn, follows obviously from the definitions of δ_λ and $\delta_\lambda\begin{pmatrix} x \\ y \end{pmatrix}$ by the series (42) and (43). Therefore the singular values are precisely the zeros of the entire function δ_λ; they are either finite or denumerable in number and possess no finite limit points.

FREDHOLM was able to apply the same method to the study of the integral equation for a singular value of λ, but his calculations, which involved minors of higher order, are less simple. Instead of reproducing them, we content ourselves with referring the reader to volume III of E. GOURSAT's Cours d'Analyse.

We note further the interesting formula

$$-\frac{d \log \delta_\lambda}{d\lambda} = \int_a^b K(x, x)dx + \lambda\int_a^b K^{(2)}(x, x)dx + \lambda^2\int_a^b K^{(3)}(x,x)\,dx + \ldots,$$

valid in the largest circle of the complex plane about the center 0 which contains no singular value in its interior. This formula follows from equation (44), by dividing it by δ_λ and substituting for

$$\frac{\delta_\lambda\begin{pmatrix} x \\ x \end{pmatrix}}{\delta_\lambda} = K_\lambda(x, x)$$

its expression as a Neumann series (18), which is uniformly convergent.

In all this theory the hypothesis that $K(x, y)$ is a *continuous* function can be replaced by less restrictive hypotheses. It suffices to assume that $K(x, y)$ is *bounded* and *summable* in the Lebesgue sense. A difficulty arises from the fact that this does not assure that $K(x, x)$ is integrable. We can avoid this by modifying the definition of $K(x, y)$, if necessary, on the diagonal $x = y$, for example by setting $K(x, x) \equiv 0$; this modification is obviously without importance for the integral equation. With this done, all the integrals appearing in series (42) exist and the convergence is established as above. As for the series (43), its terms exist for almost all x and almost all y, namely those for which $K(x, z)$ and $K(z, y)$, respectively, are summable functions of z. We thus arrive at the functions δ_λ and $\delta_\lambda\begin{pmatrix} x \\ y \end{pmatrix}$; their ratio furnishes the resolvent kernel $K_\lambda(x, y)$. Of course, the function

$$f(x) = g(x) + \lambda \int_a^b K_\lambda(x, y) g(y) dy$$

is defined only almost everywhere, and it satisfies equation (39) *almost everywhere.*

Even the hypothesis that $K(x, y)$ is a bounded function can be replaced by other more general hypotheses; it suffices that it be square-summable.[13] But the method of Fredholm loses much of its simplicity and its elegance with these generalizations.

75. Hadamard's Inequality

We give now the inequality for determinants[14] whose corollary we used in the preceding section.

If D is a determinant with complex elements $c_{hk} = a_{hk} + ib_{hk}$ $(h, k = 1, 2, \ldots, n)$, we have

$$|D| \leq A_1 A_2 \ldots A_n,$$

where

$$A_h = \left(\sum_{k=1}^{n} |c_{hk}|^2\right)^{\frac{1}{2}} = \left(\sum_{k=1}^{n} (a_{hk}^2 + b_{hk}^2)\right)^{\frac{1}{2}}.$$

Except for the obvious case when the second member is zero, equality holds only when the rows of the determinant are pairwise orthogonal, that is, when

$$\sum_{k=1}^{n} c_{ik} \overline{c_{jk}} = 0 \text{ for } i \neq j.$$

If the right member is not zero, we can reduce the problem to the case where the quantities A_h are equal to 1 by passing to the determinant with elements c_{hk}/A_h.

D is a continuous function of the $2n^2$ variables a_{hk} and b_{hk} which vary in the bounded and closed domain characterized by the conditions $A_1 = 1, A_2 = 1, \ldots, A_n = 1$;

[13] See Carleman [**1**] and, among recent contributions, Smithies [**1**].
[14] Hadamard [**1**].

hence this function is bounded and attains its maximum for certain values c_{hk}^*. This maximum $|D^*|$ is certainly ≥ 1, since the determinant with elements

$$\delta_{hk} = \begin{cases} 1 \text{ for } h = k \\ 0 \text{ for } h \neq k \end{cases}$$

satisfies our conditions.

The extremal determinant D^* has pairwise orthogonal rows. Consider, for example, the first two rows; we shall show that the hypothesis that

$$C = \sum_{k=1}^{n} c_{2k}^* \overline{c_{1k}^*}$$

is different from 0 leads to a contradiction. Starting with D^*, we construct the determinant D' whose elements c'_{hk} are identical with the elements of D^* except for those in the second row; here we define instead

$$c'_{2k} = \lambda c_{1k}^* + \mu c_{2k}^*,$$

where the quantities λ and μ are determined by the two conditions:

$$\sum_{k=1}^{n} c'_{2k} \overline{c_{1k}^*} = 0 \text{ and } \sum_{k=1}^{n} |c'_{2k}|^2 = 1.$$

From the first, we deduce that $\lambda + \mu C = 0$, from the second that

$$|\lambda|^2 + \lambda\bar{\mu}\bar{C} + \bar{\lambda}\mu C + |\mu|^2 = 1.$$

It follows that $|\mu|^2(1 - |C|^2) = 1$, hence $|\mu| > 1$. But since D' is equal to μD^*, we have $|D'| > |D^*|$, which is impossible since $|D^*|$ is the maximum of the values of $|D|$.

Since the rows of D^* are orthogonal, we have

$$D^*\bar{D}^* = \det(c_{hk}^*) \cdot \det(\overline{c_{hk}^*}) = \det\left(\sum_{j=1}^{n} c_{hj}^* \overline{c_{kj}^*}\right) = \det(\delta_{hk}) = 1,$$

which was to be proved.

We add that the inequality also has an evident geometric interpretation, at least if the elements of the determinant are real and $n = 3$: among the parallelepipeds with edges of given length, the rectangular parallelepiped has the greatest volume.

ANOTHER METHOD, BASED ON COMPLETE CONTINUITY

76. Complete Continuity

We based the proof of the Fredholm alternative for a general kernel $K(x, y)$ of the space $\boldsymbol{L}^2$ on the decomposition theorem of Sec. 72. In fact, as soon as the decomposition $K(x, y) = S(x, y) + R(x, y)$ corresponding to a singular value was obtained, the problem reduced to a problem based on the kernel of finite rank $S(x, y)$ and, consequently, to a problem of linear algebra. In order to obtain this decomposition, we had to study the properties of the transformation $(I - \lambda K)^{-1}$ as an analytic function of the parameter λ.

The following method, due to one of the authors,[15] is, by contrast, of a rather geometric character and permits at the same time a more profound study of the "principal part" $S(x, y)$ without appeal to the theory of elementary divisors. It has the further advantage that it also applies to much more general functional equations.

We start with the fundamental concept of a completely continuous transformation, which is defined as follows:

A transformation T of the space L^2 is said to be completely continuous if it transforms every infinite bounded set into a compact set, that is, if for every infinite sequence $\{f_n\}$ of elements of L^2 such that $\|f_n\| \leq C$, the sequence $\{Tf_n\}$ contains a subsequence which converges in the mean to an element of L^2.

We know that every linear transformation (in our terminology) is continuous. But there exist linear transformations which are not completely continuous, for example, the identity transformation I. In fact, if one chooses an arbitrary orthonormal sequence for $\{f_n\}$ it is impossible to select a subsequence which converges in the mean, since $\|f_m - f_n\|^2 = \|f_m\|^2 + \|f_n\|^2 = 2$ for $m \neq n$.

It follows immediately from the definition that *the sum and the product of two completely continuous linear transformations are also completely continuous; for the product, it even suffices that only one or the other of the transformations in question be completely continuous.*

The following is a less evident fact.

Every linear transformation T which can be approximated in norm arbitrarily closely by a completely continuous linear transformation is itself completely continuous.

In fact, if $\{T_i\}$ is a sequence of completely continuous linear transformations such that $\|T - T_i\| \to 0$ and if $\{f_n\}$ is a bounded sequence of elements of L^2, we can use the diagonal process to select a subsequence $\{g_n\}$ from $\{f_n\}$ with the property that $\lim_{n\to\infty} T_i g_n$ exists for $i = 1, 2, \ldots$. We shall then have for all i

$$\|Tg_n - Tg_m\| \leq \|Tg_n - T_i g_n\| + \|T_i g_n - T_i g_m\| + \|T_i g_m - Tg_m\|;$$

therefore

$$\|Tg_n - Tg_m\| \leq \|T - T_i\| (\|g_n\| + \|g_m\|) + \|T_i g_n - T_i g_m\|,$$

and consequently

$$\limsup_{n, m\to\infty} \|Tg_n - Tg_m\| = 0,$$

which implies the convergence of the sequence $\{Tg_n\}$.

[15] F. Riesz [**9**]. Further, see Hildebrandt [**3**], Banach [*] (Chapt. X), and Zaanen [**2**].

Every transformation of the form

$$Hf = (f, \psi)\varphi,$$

where φ and ψ are given elements of L^2, is completely continuous. In fact, if $\{f_n\}$ is a bounded sequence of elements of L^2, we can use the Bolzano-Weierstrass theorem to select from it a subsequence $\{g_n\}$ for which the numerical sequence $\{(g_n, \psi)\}$ is convergent; denoting its limit by c, we shall have

$$\|Hg_n - c\varphi\| = |(g_n, \psi) - c| \, \|\varphi\| \to 0.$$

But every linear transformation of finite rank is the sum of a finite number of transformations of this elementary type. It follows that *all linear transformations of finite rank as well as their uniform limits are completely continuous.*

This class of linear transformations contains, in particular, the transformations which are generated by kernels belonging to the space $\boldsymbol{L}^2$ (see Sec. 69). Hence these transformations are completely continuous, and it is on this property that our arguments will be based.

77. The Subspaces $\mathfrak{M}_n$ and $\mathfrak{N}_n$

In what follows we shall denote by K a linear transformation of the space L^2 which is generated by a kernel $K(x, y)$ belonging to $\boldsymbol{L}^2$; we set

$$T = I - K.$$

1) Consider first, for $n = 0, 1, 2, \ldots$, the set $\mathfrak{M}_n$ of elements f of L^2 for which

$$T^n f = 0,$$

where T^0 is defined to be the identity transformation I. It is clear that $\mathfrak{M}_0$ consists of the single element 0 and that $\mathfrak{M}_n$ is contained in $\mathfrak{M}_{n+1}$:

$$\mathfrak{M}_0 = (0) \subseteq \mathfrak{M}_1 \subseteq \mathfrak{M}_2 \subseteq \ldots \subseteq \mathfrak{M}_n \subseteq \mathfrak{M}_{n+1} \subseteq \ldots.$$

Since T, as well as its iterates T^n, are linear transformations and therefore also continuous, each of the $\mathfrak{M}_n$ is a *closed linear set*, that is, a *subspace* of L^2.

We shall show that the subspace $\mathfrak{M}_n$ is of *finite dimension*, that is, that its elements can be represented as linear combinations of a finite number of them.

If this were not the case for $\mathfrak{M}_1$, there would be an infinite sequence $\{g_k\}$ of elements of $\mathfrak{M}_1$ with each term linearly independent of the preceding. Let $\{f_k\}$ be the infinite orthonormal sequence which we obtain from $\{g_k\}$ by the Schmidt process. Since the elements of $\mathfrak{M}_1$ are characterized by $Tf = 0$, we should have $Kf_k = f_k$ for $k = 1, 2, \ldots$; hence by the complete continuity of K we could select a convergent subsequence from $\{f_k\}$, which is impossible since

$$\|f_k - f_i\|^2 = \|f_k\|^2 - (f_k, f_i) - (f_i, f_k) + \|f_i\|^2 = 2 \text{ for } i \neq k.$$

An analogous argument applies to $\mathfrak{M}_n$ when $n > 1$. We have only to observe that

$$T^n = (I - K)^n = I - K^{(n)},$$

where

$$K^{(n)} = nK - \binom{n}{2} K^2 + \binom{n}{3} K^3 - \ldots + (-1)^{n-1} K^n,$$

and that the transformation $K^{(n)}$ is also completely continuous.

Therefore we know that the $\mathfrak{M}_n$ are subspaces of L^2 of finite dimension. Does the number of dimensions increase indefinitely with n?

First, it is easy to see that if $\mathfrak{M}_{n+1}$ is not identical with $\mathfrak{M}_n$, then all the $\mathfrak{M}_m$ of degree $m \leq n$ are distinct. In fact, if h is an element of $\mathfrak{M}_{n+1}$ which does not belong to $\mathfrak{M}_n$, then $T^{n-m}h$ will be an element of $\mathfrak{M}_{m+1}$ which does not belong to $\mathfrak{M}_m$, since

$$T^{m+1}T^{n-m}h = T^{n+1}h = 0, \qquad T^m T^{n-m} h = T^n h \neq 0.$$

Consequently, either $\mathfrak{M}_n$ is for all n a proper subset of $\mathfrak{M}_{n+1}$, or some $\mathfrak{M}_n$ coincides with $\mathfrak{M}_{n+1}$ and also with all the following subspaces.

We shall show that the first case can not arise. If it did, we could choose in each $\mathfrak{M}_n$ an element φ_n orthogonal to $\mathfrak{M}_{n-1}$ and such that $\|\varphi_n\| = 1$. Then we should have for $n > m$:

$$K\varphi_n - K\varphi_m = \varphi_n - (\varphi_m + T\varphi_n - T\varphi_m) = \varphi_n - \varphi$$

where φ is clearly an element belonging to $\mathfrak{M}_{n-1}$ and consequently orthogonal to φ_n; therefore

$$\|K\varphi_n - K\varphi_m\|^2 = \|\varphi_n\|^2 + \|\varphi\|^2 \geq \|\varphi_n\|^2 = 1.$$

But then it would be impossible to select a convergent subsequence from the sequence $\{K\varphi_n\}$, which contradicts the property that K is completely continuous.

Therefore there exists an index $\nu \geq 0$ *such that*

$$\mathfrak{M}_0 = (0) \subset \mathfrak{M}_1 \subset \ldots \subset \mathfrak{M}_\nu = \mathfrak{M}_{\nu+1} = \mathfrak{M}_{\nu+2} = \ldots.$$

2) Now we consider, for $n = 0, 1, 2, \ldots$, the set $\mathfrak{N}_n$ of elements f which can be represented in the form

$$f = T^n g;$$

$\mathfrak{N}_n$ is the image of the entire space L^2 under the transformation T^n. It is clear that

$$\mathfrak{N}_0 = L^2 \supseteq \mathfrak{N}_1 \supseteq \mathfrak{N}_2 \supseteq \ldots \supseteq \mathfrak{N}_n \supseteq \mathfrak{N}_{n+1} \supseteq \ldots.$$

We shall show that these sets are also *subspaces* of L^2. It is obvious that they are *linear*; hence it only remains to show that they are *closed*. We give the proof for $\mathfrak{N}_1$; it also applies to $\mathfrak{N}_n$ for $n > 1$ by replacing K with the transformation $K^{(n)}$ defined above.

Let $\{f_k = Tg_k\}$ be a convergent sequence of elements of $\mathfrak{N}_1$; we have to show that its limit f^* also belongs to $\mathfrak{N}_1$. Let $g_k = u_k + v_k$ be the decomposition of g_k into the sum of an element u_k of $\mathfrak{M}_1$ and another element, v_k, orthogonal to $\mathfrak{M}_1$ (cf. Sec. 34). Then we have

$$f_k = Tu_k + Tv_k = Tv_k.$$

We shall first show that the sequence $\{v_k\}$ is bounded. If not, we could replace it by a subsequence for which $\|v_k\| \to \infty$. We set $w_k = \dfrac{v_k}{\|v_k\|}$; then $Tw_k = \dfrac{f_k}{\|v_k\|} \to 0$. Since $\|w_k\| = 1$, the sequence $\{Kw_k\}$ contains a convergent subsequence $\{Kw_{k_i}\}$. The sequence

$$w_{k_i} = Tw_{k_i} + Kw_{k_i}$$

then converges also; if its limit is denoted by w^* we have $Tw^* = \lim Tw_{k_i} = 0$, therefore w^* is an element of $\mathfrak{M}_1$. But since the w_k are orthogonal to $\mathfrak{M}_i$ we have $\|w_k - w^*\|^2 = \|w_k\|^2 + \|w^*\|^2 \geq 1$; and this contradicts the fact that $w_k \to w^*$.

Therefore the sequence $\{v_k\}$ is bounded. Consequently the sequence $\{Kv_k\}$ contains a convergent subsequence $\{Kv_{k_j}\}$. The sequence of elements

$$v_{k_j} = Tv_{k_j} + Kv_{k_j} = f_{k_j} + Kv_{k_j}$$

also converges; let its limit be v^*. Then

$$f_{k_j} = v_{k_j} - Kv_{k_j} \to v^* - Kv^*;$$

therefore $f^* = Tv^*$, that is, f^* belongs to $\mathfrak{N}_1$.

This completes the proof of the fact that the sets $\mathfrak{N}_n$ are closed, and that consequently they are subspaces of L^2.

Consider now the question of determining whether the $\mathfrak{N}_n$ are distinct. Since each of them is the image of the preceding under the transformation T, it is obvious that if an $\mathfrak{N}_n$ coincides with $\mathfrak{N}_{n+1}$, it coincides with all the $\mathfrak{N}_p$, $p > n$.

Consequently, either $\mathfrak{N}_{n+1}$ is for all n a proper subset of $\mathfrak{N}_n$, or there is an $\mathfrak{N}_n$ which coincides with $\mathfrak{N}_{n+1}$ and with all the following.

We shall show that it is always the second case which occurs. If not, we could fix an element φ_n in each $\mathfrak{N}_n$ orthogonal to $\mathfrak{N}_{n+1}$ and of norm 1. Then we could select a convergent subsequence from $\{K\varphi_n\}$, which is impossible since

$$\|K\varphi_m - K\varphi_n\| \geq 1 \text{ for } m < n;$$

this inequality results from the decomposition

$$K\varphi_m - K\varphi_n = \varphi_m - (\varphi_n + T\varphi_m - T\varphi_n) = \varphi_m - \varphi$$

where φ is clearly an element of $\mathfrak{N}_{m+1}$ and consequently orthogonal to φ_m. Therefore

$$\|\varphi_m - \varphi\|^2 = \|\varphi_m\|^2 + \|\varphi\|^2 \geq \|\varphi_m\|^2 = 1.$$

Consequently there exists an index $\mu \geq 0$ such that

$$\mathfrak{N}_0 = L^2 \supset \mathfrak{N}_1 \supset \ldots \supset \mathfrak{N}_\mu = \mathfrak{N}_{\mu+1} = \mathfrak{N}_{\mu+2} = \ldots$$

3) In this manner we have assigned to the transformation the indices ν and μ; we shall show that *these two indices are equal.*

Before beginning the proof, let us remark that the equation $T^k f = g$, where k is an integer ≥ 1 and g is an element of $\mathfrak{N}_\mu$, possesses at least one solution f belonging to $\mathfrak{N}_\mu$; this is another way of saying that the image of $\mathfrak{N}_\mu$ under the transformation T^k, that is, $\mathfrak{N}_{\mu+k}$, is identical with $\mathfrak{N}_\mu$. We shall show that this solution is *unique*, or what amounts to the same thing, that the homogeneous equation $T^k\varphi = 0$ admits no solution belonging to $\mathfrak{N}_\mu$ other than $\varphi = 0$. It suffices to show this for $k = 1$. Let φ_1 be an arbitrary solution belonging to $\mathfrak{N}_\mu$; starting with φ_1, we define the sequence $\{\varphi_n\}$ of elements of $\mathfrak{N}_\mu$ by recurrence: φ_n is to be a solution of the equation $T\varphi = \varphi_{n-1}$ ($n = 2, 3, \ldots$). Then we have, for $n > 1$:

$$T^n\varphi_n = 0 \text{ and } T^{n-1}\varphi_n = \varphi_1.$$

It follows that if we had $\varphi_1 \neq 0$ then φ_n would belong to $\mathfrak{M}_n$ without belonging to $\mathfrak{M}_{n-1}$, which is impossible as soon as $n > \nu$. Hence $\varphi_1 = 0$.

With this established, we shall prove that $\mu \geq \nu$. Let f be an element of $\mathfrak{M}_{\mu+1}$, so that $T^{\mu+1}f = 0$. The equation $T\varphi = 0$ then admits the solution $\varphi = T^\mu f$ belonging to $\mathfrak{N}_\mu$, and it follows that $\varphi = 0$, hence that f also belongs to $\mathfrak{M}_\mu$. This proves that $\mathfrak{M}_{\mu+1} = \mathfrak{M}_\mu$, therefore $\mu \geq \nu$. In particular, $\mu = 0$ implies that $\nu = 0$.

Now consider the case where $\mu \geq 1$. Let $f = T^{\mu-1}g$ be an element of $\mathfrak{N}_{\mu-1}$ which does not belong to $\mathfrak{N}_\mu$. Since $h = T^\mu g$ belongs to $\mathfrak{N}_\mu$, the equation $T^\mu\varphi = h$ has one (and only one) solution $\varphi = g'$ in $\mathfrak{N}_\mu$. Then $T^\mu(g - g') = 0$ and therefore $g - g'$ is in $\mathfrak{M}_\mu$. On the other hand, $T^{\mu-1}(g - g') = f - T^{\mu-1}g' \neq 0$, since $T^{\mu-1}g'$ is contained in $\mathfrak{N}_\mu$ whereas f is not, hence $g - g'$ is not contained in $\mathfrak{M}_{\mu-1}$. This proves that $\mathfrak{M}_{\mu-1}$ and $\mathfrak{M}_\mu$ are distinct and, therefore that $\mu \leq \nu$.

Comparing the two results, we see that $\mu = \nu$, q.e.d.

4) The next step we propose to make is to prove that *every function f belonging to the space L^2 can be decomposed in a unique fashion into the sum $f = u + v$, where u belongs to $\mathfrak{M}_\nu$ and v belongs to $\mathfrak{N}_\nu$.*

We construct this decomposition in the following manner: if $g = T^\nu f$ is an element of $\mathfrak{N}_\nu$, the equation $T^{2\nu}\varphi = g$ certainly possesses a solution φ belonging to $\mathfrak{N}_\nu$. Let us set $f' = T^\nu\varphi$, then f' belongs to $\mathfrak{N}_\nu$ and we have

$$T^\nu f' = T^{2\nu}\varphi = g = T^\nu f;$$

hence $T^\nu(f - f') = 0$, that is, $f - f'$ belongs to $\mathfrak{M}_\nu$, and the decomposition $f = (f - f') + f'$ is of the required type. As for uniqueness, it obviously suffices to show that $\mathfrak{M}_\nu$ and $\mathfrak{N}_\nu$ have only the element 0 in common. But this is equivalent to the proposition, which was already proved, that the equation $T^\nu\varphi = 0$ has no non-zero solution in $\mathfrak{N}_\nu$.

We summarize our results:

THEOREM. *The set $\mathfrak{M}_n$ of functions f for which $T^n f = 0$ and the set $\mathfrak{N}_n$ of functions f which can be written in the form $f = T^n g$ are subspaces of the space L^2; $\mathfrak{M}_n$ is of finite dimension. There exists an index $\nu \geq 0$ such that*

$$\mathfrak{M}_0 = (0) \subset \mathfrak{M}_1 \subset \ldots \subset \mathfrak{M}_\nu = \mathfrak{M}_{\nu+1} = \ldots \text{ and } \mathfrak{N}_0 = L^2 \supset \mathfrak{N}_1 \supset \ldots \supset \mathfrak{N}_\nu = \mathfrak{N}_{\nu+1} = \ldots$$

Each element f of L^2 has an unique decomposition into the sum of an element u of $\mathfrak{M}_\nu$ and an element v of $\mathfrak{N}_\nu$: $f = u + v$. Whatever be the given function g belonging to $\mathfrak{N}_\nu$ and the integer $n \geq 1$, the equation $T^n f = g$ possesses one and only one solution f belonging to $\mathfrak{N}_\nu$.

78. The Cases $\nu = 0$ and $\nu \geqq 1$. The Decomposition Theorem

There are two essentially different cases according as the index ν of the preceding theorem does or does not equal zero.

a) The case $\nu = 0$. In this case, the equation

$$Tf = g$$

has a *unique* solution f for every function g belonging to L^2. Since transformation T is linear, the solution f will depend on g in an additive and homogeneous manner. Moreover this correspondence is bounded, that is, there exists a number C such that

$$\|f\| \leq C\,\|g\|$$

for every function g in L^2 and the corresponding solution f. In the contrary case there would be a sequence $\{g_n\}$ such that the ratio $\|f_n\| : \|g_n\|$ increases indefinitely with n. Let us set

$$h_n = \frac{f_n}{\|f_n\|}; \text{ then } \|h_n\| = 1 \text{ and } Th_n = \frac{g_n}{\|f_n\|} \to 0.$$

Since the transformation K is completely continuous, the sequence $\{Kh_n\}$

contains a convergent subsequence $\{Kh_{n_i}\}$; we denote its limit by h^*. Then

$$h_{n_i} = (I - K)h_{n_i} + Kh_{n_i} = Th_{n_i} + Kh_{n_i} \to h^*,$$

and consequently, by virtue of the continuity of T,

$$Th_{n_i} \to Th^*.$$

Since on the other hand $Th_n \to 0$, we necessarily have $Th^* = 0$, and since the equation $Tf = 0$ possesses a unique solution $f = 0$, we have $h^* = 0$. But this contradicts the fact that

$$\|h^*\| = \lim \|h_{n_i}\| = 1.$$

The correspondence between g and f is therefore additive, homogeneous, and bounded, that is, it is linear.

We summarize: *When $\nu = 0$ the transformation $T = I - K$ possesses an inverse, that is, $\lambda = 1$ is a regular value with respect to K.*

b) The case $\nu \geqq 1$. In this case the equation $Tf = 0$ possesses non-zero solutions; we have only to set $f = T^{\nu-1}\varphi$, where φ is an arbitrary element which belongs to $\mathfrak{M}_\nu$ without belonging to $\mathfrak{M}_{\nu-1}$. Consequently T can not have an inverse.

But if we restrict the transformation T to the subspace $\mathfrak{N}_\nu$ it does possess an inverse. In fact, for each g in $\mathfrak{N}_\nu$ the equation

$$Tf = g$$

possesses one and only one solution f in $\mathfrak{N}_\nu$, and we have only to repeat the argument just followed in the case $\nu = 0$, but restricting ourselves this time to the subspace $\mathfrak{N}_\nu$. It follows in particular that there exists a constant C such that

$$\|f\| \leq C \, \|Tf\|$$

for all elements f of $\mathfrak{N}_\nu$, and since all the iterates $T^n f$ also belong to $\mathfrak{N}_\nu$,

$$\|f\| \leq C^\nu \, \|T^\nu f\|.$$

Let us consider the decomposition $f = u + v$ of the arbitrary element f of the space L^2 into its components in $\mathfrak{M}_\nu$ and $\mathfrak{N}_\nu$. Since $T^\nu u = 0$, and hence $T^\nu f = Tv$, we have

$$\|v\| \leq C^\nu \, \|T^\nu v\| = C^\nu \, \|T^\nu f\| \leq C^\nu \, \|T^\nu\| \, \|f\|.$$

Since also $\|u\| = \|f - v\| \leq \|f\| + \|v\|$, we conclude that there exists a constant C_1 such that

$$\|u\| \leq C_1 \, \|f\|, \quad \|v\| \leq C_1 \, \|f\|.$$

Since the components u and v are uniquely determined by f, the correspondence between f and u and the correspondence between f and v are obviously additive and homogeneous. Therefore if we set

$$Pf = u, \quad Qf = v,$$

we define two linear transformations P and Q such that $P + Q = I$; P leaves the elements of $\mathfrak{M}_\nu$ invariant and annihilates the elements of $\mathfrak{N}_\nu$, whereas for Q the situation is the reverse. This implies that $PQ = QP = O$. We can therefore say that P is *the projection of the space L^2 onto $\mathfrak{M}_\nu$ in the direction $\mathfrak{N}_\nu$* and that Q is *the projection of the space L^2 onto $\mathfrak{N}_\nu$ in the direction $\mathfrak{M}_\nu$*. Since each of the subspaces $\mathfrak{M}_\nu$, $\mathfrak{N}_\nu$ is transformed into itself by T, and therefore also by $K = I - T$, these projections are permutable with the transformation K:

$$PK = KP, \quad QK = KQ.$$

Set $S = PK$ and $R = QK$; then we have

$$S + R = (P + Q)K = K,$$

$$SR = (KP)(QK) = K(PQ)K = O, \quad RS = (KQ)(PK) = K(QP)K = O$$

and thus

$$I - K = (I - S)(I - R) = (I - R)(I - S).$$

We shall show that the index ν_R corresponding to the transformation R (which obviously is also completely continuous) is equal to 0, and that consequently $(I - R)^{-1}$ exists. In fact, in the contrary case $\nu_R \geq 1$, the equation $(I - R)f = 0$ would possess a non-zero solution f; by virtue of the equation $f = Rf$, f would belong to $\mathfrak{N}_\nu$. Then we should have

$$(I - K)f = (I - S)(I - R)f = 0,$$

which is impossible because the equation $(I - K)\varphi = 0$ has no solutions in $\mathfrak{N}_\nu$ other than $\varphi = 0$.

Since S and R are permutable, $SR = RS = 0$, we have

$$(I - K)^n = (I - S)^n (I - R)^n = (I - R)^n (I - S)^n$$

for $n = 0, 1, 2, \ldots$, and consequently

$$(I - S)^n = (I - K)^n (I - R)^{-n} = (I - R)^{-n} (I - K)^n,$$

where $(I - R)^{-n}$ denotes the n-th iterate of $(I - R)^{-1}$. These relations imply, in an obvious way, that *the subspaces $\mathfrak{M}_n$ and $\mathfrak{N}_n$ relative to the transformation S coincide with those relative to K.*

We know that the subspace $\mathfrak{M}_\nu$ is of finite dimension, say of dimension r. Its elements can therefore be represented as linear combinations of r linearly independent elements, $\varphi_1, \varphi_2, \ldots, \varphi_r$, which we can assume to be orthonormal. Since Sf belongs to $\mathfrak{M}_\nu$ for every element f, we have

$$Sf = c_1\varphi_1 + c_2\varphi_2 + \ldots + c_r\varphi_r.$$

Each coefficient

$$c_k = (Sf, \varphi_k)$$

is a linear functional in f; hence there are elements ψ_k of L^2 such that $c_k = (f, \psi_k)$ (see Sec. 30). The transformation S is therefore of finite rank.

We summarize:

In the case $\nu \geq 1$, the value $\lambda = 1$ is singular with respect to K; $\mathfrak{M}_\nu$ and $\mathfrak{N}_\nu$ are proper subspaces of L^2 and transform into themselves under K. Denoting by S and R the linear transformations which coincide with K in $\mathfrak{M}_\nu$ and $\mathfrak{N}_\nu$, respectively, and which are zero in $\mathfrak{N}_\nu$ and in $\mathfrak{M}_\nu$, respectively, we have

$$K = S + R, \quad SR = RS = O,$$

and the value $\lambda = 1$ is regular with respect to R and singular with respect to S; moreover, S is of finite rank.

Since S and R are generated by kernels, namely by

$$S(x, y) = \varphi_1(x)\overline{\psi_1(y)} + \ldots + \varphi_r(x)\overline{\psi_r(y)}$$

and

$$R(x, y) = K(x, y) - S(x, y),$$

we have arrived by an entirely different way at the decomposition theorem of Sec. 72, although with properties 3 and 4 omitted.

However, we did not use these properties when we used the theorem, in Sec. 73, to deduce *the Fredholm alternative*. Hence we have obtained a new proof.

Besides, properties 3 and 4 can also be established without difficulty. Since we are now dealing with the singular value $\lambda = 1$, we need consider only the values $\lambda \neq 1$.

All these values are *regular* with respect to the transformation of finite rank S. If this were not so, the homogeneous equation

$$(I - \lambda S)f = 0$$

would admit a solution $f \neq 0$; since

$$f = \lambda S f = \lambda P K f,$$

f would be an element of the subspace $\mathfrak{M}_\nu$. Consequently we should have

$$f = \lambda P K f = \lambda K P f = \lambda K f,$$

therefore

$$\frac{\lambda}{\lambda - 1}(I - K)f = f,$$

and hence also

$$\left(\frac{\lambda}{\lambda - 1}\right)^\nu (I - K)^\nu f = f \neq 0,$$

contrary to the fact that f belongs to $\mathfrak{M}_\nu$. This contradiction proves our assertion.

We must still show property 4, that is, that a value $\lambda \neq 1$ is regular with respect to R if and only if it is regular with respect to K. But this follows immediately from the relations

$$(I - \lambda S)(I - \lambda R) = (I - \lambda R)(I - \lambda S) = I - \lambda K,$$

where we have made use of the already established fact that $(I - \lambda S)^{-1}$ exists.

79. The Distribution of the Singular Values

We have seen in section 71 that the singular values of the transformation K are either finite or denumerable in number and have no finite limit point. We give now a new proof of this fact which is based on the complete continuity of the transformation K.

If there were a limit point we could find a bounded infinite sequence of singular values. Suppose that $\{\lambda_n\}$ is such a sequence. Each of the homogeneous equations

$$(I - \lambda_n K)\varphi = 0$$

then possesses non-zero solutions; let $\varphi_n \neq 0$ be a solution corresponding to the singular value λ_n. We thus obtain an infinite sequence of functions $\{\varphi_n\}$ and we shall show that for every m the functions $\varphi_1, \varphi_2, \ldots, \varphi_m$ are linearly independent.

In fact, if this were not true there would be a first function in the sequence $\{\varphi_n\}$ which depends linearly on the preceding, say

$$\varphi_m = c_1\varphi_1 + c_2\varphi_2 + \ldots + c_{m-1}\varphi_{m-1}. \tag{45}$$

Then we should have

$$\varphi_m = \lambda_m K\varphi_m = \lambda_m(c_1 K\varphi_1 + \ldots + c_{m-1}K\varphi_{m-1}) = \lambda_m\left(\frac{c_1}{\lambda_1}\varphi_1 + \ldots + \frac{c_{m-1}}{\lambda_{m-1}}\varphi_{m-1}\right);$$

subtracting this equation from (45), it follows that

$$\left(1 - \frac{\lambda_m}{\lambda_1}\right)c_1\varphi_1 + \ldots + \left(1 - \frac{\lambda_m}{\lambda_{m-1}}\right)c_{m-1}\varphi_{m-1} = 0.$$

But the functions $\varphi_1, \ldots, \varphi_{m-1}$ are by hypothesis linearly independent, hence we necessarily have $c_1 = \ldots = c_{m-1} = 0$, $\varphi_m = 0$, contrary to the hypothesis that $\varphi_m \neq 0$.

Let us denote by E_m the set of all the linear combinations of $\varphi_1, \varphi_2, \ldots, \varphi_m$; the infinite sequence of sets $E_1, E_2, \ldots$ is then strictly increasing. Let $\{g_n\}$ be the orthonormal sequence which we obtain from $\{\varphi_n\}$ by the Schmidt process (see Sec. 33),

$$g_n = a_{n1}\varphi_1 + a_{n2}\varphi_2 + \ldots + a_{nn}\varphi_n.$$

E_m is then the set of all the linear combinations of $g_1, g_2, \ldots, g_m$.

Since the sequence $\{\lambda_n g_n\}$ is bounded in norm, the transform $\{\lambda_n K g_n\}$ contains a subsequence which converges in the mean. But this is impossible, because

$$\|\lambda_n K g_n - \lambda_m K g_m\| \geq 1$$

for $n > m$; this inequality results from the decomposition

$$\lambda_n K g_n - \lambda_m K g_m = g_n - g$$

where

$$g = g_n - \lambda_n K g_n + \lambda_m K g_m = \sum_{i=1}^{n-1} \left(1 - \frac{\lambda_n}{\lambda_i}\right) a_{ni} \varphi_i + \lambda_m \sum_{i=1}^{m} \frac{a_{mi}}{\lambda_i} \varphi_i;$$

in fact, g is an element of E_{n-1}, hence orthogonal to g_n, and consequently

$$\|g_n - g\|^2 = \|g_n\|^2 + \|g\|^2 \geq \|g_n\|^2 = 1.$$

We have thus arrived at a contradiction; the singular values are therefore finite in number in each bounded domain of the complex plane, and there are no finite limit points.

80. The Canonical Decomposition Corresponding to a Singular Value

If $E_1, E_2, \ldots, E_m$ are linear sets in L^2, we denote by

$$E_1 + E_2 + \ldots + E_m$$

the set of sums

$$h = f_1 + f_2 + \ldots + f_m$$

of elements selected respectively from $E_1, E_2, \ldots, E_m$, and call it the *vector sum* of $E_1, E_2, \ldots, E_m$; clearly it too is a linear set. We say that $E_1, E_2, \ldots, E_m$ are *linearly independent* if the sum h is not zero unless each term is 0. Each sum h then uniquely determines its terms f_i.

In particular, let $\mathfrak{M}_\nu$ and $\mathfrak{N}_\nu$ be subspaces corresponding to the singular value 1 of the transformation $\lambda_0 K$, that is, to the singular value λ_0 of the transformation K. By what we proved in Section 77, $\mathfrak{M}_\nu$ and $\mathfrak{N}_\nu$ are linearly independent and $\mathfrak{M}_\nu + \mathfrak{N}_\nu = L^2$.

Since $\mathfrak{M}_{\nu-1}$ is properly contained in $\mathfrak{M}_\nu$, we can find a subspace $\mathfrak{Q}_\nu$ with dimension ≥ 1 such that $\mathfrak{M}_{\nu-1}$ and $\mathfrak{Q}_\nu$ are linearly independent, and $\mathfrak{M}_\nu = \mathfrak{Q}_\nu + \mathfrak{M}_{\nu-1}$; we have only to take for $\mathfrak{Q}_\nu$, for example, the set of elements of $\mathfrak{M}_\nu$ which are orthogonal to $\mathfrak{M}_{\nu-1}$. The image of $\mathfrak{Q}_\nu$ under the transformation $T = I - \lambda_0 K$, which we denote by $T\mathfrak{Q}_\nu$, is then contained in $\mathfrak{M}_{\nu-1}$, but it has only the element 0 in common with $\mathfrak{M}_{\nu-2}$. Since $T\mathfrak{Q}_\nu$ is a linear set which is contained in the finite dimensional subspace $\mathfrak{M}_{\nu-1}$, it is closed and therefore is itself a subspace (cf. Sec. 34); the same holds for $T\mathfrak{Q}_\nu + \mathfrak{M}_{\nu-2}$. Since the latter is contained in $\mathfrak{M}_{\nu-1}$, we can find a subspace $\mathfrak{Q}_{\nu-1}$ such that $\mathfrak{Q}_{\nu-1} + T\mathfrak{Q}_\nu + \mathfrak{M}_{\nu-2} = \mathfrak{M}_{\nu-1}$ and that $\mathfrak{Q}_{\nu-1}$, $T\mathfrak{Q}_\nu$, and $\mathfrak{M}_{\nu-2}$ are linearly independent; it might happen that $\mathfrak{Q}_{\nu-1}$ consists of the single element 0. We successively construct the decompositions

$$\mathfrak{M}_\nu = \mathfrak{Q}_\nu + \mathfrak{M}_{\nu-1},$$

$$\mathfrak{M}_{\nu-1} = \mathfrak{Q}_{\nu-1} + T\mathfrak{Q}_\nu + \mathfrak{M}_{\nu-2},$$

$$\mathfrak{M}_{\nu-2} = \mathfrak{Q}_{\nu-2} + T\mathfrak{Q}_{\nu-1} + T^2\mathfrak{Q}_\nu + \mathfrak{M}_{\nu-3}, \text{ etc.},$$

always with linearly independent terms. From them we obtain a decomposition of $\mathfrak{M}_\nu$ into linearly independent terms:

$$\mathfrak{M}_\nu = \sum_{n=0}^{\nu-1} T^n \mathfrak{Q}_\nu + \sum_{n=0}^{\nu-2} T^n \mathfrak{Q}_{\nu-1} + \ldots + \sum_{n=0}^{1} T^n \mathfrak{Q}_2 + \mathfrak{Q}_1;$$

the $\mathfrak{Q}_\mu$ are subspaces of finite dimension which except for $\mathfrak{Q}_\nu$ may reduce to the single element 0.

When $\mathfrak{Q}_\mu \neq (0)$, we choose a system of linearly independent functions $\{\varphi_{\mu i}^{(0)}\}$ $(i = 1, 2, \ldots, r)$ in $\mathfrak{Q}_\mu$ which determine $\mathfrak{Q}_\mu$. We set

$$\varphi_{\mu i}^{(\varkappa)} = \left(-\frac{1}{\lambda_0} T\right)^\varkappa \varphi_{\mu i}^{(0)} \text{ for } i = 1, 2, \ldots, r_\mu;\ \varkappa = 1, 2, \ldots, \mu - 1.$$

The functions $\varphi_{\mu i}^{(\varkappa)}$ $(i = 1, 2, \ldots, r_\mu)$ obviously determine $T^\varkappa\mathfrak{Q}_\mu$; moreover, they are linearly independent. In fact, if

$$\sum_i c_i \varphi_{\mu i}^{(\varkappa)} = 0,$$

that is, if $T^\varkappa \psi = 0$, where

$$\psi = \sum_i c_i \varphi_{\mu i}^{(0)},$$

ψ belongs to $\mathfrak{M}_\varkappa$ and hence necessarily to $\mathfrak{M}_{\mu-1}$, and since ψ also belongs to $\mathfrak{Q}_\mu$, we must have $\psi = 0$; therefore $c_1 = c_2 = \ldots = c_{r_\mu} = 0$. It follows that *all* the functions

$$\varphi_{\mu i}^{(\varkappa)} \quad (\mu = 1, 2, \ldots, \nu;\ i = 1, 2, \ldots, r_\mu;\ \varkappa = 1, 2, \ldots, \mu - 1)$$

are linearly independent and that they determine the subspace $\mathfrak{M}_\mu$.

We obviously have

$$\varphi_{\mu i}^{(\varkappa+1)} = -\frac{1}{\lambda_0} T\varphi_{\mu i}^{(\varkappa)} = \left(K - \frac{1}{\lambda_0} I\right) \varphi_{\mu i}^{(\varkappa)} \quad \text{for} \quad \varkappa = 0, 1, \ldots, \mu - 2$$

and

$$T\varphi_{\mu i}^{(\mu-1)} = (I - \lambda_0 K)\varphi_{\mu i}^{(\mu-1)} = 0,$$

since $\varphi_{\mu i}^{(\mu-1)}$ belongs to $\mathfrak{M}_1$. Therefore

$$(46)\qquad \begin{aligned} K\varphi_{\mu i}^{(\varkappa)} &= \frac{1}{\lambda_0}\,\varphi_{\mu i}^{(\varkappa)} + \varphi_{\mu i}^{(\varkappa+1)} \quad \text{for} \quad \varkappa = 0, 1, \ldots, \mu-2,\\ K\varphi_{\mu i}^{(\mu-1)} &= \frac{1}{\lambda_0}\,\varphi_{\mu i}^{(\mu-1)}. \end{aligned}$$

Let $\mathfrak{L}_{\mu i}$ be the subspace determined by $\varphi_{\mu i}^{(0)}, \varphi_{\mu i}^{(1)}, \ldots, \varphi_{\mu i}^{(\mu-1)}$; we have $\mathfrak{M}_\nu = \sum_{\mu, i} \mathfrak{L}_{\mu i}$. It follows from (46) that each $\mathfrak{L}_{\mu i}$ is transformed by K into itself.

We summarize our results:

THEOREM. *When λ_0 is a singular value with respect to the transformation K, the space L^2 can be decomposed into the vector sum of linearly independent subspaces $\mathfrak{L}_1, \mathfrak{L}_2, \ldots, \mathfrak{L}_p, \mathfrak{N}$, each of which is transformed by K into itself; this can be done in such a way that $I - \lambda_0 K$ possesses an inverse in $\mathfrak{N}$, and that in each of the subspaces $\mathfrak{L}$ one can choose a complete finite system of linearly independent elements, say $\varphi_1, \varphi_2, \ldots, \varphi_\varrho$, such that:*

$$K\varphi_\varkappa = \frac{1}{\lambda_0}\,\varphi_\varkappa + \varphi_{\varkappa+1} \quad (\varkappa = 1, 2, \ldots, \varrho - 1) \textit{ and } K\varphi_\varrho = \frac{1}{\lambda_0}\,\varphi_\varrho.$$

APPLICATIONS TO POTENTIAL THEORY

81. The Dirichlet and Neumann Problems. Solution by Fredholm's Method[16]

Among the various applications of integral equations to problems of mathematics and physics, we find it appropriate to mention particularly two problems from potential theory which bear the names of DIRICHLET and NEUMANN. Indeed, it is to the study of these problems that we owe the rapid development of the theory of integral equations by C. NEUMANN, H. POINCARÉ, and I. FREDHOLM.

In the Dirichlet problem one seeks a harmonic function on a domain which is continuous on the domain and its boundary, and which reduces to a given continuous function on the boundary. Neumann's problem is an analogous problem in which instead of prescribing the value of the solution on the boundary, we prescribe the value of its normal derivative there.

Let us limit ourselves to the case of a plane domain which is bounded by a simple closed curve C with continuous curvature. We speak of the *interior problem* or of the *exterior problem* according as the domain under consideration

[16] FREDHOLM [**1**]. We refer the reader to the book by GOURSAT [*] for a detailed discussion.

is equal to the interior D_i or to the exterior D_e of C. The points of C are identified by the arc length s calculated from a fixed point, $0 \leq s \leq S$.

1) In the interior Dirichlet problem we seek a function $u(P)$ which is harmonic in D_i and whose limit $u_i(s)$, when the point P approaches the point s on C while remaining in D_i, is equal to the given continuous function $g(s)$ on C:

$$u_i(s) = g(s). \tag{47}$$

Following a classical method due to C. NEUMANN, we try to find the function u in the form

$$u(P) = \int_C \mu(t) \frac{\partial}{\partial n_t} \log \frac{1}{r_{Pt}} dt = \int_C \mu(t) \frac{\cos(r_{Pt}, n_t)}{r_{Pt}} dt, \tag{48}$$

that is, as the potential of a double layer $\mu(t)$ distributed over C; here r_{Pt} denotes the distance from the point P to the point t of C and n_t denotes the interior normal at the point t of C. When this double layer is a continuous function, we know that the potential (48) is harmonic in D_i and in D_e, and that it is discontinuous when we cross C; the value of u at a point s of C, the interior limit u_i, and the exterior limit u_e are related to one another by the following equations:

$$u_i(s) = u(s) + \pi\mu(s), \quad u_e(s) = u(s) - \pi\mu(s). \tag{49}$$

On the other hand, the normal derivative is continuous:

$$\left(\frac{\partial u}{\partial n}\right)_i = \left(\frac{\partial u}{\partial n}\right)_e.$$

Therefore the function μ must satisfy the integral equation

$$\frac{1}{\pi} g(s) = \mu(s) + \int_C K(s, t)\mu(t)dt, \tag{50}$$

where

$$K(s, t) = \frac{1}{\pi} \frac{\partial}{\partial n_t} \log \frac{1}{r_{st}} = \frac{1}{\pi} \frac{\cos(r_{st}, n_t)}{r_{st}}.$$

The kernel $K(s, t)$ is continuous not only for $s \neq t$, but also for $s = t$. In fact, if we denote the rectangular coordinates of the point s on C by $x(s)$, $y(s)$, these functions possess continuous derivatives of the first *two* orders by hypothesis and we have

$$K(s, t) = \frac{[y(s)-y(t)]x'(t)-[x(s)-x(t)]y'(t)}{[x(s)-x(t)]^2+[y(s)-y(t)]^2} \to \frac{y''(s_0)x'(s_0)-x''(s_0)y'(s_0)}{2} = k(s_0)$$

when s and t tend to the same limit s_0; $k(s_0)$ is equal to the curvature of C at the point s_0.

Therefore we can apply the Fredholm theory. It follows that either *the non-homogeneous equation* (50) possesses a (necessarily continuous) solution $\mu(s)$, whatever be the given continuous function $g(s)$, or else *the homogeneous equation*

(51) $$\nu(s) + \int_C K(s, t)\nu(t)dt = 0$$

possesses a (necessarily continuous) solution $\nu(s) \not\equiv 0$. But the existence of such a solution $\nu(s) \not\equiv 0$ can be excluded: It follows from (47), (50) and (51) that for the potential $v(P)$ corresponding to the double layer $\nu(s)$ we have $v_i(s) \equiv 0$, which implies that $v(P) \equiv 0$ in D_i, because a function harmonic in a domain takes on its extremal values on the boundary. Therefore $\left(\frac{\partial v}{\partial n}\right)_i \equiv 0$, and consequently $\left(\frac{\partial v}{\partial n}\right)_e \equiv 0$. Since the potential v is harmonic in D_e, including the point at infinity, we can apply Green's formula

(52) $$\iint_{D_e} (v_x^2 + v_y^2)dx\,dy = -\int_C v_e\left(\frac{\partial v}{\partial n}\right)_e dt,$$

which implies that $v_x \equiv 0$, $v_y \equiv 0$ in D_e; therefore v is constant in D_e. Since v is zero at infinity, this constant value is nessarily 0. It follows that $v_e(s) \equiv 0$, and also, by (49), that

$$\nu(s) = \frac{1}{2\pi}(v_i(s) - v_e(s)) \equiv 0,$$

that is, that the homogeneous equation possesses the single solution $\nu(s) \equiv 0$. Therefore:

The interior Dirichlet problem possesses a solution for every continuous function $g(s)$ given on the boundary.

The exterior Dirichlet problem leads in the same manner to the equation

(53) $$\frac{1}{\pi}g(s) = \mu(s) - \int_C K(s, t)\mu(t)dt.$$

We see by reasoning analogous to that above that the homogeneous equation possesses only constant solutions, but these are not necessarily zero, as is seen by observing that

$$\int_C K(s, t)dt = 1.$$

The number of linearly independent solutions is therefore equal to 1. The same is then true for the adjoint equation

$$\varrho(s) - \int_C K(t, s)\varrho(t)dt = 0,$$

that is, the adjoint equation possesses a solution $\varrho_0(s) \not\equiv 0$, and all the other solutions are numerical multiples of $\varrho_0(s)$. Therefore, in order that equation (53) possess a solution, it is necessary and sufficient that we have

$$\int_C g(s)\varrho_0(s)ds = 0.$$

We now determine the constant c so that $g_1(s) = g(s) - c$ is orthogonal to $\varrho_0(s)$; denoting by u_1 the potential of the double layer which by equation (53) corresponds to g_1, the exterior Dirichlet problem has $u = u_1 + c$ for solution. Therefore:

The exterior Dirichlet problem possesses a solution for every continuous function $g(s)$ given on the boundary.

2) For *the Neumann problem,* we express the solution as the potential of a continuous, single layer density $\varrho(s)$,

$$u(P) = \int_C \varrho(t) \log \frac{1}{r_{Pt}} dt.$$

This potential is harmonic in D_i and in D_e, it is even continuous on the boundary C, but its normal derivative has discontinuities there; we have:

$$u_i = u_e,$$

$$\left(\frac{\partial u}{\partial n_s}\right)_i + \pi\varrho(s) = \left(\frac{\partial u}{\partial n_s}\right)_e - \pi\varrho(s) = \int_C \varrho(t)\left(\frac{\partial}{\partial n_s} \log \frac{1}{r_{st}}\right) dt.$$

The interior Neumann problem, expressed by the equation

$$\left(\frac{\partial u}{\partial n_s}\right)_i = h(s),$$

therefore leads to the integral equation

$$(54) \qquad -\frac{1}{\pi} h(s) = \varrho(s) - \int_C \varrho(t)K(t, s)dt$$

whose kernel $K(t, s)$ is the adjoint of the one we have just encountered in the course of the solution of the Dirichlet problem. Now we have already remarked that the homogeneous equation

$$\mu(s) - \int_C K(s, t)\mu(t)dt = 0$$

possesses only constant solutions, hence equation (54) possesses a solution if and only if $h(s)$ is orthogonal to 1. Therefore:

The interior Neumann problem possesses a solution for every continuous function $h(s)$ such that

$$\int_C h(s)ds = 0.$$

Moreover, this condition is also necessary, since we have from Green's formula

$$\int_C \left(\frac{\partial u}{\partial n}\right)_i dt = \iint_{D_i} \Delta u \, dx \, dy = 0.$$

The exterior Neumann problem also leads to equation (54), but with the $+$ sign instead of $-$ in the second member. The corresponding homogeneous equation has no non-zero solutions since, as we saw above, the adjoint homogeneous equation has none. Therefore:

The exterior Neumann problem has a solution for every given continuous function $h(s)$.

The Dirichlet and Neumann problems are only special cases of the "third" limit problem, in which the limit values of a certain linear combination $hu + k\frac{\partial u}{\partial n}$ are prescribed, where the coefficients h and k are given continuous functions on C. In attempting to find the solution in the form of the sum of the potentials of a single and a double layer, we arrive at an integral equation analogous to those considered above.

The same method can also be applied to analogous problems in space or even in spaces of a higher dimension. The only difference is that in these cases the kernel will not in general be bounded. But one can show that a certain iterate of the kernel is bounded, which ensures that the Fredholm theorems apply.[17]

[17] Cf. Sec. 151.

CHAPTER V

HILBERT AND BANACH SPACES

HILBERT SPACE

82. Hilbert Coordinate Space

Integral equations of the type considered in the preceding chapter can be reduced to systems of linear equations in an infinite number of unknowns[1] by means of complete orthonormal sequences. In fact, let $f - Kf = g$ be an integral equation in L^2 whose kernel $K(x, y)$ belongs to $\mathbf{L}^2$, and let $\{\varphi_n\}$ be a complete orthonormal sequence in L^2. Let

$$\sum_k x_k \varphi_k, \quad \sum_i y_i \varphi_i, \quad \sum_i K_{ik} \varphi_i$$

be the developments of the elements f, g, $K\varphi_k$ in terms of $\{\varphi_n\}$. Then we have

$$Kf = K \sum_k x_k \varphi_k = \sum_k x_k K\varphi_k;$$

hence

$$(Kf, \varphi_i) = \sum_k x_k (K\varphi_k, \varphi_i) = \sum_k x_k K_{ik}$$

and consequently

$$f - Kf = \sum_i (x_i - \sum_k K_{ik} x_k) \varphi_i.$$

Comparing the coefficients with those of the development of g, we obtain the infinite system of equations

$$x_i - \sum_k K_{ik} x_k = y_i \qquad (i = 1, 2, \ldots). \tag{1}$$

Here the given quantities y_i form a sequence such that $\sum |y_i|^2$ converges; in fact, by Parseval's formula (Sec. 33) this series has the sum $\|g\|^2$. For the same reason the desired solution $\{x_i\}$ must be a sequence such that $\sum |x_i|^2$ converges.

We shall call a sequence $\{a_k\}$ for which the series $\sum |a_k|^2$ converges an element of *Hilbert coordinate space*, or *the space* l^2. In this space, which is the

[1] HILBERT [*] (Notes 4 and 5); E. SCHMIDT [**3**]; F. RIESZ [*] (Chapters III–IV).

immediate generalization of the complex vector space of n coordinates, the fundamental operations are defined in the following manner·

$$\{a_k\} + \{b_k\} = \{a_k + b_k\}, \qquad c\{a_k\} = \{ca_k\},$$

$$(\{a_k\}, \{b_k\}) = \sum_k a_k \overline{b_k}, \qquad \|\{a_k\}\| = [\sum_k |a_k|^2]^{\frac{1}{2}}.$$

The complete orthonormal sequence $\{\varphi_n\}$ establishes a one-to-one correspondence between the elements f of L^2 and the elements $\{a_k\}$ of l^2, by means of the formula $f = \sum a_k \varphi_k$ which expresses that the a_k are the "coordinates of f in the orthonormal system of the φ_k" (cf. Sec. 33). This correspondence preserves *linear* structure, that is, to linear combinations of the f correspond linear combinations of the $\{a_k\}$, and it also preserves *metric* structure, since by virtue of the Parseval formulas, scalar products and norms of the elements of L^2 are equal to those of the corresponding elements in l^2.

It is clear that everything which has just been said also holds for the space L^2 of functions defined on an arbitrary measurable domain D which lies on a straight line or in a space of several dimensions; the integral can be either the Lebesgue integral or the Stieltjes-Lebesgue integral taken with respect to a distribution of positive masses. Since all these L^2 spaces are separable, we can construct in each a complete orthonormal sequence which is denumerably infinite provided that the masses are not concentrated in D at a finite number of points. Except for this obvious case, every space L^2 is therefore of the same linear and metric structure as the space l^2. Considered from this general point of view, the space l^2 even appears as a particular case of an L^2 space, namely, that corresponding to a domain D which consists of an infinite sequence of points, each supporting a mass of 1.

We return to our integral equation or rather to the system of linear equations with an infinite number of unknowns (47) which we derived from it. The double series

$$\sum_{i,k} |K_{ik}|^2, \tag{2}$$

formed with the coefficients of this system, is convergent. In fact, denoting the product $\varphi_i(x)\overline{\varphi_k(y)}$ by $\Phi_{ik}(x, y)$, which is an element of $\boldsymbol{L}^2$, we have

$$K_{ik} = (K\varphi_k, \varphi_i) = (K, \Phi_{ik}),$$

and since these products obviously form an orthonormal system in $\boldsymbol{L}^2$, we have by Bessel's inequality (cf. Sec. 33):

$$\sum_{i,k} |(K, \Phi_{ik})|^2 \leq \|K\|^2.$$

Conversely, every infinite system of linear equations of type (1) for which series (2) converges generates an integral equation in L^2 given by

$$g = \sum_i y_i \varphi_i \text{ and } K(x, y) = \sum_{i,k} K_{ik} \Phi_{ik}(x, y),$$

where these series converge in the mean in L^2 and in $\mathbf{L}^2$ respectively.

In passing from the integral equation to the infinite system of linear equations, a new method of solution is suggested. We replace the infinite system (1) by the system reduced to n unknowns

$$\text{(3)} \qquad x_i - \sum_{k=1}^{n} K_{ik} x_k = y_i \qquad (i = 1, 2, \ldots, n);$$

denoting its solution, when it exists, by $(x_1^{(n)}, \ldots, x_n^{(n)})$, we expect that it converges, for $n \to \infty$, to a solution of the infinite system (1). As HILBERT showed, this method applies successfully not only to the case where series (2) converges, but also to more general infinite matrices (K_{ik}), namely, to all those which correspond to completely continuous transformations of the space l^2. We shall not go into this method, but we shall return in a moment to the problem of these transformations.

83. Abstract Hilbert Space

The considerations of the preceding section lead us to recognize that the spaces L^2 and l^2 have the same linear and metric structure, and that consequently they can be considered as realizations of the same abstract space, which is defined by precisely these common properties: linearity, existence of the scalar product and the norm, and finally the validity of the Cauchy criterion for mean convergence (the Riesz-Fischer theorem in the space L^2).

Hence here is the definition:

A set of abstract elements $f, g, h, \ldots$ which possesses the following properties **A**, **B** and **C** *will be called an abstract Hilbert space and will be denoted by* $\mathfrak{H}$:[2]

A. $\mathfrak{H}$ is *a linear space*, that is, the operations of addition and of multiplication by real or complex numbers are defined for its elements, and these operations obey the usual rules of the algebra of vectors; in particular, there exists an element 0 which is equal to $0 \cdot f$ for all elements f of $\mathfrak{H}$.

B. $\mathfrak{H}$ is *a metric space whose metric is derived from a scalar product.* This means that to every pair of elements f, g there is associated a real or complex number, called their scalar product and denoted by (f, g), in such a way

[2] Cf. J. v. NEUMANN [**1**]; this author states two more axioms requiring the space to be separable and of infinite dimension (hence, of denumerably infinite dimension). We prefer not to exclude in the beginning spaces of finite dimension and non-separable spaces. For non-separable spaces, see LÖWIG [**1**], RIESZ [**15**], and RELLICH [**1**].

that the following rules are satisfied:

$$(af, g) = a(f, g) \text{ for every number } a; \quad (f + g, h) = (f, h) + (g, h);$$

$$(f, g) = \overline{(g, f)}; \quad (f, f) > 0 \text{ for } f \neq 0; \quad (f, f) = 0 \text{ for } f = 0.$$

The *norm* of an element f is then defined by

$$\|f\| = (f, f)^{\frac{1}{2}},$$

and the *distance* between the elements f and g by

$$\|f - g\|.$$

C. $\mathfrak{H}$ is a *complete space* in the sense that if a sequence of elements $\{f_n\}$ of $\mathfrak{H}$ satisfies the condition $\|f_n - f_m\| \to 0$ for $m, n \to \infty$, then there exists an element f^* of $\mathfrak{H}$ such that $\|f_n - f^*\| \to 0$ for $n \to \infty$.

We remark that we could have postulated only *real* numerical multipliers and have required, at the same time, that the scalar products possess only real values; the result would have been *real* Hilbert space, for which the space L^2 of functions with real values is a realization. Our considerations on integral equations, in particular the analytic properties of the resolvent transformation, show that it is preferable to consider complex spaces; furthermore, most of the results obtained hold, with evident modifications, for real spaces.

Just as in the concrete space L^2, we can define two types of convergence in the space $\mathfrak{H}$, *strong* and *weak*. The sequence of elements $\{f_n\}$ converges to the element f^* *strongly* if

$$\|f_n - f^*\| \to 0,$$

and *weakly* if

$$(f_n, g) \to (f, g)$$

for each element g of $\mathfrak{H}$. We shall denote *strong* convergence by the sign "$\to$" and *weak* convergence by the sign "$\rightharpoonup$". In order to abbreviate, we shall often omit the adjective "strong" by saying "convergence" instead of "strong convergence"; this notion of convergence is the abstraction of mean convergence in the space L^2.

Let f and g be two arbitrary elements of $\mathfrak{H}$. We have, for all real λ,

$$0 \leq (f + \lambda(f, g)g, f + \lambda(f, g)g) = (f, f) + 2\lambda.|(f, g)|^2 + \lambda^2\,|(f, g)|^2\,(g, g);$$

therefore this polynomial in λ, which is of second degree and has real coefficients, can not have two real distinct zeros, which implies that

$$|(f, g)|^4 - (f, f)\,|(f, g)|^2\,(g, g) \leq 0;$$

hence (even if $(f, g) = 0$)

$$|(f, g)|^2 \leq (f, f)\,(g, g)$$

or

$$|(f, g)| \leq \|f\|\,\|g\|.$$

This is *Schwarz's inequality*, which is therefore valid in the abstract space $\mathfrak{H}$; we also see that, apart from the evident cases $f = 0$ or $g = 0$, the equality sign holds only if $f = -\lambda(f, g)g$ for some value of λ, that is, only if f is a numerical multiple of g.

The Minkowski inequality

$$\|f + g\| \leqq \|f\| + \|g\|$$

follows from that of Schwarz just as in the space L^2 (Sec. 21); it also follows that

$$\|f_1 - f_3\| \leqq \|f_1 - f_2\| + \|f_2 - f_3\|,$$

that is, the notion of distance which we introduced satisfies *the triangle inequality*.

An immediate consequence of Schwarz's inequality is that the scalar product is a continuous function of its two factors, that is, that

$$f_n \to f, \quad g_n \to g \quad \text{imply} \quad (f_n, g_n) \to (f, g);$$

in fact,

$$|(f_n, g_n) - (f, g)| = |(f_n - f, g_n - g) + (f_n - f, g) + (f, g_n - g)| \leqq$$
$$\leqq \|f_n - f\| \, \|g_n - g\| + \|f_n - f\| \, \|g\| + \|f\| \, \|g_n - g\| \to 0.$$

It follows, in particular, that strong convergence implies weak convergence.

In addition, such concepts as orthogonality, subspaces, linear functionals, etc., as well as the fundamental theorems proved in sections 30 to 35 for the space L^2, extend without modification to the case of the abstract space $\mathfrak{H}$. The only exception concerns the separability of the space L^2 (Sec. 32); furthermore, the decomposition theorem and the theorem of choice could have been proved without recourse to this property of L^2, and consequently they too remain valid in the space $\mathfrak{H}$.

The *dimension* of the space $\mathfrak{H}$ is defined as *the smallest cardinal number equivalent to a complete subset of* $\mathfrak{H}$, that is, to a set of elements of $\mathfrak{H}$ whose finite linear combinations are everywhere dense in $\mathfrak{H}$. The dimension of a separable space is therefore either finite or denumerably infinite. When it is finite and equal to n, we call $\mathfrak{H}$ a *unitary space of dimension* n. When it is denumerably infinite, we call $\mathfrak{H}$ a *true Hilbert space*. In this space there always exists an infinite complete orthonormal sequence: we have only to apply the SCHMIDT orthogonalization process (Sec. 33) to a complete denumerable subset of $\mathfrak{H}$.

84. Linear Transformations of Hilbert Space. Fundamental Concepts

The *linear transformations* of the space $\mathfrak{H}$ will be defined, just as in the case of L^2, by the three properties of being additive, homogeneous, and bounded; the *norm* is the smallest of the bounds. Sums, products, and inverses, as well the different types of convergence (weak, strong, in norm) of linear transformations are defined without any modification (cf. Secs. 66—67). We denote these three types of convergence by the respective signs "$\rightharpoonup$", "$\rightarrow$", and "$\Rightarrow$". Instead of "strong convergence" we shall also say simply "convergence" and denote the strong limit of $\{T_n\}$ by $\lim T_n$; instead of "convergence in the norm" we shall also say "uniform convergence."

In these definitions, even in the definition of weak convergence, there is no need to assume in advance that the limit T of the sequence of linear transformations T_n is itself linear, since this is automatically assured. Additivity and homogeneity of the transformation T result immediately from the fact that the transformations T_n have these properties; therefore we have only to prove that T is also bounded. This is done easily if we know that the linear transformations T_n are uniformly bounded, that is, that there exists a constant C such that

$$\|T_n\| \leq C \qquad (n = 1, 2, \ldots);$$

in fact we then obtain, by virtue of the semi-continuity of the norm with respect to weak convergence (cf. Sec. 29),

$$\|Tf\| \leq \liminf \|T_n f\| \leq C\,\|f\|, \text{ hence } \|T\| \leq C.$$

But the existence of such a constant C is ensured even by the hypothesis that the sequence $\{T_n\}$ is weakly convergent. In fact, since we know that for arbitrary fixed f the sequence of elements $\{T_n f\}$ is weakly convergent, hence also bounded (cf. the end of Sec. 31), we have only to apply the following theorem, which is proved exactly as was its analogue for sequences of linear functionals (Sec. 31):

THEOREM. *The sequence $\{T_n\}$ of linear transformations cannot converge or even remain bounded for each element f of the space $\mathfrak{H}$ unless it is uniformly bounded, that is, unless the sequence of norms $\|T_n\|$ is bounded.*

We can operate with convergent sequences of linear transformations in the same way as with numerical sequences: *the sum of two convergent sequences converges to the sum of the two limits, and the product to the product,* with the same type of convergence (weak, strong, uniform) except in the case of the product of two weakly convergent sequences. The assertion concerning the sum is evident; those concerning the product of strongly or uniformly convergent sequences follow from the relations

$$\|T_n T'_n f - TT'f\| = \|T_n(T'_n - T')f + (T_n - T)T'f\| \leq$$
$$\leq C\,\|(T'_n - T')f\| + \|(T_n - T)T'f\|$$

and

$$\|T_n T'_n - TT'\| \leq C\,\|T'_n - T'\| + \|T_n - T\|\,\|T'\|,$$

where $T = \lim T_n$, $T' = \lim T'_n$, and C is a bound of the sequence $\{\|T_n\|\}$.

In the case of weak convergence we have only the following:

$$T_n \rightharpoonup T \text{ and } T'_n \to T' \text{ imply } T_n T'_n \rightharpoonup TT',$$

a consequence of the fact that

$$|(T_n T'_n f - TT'f, g)| \leq |(T_n(T'_n - T')f, g)| + |((T_n - T)T'f, g)| \leq$$
$$\leq C\,\|(T'_n - T')f\|\,\|g\| + |(T_n T'f - TT'f, g)|.$$

Everything that was said in Sec. 67 about regular and singular values of a linear transformation and about the analytic behavior of the resolvent transformation obviously remains valid in the abstract case.

On the other hand, we used the kernels in order to define adjoint transformations: for a transformation T generated by a kernel $T(x, y)$ we defined the adjoint transformation T^* to be the transformation generated by the kernel $\overline{T(y, x)}$. However, we obtained the relation

$$(Tf, g) = (f, T^*g) \tag{4}$$

for every pair of elements f, g of L^2, and this relation can serve as the *definition* of the *adjoint transformation* T^*, for any linear transformation T of the space L^2 or of the abstract space $\mathfrak{H}$.

In fact, (Tf, g) is for fixed g a linear functional operating on the variable element f; hence by the theorem proved in Sec. 30, there exists a uniquely determined element g^* such that

$$(Tf, g) = (f, g^*)$$

for all f. We set $T^*g = g^*$; the transformation T^* thus defined is obviously additive and homogeneous; let us show that it is also bounded and that its norm equals the norm of the transformation T. Setting $f = T^*g$ in (4), we obtain

$$(T^*g, T^*g) = (TT^*g, g) \leq \|TT^*g\|\,\|g\| \leq \|T\|\,\|T^*g\|\,\|g\|,$$

hence

$$\|T^*g\| \leq \|T\|\,\|g\|, \quad \|T^*\| \leq \|T\|.$$

Setting $g = Tf$ in (4), we obtain in the same manner $\|T\| \leq \|T^*\|$. Therefore

$$\|T^*\| = \|T\|. \tag{5}$$

Since relation (4) is symmetric in T and T^*, we have

$$(T^*)^* = T.$$

The transformations I and O — the identity and the transformation carrying every element into the element 0 — coincide with their adjoints: $I^* = I$, $O^* = O$.

The following relations are obvious consequences of the definition of adjoint transformation:

$$(aT)^* = \bar{a}T^*, \quad (T_1 + T_2)^* = T_1^* + T_2^*, \quad (T_1 T_2)^* = T_2^* T_1^*.$$

It follows from the equality of the norms (5) that $T_n \Rightarrow T$ implies $T_n^* \Rightarrow T^*$. On the other hand, it is obvious that $T_n \rightharpoonup T$ implies $T_n^* \rightharpoonup T$. However, in general $T_n \to T$ does *not* imply that $T_n^* \to T^*$.[3]

When T possesses an *inverse*, that is, when there is a linear transformation T^{-1} such that $T^{-1}T = TT^{-1} = I$, we have

$$T^*(T^{-1})^* = (T^{-1})^*T^* = I^* = I;$$

therefore T* also possesses an inverse and

$$(T^*)^{-1} = (T^{-1})^*.$$

The first member of relation (4) is a *bilinear form* of the two variables f and g, which means that it possesses the following properties [denoting it for the moment by $(f|g)$]:

$$(f_1 + f_2|g) = (f_1|g) + (f_2|g), \quad (f|g_1 + g_2) = (f|g_1) + (f|g_2),$$
$$(af|g) = a(f|g), \quad (f|ag) = \bar{a}(f|g);$$

furthermore, it is *bounded*, that is,

$$|(f|g)| \leq M \,\|f\|\, \|g\|,$$

and the smallest possible constant M is obviously equal to $\|T\|$. Conversely, *to every bounded bilinear form* $(f|g)$ *there corresponds a linear transformation* T *such that*

$$(f|g) = (Tf, g),$$

which we see by an argument analogous to that we used in the construction of the adjoint transformation.

[3] Let T_n be the linear transformation of the space l^2 defined by $T_n\{x_1, x_2, \dots\} = \{x_{n+1}, x_{n+2}, \dots\}$. Then we have $T^*\{x_1, x_2, \dots\} = \{0, \dots, 0, x_1, x_2, \dots\}$, where there are n zeros before x_1. It is easy to see that, for every vector x, $T_n x \to 0$, while $\|T_n^* x\| = \|x\|$.

85. Completely Continuous Linear Transformations

Let us consider now the equations

$$f - Kf = g \text{ and } f - K^*f = g,$$

which correspond, in the abstract Hilbert space $\mathfrak{H}$, to our integral equations with adjoint kernels in L^2; of course, it is no longer a question of either integrals or kernels, since K and K^* are two arbitrary linear transformations of the space $\mathfrak{H}$, adjoint to one another in the sense defined in the preceding section.

When the transformation K is of *finite rank*,

$$Kf = \sum_{i=1}^{r} (f, \psi_i)\varphi_i,$$

we have

$$K^*f = \sum_{i=1}^{r} (f, \varphi_i)\psi_i,$$

and the problem of our equations reduces to a problem of linear algebra, just as in the case of integral equations; see Sec. 70. We remark that, as we observed with regard to a particular problem in Sec. 78, the linear transformations of finite rank of the space $\mathfrak{H}$ can be defined as those which transform the entire space $\mathfrak{H}$ into a subspace of finite dimension.

For a linear transformation K of the most general type our methods do not suffice, which is not astonishing since in general the Fredholm alternative is no longer valid. In fact, consider the following linear transformation of the space L^2 of functions defined on the interval (0, 1):

$$Kf(x) = (1 - x)f(x);$$

this transformation is equal to its adjoint. But the equation $(I - K)f(x) = g(x)$, that is, the equation $xf(x) = g(x)$, cannot be solved in L^2 for all given functions $g(x)$ belonging to L^2, despite the fact that the homogeneous adjoint equation $xf(x) = 0$ possesses only the single solution $f(x) = 0$.

However, if we restrict ourselves to the consideration of *completely continuous* linear transformations K, the method given in sections 77 to 80 for the space L^2 applies word for word to the abstract space $\mathfrak{H}$. We prove in particular the decomposition theorem (Sec. 78) and then, applying it, the Fredholm alternative (Sec. 73). Of course, we no longer pass from the equations $KK_1 = K_1K = K_1 - K$ to the equations $K_1^*K^* = K^*K_1^* = K_1^* - K^*$ by means of kernels (which no longer has meaning), but directly by basing our arguments on the definition and on the properties of adjoint transformations given in Sec. 84.

Instead of this "geometric" method, we can also use the "analytic" method of sections 71 and 72, at least for linear transformations K which can be approximated arbitrarily closely, in norm, by transformations of finite rank. Now this is true for all completely continuous linear transformations K. In fact, assuming at first that we are dealing with a separable Hilbert space and that $\{\varphi_k\}$ is a complete orthonormal sequence in it, we have the following theorem:

THEOREM. *If K is a completely continuous linear transformation, the "reduced" transformations K_n, defined by*

$$K_n f = \sum_{i,j=1}^{n} (f, \varphi_i)\,(K\varphi_i, \varphi_j)\varphi_j,$$

tend uniformly to the transformation K when $n \to \infty$.

We observe first that the transformation K_n can be written in the form $K_n = P_n K P_n$, where P_n denotes the orthogonal projection onto the subspace determined by $\varphi_1, \varphi_2, \ldots, \varphi_n$, that is

$$P_n f = \sum_{i=1}^{n} (f, \varphi_i)\varphi_i.$$

We note the following relations which we shall use:

$$P_n^* = P_n, \quad \|P_n\| \leq 1, \quad \|I - P_n\| \leq 1, \quad P_n \to I.$$

If the theorem were not true, that is, if $\|K - K_n\|$ did not tend to 0 with $1/n$, we could choose a sequence of elements f_n such that

$$\|f_n\| = 1 \text{ and } \|(K - K_n)f_n\| \geq q, \tag{6}$$

where q is some positive quantity independent of n. In view of the complete continuity of K, we can assume without loss of generality that the sequences $\{Kf_n\}$ and $\{K(I - P_n)f_n\}$ are convergent; denote their limits by g and h respectively. Consider the decomposition

$$K - K_n = K - P_n K P_n = (I - P_n)K + P_n K(I - P_n);$$

we have

$$\|(I-P_n)Kf_n\| \leq \|(I-P_n)(Kf_n-g)\| + \|(I-P_n)g\| \leq \|Kf_n-g\| + \|g-P_n g\| \to 0,$$

$$\|P_n K(I-P_n)f_n\| \leq K(I-P_n)f_n\| \to \|h\|,$$

$$(h, h) = \lim_n\, (K(I-P_n)f_n, h) = \lim_n\, (f_n, (I-P_n)K^*h) \geq \lim_n\, \|(I-P_n)K^*h\| = 0,$$

and consequently

$$\|(K - K_n)f_n\| \to 0,$$

which is a contradiction of (6). Hence the theorem is proved.

We could just as well have used the transformations KP_n or $P_n K$ instead of $K_n = P_n K P_n$. The advantage of K_n is that it transforms the subspace determined by $\varphi_1, \varphi_2, \ldots, \varphi_n$ into itself and is zero on the orthogonal complement;

hence it is essentially a transformation of an n-dimensional space into itself.

The case of a completely continuous linear transformation K of a *non-separable* space $\mathfrak{H}$ reduces to the preceding case by virtue of the fact that K is essentially the transformation of a separable subspace, that is, *there exists in $\mathfrak{H}$ a separable subspace $\mathfrak{H}_0$ which is transformed by K into itself and whose orthogonal complement is transformed by K into the element* 0.

In fact, since the transformation $A = K^*K$ is completely continuous and has the property that $A^* = A$, there exists, as we shall see in the following chapter, a sequence of elements $\varphi_1, \varphi_2, \ldots$ (the characteristic elements of A corresponding to non-zero characteristic values), such that for every element f orthogonal to all the φ_n we have $Af = 0$, and consequently

$$(Kf, Kf) = (K^*Kf, f) = (Af, f) = 0, \quad Kf = 0.$$

The denumerable set of elements

$$K^m \varphi_n \qquad (n = 1, 2, \ldots; m = 0, 1, \ldots)$$

then determines a separable subspace $\mathfrak{H}_0$ of $\mathfrak{H}$ which obviously is transformed by K into itself; since the orthogonal complement of $\mathfrak{H}_0$ is, in particular, orthogonal to all the φ_n, it is transformed by K into the element 0.

The "analytic" method has the further advantage that it can also be applied, at least partially, to linear transformations K which *are not completely continuous*. Let us define the *Fredholm radius* of the linear transformation K to be the least upper bound Ω of the values $\omega > 0$ for which there exists a linear transformation of finite rank L such that

$$\|K - L\| \leq \frac{1}{\omega}.$$

As we have just seen, $\Omega = \infty$ for completely continuous linear transformations. Choosing $L = 0$, we see that for every transformation

$$\Omega \geq \frac{1}{\|K\|}.$$

According to sections 71—73, it follows that the resolvent transformation K_λ behaves in the interior of the circle

$$|\lambda| = \Omega$$

exactly as if K were completely continuous: hence it has only polar singularities which cannot have an accumulation point in the interior of this circle, and the Fredholm alternative holds for the functional equations

$$f - \lambda Kf = g, f' - \bar{\lambda} K^* f' = g'.$$

*

Let us mention several other variants of the definition of *complete continuity*. Recall the definition given in Sec. 76:[4]

DEFINITION 1. *A linear transformation K is said to be completely continuous if it transforms every infinite and bounded set into a compact set, that is, if for every infinite sequence of elements f_n such that $\|f_n\| \leq C$, the sequence $\{Kf_n\}$ contains a subsequence which converges in the strong sense to an element of the space $\mathfrak{H}$.*

Originally, HILBERT introduced the notion of complete continuity for numerical-valued functions $F(f, g, \ldots, v)$, where $f, g, \ldots, v$ are variable elements of the space $\mathfrak{H}$; the function F is completely continuous if

$$F(f_n, g_n, \ldots, v_n) \to F(f, g, \ldots, v)$$

when the elements $f_n, g_n, \ldots, v_n$ tend *weakly* to the elements $f, g, \ldots, v$.[5]

With the aid of this notion we can define complete continuity for a linear transformation in the following manner:

DEFINITION 2. *A linear transformation K is said to be completely continuous if the bilinear form $(f|g) = (Kf, g)$ is a weakly continuous function of the elements f and g: $f_n \rightharpoonup f, g_n \rightharpoonup g$ imply $(f_n|g_n) \to (f|g)$.*

Two other definitions, which are more convenient in certain applications, are:

DEFINITION 3. *A linear transformation K is said to be completely continuous if it transforms every weakly convergent sequence of elements into a strongly convergent sequence, that is, if*

$$f_n \rightharpoonup f \quad \textit{implies} \quad Kf_n \to Kf.$$

DEFINITION 4. *A linear transformation K is said to be completely continuous if from every bounded infinite sequence of elements we can select a subsequence $\{f_n\}$ for which*

$$(f_n - f_m|f_n - f_m) = (K(f_n - f_m), f_n - f_m) \to 0 \text{ for } m, n \to \infty.$$

We shall show that all these definitions are equivalent.

$1 \to 2$. We assume that K is completely continuous according to definition 1, and show that then $f_n \rightharpoonup f$, $g_n \rightharpoonup g$ imply $(Kf_n, g_n) \to (Kf, g)$. If this were not the case, there would be a positive quantity q such that

$$|(Kf_n, g_n) - (Kf, g)| \geq q$$

for an infinite number of indices $n = n_1, n_2, \ldots$. Since the sequence $\{f_n\}$ is weakly convergent and hence bounded, we can also require, without loss of

[4] Cf. F. RIESZ [9] (p. 74).

[5] HILBERT [*] (Note 4) and RIESZ [*] (p. 96).

generality, that the sequence $\{Kf_{n_k}\}$ be convergent in the strong sense. On the other hand, $f_n \rightharpoonup f$ implies $Kf_n \rightharpoonup Kf$, since

$$(Kf_n, h) = (f_n, K^*h) \to (f, K^*h) = (Kf, h),$$

for every element h; hence we necessarily have $Kf_{n_k} \to Kf$. From this it follows that

$$|(Kf_{n_k}, g_{n_k}) - (Kf, g)| = |(Kf_{n_k} - Kf, g_{n_k}) + (Kf, g_{n_k} - g)| \leq$$
$$\leq \|Kf_{n_k} - Kf\| \, \|g_{n_k}\| + |(Kf, g_{n_k} - g)| \to 0$$

for $k \to \infty$. Since on the other hand the first member is $\geq q > 0$, we have encountered a contradiction. Therefore K is also completely continuous in the sense of Definition 2.

$2 \to 3$. We have seen that $f_n \rightharpoonup f$ implies $Kf_n \rightharpoonup Kf$ for every linear transformation K. When in addition K is completely continuous in the sense of definition 2 we have that $h_n = f_n - f \rightharpoonup 0$ and $g_n = Kf_n - Kf \rightharpoonup 0$ imply

$$\|Kf_n - Kf\|^2 = (Kh_n, g_n) \to 0;$$

hence

$$Kf_n \to Kf,$$

which proves that K is also completely continuous in the sense of Definition 3.

$3 \to 4$. Let $\{h_n\}$ be a bounded infinite sequence of elements of $\mathfrak{H}$, $\|h_n\| \leq C$. By virtue of the theorem of choice proved in sections 32 and 35, we can select a weakly convergent subsequence $\{f_n\}$ from the sequence $\{h_n\}$. Since the transformation K is assumed completely continuous in the sense of Definition 3, the sequence $\{Kf_n\}$ will be strongly convergent and hence

$$|(K(f_n - f_m), f_n - f_m)| \leq \|Kf_n - Kf_m\| \cdot 2C \to 0$$

when $m, n \to \infty$. Therefore K is also completely continuous in the sense of Definition 4.

$4 \to 1$. We assume that the linear transformation K is completely continuous in the sense of Definition 4. Let $\{h_n\}$ be a bounded infinite sequence of elements of $\mathfrak{H}$; the sequences

$$h_{1n} = h_n + Kh_n, \quad h_{2n} = h_n - Kh_n, \quad h_{3n} = h_n + iKh_n, \quad h_{4n} = h_n - iKh_n$$

are then also bounded. Hence we can determine a sequence of integers $\{n_k\}$ such that, denoting h_{n_k} by f_k and h_{rn_k} by f_{rk} $(r = 1, 2, 3, 4)$, we have

$$(K(f_{rk} - f_{rj}), f_{rk} - f_{rj}) \to 0 \qquad (j, k \to \infty;\ r = 1, 2, 3, 4).$$

Inasmuch as

$$(K(f_{1k} - f_{1j}), f_{1k} - f_{1j}) - (K(f_{2k} - f_{2j}), f_{2k} - f_{2j}) +$$
$$+ i(K(f_{3k} - f_{3j}), f_{3k} - f_{3j}) - i(K(f_{4k} - f_{4j}), f_{4k} - f_{4j}) =$$
$$= 4(K(f_k - f_j), K(f_k - f_j)),$$

we shall have

$$\|Kf_k - Kf_j\| \to 0 \quad \text{when} \quad j, k \to \infty;$$

therefore the sequence $\{Kf_k\}$ is (strongly) convergent. The transformation K is consequently also completely continuous in the sense of Definition 1.

This completes the proof of the equivalence of the four definitions.

86. Biorthogonal Sequences. A Theorem of Paley and Wiener

We say that the sequences $\{f_n\}$, $\{g_n\}$ of elements of the Hilbert space $\mathfrak{H}$ form a *normalized biorthogonal system* if

$$(f_n, g_m) = 0 \text{ for } n \neq m \text{ and } (f_n, g_n) = 1.$$

This biorthogonal system is said to be *complete* if each of the systems $\{f_n\}$, $\{g_n\}$ is complete in $\mathfrak{H}$, that is, if the linear combinations of the f_n, as well as those of the g_n, are everywhere dense in $\mathfrak{H}$. Then we have the biorthogonal developments

$$f = \sum_{n=1}^{\infty} (f, g_n) f_n, \qquad f = \sum_{n=1}^{\infty} (f, f_n) g_n,$$

valid whenever the series in the second member converges. It clearly suffices to consider the first series. Denoting its sum by f', we shall have

$$(f', g_m) = \sum_{n=1}^{\infty} (f, g_n)\,(f_n, g_m) = (f, g_m),$$

hence

$$(f' - f, g_m) = 0 \qquad (m = 1, 2, \ldots),$$

which implies that $f' - f = 0$, $f' = f$.

The following theorem is very useful in the theories of various different series of functions. Its proof will be based on the fact that, for every linear transformation K such that $\|K\| < 1$, the transformation $(I - K)^{-1}$ exists.

THEOREM.[6] *Assume that the sequence $\{f_n\}$ differs only slightly from the complete orthonormal sequence $\{\varphi_n\}$, in the sense that there exists a constant θ, $0 \leq \theta < 1$, such that*

$$\left\| \sum a_n(\varphi_n - f_n)\right\|^2 \leq \theta^2 \sum |a_n|^2$$

for every finite sequence $\{a_n\}$ of complex numbers. Then there exists a sequence $\{g_n\}$ which, with $\{f_n\}$, forms a complete normalized biorthogonal system; furthermore, every element f of Hilbert space has convergent developments

$$f = \sum (f, g_n) f_n, \; f = \sum (f, f_n) g_n$$

[6] PALEY and WIENER [*] (p. 100). The above proof is due to SZ.-NAGY [5]; moreover, a more general theorem is to be found there.

and we have

$$(1+\theta)^{-1}\|f\| \leq \left(\sum |(f, g_n)|^2\right)^{\frac{1}{2}} \leq (1-\theta)^{-1}\|f\|,$$
$$(1-\theta)\|f\| \leq \left(\sum |(f, f_n)|^2\right)^{\frac{1}{2}} \leq (1+\theta)\|f\|.$$

To prove this theorem, we observe first that under the given hypothesis, the series

$$\sum (f, \varphi_n)(\varphi_n - f_n)$$

converges for every f and, denoting its sum by Kf, that the transformation K thus defined is linear and bounded, $\|K\| \leq \theta$. In fact, we have

$$\|\sum_{k=m}^{n} (f, \varphi_k)(\varphi_k - f_k)\|^2 \leq \theta^2 \sum_{k=m}^{n} |(f, \varphi_k)|^2 \to 0$$

for $m, n \to \infty$, which assures the existence of Kf; the linearity of K is evident; finally, since

$$\|Kf\|^2 = \lim_{n\to\infty} \|\sum_{k=1}^{n} (f, \varphi_k)(\varphi_k - f_k)\|^2 \leq \lim_{n\to\infty} \theta^2 \sum_{k=1}^{n} |(f, \varphi_k)|^2 = \theta^2 \|f\|^2,$$

we have $\|K\| \leq \theta$ and hence also $\|K^*\| \leq \theta$.

The transformation $T = I - K$ therefore has an inverse and we have

$$(1-\theta)\|f\| \leq \|Tf\| \leq (1+\theta)\|f\|, \quad (1-\theta)\|f\| \leq \|T^*f\| \leq (1+\theta)\|f\|,$$
$$(1-\theta)\|T^{-1}g\| \leq \|g\| \leq (1+\theta)\|T^{-1}g\|.$$

Since $T\varphi_n$ is clearly equal to f_n, we show that the elements $g_n = (T^{-1})^*\varphi_n$ satisfy the theorem. In fact, we have

$$(f_n, g_m) = (T\varphi_n, (T^{-1})^*\varphi_m) = (T^{-1}T\varphi_n, \varphi_m) = (\varphi_n, \varphi_m) = \delta_{nm}$$

and, whatever be the element f of $\mathfrak{H}$,

$$f = T(T^{-1}f) = T\sum_n (T^{-1}f, \varphi_n)\varphi_n = \sum_n (f, (T^{-1})^*\varphi_n)T\varphi_n = \sum_n (f, g_n)f_n,$$

$$f = (T^*)^{-1}T^*f = (T^*)^{-1}\sum_n (T^*f, \varphi_n)\varphi_n = \sum_n (f, T\varphi_n)(T^*)^{-1}\varphi_n = \sum_n (f, f_n)g_n;$$

consequently the systems $\{f_n\}$ and $\{g_n\}$ are complete. Furthermore, we have

$$\sum_n |(f, g_n)|^2 = \sum_n |(T^{-1}f, \varphi_n)|^2 = \|T^{-1}f\|^2,$$
$$\sum_n |(f, f_n)|^2 = \sum_n |(T^*f, \varphi_n)|^2 = \|T^*f\|^2,$$

which completes the proof.

We give as an example the following application of the theorem to *non-harmonic Fourier series*. Consider, in the space $L^2(-\pi, \pi)$, the functions

$$f_n(x) = \frac{1}{\sqrt{2\pi}} e^{i\lambda_n x} \qquad (n = 0, \pm 1, \pm 2, \ldots)$$

and assume they differ only slightly from the functions

$$\varphi_n(x) = \frac{1}{\sqrt{2\pi}} e^{inx},$$

which form a complete orthonormal sequence, in the sense that

$$M = \max_n |\lambda_n - n| < \frac{\log 2}{\pi}.$$

Inasmuch as

$$e^{i\lambda_n x} - e^{inx} = \sum_{k=1}^{\infty} \frac{[i(\lambda_n - n)]^k}{k!} x^k e^{inx},$$

and

$$\|x^k g(x)\| \leq \pi^k \|g(x)\|,$$

we have for every finite sequence $\{a_n\}$

$$\|\sum_n a_n(f_n - \varphi_n)\| = \left\| \sum_n a_n \sum_{k=1}^{\infty} \frac{[i(\lambda_n - n)]^k}{k!} x^k \varphi_n(x) \right\| \leq$$

$$\leq \sum_{k=1}^{\infty} \frac{\pi^k}{k!} \|\sum_n a_n[i(\lambda_n - n)]^k \varphi_n(x)\| = \sum_{k=1}^{\infty} \frac{\pi^k}{k!} \{\sum_n |a_n|^2 |\lambda_n - n|^{2k}\}^{\frac{1}{2}} \leq$$

$$\leq \sum_{k=1}^{\infty} \frac{\pi^k}{k!} M^k \{\sum_n |a_n|^2\}^{\frac{1}{2}} = (e^{M\pi} - 1) \{\sum_n |a_n|^2\}^{\frac{1}{2}}.$$

Setting $\theta = e^{M\pi} - 1$, we have $\theta < 1$ and the theorem can be applied to the sequence $\{f_n(x)\}$.[7]

BANACH SPACES

87. Banach Spaces and Their Conjugate Spaces

The functional spaces L^p where $1 \leq p \leq \infty$ (see Sec. 36) and the space of continuous functions C (see Sec. 50) are, just as the space L^2, *linear* and *metric* spaces; while the definition of the norm of an element is different, we always have the inequality

$$\|f + g\| \leq \|f\| + \|g\|,$$

and in each space convergence with respect to the distance $\|f - g\|$ satisfies the *Cauchy criterion*.

[7] Cf. DUFFIN and EACHUS [1]. Under the less precise condition $M < \frac{1}{\pi^2}$ this result had appeared in PALEY and WIENER [*] (p. 108).

But the spaces L^p ($p \neq 2$) and C differ from the space L^2 in that in these spaces the norm is not derived from a scalar product. By this we mean that there is no scalar product (f, g) defined for all pairs of elements f, g such that $\|f\| = (f, f)^{\frac{1}{2}}$, that is, we can not make Hilbert spaces from these spaces with the same definition of the norm.

In fact, in Hilbert space we always have

$$\|f + g\|^2 + \|f - g\|^2 = 2\,\|f\|^2 + 2\,\|g\|^2, \tag{7}$$

as we deduced in Sec. 30. But in L^p ($p \neq 2$) and in C this formula is not valid; we see this easily by, for example, taking f and g to be two functions with norms equal to 1 and such that $f(x) \cdot g(x) \equiv 0$.

The spaces L^p and C are particular cases of a class of abstract spaces called Banach spaces which are defined in the following manner:[8]

A Banach space is a set $\mathfrak{B}$ *of abstract elements* $f, g \ldots$, *which satisfies conditions* **A** *and* **C** *of the definition of Hilbert space* (*Sec.* 83), *and instead of* **B** *the less stringent condition*:

B′. *To every element* f *of* $\mathfrak{B}$ *there is attached a number* $\|f\| \geq 0$, *its norm, with the properties*

$$\|f + g\| \leq \|f\| + \|g\|, \quad \|af\| = |a|\,\|f\|,$$

and $\|f\| = 0$ *if and only if* $f = 0$.

We distinguish between *real* and *complex* Banach spaces according to whether the multiplication af is defined only for real numbers a or also for complex numbers a.

We note without proof that relation (7) is not only necessary for the Banach space to be a Hilbert space, but also sufficient: when it is satisfied for arbitrary f and g, we can define a bilinear function (f, g) in the space which is a scalar product and satisfies $(f, f) = \|f\|^2$; we have only to set

$$(f, g) = \left\|\frac{f+g}{2}\right\|^2 - \left\|\frac{f-g}{2}\right\|^2$$

or

$$(f, g) = \left\|\frac{f+g}{2}\right\|^2 - \left\|\frac{f-g}{2}\right\|^2 + i\left\|\frac{f+ig}{2}\right\|^2 - i\left\|\frac{f-ig}{2}\right\|^2,$$

depending on whether we are dealing with a real or complex space.[9]

Linear functionals (or *operators*) and their norms are defined on an arbitrary space $\mathfrak{B}$ just as in the particular cases already considered (Secs. 30 and 50). The theorem on sequences of linear functionals which was proved in Sec. 31 for the space L^2 extends immediately to the general case.

[8] BANACH [*], cf. also HAUSDORFF [**1**], DIEUDONNÉ [**1**].

[9] JORDAN–VON NEUMANN [**1**]. A list of other characterizations of this type of Hilbert space, due to other authors, is found in the article by LORCH [**5**].

The fundamental theorem on the *extension* of functionals which we proved in Sec. 35 for L^2 and in Sec. 52 for the space C also extends to an arbitrary space $\mathfrak{B}$, but with more serious difficulties. In the proof of this theorem for the space L^2 we made use of the theorem on orthogonal decomposition of L^2 (Sec. 34), hence of a theorem which is valid for all Hilbert spaces (separable or not), but not for the other Banach spaces, where there is no concept of orthogonality. On the other hand, the proof of the theorem given for the space C is valid for an arbitrary *separable* space $\mathfrak{B}$, that is, a space in which there is a finite or denumerably infinite sequence of elements whose linear combinations are everywhere dense in $\mathfrak{B}$. In the case of a non-separable space $\mathfrak{B}$, we could use ZERMELO's theorem that every set can be well-ordered, a theorem which depends on Zermelo's axiom of choice. If the values of the functional are given on the subset E of the space $\mathfrak{B}$, it suffices to well-order the set $\mathfrak{B} - E$ in order to arrive, by a transfinite repetition of the extension process described in Sec. 52, at the extension of the functional to the entire space $\mathfrak{B}$.

In particular, it follows from the extension theorem that *there exist linear functionals defined on the entire Banach space* $\mathfrak{B}$ *which are not identically zero.* We have even more:

THEOREM. *For every given element* f_0 *of the space* $\mathfrak{B}$, *we can construct a linear functional* Ff, *defined on* $\mathfrak{B}$ *and such that*

$$Ff_0 = \|f_0\| \text{ and } \|F\| = 1.$$

To prove this theorem, we first define Ff on the set E of elements of the form $f = cf_0$ by setting $Ff = c\,\|f_0\|$, then we extend F to the entire space $\mathfrak{B}$, keeping its norm equal to 1.

All these theorems hold for real as well as for complex spaces.

We now introduce the important notion of *conjugate* space. For reasons of simplicity, we restrict ourselves at first to the case of a *real* space $\mathfrak{B}$.

Denote by $\mathfrak{B}^*$ the set of all the real-valued linear functionals defined on $\mathfrak{B}$. The set $\mathfrak{B}^*$ is *linear*; the addition of its elements and their multiplication by real numbers are defined in an obvious manner. It is also *normed*, and it is easy to see that the norm satisfies condition **B′**. Finally, condition **C** is also satisfied, that is, $\mathfrak{B}^*$ is a *complete* metric space. In fact, if the sequence $\{F_n\}$ of elements of $\mathfrak{B}^*$ is such that $\|F_n - F_m\| \to 0$ for $m, n \to \infty$, we have

$$|F_nf - F_mf| \leq \varepsilon\,\|f\| \tag{8}$$

for $m, n \geq n_0(\varepsilon)$ and for all the elements f of $\mathfrak{B}$. It follows that the numerical sequence $\{F_nf\}$ converges for every f; denoting its limit by Ff, we have defined a functional F which is clearly additive and homogeneous since the functionals F_n are so. When $n \to \infty$, it follows from (8) that

$$|Ff - F_mf| \leq \varepsilon\,\|f\| \text{ for } m \geq n_0(\varepsilon),$$

which implies that

$$\|F\| \leq \|F_m\| + \varepsilon \text{ for } m \geq n_0(\varepsilon)$$

and

$$\|F - F_m\| \to 0 \text{ for } m \to \infty.$$

Therefore the functional F is also bounded, and hence is an element of $\mathfrak{B}^*$; the sequence $\{F_n\}$ converges to F in the metric of $\mathfrak{B}^*$.

It is this space $\mathfrak{B}^*$ which we call the *conjugate* or the *dual* of the space $\mathfrak{B}$.

In particular, the space $(L^p)^*$ where $1 \leq p < \infty$ can be identified with the space L^q where $q = p/(p - 1)$, since by the theorem proved in Sec. 36 there exists a one-to-one correspondence between the linear functionals of L^p and the elements of L^q, a correspondence which preserves linear structure and also norm. Therefore the space L^2, and more generally, every Hilbert space, can be identified with its conjugate space.

The space C^*, in its turn, can be identified, according to Secs. 50, 51, with the space V of functions of bounded variation $\alpha(x)$ which are zero at the left endpoint of the interval (a, b) in question and satisfy $\alpha(x) = \frac{1}{2}[\alpha(x - 0) + \alpha(x + 0)]$ in the interior of (a, b); the *norm* $\|\alpha\|$ is defined to be the total variation of $\alpha(x)$ in (a, b).

Let us write the linear functional Ff, defined on the space $\mathfrak{B}$, in the form (F, f) or (f, F); then by the definition of $\|F\|$,

$$(9) \qquad |(F, f)| \leq \|F\| \cdot \|f\|.$$

If we fix f and let F vary in the conjugate space $\mathfrak{B}^*$, then (F, f) will obviously be a linear functional on the space $\mathfrak{B}^*$, hence an element of the space $\mathfrak{B}^{**} = (\mathfrak{B}^*)^*$; in view of (9) its norm in $\mathfrak{B}^{**}$ will not exceed the norm of f in $\mathfrak{B}$. But these two norms are actually equal! In order to see this, it suffices to find an element F of $\mathfrak{B}^*$ for which

$$(F, f) = \|f\| \quad \text{and} \quad \|F\| = 1.$$

But we established above the existence of an F fulfilling these conditions as a consequence of the theorem on the extension of functionals.

To each element f of $\mathfrak{B}$ we have thus assigned a well-defined element Φ_f of $\mathfrak{B}^{**}$ in such a way as to preserve the linear structure and the metric, that is, such that

$$\Phi_{af+bg} = a\Phi_f + b\Phi_g, \quad \|f\| = \|\Phi_f\|.$$

Since we have, in particular,

$$\|f - g\| = \|\Phi_f - \Phi_g\|,$$

Φ_f and Φ_g do not coincide unless $f = g$. This permits us to *identify* Φ_f with f and in this manner to embed the space $\mathfrak{B}$ in the space $\mathfrak{B}^{**}$. This is expressed by writing:

$$\mathfrak{B} \subseteq \mathfrak{B}^{**}.$$

It can happen that we have

$$\mathfrak{B} = \mathfrak{B}^{**};$$

we then say the space $\mathfrak{B}$ is *regular* or *reflexive*.[10] The spaces L^p ($p > 1$) are examples of reflexive spaces, for $(L^p)^{**} = (L^q)^* = L^p$.

On the other hand, the space C is not reflexive, that is, the space $C^{**} = V^*$ is larger than the space C. We see this as follows: to each function of bounded variation $\alpha(x)$ we make correspond the sum of its jumps:

$$\sum_x [\alpha(x+0) - \alpha(x-0)].$$

We have obviously defined a linear functional on the space V, but it can not be represented in the form of an integral

$$\int_a^b f(x) d\alpha(x)$$

with a fixed continuous function $f(x)$.[11] Hence this functional can not be identified with any element of C.

The space L^1 is not reflexive either. We know that $(L^1)^* = L^\infty$, but the converse is not true: there exist linear functionals Ff in the space L^∞ which can not be put in the integral form

$$\int_a^b f(x)g(x)dx$$

with a fixed summable function $g(x)$. In fact, fixing a point x_0 of the interval (a, b), let us at first define the functional Ff for continuous and bounded functions by

$$Ff = f(x_0);$$

then we extend this functional (which obviously is linear in the subspace C of L^1) to a linear functional of the entire space L^1. This resulting functional can not be put in the required form, even for continuous functions $f(x)$.

We conclude with a few words about complex spaces. We can form the space of linear functionals with complex values, but there are certain advantages in defining the conjugate space $\mathfrak{B}^*$ to be rather the space of functionals Ff which are the complex conjugates of the linear functionals. Denoting Ff by (F, f) or $\overline{(f, F)}$, we shall have, for F in $\mathfrak{B}^*$, f in $\mathfrak{B}$, and arbitrary complex

[10] HAHN [**2**], LORCH [**2**].

[11] In fact, considering the function $\alpha(x)$ which is equal to 0 for $x < \xi$ and to 1 for $x \geq \xi$, it is seen that $f(x)$ would be equal to 1 at the point ξ; since ξ was arbitrary, it would be equal to 1 everywhere. But if $f(x) \equiv 1$, the integral reduces to $\alpha(b) - \alpha(a)$, which is equal to the sum of the jumps of $\alpha(x)$ only if $\alpha(x)$ is a monotonic saltus function.

numbers a, b:

$$(aF, bf) = a\bar{b}(F, f),$$

which is similar to the situation for complex Hilbert space. The space $\mathfrak{B}^*$ will clearly be itself a complex space, as will $\mathfrak{B}^{**}$. All the above considerations adapt themselves immediately to the complex case.

88. Linear Transformations and Their Adjoints

The *linear transformations* T of the Banach space $\mathfrak{B}$, the norm $\|T\|$, as well as addition, multiplication, inversion, (strong) convergence and convergence in norm (or uniform convergence) of linear transformations are all defined just as in the case of Hilbert space.

We can also define the *adjoint transformation* T^* by the formula

$$(T^*F, f) = (F, Tf);$$

T^* will therefore be a transformation, obviously additive and homogeneous, of the space $\mathfrak{B}^*$ into itself. It follows from the definition that

$$|(T^*F, f)| \leq \|F\| \cdot \|T\| \cdot \|f\|$$

and, consequently,

$$\|T^*F\| \leq \|F\| \cdot \|T\|;$$

hence T^* is also bounded and $\|T^*\| \leq \|T\|$. We even have $\|T^*\| = \|T\|$; in fact, since we can construct for every element f of $\mathfrak{B}$ an element F of $\mathfrak{B}^*$ such that

$$(F, Tf) = \|Tf\| \quad \text{and} \quad \|F\| = 1,$$

we have

$$\|Tf\| = (F, Tf) = (T^*F, f) \leq \|T^*\| \cdot \|F\| \cdot \|f\| = \|T^*\| \cdot \|f\|$$

and consequently $\|T\| \leq \|T^*\|$, which proves that

$$\|T^*\| = \|T\|.$$

For the transformation of finite rank

$$Tf = \sum_{k=1}^{r} (f, \Psi_k)\varphi_k, \tag{10}$$

where the φ_k and the Ψ_k are elements of $\mathfrak{B}$ and of $\mathfrak{B}^*$ respectively, we have

$$(T^*F, f) = (F, Tf) = \sum (\Psi_k, f)\,(F, \varphi_k) = \sum (F, \varphi_k)\,(\Psi_k, f),$$

hence

$$T^*F = \sum_{k=1}^{r} (F, \varphi_k)\Psi_k. \tag{11}$$

It will be useful for us to note that, just as for the space L^2 (cf. Sec. 78) and even for abstract Hilbert space, *every linear transformation T which transforms the entire space $\mathfrak{B}$ into a linear subset of finite dimension is necessarily of the form* (10).

In fact, if this linear subset is determined by the linearly independent elements $\varphi_1, \varphi_2, \ldots, \varphi_r$, each element of the form Tf has one and only one development

$$Tf = c_1\varphi_1 + c_2\varphi_2 + \ldots + c_r\varphi_r$$

where the coefficients $c_k = c_k(f)$ are obviously additive and homogeneous functions of the variable element f. In order to prove that $c_k(f)$ is a linear functional, or equivalently, that there exists an element Ψ_k of $\mathfrak{B}^*$ such that $c_k(f) = (f, \Psi_k)$, it therefore only remains to show that $c_k(f)$ is also bounded, that is, that there exists a constant M such that

$$|c_k(f)| \leq M \|f\| \qquad (k = 1, 2, \ldots, r).$$

But this follows from the following lemma.

LEMMA 1. *To every finite system of linearly independent elements $\varphi_1, \varphi_2, \ldots, \varphi_r$ of Banach space we can assign a constant M such that*

$$|c_1| + |c_2| + \ldots + |c_r| \leq M \|c_1\varphi_1 + c_2\varphi_2 + \ldots + c_r\varphi_r\|,$$

whatever be the coefficients c_k.

In view of the homogeneity it suffices to prove that if the c_k vary under the condition $\sum |c_k| = 1$, then $\| \sum c_k \varphi_k\|$ has a *positive* minimum. Let

$$g_n = \sum_k c_k^{(n)} \varphi_k$$

be a minimizing sequence. By the Bolzano-Weierstrass theorem we can select from it a subsequence

$$g_{n_i} = \sum_k d_k^{(i)} \varphi_k$$

for which each of the numerical sequences $d_k^{(i)}$ $(k = 1, 2, \ldots, r)$ is convergent. Let c_k^* $(k = 1, 2, \ldots, r)$ be the respective limits and set $g^* = \sum_k c_k^* \varphi_k$.

Since we also have $\sum |c_k^*| = 1$, the minimum in question is *attained* and its value is equal to $\|g^*\|$. Now $g^* \neq 0$, $\|g^*\| \neq 0$, since the coefficients c_k^* are not all zero. This completes the proof.

89. Functional Equations

Let K be a linear transformation of the Banach space $\mathfrak{B}$ and consider the functional equations

$$(12) \qquad (I - K)f = g \quad \text{and} \quad (I - K^*)F = G,$$

in $\mathfrak{B}$ and in $\mathfrak{B}^*$ respectively.

When $\|K\| < 1$, the series $I + K + K^2 + \ldots$ is uniformly convergent and its sum is equal to the inverse of $I - K$. Hence in this case $(I - K)^{-1}$ and $(I - K^*)^{-1} = ((I - K)^{-1})^*$ exist and the equations (12) possess unique solutions f and F for every g and G.

When K is a transformation of finite rank, we can apply the method of Sec. 70 without modification. The Fredholm alternative therefore also holds in this general case; of course, the orthogonality of an element g of $\mathfrak{B}$ to an element F of $\mathfrak{B}^*$ means that $Fg = 0$.

For an arbitrary linear transformation K, we define the Fredholm radius Ω just as it was defined in Hilbert space (cf. Sec. 85); we can then apply the method of Secs. 71 to 73. It follows that for $|\lambda| < \Omega$ the Fredholm alternative holds for the equations (12), written with δK instead of K; moreover, the resolvent transformation K_λ has only poles for singularities; these are finite in number in the interior of every circle $|\lambda| = \omega$ with $\omega < \Omega$. We always have $\Omega \geq \dfrac{1}{\|K\|}$, but *it is unknown* whether or not the equation $\Omega = \infty$ holds for every completely continuous transformation K. However, it can be shown that in the particular spaces L^p and C this is always the case; for the space C, see Sec. 90.

On the other hand, the method discussed in Secs. 77 to 80 applies, after some modifications,[12] to every completely continuous transformation K of an arbitrary Banach space. These modifications are for the purpose of eliminating the use of the notion of scalar product and in particular the notion of the orthogonality of two elements of $\mathfrak{B}$.

The proof of the fact that $\mathfrak{M}_n$ is of finite dimension was based on the fact that a subspace of infinite dimension contains an orthonormal sequence $\{f_n\}$. However, we actually used only the following properties of the sequence $\{f_n\}$:

$$\|f_n\| = 1 \quad \text{and} \quad \|f_k - f_i\| = \sqrt{2} \quad \text{for} \quad k \neq i.$$

It would even have sufficed to know that

$$\|f_n\| = 1 \quad \text{and} \quad \|f_k - f_i\| \geq \delta \quad \text{for} \quad k \neq i,$$

where δ is a fixed positive quantity. But a sequence possessing these properties can be constructed in every infinite-dimensional subspace of an arbitrary

[12] In the paper by F. RIESZ [**9**] the theory is already discussed in all its generality except for that which concerns the study of the adjoint transformation where one is restricted to the case of a transformation of integral type in the space L^p or C. The theory was completed on this point by HILDEBRANDT [**3**] and SCHAUDER [**1**]; cf. also BANACH [*] (Chapt. X). It is in SCHAUDER that the notion of adjoint transformation appears for the first time in its general form given above. He proved, among other things, that if a linear transformation is completely continuous the same is true of its adjoint.

Banach space; we have only to apply repeatedly the following lemma:

LEMMA 2. *If E_1 and E_2 are two subspaces of a Banach space of which the first is properly contained in the second, and if δ is a given positive number less than* 1, *then there exists in E_2 an element f such that $\|f\| = 1$ and $\|f - g\| \geq \delta$ for all elements g of E_1.*

To prove this lemma, let us choose an element h of E^2 which does not belong to E_1. Since E_1 is a closed set, the distance from h to the elements of E_1 possesses a positive minimum d. Since $\frac{d}{\delta} > d$, there exists an element g^* in E_1 for which $\|h - g^*\| < \frac{d}{\delta}$. Then we have, for all the elements g of E_1:

$$\left\| \frac{h - g^*}{\|h - g^*\|} - g \right\| = \frac{1}{\|h - g^*\|} \|h - (g^* + \|h - g^*\|g)\| \geq \frac{\delta}{d} d = \delta,$$

since $g^* + \|h - g^*\|g$ is an element of E_1. The element

$$f = \frac{h - g^*}{\|h - g^*\|}$$

thus satisfies the requirements of the lemma.

The proof of the fact that the $\mathfrak{M}_n$, and also the $\mathfrak{N}_n$, coincide starting with some index, and then the reasoning of Sec. 79 (distribution of singular values), can be based on the same lemma instead of on the theorem affirming the existence of an orthogonal element.

In Sec. 77 we once again made use of orthogonality, namely, in the proof of the fact that $\mathfrak{N}_1$ is a closed set. There we considered a sequence $\{g_k\}$ whose transform $\{Tg_k\}$ is convergent and we showed that the limit element is of the form Tv^*. In doing so we made use of the decomposition $g_k = u_k + v_k$, where u_k is an element of $\mathfrak{M}_1$ and v_k is orthogonal to $\mathfrak{M}_1$. But it would have sufficed to choose u_k and $v_k = g_k - u_k$ so that u_k is an element of $\mathfrak{M}_1$ whose distance to g_k, namely $\|v_k\|$, does not exceed twice the distance d_k from g_k to $\mathfrak{M}_1$. The sequence $\{v_k\}$ is then bounded; in fact, if it were not we could conclude, just as in Sec. 77, that the sequence $w_k = \frac{v_k}{\|v_k\|}$ contains a subsequence converging to an element w^* of $\mathfrak{M}_1$, which is impossible since

$$\|w_k - w^*\| = \frac{1}{\|v_k\|} \|g_k - (u_k + \|v_k\|w^*)\| \geq \frac{d_k}{\|v_k\|} \geq \tfrac{1}{2}.$$

The proof is completed as in Sec. 77.

In Sec. 78, we made use of orthogonality to obtain the canonical representation of the transformation S. But, as we saw in Sec. 88, we can arrive at the analogous representation (10) even in the space $\mathfrak{B}$.

Finally, in order to adapt the reasoning of Sec. 80 to the general case of a space $\mathfrak{B}$, it is still necessary to show that:

1) every finite-dimensional linear subset E of $\mathfrak{B}$ is closed and hence is a subspace;

2) if E_1 is a proper subspace of the finite-dimensional subspace E, then there is a subspace E_2 such that E_1 and E_2 are linearly independent and $E_1 + E_2 = E$.

In order to prove 1), consider a basis $\varphi_1, \varphi_2, \ldots, \varphi_r$ of E[13] and a convergent sequence of elements

$$g_n = \sum_i c_i^{(n)} \varphi_i.$$

It follows from Lemma 1, Sec. 88, that there exists a number C such that

$$\sum_i |c_i^{(n)} - c_i^{(m)}| \leq C \, \|g_n - g_m\| \text{ for } n, m = 1, 2, \ldots.$$

This makes it clear that each of the numerical sequences $\{c_i^{(n)}\}$ is convergent. Setting

$$c_i^* = \lim c_i^{(n)} \text{ and } g^* = \sum_i c_i^* \varphi_i,$$

we obviously have $g^* = \lim g_n$. Therefore the limit of $\{g_n\}$ also belongs to the set E, q. e. d.

As to 2), we have only to choose a basis of E_1 and complete it by suitable elements $\varphi_{r+1}, \ldots, \varphi_t$ into a basis of E; the linear combinations of the φ then form the subspace E_2 sought.

Summarizing, all the arguments discussed in Secs. 77 to 80 apply to arbitrary Banach spaces. Hence:

The Fredholm alternative is valid for an arbitrary completely continuous linear transformation of Banach space.

90. Transformations of the Space of Continuous Functions[14]

Let T be a linear transformation of the space C of continuous real-valued functions defined in the interval $a \leq x \leq b$. Since for each fixed value of x, $Tf(x)$ is a linear functional acting on f whose norm is at most equal to that of T, there exists a function of bounded variation $\tau_x(y)$ such that

$$Tf(x) = \int_a^b f(y) d\tau_x(y) \tag{13}$$

[13] That is, a system of linearly independent elements which determine E.

[14] The results of this section are due to RADON [**2**].

and

$$\int_a^b |d\tau_x(y)| \leq \|T\|.$$

We can assume that $\tau_x(a) = 0$. Since $Tf(x)$ is a continuous function of x for every element f of C and in particular for $f(x) = 1$ and for $f(x) = (\xi - x)^+$, it follows, integrating by parts, that

$$\tau_x(b) \text{ and } \int_a^\xi \tau_x(y)dy \tag{14}$$

are continuous functions of x for every fixed value of ξ between a and b.

It follows from our previous results on the convergence of linear functionals in the space C (Sec. 55) that these conditions are also sufficient for integral (13) to be a continuous function of x. Hence we have the theorem:

THEOREM. *Let $\tau_x(y)$ be a function defined for $a \leq x \leq b$, $a \leq y \leq b$ whose total variation with respect to y is less than some finite value independent of x, for which $\tau_x(a) = 0$, and for which the expressions* (14) *are continuous functions of x. Then $\tau_x(y)$ generates, by formula* (13), *a linear transformation T of the space C into itself. Conversely, every linear transformation of C admits a representation of this type.*

Now let us seek an analytic representation of the adjoint transformation $\alpha^* = T^*\alpha$ of the space $C^* = V$. By definition we have

$$(\alpha^*, f) = (T^*\alpha, f) = (\alpha, Tf),$$

that is,

$$\int_a^b f(x)d\alpha^*(x) = \int_a^b [\int_a^b f(y)d\tau_x(y)]d\alpha(x)$$

for every continuous function $f(x)$. We let $f(x)$ run through a sequence $\{f_n(x)\}$ which tends to 1 for $x \leq \xi$ and to 0 for $x > \xi$ and satisfies $|f_n(x)| \leq 1$. Then, by the Lebesgue theorem, the integrals with respect to $\alpha^*(x)$ and $\tau_x(y)$ tend to $\alpha^*(\xi)$ and $\tau_x(\xi)$ respectively. Since, furthermore,

$$|\int_a^b f_n(y)d\tau_x(y)|$$

is less than $\|T\|$, we can also apply the Lebesgue theorem to the integral with respect to $\alpha(x)$ and obtain finally

$$\alpha^*(\xi) = \int_a^b \tau_x(\xi)d\alpha(x)$$

or, changing the notation of the variables,

$$\alpha^*(x) = \int_a^b \tau_k(x)d\alpha(y), \tag{15}$$

where the integral is taken in the Stieltjes-Lebesgue sense.

Let us consider $\tau_x(y)$ as a element of the space V, depending on the parameter x. In V, the norm of a function is its total variation; we shall denote it by $\|\ldots\|_V$. We shall see that *τ_x is a continuous function of x in the sense of the metric of the space V — that is, that $\|\tau_\xi - \tau_x\|_V \to 0$ when $\xi \to x$ — if and only if the corresponding transformation T is completely continuous.*

We assume first that T is completely continuous. If τ_x were not continuous at the point x_0, there would be a sequence $\{x_k\}$ tending to x_0 for which $\|\tau_{x_k} - \tau_{x_0}\|_V$ remains larger than some positive quantity ε_0. Since $\|\tau_{x_k} - \tau_{x_0}\|_V$ is equal to the norm of the functional $Tf(x_k) - Tf(x_0)$, there exists an element f_k of C, $\|f_k\| \leq 1$, for which this functional approaches within $\frac{\varepsilon_0}{2}$ of the value $\|\tau_{x_k} - \tau_{x_0}\|_V$. Setting $g_k(x) = Tf_k(x)$, we shall therefore have

$$|g_k(x_k) - g_k(x_0)| > \frac{\varepsilon_0}{2}.$$

Since the transformation T is completely continuous, we can select from $\{g_k(x)\}$ a subsequence $\{g_{k_i}(x)\}$ which is convergent in the metric of C, that is, uniformly in every interval $a \leq x \leq b$; denote its limit by $g(x)$. For sufficiently large indices i we shall then have

$$|g_{k_i}(x) - g(x)| < \frac{\varepsilon_0}{6}$$

in the entire interval $a \leq x \leq b$, and by virtue of the continuity of $g(x)$, also

$$|g(x_{k_i}) - g(x_0)| < \frac{\varepsilon_0}{6};$$

hence

$$\frac{\varepsilon_0}{2} < |g_{k_i}(x_{k_i}) - g_{k_i}(x_0)| \leq |g_{k_i}(x_{k_i}) - g(x_{k_i})| + |g(x_{k_i}) - g(x_0)| + |g(x_0) - g_{k_i}(x_0)| <$$

$$< \frac{\varepsilon_0}{6} + \frac{\varepsilon_0}{6} + \frac{\varepsilon_0}{6} = \frac{\varepsilon_0}{2},$$

which yields a contradiction. Therefore the function τ_x is continuous.

Now we assume that τ_x is continuous in $a \leq x \leq b$, and show that T is then completely continuous. We note first that τ_x is uniformly continuous in every interval $a \leq x \leq b$, which one sees just as for ordinary functions. To every $\varepsilon > 0$ we can therefore find points

$$a = x_0 < x_1 < x_2 < \ldots < x_n = b$$

such that $\|\tau_x - \tau_\xi\|_V$ is less than ε whenever x and ξ belong to the same closed interval $[x_{k-1}, x_k]$. We assign to this decomposition of $[a, b]$ a system of continuous functions

$$h_0(x), h_1(x), \ldots, h_n(x),$$

for which $h_m(x)$ is zero at all the points x_k except at x_m, where it takes on the value 1, and is linear in each subinterval (x_{k-1}, x_k). The function

$$\lambda_x(y) = \sum_{m=0}^{n} h_m(x)\tau_{x_m}(y)$$

coincides with $\tau_x(y)$ for $x = x_0, x_1, \ldots, x_n$, and for a point x of the subinterval (x_{k-1}, x_k) it is the arithmetic mean of the functions $\tau_{x_{k-1}}(y)$ and $\tau_{x_k}(y)$ with weights depending only on x. It follows that

$$\|\lambda_x - \tau_x\| < \varepsilon$$

in every interval $a \leq x \leq b$. Denoting by L the transformation of finite rank

$$Lf(x) = \sum_{m=0}^{n} h_m(x) \cdot Tf(x_m) = \int_a^b f(y)d\lambda_x(y),$$

we therefore have

$$\|L - T\| < \varepsilon.$$

Since ε was arbitrary, T can be approximated arbitrary closely in the norm by transformations of finite rank, hence by completely continuous transformations; from this it follows that T is also completely continuous, completing the proof.

In the course of the proof we have obtained the following result:

THEOREM. *Every completely continuous linear transformation of the space C can be approximated arbitrarily closely in norm by linear transformations of finite rank.*

By the same reasoning we prove the following theorem:

THEOREM. *If T is a linear transformation of the space C represented by the integral* (13), *we can approximate T to within $\omega + \varepsilon$ by a linear transformation of finite rank, where*

$$\omega = \limsup_{x-\xi\to 0} \|\tau_x - \tau_\xi\|_V$$

and ε is an arbitrarily small positive quantity; in other words: the Fredholm radius of T is $\geq 1/\omega$.

The integral representation (13) becomes simpler if the functions of bounded variation $\tau_x(y)$ are absolutely continuous with respect to the same non-decreasing function $\sigma(y)$, that is, if they admit the representation

$$\tau_x(y) = \int_a^y K(x, z)d\sigma(z),$$

where for each fixed value of x the kernel $K(x, y)$ is a summable function with respect to $\sigma(y)$. We then have

$$\|\tau_x\|_V = \int_a^b |K(x, z)| \, d\sigma(z)$$[15]

and

$$\|\tau_x - \tau_\xi\|_V = \int_a^b |K(x, z) - K(\xi, z)| \, d\sigma(z).$$

In order that the transformation in question be completely continuous, it is therefore necessary and sufficient that the last integral tend to 0 when ξ tends to x.

The problem arises of knowing if every completely continuous linear transformation T admits a representation with a nondecreasing function $\sigma(y)$ and a kernel $K(x, y)$. We shall show that the answer is in the affirmative.

Let $\{\tau_x(y)\}$ be the family of functions associated with the transformation T. The problem is to find a nondecreasing function $\sigma(y)$ with respect to which all these functions are absolutely continuous. We shall solve this problem by considering first a denumerable subset of this family of functions obtained by letting the parameter x run through a denumerable set of values which is everywhere dense in (a, b), say $x = r_1, r_2, \ldots$; we shall pass to an arbitrary value of x by making use of the fact established above that, for T completely continuous, τ_x depends on x in a continuous manner.

Denoting by $\sigma_x(y)$ the indefinite total variation of $\tau_x(y)$ for which $\sigma_x(a) = 0$, we shall determine the sequence $\{c_n\}$ of positive numbers such that the series

$$\sum_{n=1}^{\infty} c_n \sigma_{r_n}(y)$$

is convergent even for $y = b$. The sum of this series, $\sigma(y)$, is a nondecreasing function of y with respect to which all the functions $\tau_{r_n}(y)$ are obviously absolutely continuous. We shall show that the same holds for $\tau_x(y)$, whatever be the value of x.

Given $\varepsilon > 0$, we choose a value r belonging to the everywhere dense sequence $\{r_n\}$ and sufficiently close to the value x that

$$\|\tau_x - \tau_r\|_V < \frac{\varepsilon}{2}.$$

For every system of non-overlapping intervals $\{(a_k, b_k)\}$ we shall have, a

[15] The total variation of an indefinite integral is equal to the integral of the absolute value of the function in question. This fact was proved in Sec. 23 for the ordinary Lebesgue integral and the same proof is also valid for the Stieltjes-Lebesgue integral with respect to a nondecreasing function.

fortiori,

$$\sum_k |[\tau_x(b_k) - \tau_r(b_k)] - [\tau_x(a_k) - \tau_r(a_k)]| < \frac{\varepsilon}{2},$$

from which it follows that the sum

$$S_x = \sum_k |\tau_x(b_k) - \tau_x(a_k)|$$

differs from the analogous sum S_r by at most $\frac{\varepsilon}{2}$. But the function $\tau_r(y)$ is absolutely continuous with respect to $\sigma(y)$, and consequently $S_r < \frac{\varepsilon}{2}$ whenever the sum

$$\sum_k [\sigma(b_k) - \sigma(a_k)]$$

is small enough. We then have $S_x < \varepsilon$, which proves that $\tau_x(y)$ is also absolutely continuous with respect to $\sigma(y)$.

Therefore we have the following result:

THEOREM. *A necessary and sufficient condition that the linear transformation $g = Tf$ of the space C into itself be completely continuous is that it be representable in the form*

$$g(x) = \int_a^b K(x, y)f(y)d\sigma(y),$$

where $\sigma(y)$ is a nondecreasing function and $K(x, y)$ is a summable function with respect to $\sigma(y)$ for every fixed value of x, and such that

$$\int_a^b |K(x, y) - K(\xi, y)|\, d\sigma(y) \to 0 \qquad \textit{when } \xi \to x.$$

91. A Return to Potential Theory

Let us return to the Dirichlet and Neumann problems in potential theory and to their solution by the FREDHOLM method, Sec. 81. Despite the elegance of this method, it has the inconvenience of being applicable only under very restrictive hypotheses concerning the boundary of the domains considered. These restrictions are inherent in the method; in fact, since the double layers are made up of oriented dipoles, they can not be distributed over any but sufficiently regular curves (or surfaces).

Nevertheless, we can somewhat weaken the hypotheses set down in the preceding section and thus extend the Fredholm method to domains of more general type. In the case of the plane, RADON[16] has extended the generality

[16] RADON [**3**].

of the method to what seems to us to be its natural limit. The domains considered by RADON are those bounded by curves of "bounded rotation."

For a curve whose tangent varies continuously, we can determine the angle which the tangent makes with a fixed straight line so that this angle is a continuous function of the point moving along the curve; the "total rotation" of the curve is then defined as the total variation of this function. If it is finite, we say that the curve is of "bounded rotation."

We can also extend this definition, as RADON showed, to curves whose tangent does not vary continuously, if we agree to determine the angle which the half-tangent makes with the fixed direction in such a way that the jumps do not exceed π in absolute value. Furthermore, as PAATERO[17] observed later, we can also define the total rotation of the boundary of a simply connected domain by the lower limit of the total rotation of curves with continuous tangent, traced in this domain and tending to its boundary.

The double layer distributed on the curve C of bounded rotation, whose "moment" $\mu(s)$ is a continuous function on C, has, at the point P, the potential

$$u(P) = \int_C \mu(t) d\tau_P(t),$$

where $\tau_P(t)$ denotes the angle which the half-line $\overrightarrow{Pt}$ makes with the given direction. We thus arrive, for the interior and exterior Dirichlet problems, at the functional equation

$$\frac{1}{\pi} g = \mu - \lambda K \mu$$

with $\lambda = -1$ and $\lambda = 1$ respectively, where we have set

$$K\mu(s) = \frac{1}{\pi} \int_C \mu(t) d\tau_s(t).$$

We easily see that we are dealing with a linear transformation of the space of continuous functions defined on the curve C (cf. Sec. 89). In particular, the function $\tau_P(s)$ is of bounded variation, and we can show that its total variation does not exceed the total rotation of the curve C, which is a quantity independent of P.

In order that the Fredholm alternative be valid for the values ± 1 of the parameter λ, it is necessary that the Fredholm radius of the transformation K be > 1. Since by Sec. 89 this radius is the reciprocal of the quantity

$$\omega = \limsup_{s-\sigma\to 0} \int_C |d[\tau_s(t) - \tau_\sigma(t)]|,$$

the alternative is valid if $\omega < 1$. But we can show that ω equals $\frac{\vartheta}{\pi}$, where

[17] PAATERO [1], [2].

ϑ ($\leq \pi$) is the largest (in modulus) of the jumps which the half-tangent makes when it traverses C. Therefore if $\vartheta < \pi$, that is, if *the curve C has no "cusp" (directed toward the interior or toward the exterior), we can apply the method of Fredholm.*

The Neumann problem can also be generalized to this case, but not without modification. We cannot require that the normal derivative assume everywhere on C values which are prescribed, since at the vertices there is not even a defined normal. Instead, we can require that the "flux" on C of the desired harmonic function u be equal to a given function $\chi(s)$ of bounded variation, where by this we mean the following:

Let $\varphi(P)$ be an arbitrary function which is continuous in the domain considered as well as on its boundary C, and let its values on C be denoted by $\varphi(s)$. Let c be a closed curve with continuous curvature which lies in the interior of the domain, and let us form the integral of the product $\frac{\partial u}{\partial n} \varphi$ along c, where of course the normal to be understood here is that of c. We then require that when we let c tend to C so that the total rotation of c tends to that of C, this integral, taken on c, tends to the Stieltjes integral

$$\int_C \varphi(s) d\chi(s),$$

for every function φ of the type considered. With the problem thus posed, we also seek the solution in the form of a Stieltjes integral

$$u_P = \frac{1}{\pi} \int_C \log \frac{1}{r_{Pt}} \, da(t),$$

that is, as the potential of a single layer of charges distributed over C. We thus arrive at the functional equation

$$-\frac{1}{\pi} \chi = a - \lambda K^* a$$

with $\lambda = \pm 1$, where K^*, the adjoint of the transformation K, is a transformation from the conjugate space to the space of continuous functions on C, that is, from the space of functions of bounded variation on C. Since the Fredholm alternative also holds for this equation, we obtain in this way a solution of the Neumann problem for every curve C of bounded rotation and without cusps.

For details, we refer the reader to the paper of Radon cited above.

We note that the analogue of the notion of curves of bounded rotation has not yet been found for space.

CHAPTER VI

COMPLETELY CONTINUOUS SYMMETRIC TRANSFORMATIONS OF HILBERT SPACE

EXISTENCE OF CHARACTERISTIC ELEMENTS. THEOREM ON SERIES DEVELOPMENT

92. Characteristic Values and Characteristic Elements. Basic Properties of Symmetric Transformations.

Given a linear transformation T of Hilbert space or, more generally, of a Banach space, one of the first problems is to determine all the elements f of the space which remain invariant or which at least do not change direction under T, that is, which transform into a certain numerical multiple

$$Tf = \mu f. \tag{1}$$

Each of these elements f which differs from 0 is called a *characteristic element*, and the corresponding number μ, a *characteristic value* (we admit the value $\mu = 0$). The solutions f of (1) which correspond to the same characteristic value μ obviously form a linear set; since T is continuous (Cf. Sec. 66), this set is also closed, that is, it is a subspace: the characteristic subspace corresponding to the characteristic value μ. The *multiplicity* of a characteristic value is defined to be the dimension of the corresponding characteristic subspace; in particular, we say that the characteristic value is *simple* if the corresponding characteristic elements are all numerical multiples of one from among them.

The reciprocal of a characteristic value $\mu \neq 0$ is a singular value in the sense of Sec. 67; in fact, since the equation $\left(I - \frac{1}{\mu} T\right)f = 0$ admits a solution $f \neq 0$, $I - \frac{1}{\mu} T$ can not have an inverse.

For a linear transformation $y_i = \sum_{k=1}^{n} t_{ik} x_k$ of the n-dimensional space of complex vectors $x = (x_1, x_2, \ldots, x_n)$, equation (1) is equivalent to the system of linear equations

$$\sum_{k=1}^{n} (t_{ik} - \mu\delta_{ik})x_k = 0 \qquad (i = 1, 2, \ldots, n).$$

In order that this system admit a non-zero solution it is necessary and sufficient that its determinant be zero. But since this determinant is a polynomial in μ of precisely the degree n, it becomes zero for at least one real or complex value. *A linear transformation of a finite-dimensional space therefore always has at least one characteristic value.* In general we can not say more; for example, the transformation

$$y_i = x_i + x_{i+1} \ (i = 1, 2, \ldots, n-1), \ y_n = x_n$$

has only the characteristic value $\mu = 1$, and moreover this is a simple characteristic value.

But we know that if the matrix (t_{ik}) of the transformation T is *Hermitian symmetric*: $t_{ki} = \overline{t_{ik}}$, or equivalently, if we have

$$(Tx, x') = (x, Tx')$$

for all vectors x and x', there exist n *characteristic vectors*

$$e_i = (e_{i1}, e_{i2}, \ldots, e_{in}) \qquad (i = 1, 2, \ldots, n)$$

corresponding to the characteristic values μ_i and *forming an orthonormal system.* Denoting by z_i the components of the vector x with respect to the new system of coordinates formed with respect to these vectors e_i, that is, writing $x = z_1e_1 + z_2e_2 + \ldots + z_ne_n$, *the quadratic (Hermitian) form*

$$(Tx, x) = \sum_{i,k=1}^{n} t_{ik}\overline{x_i}x_k \qquad (t_{ki} = \overline{t_{ik}})$$

transforms into the purely quadratic form

$$\sum_{i=1}^{n} \mu_i \bar{z}_i z_i = \sum_{i=1}^{n} \mu_i \, |z_i|^2.$$

This is the so-called *theorem of principal axes* from algebra.

Passing to the more general case of a Hilbert space $\mathfrak{H}$, we are going to consider the analogues of these transformations, that is, the linear transformations T for which

$$(Tf, g) = (f, Tg)$$

for every pair f, g of elements of $\mathfrak{H}$; we shall call them *symmetric transformations.* These are therefore the linear transformations for which

$$T^* = T,$$

and this is why they are also called *self-adjoint* ("auto-adjointes" in French and "selbstadjungiert" in German). Self-adjoint and symmetric are therefore equivalent, at least for the linear transformations as we have defined them in Secs. 66 and 84, that is, for transformations which are additive, homogeneous, and *bounded.* However, when we also consider *unbounded* transformations we shall have to distinguish between these two notions (see Chapter VIII).

For a symmetric transformation A, the "quadratic form" (Af, f) is always real-valued; in fact:

$$(Af, f) = (f, A^*f) = (f, Af) = \overline{(Af, f)}.$$

Conversely, if the linear transformation A is such that the quadratic form (Af, f) is real-valued, A is necessarily symmetric. In order to see this we start with the identity, valid for *every linear transformation* T,

$$(T(f+g), f+g) - (T(f-g), f-g) + i(T(f+ig), f+ig) - \\ - i(T(f-ig), f-ig) = 4(Tf, g), \tag{2}$$

and from the analogous relation

$$(f+g, T(f+g)) - (f-g, T(f-g)) + i(f+ig, T(f+ig)) - \\ - i(f-ig, T(f-ig)) = 4(f, Tg), \tag{2'}$$

which follows from (2) by interchanging f and g and taking the complex conjugates. But if the quadratic form (Af, f) is always real-valued, we have $(f, Af) = \overline{(Af, f)} = (Af, f)$, and thus the first members of relations (2) and (2′), written for $T = A$, are equal, and consequently $(Af, g) = (f, Ag)$; hence $A^* = A$.

The proposition just proved, as well as relations (2) and (2′), holds only for complex Hilbert space and not for real Hilbert space.

The fact that the quadratic form (Af, f) is real implies that all the characteristic values of A are real; in fact, if μ is a characteristic value and f a corresponding characteristic element, we have

$$\mu = (Af, f)/(f, f).$$

Two characteristic elements f and g corresponding to different characteristic values μ and ν are orthogonal to one another; in fact

$$(Af, g) = (\mu f, g) = \mu(f, g), \quad (f, Ag) = (f, \nu g) = \nu(f, g);$$

hence

$$\mu(f, g) = \nu(f, g), \quad (f, g) = 0.$$

We consider again the quadratic form (Af, f). We obviously have

$$|(Af, f)| \leq \|Af\| \, \|f\| \leq \|A\| \, \|f\|^2;$$

hence if we denote by N_A the smallest constant for which the inequality

$$|(Af, f)| \leq N_A \|f\|^2$$

is verified for every element f, we have

$$N_A \leq \|A\|$$

This is true even if A is not symmetric. We shall show that for A symmetric the two quantities are equal.

In fact, inasmuch as $(A^2f, f) = (Af, Af)$, we have for arbitrary $\lambda > 0$

$$\|Af\|^2 = \tfrac{1}{4}\left[\left(A(\lambda f + \frac{1}{\lambda}Af), \lambda f + \frac{1}{\lambda}Af\right) - \left(A(\lambda f - \frac{1}{\lambda}Af), \lambda f - \frac{1}{\lambda}Af\right)\right] \leq$$

$$\leq \tfrac{1}{4}\left[N_A \|\lambda f + \frac{1}{\lambda}Af\|^2 + N_A \|\lambda f - \frac{1}{\lambda}Af\|^2\right] = \tfrac{1}{2} N_A \left[\lambda^2 \|f\|^2 + \frac{1}{\lambda^2}\|Af\|^2\right];$$

when $\|Af\| \neq 0$ the minimum of the last member will be attained when

$$\lambda^2 = \frac{\|Af\|}{\|f\|},$$

which yields

$$\|Af\|^2 \leq N_A \|Af\| \, \|f\|, \quad \|Af\| \leq N_A \|f\|;$$

these inequalities are obviously also valid when $\|Af\| = 0$. This proves that $\|A\| \leq N_A$. Comparing with the inequality obtained above, we see that $N_A = \|A\|$.

We summarize the properties proved:

THEOREM. *For a symmetric transformation A of the space $\mathfrak{H}$ the characteristic values are real, characteristic elements corresponding to distinct characteristic values are orthogonal, the quadratic form (Af, f) is real-valued, and finally, the smallest constant N_A for which*

$$|(Af, f)| \leq N_A \|f\|^2,$$

equals $\|A\|$.

The theorem on principal axes does not remain valid, at least not in its original form, for an arbitrary symmetric transformation of Hilbert space. For example, the transformation of the space L^2 of functions $f(x)$ defined in the interval $(0, 1)$ which carries $f(x)$ into $x \cdot f(x)$ is symmetric, but has no characteristic values, since the equation $xf(x) = \mu f(x)$ admits no solution other than $f(x) = 0$.

We can nevertheless formulate this theorem in an equivalent form by using Stieltjes integrals; it will then remain valid in Hilbert space of any dimension, as we shall show in the following chapter.

In the present chapter we shall consider only *completely continuous* symmetric transformations. For them, the theorem of principal axes holds in its original form almost word for word; in particular, a completely continuous symmetric transformation always has characteristic values.

93. Completely Continuous Symmetric Transformations

It follows from what we have just proved that for an arbitrary symmetric transformation A and for an element f varying under the condition $\|f\| = 1$, the quantities $|(Af, f)|$ and $\|Af\|$ have the same least upper bounds, namely

$$\sup_f |(Af, f)| = \sup_f \|Af\| = \|A\|.$$

Let $\{f_n\}$ be a sequence of elements such that

$$\|f_n\| = 1 \quad \text{and} \quad |(Af_n, f_n)| \to \|A\|.$$

Assume that it is already chosen in such a manner that the sequence (Af_n, f_n) is itself convergent,

$$(Af_n, f_n) \to \mu_1;$$

μ_1 is then equal to either $\|A\|$ or to $-\|A\|$. We write:

$$0 \leq \|Af_n - \mu_1 f_n\|^2 = \|Af_n\|^2 - 2\mu_1(Af_n, f_n) + \mu_1^2 \|f_n\|^2.$$

As n increases indefinitely the second member of the equation becomes less than any positive number, since

$$\|Af_n\|^2 \leq \|A\|^2 = \mu_1^2, \quad (Af_n, f_n) \to \mu_1, \quad \text{and} \quad \|f_n\| = 1.$$

Hence we have for $n \to \infty$:

$$Af_n - \mu_1 f_n \to 0, \tag{3}$$

which we can express by saying that the equation

$$Af = \mu_1 f \tag{3a}$$

has the elements f_n for "approximate solutions."

Now assume that the symmetric transformation A is *completely continuous*, that is, has the property that every bounded infinite sequence of elements of $\mathfrak{H}$ contains a subsequence whose transform under A converges to an element of $\mathfrak{H}$. We shall show that equation (3a) then has an exact solution.

In fact, the sequence $\{Af_n\}$ then contains a convergent subsequence $\{Af_{n_k}\}$; by virtue of (3), the sequence $\{f_{n_k}\}$ itself is convergent. Denoting the limit of the latter by f, we have

$$Af = \lim Af_{n_k}, \quad \|f\| = \lim \|f_{n_k}\| = 1$$

and, by (3),

$$Af - \mu_1 f = 0, \quad Af = \mu_1 f;$$

furthermore,

$$|(Af, f)| = |(\mu_1 f, f)| = |\mu_1| = \|A\|, \; \|Af\| = \|\mu_1 f\| = |\mu_1| = \|A\|.$$

Hence we have the

THEOREM.[1] *If A is a completely continuous symmetric transformation, the extremal problem:*

$$\text{“}|(Af, f)| = \textit{maximum under the condition } \|f\| = 1\text{”}$$

admits solutions; every solution $f = \varphi$ is a characteristic element of the transformation A. The corresponding characteristic value μ_1 equals in absolute value the maximum in question, or even the maximum of $\|Af\|$, where f varies under the condition $\|f\| = 1$; that is, $|\mu_1| = \|A\|$. Every completely continuous symmetric transformation $A \neq O$ therefore has at least one characteristic value different from 0.

We fix one of the solutions φ, say φ_1, and try to construct new characteristic elements which are orthogonal to φ_1. To do this, we consider the subspace $\mathfrak{H}_1$ of $\mathfrak{H}$ which consists of all the elements which are orthogonal to the characteristic element φ_1. This subspace is transformed by A into itself; in fact, if f is an element of $\mathfrak{H}_1$, we have

$$(Af, \varphi_1) = (f, A\varphi_1) = (f, \mu_1\varphi_1) = \mu_1(f, \varphi_1) = 0;$$

therefore Af also belongs to $\mathfrak{H}_1$. Considered as a transformation of $\mathfrak{H}_1$, A remains symmetric and completely continuous; hence we can apply the result just found. Therefore there exists a characteristic element φ_2 belonging to $\mathfrak{H}_1$, $\|\varphi_2\| = 1$, such that the corresponding characteristic value μ_2 is equal in absolute value to the maximum of $|(Af, f)|$ or, equivalently, of $\|Af\|$, when f varies in $\mathfrak{H}_1$ under the condition $\|f\| = 1$.

Repeating this procedure for the subspace $\mathfrak{H}_2$ of elements simultaneously orthogonal to φ_1 and to φ_2, we obtain a characteristic element φ_3 and a characteristic value μ_3, and so on. In general, *when the characteristic elements φ_1, $\varphi_2, \ldots, \varphi_{n-1}$ are determined, the characteristic element φ_n is obtained as the solution of the extremal problem:*

$$\text{“}|(Af, f)| = \text{maximum}$$

$$\textit{under the conditions } \|f\| = 1,\ (f, \varphi_i) = 0\ (i = 1, 2, \ldots, n-1),\text{”}$$

and the corresponding characteristic value μ_n is equal in absolute value to the maximum in question or even to that of $\|Af\|$ under the same conditions.

Apart from the obvious case where $\mathfrak{H}$ is of finite dimension, this procedure furnishes an infinite sequence of characteristic elements $\{\varphi_n\}$ which form an orthonormal system; for the corresponding characteristic values μ_n we evidently have:

$$|\mu_1| \geq |\mu_2| \geq \ldots.$$

[1] HILBERT [*] (Note 4); the above direct proof is due to F. RIESZ [**6**] (14). See also F. RELLICH [**1**].

We assert that $\mu_n \to 0$. If not, the sequence $\left\{\frac{1}{\mu_n}\varphi_n\right\}$ would be bounded and its transform by A, $\{\varphi_n\}$, would thus contain a convergent subsequence, which is impossible, since $\|\varphi_i - \varphi_k\|^2 = 2$ for $i \neq k$.

Let f be an arbitrary element of $\mathfrak{H}$ and set

$$g_n = f - \sum_{i=1}^{n} (f, \varphi_i)\varphi_i.$$

Since g_n obviously belongs to the subspace $\mathfrak{H}_n$ of elements orthogonal to $\varphi_1, \varphi_2, \ldots, \varphi_n$, we have

$$\|Ag_n\| \leq |\mu_{n+1}|\ \|g_n\|.$$

Since, moreover,

$$\|g_n\|^2 = \|f\|^2 - \sum_{i=1}^{n} |(f, \varphi_i)|^2 \leq \|f\|^2 \text{ and } \mu_{n+1} \to 0,$$

it follows that

$$Ag_n = Af - \sum_{i=1}^{n} (f, \varphi_i)A\varphi_i \to 0 \text{ for } n \to \infty,$$

which can also be written in the form:

$$Af = \sum_{i=1}^{\infty} \mu_i\ (f,\varphi_i)\varphi_i = \sum_{i=1}^{\infty} (f, A\varphi_i)\varphi_i, \tag{4}$$

or also

$$Af = \sum_{i=1}^{\infty} (Af, \varphi_i)\varphi_i.$$

This development reduces, of course, to a finite number of terms when the μ_i are zero after some index.

The sequence (finite or infinite) of the $\mu_i \neq 0$ contains all the characteristic values $\neq 0$ of A, each as many times as its multiplicity indicates. In the contrary case, there would be a characteristic value $\mu \neq 0$ and a corresponding characteristic element φ which would be orthogonal to all the φ_i. Applying (4) to $f = \varphi$, we encounter a contradiction.

It follows in particular that every characteristic value $\mu \neq 0$ is of finite multiplicity. Hence we have the theorem:

THEOREM.[2] *If A is a completely continuous symmetric transformation, we obtain by the above extremal method all the characteristic values of A different from 0, each as many times as its multiplicity indicates. They are of finite multiplicity and either finite or denumerably infinite in number; in the latter case they*

[2] For transformations of integral type, cf. HILBERT [*] (Note 1), E. SCHMIDT [**1**]. Cf. also RELLICH [**1**].

form a sequence tending to 0. *Every element of the form* Af *can be developed in terms of the orthonormal system* $\{\varphi_i\}$ *of corresponding characteristic elements*:

$$Af = \sum_{i=1}^{\infty} (Af, \varphi_i)\varphi_i = \sum_{i=1}^{\infty} \mu_i(f, \varphi_i)\varphi_i. \tag{4'}$$

It will be useful to note that in our proofs we have not made use of property **C** of Hilbert space, that it is *complete*. In fact, every time we concluded the existence of a limit element of a certain sequence, this was not by means of the Cauchy condition, but because it was the transform of a bounded sequence by the completely continuous transformation A; the existence of a limit element was assured by the very definition of complete continuity. *Therefore our theorems hold also for spaces satisfying only the conditions* **A** *and* **B** (Sec. 83), which spaces we shall call *incomplete Hilbert spaces.* An incomplete space can always be completed by the adjunction of certain ideal elements in the same manner that we derive the system of real numbers from that of the rational numbers, namely by the Cantor method of fundamental sequences.

In the case of a *complete* Hilbert space $\mathfrak{H}$, the series

$$\sum_i (g, \varphi_i)\varphi_i$$

is convergent, whatever be the element g of $\mathfrak{H}$, and has for sum an element f of $\mathfrak{H}$. We have on the one hand, applying the transformation A term by term,

$$Af = \sum_i \mu_i(g, \varphi_i)\varphi_i,$$

and on the other hand, according to the theorem just proved,

$$Ag = \sum_i \mu_i(g, \varphi_i)\varphi_i,$$

hence $Af = Ag$. Setting $h = g - f$, we have $Ah = 0$ and

$$g = h + f = h + \sum_i (g, \varphi_i)\varphi_i;$$

the element h is obviously orthogonal to all the φ_i. The following proposition follows from this immediately:

THEOREM. *In order that the sequence of characteristic elements* φ_i *of* A *corresponding to the characteristic values* μ_i *different from* 0 *form a complete orthonormal sequence in the* (*complete*) *space* $\mathfrak{H}$, *it is necessary and sufficient that* 0 *not be a characteristic value of the transformation* A, *that is, that we have* $Af \neq 0$ *when* $f \neq 0$.

In concluding this section, we observe further that *the representation* (4') *with* $\mu_i \to 0$ *is characteristic of completely continuous symmetrictrans formations.* In fact, let A be a linear transformation such that (4') holds for an orthonormal sequence $\{\varphi_k\}$ and a sequence of real numbers $\{\mu_k\}$ tending to 0. It is clear that A is symmetric; we shall show that it is also completely continuous. Let $\{f\}$ be a bounded infinite set, $\|f\| \leq C$. Let $\{f_n\}$ be a weakly convergent sequence selected from this set. We have

$$\|A(f_n - f_m)\|^2 = \sum_{k=1}^{\infty} \mu_k^2 |(f_n - f_m, \varphi_k)|^2 = \sum_{k=1}^{r} + \sum_{k=r+1}^{\infty} = S_r + R_r.$$

Choosing r sufficiently large, R_r will (by virtue of the fact that $\mu_k \to 0$) be less than

$$\varepsilon^2 \sum_{k=r+1}^{\infty} |(f_n - f_n, \varphi_k)|^2 \leq \varepsilon^2 \|f_n - f_m\|^2 \leq \varepsilon^2(2C)^2;$$

having fixed r, we can find (by the weak convergence of $\{f_n\}$) an integer N such that, for $m, n \geq N$, each term of the sum S_r is less than ε^2/r. Then we shall have

$$\|A(f_n - f_m)\| \leq \varepsilon(1 + 4C^2)^{\frac{1}{2}},$$

hence $\{Af_n\}$ converges. This proves that A is completely continuous.

94. Solution of the Functional Equation $f - \lambda Af = g$

If the characteristic values $\mu_i \neq 0$ and the corresponding characteristic elements φ_i of the completely continuous symmetric transformation A are known, the functional equation

(5) $$f - \lambda Af = g$$

is solved in the following manner, which is immediate from the formal point of view.

If for a given element g (5) admits a solution f, we have by the preceding theorem:

(6) $$f = g + \lambda Af = g + \lambda \sum_i \mu_i(f, \varphi_i)\varphi_i,$$

therefore

$$(f, \varphi_k) = (g, \varphi_k) + \lambda\mu_k(f, \varphi_k),$$

(7) $$(1 - \lambda\mu_k)(f, \varphi_k) = (g, \varphi_k) \qquad (k = 1, 2, \ldots).$$

We consider first the case where λ is not equal to the reciprocal of any characteristic value; then we have

$$(f, \varphi_k) = \frac{1}{1 - \lambda\mu_k}(g, \varphi_k) \qquad (k = 1, 2, \ldots),$$

and, by (6),

(8) $$f = g + \lambda \sum_i \frac{\mu_i}{1 - \lambda\mu_i}(g, \varphi_i)\varphi_i.$$

Conversely, when series (8) converges its sum f is obviously a solution of (5). But convergence follows, for a complete space $\mathfrak{H}$, from the fact that the partial sums u_n of (8) satisfy the Cauchy condition:

$$\|u_n - u_m\|^2 = \left|\sum_{m+1}^{n}\left|\frac{\lambda\mu_i}{1-\lambda\mu_i}(g, \varphi_i)\right|^2 \leq \alpha^2 \sum_{m+1}^{n} |(g, \varphi_i)|^2 \to 0 \quad \text{for} \quad m, n \to \infty,$$

where α denotes the least upper bound of the quantities

$$\left|\frac{\lambda\mu_i}{1-\lambda\mu_i}\right| \qquad (i = 1, 2, \ldots),$$

which exists since $\lambda\mu_i \neq 1$ and $\mu_i \to 0$.

When we do not assume that $\mathfrak{H}$ is complete we can reason as follows. The sequence of elements

$$v_n = \sum_{i=1}^{n} \frac{1}{1-\lambda\mu_i}(g, \varphi_i)\varphi_i$$

is bounded, since

$$\|v_n\|^2 = \sum_{i=1}^{n}\left|\frac{1}{1-\lambda\mu_i}(g, \varphi_i)\right|^2 \leq \beta^2 \sum_{i=1}^{n} |(g, \varphi_i)|^2 \leq \beta^2 \|g\|^2,$$

where β denotes the least upper bound of the quantities

$$\left|\frac{1}{1-\lambda\mu_i}\right| \qquad (i = 1, 2, \ldots).$$

The sequence $\{u_n = g + \lambda A v_n\}$ therefore contains a subsequence $\{u_{n_k}\}$ converging to an element u of $\mathfrak{H}$. Since

$$\|u - u_k\| \leq \|u - u_{n_k}\| + \|u_{n_k} - u_k\| \to 0 \quad \text{for} \quad k \to \infty,$$

u is also the limit of the original sequence.

Now consider the case where λ is the reciprocal of a characteristic value of multiplicity r, say

$$\lambda = \frac{1}{\mu_k} \qquad (k = s, s+1, \ldots, s+r-1).$$

By (7), g must then be orthogonal to $\varphi_s, \varphi_{s+1}, \ldots, \varphi_{s+r-1}$, and hence to every characteristic element corresponding to the characteristic value $\dfrac{1}{\lambda}$; the coefficients $c_k = \lambda\mu_k(f, \varphi_k)$ are indeterminate for $k = s, s+1, \ldots, s+r-1$, while for $k < s$ and for $k \geq s+r$ we have $c_k = \dfrac{\lambda\mu_k}{1-\lambda\mu_k}(g, \varphi_k)$.

One shows that conversely, when g satisfies the orthogonality condition and the c_k are chosen in the indicated manner, the series

$$f = g + \sum_k c_k\varphi_k$$

converges and furnishes a solution of equation (5).

95. Direct Determination of the n-th Characteristic Value of Given Sign

It follows from the development by characteristic functions that we have

$$(9) \qquad (Af, f) = \sum_i \mu_i |(f, \varphi_i)|^2,$$

for every element f of $\mathfrak{H}$.

We say the transformation A is *positive*, written $A \geq O$, when $(Af, f) \geq 0$ for all f. It follows from formula (9) that *the transformation A is positive if and only if all its characteristic values are* ≥ 0.

Up to now we have arranged the characteristic values μ_i according to their absolute values without considering their signs. Now it follows from Bessel's inequality that the series development (9) converges *absolutely*, since the sequence of factors μ_i is bounded. Let us arrange the positive and negative terms of the sequence $\{\mu_i\}$ into two separate sequences:

$$\mu_1^+, \mu_2^+, \ldots \quad \text{and} \quad \mu_1^-, \mu_2^-, \ldots,$$

of which the first is nonincreasing and the second nondecreasing. Each of the two sequences can be finite or infinite, or even empty. Let

$$\varphi_1^+, \varphi_2^+, \ldots \quad \text{and} \quad \varphi_1^-, \varphi_2^-, \ldots$$

be the corresponding normalized characteristic elements. By (9) we have

$$(Af, f) = \sum_i \mu_i^+ |(f, \varphi_i^+)|^2 + \sum_i \mu_i^- |(f, \varphi_i^-)|^2,$$

from which we deduce the following sharpened formulation of one of the theorems of Sec. 93:

THEOREM.[3] *The n-th positive characteristic value μ_n^+ equals the maximum of the quadratic form (Af, f) when f varies under the conditions*

$$\|f\| = 1 \quad \textit{and} \quad (f, \varphi_i^+) = 0 \quad (i = 1, 2, \ldots, n-1).$$

This maximum is attained for the characteristic element φ_n^+ corresponding to μ_n^+. Similarly, the n-th characteristic value μ_n^- equals the minimum of (Af, f) under the conditions

$$\|f\| = 1 \quad \textit{and} \quad (f, \varphi_i^-) = 0 \quad (i = 1, 2, \ldots, n-1);$$

this minimum is attained for φ_n^-.

This characterization of the n-th characteristic value and the n-th characteristic element has the inconvenience that it requires knowledge of all the characteristic elements of lower index. Here is a direct characterization which is due — for the case of quadratic forms in spaces of finite dimension — to E. FISCHER:[4]

[3] HILBERT [*] (Note 1, Chapters V and XIV).

[4] Cf. FISCHER [**2**]; COURANT [**2**]; COURANT–HILBERT [*] (pp. 38, 112–113).

THEOREM. *Let $h_1, h_2, \ldots, h_{n-1}$ be $n-1$ elements of $\mathfrak{H}$, and denote by $\nu = \nu(h_1, h_2, \ldots, h_{n-1})$ the maximum of (Af, f) under the conditions*

$$\|f\| = 1 \quad \textit{and} \quad (f, h_i) = 0 \qquad (i = 1, 2, \ldots, n-1).$$

If we let the h_i vary, the minimum of ν will equal μ_n^+ and it will be attained for

$$h_i = \varphi_i^+ \qquad (i = 1, 2, \ldots, n-1).$$

For μ_n^- we have an analogous theorem, but the rôles of maximum and minimum must be interchanged.

It suffices to prove the proposition concerning μ_n^+. We know that

$$\nu(\varphi_1^+, \varphi_2^+, \ldots, \varphi_{n-1}^+) = \mu_n^+.$$

Hence, we have only to show that, for $h_1, h_2, \ldots, h_{n-1}$ arbitrary, $\nu \geq \mu_n^+$. To do this, we choose a non-zero solution of the system of linear equations

$$\sum_{i=1}^{n} x_i(\varphi_i^+, h_k) = 0 \qquad (k = 1, 2, \ldots, n-1)$$

which we normalize by the condition

$$\sum_{i=1}^{n} |x_i|^2 = 1.$$

For

$$\varphi = \sum_{i=1}^{n} x_i \varphi_i^+$$

we shall have

$$\|\varphi\| = 1 \quad \text{and} \quad (\varphi, h_k) = 0 \quad (k = 1, 2, \ldots, n-1),$$

hence

$$(A\varphi, \varphi) \leq \nu(h_1, h_2, \ldots, h_{n-1}).$$

On the other hand,

$$(A\varphi, \varphi) = \sum_{i,j=1}^{n} x_i \bar{x}_j (A\varphi_i^+, \varphi_j^+) = \sum_{i=1}^{n} \mu_i^+ |x_i|^2 \geq \mu_n^+ \sum_{i=1}^{n} |x_i|^2 = \mu_n^+,$$

from which it follows that $\nu \geq \mu_n^+$, which completes the proof.

This theorem gives a method of comparing the characteristic values of several transformations. Here are some results of this type, due to H. WEYL and to R. COURANT.[5]

THEOREM. *Let A_1 and A_2 be two completely continuous symmetric transformations and let $A = A_1 + A_2$. Denoting the n-th positive characteristic values of A_1, A_2, and A by respectively μ_{1n}^+, μ_{2n}^+,* and *μ_n^+, and their n-th negative charac-*

[5] WEYL [**2**], [**3**]; COURANT [**2**].

teristic values by $\overset{-}{\mu}_{1n}$, $\overset{-}{\mu}_{2n}$, *and* $\overset{-}{\mu}_{n}$, *we shall have*

(10) $$\overset{+}{\mu}_{p+q-1} \leqq \overset{+}{\mu}_{1p} + \overset{+}{\mu}_{2q},$$

(11) $$\overset{-}{\mu}_{p+q-1} \geqq \overset{-}{\mu}_{1p} + \overset{-}{\mu}_{2q},$$

and, if A_2 *has only a finite number* N^+ *of positive characteristic values, or a finite number* N^- *of negative characteristic values, we have*

(12) $$\overset{+}{\mu}_{p+N^+} \leqq \overset{+}{\mu}_{1p},$$

and

(13) $$\overset{-}{\mu}_{p+N^-} \geqq \overset{-}{\mu}_{1p},$$

respectively.

Characteristic values with the same index, positive or negative, of the transformations A_1 *and* $A = A_1 + A_2$ *differ by at most* $\|A_2\|$. *If* $A_2 \geq 0$, *each characteristic value of* A *is greater than or equal to the characteristic value of* A_1 *with the same index. If* $A_2 \leq 0$, *that is, if* $-A_2 \geq 0$, *each characteristic value of* A *is less than or equal to the characteristic value of* A_1 *with the same index.*

The proof of (10) follows from the obvious inequalities

$$\overset{+}{\mu}_{p+q-1} \leq \max_{12}(Af, f) \leq \max_{12}(A_1 f, f) + \max_{12}(A_2 f, f) \leq$$
$$\leq \max_1(A_1 f, f) + \max_2(A_2 f, f) = \overset{+}{\mu}_{1p} + \overset{+}{\mu}_{2q},$$

where the indices 1, 2, or 12 of the sign "max" indicate that the admissible elements f are restricted, not only by the condition $\|f\| = 1$, but also by the condition of being orthogonal to the characteristic elements

$$\overset{+}{\varphi}_{11}, \overset{+}{\varphi}_{12}, \ldots, \overset{+}{\varphi}_{1,p-1}$$

of A_1 corresponding to the characteristic values

$$\overset{+}{\mu}_{11}, \overset{+}{\mu}_{12}, \ldots, \overset{+}{\mu}_{1,p-1};$$

respectively to the characteristic elements

$$\overset{+}{\varphi}_{21}, \overset{+}{\varphi}_{22}, \ldots, \overset{+}{\varphi}_{2,q-1}$$

of A_2 corresponding to the characteristic values

$$\overset{+}{\mu}_{21}, \overset{+}{\mu}_{22}, \ldots, \overset{+}{\mu}_{2,q-1};$$

respectively to all these elements.

Leaving the task of proving the other propositions to the reader, we note the following consequence of inequalities (12) and (13):

If A_2 *has only* N *characteristic values different from* 0, *and if we denote by* μ_{1n} *and* μ_n *the characteristic values of* A_1 *and* $A = A_1 + A_2$, *arranged according to decreasing absolute values, we have*

(14) $$|\mu_{p+N}| \leq |\mu_{1p}| \qquad (p = 1, 2, \ldots).$$

In fact, let for example

$$\mu_{1p} = \mu_{1r}^{+},$$

then obviously

$$\mu_{1p} \geq |\mu_{1,p-r+1}^{-}|.$$

In view of (12) the number of μ_k^+ greater than μ_{1r}^+ does not exceed the number $r + N^+ - 1$, and by (13) the number of μ_k^- less than $\mu_{1,p-r+1}^-$ does not exceed the number $p - r + N^-$; consequently, the number of μ_k such that $|\mu_k| > |\mu_{1p}|$ does not exceed the number

$$(r + N^+ - 1) + (p - r + N^-) = p + N^+ + N^- - 1 = p + N - 1,$$

which was to be proved.

96. Another Method of Constructing Characteristic Values and Characteristic Elements

The following method of constructing the characteristic values different from 0 of a completely continuous symmetric transformation $A \neq 0$ is due to KELLOGG [**1**].

Let f_0 be an element such that $Af_0 \neq 0$, but otherwise arbitrary. Set

$$f_n = A^n f_0 \qquad (n = 1, 2, \ldots).$$

The relation

$$(15) \qquad (f_n, f_n) = (Af_{n-1}, f_n) = (f_{n-1}, Af_n) = (f_{n-1}, f_{n+1}) \qquad (n = 1, 2, \ldots)$$

makes it clear that if $f_n \neq 0$, then also $f_{n+1} \neq 0$. Since $f_0 \neq 0$ and $f_1 \neq 0$, we have for all n: $f_n \neq 0$.

Denoting the ratio $\|f_n\|/\|f_{n-1}\|$ by r_n and setting $\|f_0\| = r_0$, we have

$$f_n = (r_0 r_1 \ldots r_n) g_n,$$

where g_n is an element of $\mathfrak{H}$ of norm 1. We have the equations

$$(16) \qquad Ag_n = r_{n+1} g_{n+1} \text{ and } A^2 g_n = r_{n+1} r_{n+2} g_{n+2}.$$

It follows from the first equation that $r_{n+1} \leq \|A\|$. On the other hand, (15) implies that $r_n = (g_{n-1}, g_{n+1}) r_{n+1}$, hence $r_n \leq r_{n+1}$. Consequently the sequence $\{r_n\}$ converges to some positive quantity r.

Making use of equations (16), we obtain

$$\|g_n - g_{n+2}\|^2 = 2 - (g_n, g_{n+2}) - (g_{n+2}, g_n) = 2 - \left(g_n, \frac{A^2 g_n}{r_{n+1} r_{n+2}}\right) - \left(\frac{A^2 g_n}{r_{n+1} r_{n+2}}, g_n\right) =$$

$$= 2 - 2\frac{(Ag_n, Ag_n)}{r_{n+1} r_{n+2}} = 2 - 2\frac{r^2_{n+1}}{r_{n+1} r_{n+2}} = 2\left(1 - \frac{r_{n+1}}{r_{n+2}}\right) \to 0;$$

hence

$$g_n - g_{n+2} = g_n - \frac{1}{r_{n+1} r_{n+2}} A^2 g_n \to 0 \quad \text{as} \quad n \to \infty. \tag{17}$$

Since the sequence of elements

$$h_n = \frac{1}{r_{n+1} r_{n+2}} g_n$$

is bounded, it contains a subsequence for which $A^2 h_n$ converges to some element φ of $\mathfrak{H}$. The corresponding subsequence of $\{g_n\}$ converges, by (17), to the same element φ, and that of the sequence

$$\left\{g_n - \frac{1}{r_{n+1} r_{n+2}} A^2 g_n\right\}$$

to

$$\varphi - \frac{1}{r^2} A^2 \varphi;$$

therefore we have

$$\varphi - \frac{1}{r^2} A^2 \varphi = 0, \quad A^2 \varphi = r^2 \varphi.$$

Since φ is the limit of the sequence $\{g_n\}$ of elements of norm 1, φ has norm 1 also. Therefore φ is a characteristic element of A^2. We obtain a characteristic element of A from it in the following manner. We set

$$\psi = \varphi + \frac{1}{r} A\varphi, \quad \chi = \varphi - \frac{1}{r} A\varphi;$$

then

$$A\psi = r\psi, \quad A\chi = -r\chi.$$

Since $\psi + \chi = 2\varphi \neq 0$, the elements ψ and χ can not be zero simultaneously, hence at least one of the values r and $-r$ is a characteristic value of A.

A more profound study of the sequence $\{g_n\}$, as well as an application of the same method to certain *non-completely continuous* symmetric transformations, is found in the Notes [1], [2] of R. Wavre. The same method can also be applied with success to *nonsymmetric* completely continuous linear transformations which can be made symmetric if multiplied on the left by a suitable positive symmetric transformation (*symmetrizable* transformations).[6]

[6] Cf. Zaanen [1]. Integral equations with "symmetrizable" kernels have been studied by several authors, starting with Hilbert (Note 5); see the article in the Enzyklopädie by Hellinger and Toeplitz [*] (pp. 1536—1543) and, among recent contributions, see Reid [1], where bibliographical references are also found.

TRANSFORMATIONS WITH SYMMETRIC KERNEL

97. Theorems of Hilbert and Schmidt

Let us apply the theory of completely continuous symmetric transformations to the particular case of a transformation A of the functional space $L^2(a, b)$ which is generated by a *symmetric* or *Hermitian* kernel $A(x, y)$, that is, a kernel such that

$$A(x, y) = \overline{A(y, x)},$$

which belongs to the space $\boldsymbol{L}^2$ of square-summable functions in the plane domain $a \leq x \leq b$, $a \leq y \leq b$. By the remark made in Sec. 69, the transformation A can not be zero for all the elements of L^2 unless the kernel $A(x, y)$ *itself* is zero almost everywhere. Therefore we have the

THEOREM. *If the kernel $A(x, y)$ is not zero almost everywhere, the transformation A has a least one characteristic value different from 0, and each of its characteristic values is of finite multiplicity. There is an orthonormal sequence (finite or infinite) of characteristic functions $\varphi_i(x)$ of A corresponding to the characteristic values $\mu_i \neq 0$, and every function $g(x)$ belonging to L^2 admits the development, convergent in the mean:*

$$g(x) = h(x) + \sum_i (g, \varphi_i)\varphi_i(x) \tag{18}$$

where $h(x)$ is a function (depending on $g(x)$) such that $Ah(x) = 0$; consequently,

$$Ag(x) = \sum_i \mu_i(g, \varphi_i)\varphi_i(x). \tag{19}$$

We shall complete this theorem by several statements which are peculiar to the transformations being considered.

Since the functions

$$\Phi_i(x, y) = \varphi_i(x)\overline{\varphi_i(y)} \qquad (i = 1, 2, \ldots)$$

form an orthonormal system in $\boldsymbol{L}^2$, the series

$$\sum_i (A, \Phi_i)\Phi_i(x, y)$$

converges in the mean to some function $S(x, y)$ belonging to $\boldsymbol{L}^2$. We have

$$(A, \Phi_i) = \int_a^b\int_a^b A(x, y)\varphi_i(y)\overline{\varphi_i(x)}dxdy = (A\varphi_i, \varphi_i) = \mu_i;$$

hence

$$S(x, y) = \sum \mu_i \Phi_i(x, y).$$

For any f and g in L^2, the product $F(x, y) = g(x)\overline{f(y)}$ belongs to $\boldsymbol{L}^2$ and we have

$$(S, F) = \sum_i \mu_i(\Phi_i, F).$$

On the other hand, since $Ah = 0$ implies $(Af, h) = (f, Ah) = 0$, (18) furnishes:

$$(Af, g) = \sum_i (Af, \varphi_i)(\varphi_i, g) = \sum_i (f, A\varphi_i)(\varphi_i, g) = \sum \mu_i (f, \varphi_i)(\varphi_i, g);$$

hence

$$(A, F) = (Af, g) = \sum_i \mu_i(\Phi_i, F),$$

and consequently

$$(A, F) = (S, F).$$

The difference $A(x, y) - S(x, y)$ is therefore orthogonal to all these functions $F(x, y)$ and hence also to all the kernels of finite rank, and since these are everywhere dense in $\boldsymbol{L}^2$, it follows that $A(x, y) - S(x, y) = 0$ almost everywhere. From this we have the

THEOREM.[7] *Every function $A(x, y)$ which is symmetric and square-summable can be developed, in the sense of convergence in the mean, into the series*

$$A(x, y) = \sum_i \mu_i \varphi_i(x)\overline{\varphi_i(y)}, \tag{20}$$

where $\{\varphi_i(x)\}$ *denotes the orthonormal sequence of characteristic functions, and* $\{\mu_i\}$ *the sequence of corresponding characteristic values, of the transformation A generated by the kernel* $A(x, y)$.

It follows in particular that

$$\| A(x, y) \|^2 = \sum_i \mu_i^2, \tag{21}$$

and further, denoting by $S_N(x, y)$ the N-th partial sum of the series (20), that

$$\| A(x, y) - S_N(x, y) \|^2 = \sum_{i>N} \mu_i^2.$$

Up to now we have made no hypothesis concerning the arrangement of the characteristic values μ_i. If they are arranged in such a way that their absolute values form a nondecreasing sequence, that is,

$$|\mu_1| \geqq |\mu_2| \geqq |\mu_3| \geqq \ldots,$$

the partial sum $S_N(x, y)$ which is obviously a kernel of finite rank, possesses a remarkable minimum property:

For every symmetric kernel $A_N(x, y)$ of rank N, we have

$$\| A(x, y) - A_N(x, y) \| \geqq \| A(x, y) - S_N(x, y) \|.$$

In fact, since the number of characteristic values of A_N which are different from 0 is at most equal to N, the characteristic value of degree $p + N$ of

[7] E. SCHMIDT [**1**].

$A = (A - A_N) + A_N$, that is, μ_{p+N}, can not exceed in absolute value the characteristic value of degree p of $A - A_N$, which we shall denote by χ_p (see Sec. 95, inequality (14)). Applying relation (21) to $A - A_N$ instead of A, we obtain

$$|A - A_N|^2 = \sum_{p=1}^{\infty} \chi_p^2 \geq \sum_{p=1}^{\infty} \mu_{p+N}^2 = |A - S_N|^2,$$

which was to be proved.

Let us return to the development (19), which is valid in the sense of convergence in the mean for all functions of the form

$$Ag(x) = \int_a^b A(x, y)g(y)dy.$$

It is important to know if, in certain cases, this development is also convergent in the ordinary sense or even absolutely and uniformly convergent. This will be the case in particular if there exists a constant C such that

$$\int_a^b |A(x, y)|^2 dy < C^2 \tag{22}$$

for all values of x. (Example: $A(x, y) = |x - y|^{-\alpha}$, $\alpha < \frac{1}{2}$.)

In fact, condition (22) implies that if the sequence $\{f_n(x)\}$ converges in the mean to $f(x)$, its transform, $\{Af_n(x)\}$, converges uniformly to $Af(x)$, since

$$|Af(x) - Af_n(x)|^2 = |\int_a^b A(x, y) [f(y) - f_n(y)]dy|^2 \leq$$

$$\leq \int_a^b |A(x, y)|^2 dy \cdot \int_a^b |f(y) - f_n(y)|^2 dy \leq C^2 \|f - f_n\|^2.$$

But since the development (18) is convergent in the mean, development (19), which is derived from it by applying the transformation A to both sides, is uniformly convergent in the interval $a \leq x \leq b$.

Moreover, the convergence of development (19) is also absolute, that is, we can rearrange it in any arbitrary manner. In fact, ordinary convergence is, as we have just established, a consequence of mean convergence; but since mean convergence of a series with orthogonal terms $\Sigma\psi_i$ is equivalent to the convergence of the numerical series with non-negative terms $\Sigma\|\psi_i\|^2$, it does not depend on the arrangement of its terms.

Thus we have obtained the

THEOREM. *If the symmetric kernel* $A(x, y)$ *satisfies condition* (22), *the development* (19) *converges absolutely and uniformly, whatever the function* $g(x)$ *belonging to* L^2.

98. Mercer's Theorem

Development (20) of the kernel $A(x, y)$ is not necessarily uniformly convergent, even for continuous $A(x, y)$. However, for an important class of continuous kernels the convergence is uniform. We have namely the theorem of Mercer [**1**]:

THEOREM. *If the transformation A generated by the continuous symmetric kernel $A(x, y)$ is positive, that is, if $(Af, f) \geq 0$ for all f, or, equivalently, if all the characteristic values $\mu_i \neq 0$ are positive, the development* (20) *is uniformly convergent.*

This theorem extends immediately to the case where all but a finite number of the $\mu_i \neq 0$ are of the same sign, positive or negative.

We observe first that since the kernel $A(x, y)$ is continuous, all the image functions

$$Af(x) = \int_a^b A(x, y)f(y)dy$$

are continuous; therefore, in particular, all the characteristic functions $\varphi_i(x) = \dfrac{1}{\mu_i} A\varphi_i(x)$ are continuous. Consequently the "remainders"

$$A_n(x, y) = A(x, y) - \sum_{i=1}^{n} \mu_i\varphi_i(x)\overline{\varphi_i(y)} \qquad (n = 1, 2, \ldots),$$

are also continuous functions. Since we have

$$A_n(x, y) = \sum_{i=n+1}^{\infty} \mu_i\varphi_i(x)\overline{\varphi_i(y)}$$

in the sense of mean convergence, it follows that

$$\int_a^b\int_a^b A_n(x, y)f(y)\overline{f(x)}dxdy = \sum_{i=n+1}^{\infty} \mu_i(\varphi_i, f)\,(f, \varphi_i) \geq 0 \tag{23}$$

for every element f of L^2.

From this we deduce that $A_n(x, x) \geq 0$. In fact, if we had $A_n(x_0, x_0) < 0$, we should have by continuity $A_n(x, y) < 0$ in a neighborhood

$$x_0 - \delta < x < x_0 + \delta, \quad x_0 - \delta < y < x_0 + \delta$$

of the point (x_0, x_0). Setting $f(x) = 1$ for $x_0 - \delta < x < x_0 + \delta$ and $f(x) = 0$ elsewhere, integral (23) would become negative, a contradiction.

Hence we have

$$A_n(x, x) = A(x, x) - \sum_{i=1}^{n} \mu_i\varphi_i(x)\overline{\varphi_i(x)} \geq 0$$

for $n = 1, 2, \ldots$. From this we conclude that the series of positive terms

$$\sum_{i=1}^{\infty} \mu_i\varphi_i(x)\overline{\varphi_i(x)}$$

is convergent and that its sum is $\leq A(x, x)$. Denoting by M the maximum of the continuous function $A(x, x)$, we have by Cauchy's inequality:

$$\left|\sum_{i=m}^{n} \mu_i\varphi_i(x)\overline{\varphi_i(y)}\right|^2 \leq \sum_{i=m}^{n} \mu_i\,|\varphi_i(x)|^2 \sum_{i=m}^{n} \mu_i\,|\varphi_i(y)|^2 \leq M \sum_{i=m}^{n} \mu_i\,|\varphi_i(x)|^2. \tag{24}$$

From this it follows that the series

$$\sum_{i=1}^{\infty} \mu_i \varphi_i(x)\overline{\varphi_i(y)} \tag{25}$$

converges, for every fixed value of x, uniformly in y; its sum $B(x, y)$ is therefore a continuous function of y, and for every continuous function $f(y)$ we have

$$\int_a^b B(x, y)f(y)dy = \sum_{i=1}^{\infty} \mu_i\varphi_i(x) \int_a^b \overline{\varphi_i(y)}f(y)dy.$$

Now by one of the theorems proved in the preceding section, the series in the second member converges to $Af(x)$. Hence we have

$$\int_a^b [B(x, y) - A(x, y)]f(y)dy = 0;$$

setting in particular $f(y) = \overline{B(x, y) - A(x, y)}$ (for a fixed value of x), it follows that $B(x, y) - A(x, y) = 0$ for $a \leq y \leq b$, hence

$$A(x, x) = B(x, x) = \sum_{i=1}^{\infty} \mu_i \,|\, \varphi_i(x) \,|^2.$$

Since the terms of this series are positive continuous functions of x and its sum $A(x, x)$ is a continuous function, it follows from a known theorem of Dini that the series converges *uniformly*. Applying Cauchy's inequality (24) again, we deduce from this that series (25) converges uniformly with respect to its two variables x and y simultaneously, which was to be proved.

Whatever be the continuous symmetric kernel $A(x, y)$, its iterate

$$A_2(x, y) = \int_a^b A(x, z)A(z, y)dz$$

is continuous and of positive type. In fact,

$$(A_2 f, f) = (A^2 f, f) = (Af, Af) \geq 0.$$

The characteristic functions $\varphi_i(x)$ of A are also characteristic functions for A^2, but they correspond to the squares of the characteristic values μ_i of A:

$$A^2\varphi_i = A(A\varphi_i) = A(\mu_i\varphi_i) = \mu_i^2\varphi_i.$$

The sequence $\mu_1^2, \mu_2^2, \ldots$ contains all the characteristic values of A^2 different from 0, each as many times as its multiplicity indicates. If not, there would be a characteristic function φ corresponding to a characteristic value $\mu \neq 0$ of A^2 and orthogonal to all the φ_i. This would be in contradiction to the fact that

$$\mu\varphi = A^2\varphi = \sum_{i=1}^{\infty} (A^2\varphi, \varphi_i)\varphi_i = \sum_{i=1}^{\infty} (\varphi, A^2\varphi_i)\varphi_i = \sum_{i=1}^{\infty} \mu_i^2(\varphi, \varphi_i)\varphi_i = 0.$$

By the theorem of Mercer we therefore have, for the iterate of an arbitrary continuous kernel $A(x, y)$, the *uniformly convergent* development:

$$A_2(x, y) = \sum_{i=1}^{\infty} \mu_i^2\varphi_i(x)\overline{\varphi_i(y)}.$$

APPLICATIONS TO THE VIBRATING STRING PROBLEM AND TO ALMOST PERIODIC FUNCTIONS

99. The Vibrating String Problem. The Spaces *D* and *H*

Of the numerous applications which the theory we have just discussed has to problems of mathematical physics, we shall consider only one of the simplest, which is nevertheless characteristic; this is the application to the problem of small planar transversal vibrations of a homogeneous or non-homogeneous string.[8]

We choose the (x, y) plane as the plane of vibration, and assume that the extremities of the string are fixed at the points $(0, 0)$ and $(1, 0)$. The movement of the string will be described by a function $y(x, t)$ giving the ordinate at the time t.

The kinetic energy is expressed by the Stieltjes integral

$$E_c = \tfrac{1}{2}\int_0^1 \dot{y}^2\, dm(x),^{9}$$

where we have denoted by $m(x)$ the mass borne by the segment $(0, x)$ of the string. The potential energy E_p is proportional to the increase of the length of the string, that is, to

$$\Delta = \int_0^1 \sqrt{1 + y'^2}\, dx - 1;$$

the coefficient of proportionality is equal to the tension, which for simplicity of writing we assume equal to unity. Let us restrict ourselves to the study of small vibrations, that is, assume that in the above integral $\sqrt{1 + y'^2}$ can be replaced by its approximate value $1 + \frac{1}{2}y'^2$. Then we shall have

$$E_p = \tfrac{1}{2}\int_0^1 y'^2\, dx.$$

The functions $y(x, t)$ corresponding to the movements which the string is capable of making in the time interval $0 \leq t \leq T$ are, according to Hamilton's principle, those which render the integral

$$\int_0^T (E_c - E_p)dt = \int_0^T dt\, [\tfrac{1}{2}\int_0^1 \dot{y}^2\, dm(x) - \tfrac{1}{2}\int_0^1 y'^2\, dx] \tag{26}$$

stationary, when we assume the configurations given at the initial and final

[8] We follow the discussion due to Sz.-Nagy [**6**].

[9] We shall write $\dot{y}$ instead of $\dfrac{\partial y}{\partial t}$ and y' instead of $\dfrac{\partial y}{\partial x}$.

instants: $y(x, 0)$ and $y(x, T)$. Hence we have

$$\frac{d}{d\varepsilon}\int_0^T dt\,[\tfrac{1}{2}\int_0^1 (\dot{y} + \varepsilon\dot{\eta})^2 dm(x) - \tfrac{1}{2}\int_0^1 (y' + \varepsilon\eta')^2 dx] = 0$$

for $\varepsilon = 0$, that is,

$$\int_0^T dt\,[\int_0^1 \dot{y}\dot{\eta}\, dm(x) - \int_0^1 y'\eta' dx] = 0 \tag{27}$$

for every *admissible* function $\eta(x, t)$ such that

$$\eta(x, 0) = 0, \quad \eta(x, T) = 0. \tag{28}$$

We say that a function $y(x, t)$ is admissible when it satisfies certain conditions imposed by the very nature of the problem, namely the *limit conditions*

$$y(0, t) = 0, \quad y(1, t) = 0, \tag{29}$$

and the conditions assuring that the derivatives and the integrals appearing in (26) have meaning.

In order to be able to make these conditions precise, we introduce two functional spaces, D and H, which are realizations of real abstract Hilbert space.

The space D consists of the real-valued functions $u(x)$ which are defined in the interval $0 \leq x \leq 1$, absolutely continuous, zero at the points 0 and 1, and for which $u'(x)$ — existing almost everywhere — is square-summable; scalar products and norms are defined by means of the derivatives as

$$(u, v)_D = \int_0^1 u'(x)v'(x)dx, \quad \|u\|_D = \sqrt{(u, u)_D}.$$

One verifies without difficulty that D satisfies the axioms of (real) abstract Hilbert space, in particular, that it is complete.

Let H be the space L^2 of real-valued functions $f(x)$ defined in the interval $0 \leq x \leq 1$, the integral being taken with respect to the strictly increasing function $m(x)$; scalar products and norms in H are therefore defined by

$$(f, g)_H = \int_0^1 f(x)g(x)dm(x), \quad \|f\|_H = \sqrt{(f, f)_H}.$$

The *admissibility conditions* are now expressed as follows:

a) *$y(x, t)$ is an element of the space D and depends on the parameter t in a weakly continuous manner, that is, $(y, u)_D$ is a continuous function of t for every element u of D;*

b) *$y(x, t)$, considered as an element of the space H depending on the parameter t, has a weak derivative which is weakly continuous with respect to t, that is, $(y, f)_H$ has a continuous derivative with respect to t for every element f of H.*

The admissible functions obviously form a linear set. It follows from a) that $\|y\|_D^2$ is an upper semi-continuous function of t and is therefore summable in every finite interval $0 \leq t \leq T$; consequently,

$$(y_1, y_2)_D = \tfrac{1}{4} \|y_1 + y_2\|_D^2 - \tfrac{1}{4} \|y_1 - y_2\|_D^2$$

is also summable, for all admissible functions $y_1(x, t)$ and $y_2(x, t)$. From b) it follows that the quotient

$$z(x; t, h) = \frac{y(x, t+h) - y(x, t)}{h},$$

considered as an element of H, satisfies the Cauchy condition for weak convergence when $h \to 0$; therefore there exists a weak limit in H which we shall denote $\dot{y}(x, t)$. Since we have by definition

$$(\dot{y}, f)_H = [(y, f)_H]^{\cdot},$$

this element $\dot{y}$ is a weakly continuous function of the parameter t. Consequently $\|\dot{y}\|_H^2$ and $(\dot{y}_1, \dot{y}_2)_H$ are summable functions of t in every finite interval $0 \leq t \leq T$.

We shall need the following inequalities for functions $u(x)$ belonging to the space D—simple consequences of Schwarz's inequality:

$$\max |u(x)| \leq \tfrac{1}{2} \|u\|_D, \tag{30}$$

$$|u(x_1) - u(x_2)| \leq \sqrt{|x_1 - x_2|}\, \|u\|_D. \tag{31}$$

We have, in fact,

$$2\,|u(x)| = \left|\int_0^x u'(\xi)d\xi - \int_x^1 u'(\xi)d\xi\right| = \left|\int_0^1 [\operatorname{sgn}(\xi - x)]\, u'(\xi)d\xi\right| \leq$$

$$\leq \left\{\int_0^1 [\operatorname{sgn}(\xi - x)]^2\, d\xi \cdot \int_0^1 [u'(\xi)]^2\, d\xi\right\}^{\frac{1}{2}} = \|u\|_D$$

and

$$|u(x_1) - u(x_2)| = \left|\int_{x_2}^{x_1} u'\, dx\right| \leq \left\{\left|\int_{x_2}^{x_1} dx\right| \cdot \int_0^1 u'^2 dx\right\}^{\frac{1}{2}} = \sqrt{x_1 - x_2}\, \|u\|_D.$$

It follows from inequality (30) that the convergence of a sequence $\{u_n(x)\}$ in the metric of the space D implies its ordinary convergence, even its uniform convergence, in the entire interval $0 \leq x \leq 1$, and that the functions lying in a "ball" $\|u\|_D \leq C$ of the space D are also uniformly bounded in the ordinary sense; moreover, by (31) these functions are uniformly equi-continuous.

But a known theorem of ARZELÀ [**2**] asserts that from every sequence of functions which are uniformly bounded and uniformly equi-continuous it is possible to select a uniformly convergent subsequence. Therefore *from every* *sequence $\{u_k(x)\}$ which is bounded in the sense of the metric of the space D it is possible to select a uniformly convergent subsequence.*

For two arbitrary elements u and v of D we have, by (30):

$$|(u, v)_H| = |\int_0^1 uv\, dm(x)| \leq \max |u| \cdot \max |v| \cdot M \leq \frac{M}{4} \|u\|_D \|v\|_D,$$

where M denotes the total mass of the string, $M = m(1)$. The expression $(u, v)_H$, which is symmetric in u and v, is therefore a bounded bilinear form on the space D in the sense of Sec. 84; consequently there exists an (obviously symmetric) linear transformation A of the space D into itself such that

$$(u, v)_H = (Au, v)_D.$$

Moreover, this transformation is *completely continuous.* In fact, from every sequence $\{u_n(x)\}$ bounded in the sense of the metric of D it is possible to select a uniformly convergent subsequence $\{w_n(x)\}$, and we obviously have

$$(A(w_n - w_m), w_n - w_m)_D = (w_n - w_m, w_n - w_m)_H = \int_0^1 (w_n - w_m)^2 dm(x) \to 0$$

as $m, n \to \infty$, which is simply the fourth criterion for complete continuity given in Sec. 85.

We can therefore apply the results of Sec. 93:

There exists an orthonormal sequence $\{u_n\}$ of elements of D and a numerical sequence $\{\mu_n\}$ such that, for every pair u, v of elements of D,

$$(u, v)_H = \sum \mu_n(u, u_n)_D (u_n, v)_D. \tag{32}$$

Since $(u, u)_H \geq 0$, the quantities μ_n are all positive, $\mu_n = \frac{1}{\varkappa_n^2}$ and as $(u, u)_H$ is zero only for $u(x) \equiv 0$, since the function $m(x)$ is strictly increasing, the orthonormal sequence $\{u_n\}$ is complete, that is, every function $u(x)$ belonging to D admits the development

$$u(x) = \sum a_n u_n(x) \text{ where } a_n = (u, u_n)_D;$$

this development converges in the metric of D and, a fortiori, converges uniformly. Since the arrangement of the terms is arbitrary, the convergence is also absolute.

We deduce from (32) in particular that

$$(u_n, u_n)_H = \mu_n \text{ and } (u_n, u_m)_H = 0 \text{ for } n \neq m,$$

that is, that *the functions*

$$\varphi_n(x) = \varkappa_n u_n(x)$$

form an orthonormal sequence in the space H. This sequence is also *complete.* In fact, we know that the continuous functions form an everywhere dense set in H; it even suffices to consider only the functions which are zero at the points 0 and 1, and whose graphs consist of a finite number of line segments. But since these functions also belong to D, they can be approximated uniformly and, a fortiori, in the metric of H by linear combinations of the u_n or of the φ_n, which proves our assertion.

Summarizing: *Every function $f(x)$ belonging to the space H has the development, convergent in the metric of H,*

$$f(x) = \sum b_n \varphi_n(x), \textit{ where } b_n = (f, \varphi_n)_H.$$

100. The Vibrating String Problem. Characteristic Vibrations

We return to our problem, which consists in determining the possible movements $y(x, t)$ of the string. Let

$$y(x, t) = \sum c_k(t) u_k(x), \quad \dot{y}(x, t) = \sum d_k(t) \varphi_k(x)$$

be the developments converging in the metric of D and H respectively; the coefficients

$$c_k(t) = (y, u_k)_D \text{ and } d_k(t) = (\dot{y}, \varphi_k)_H$$

are continuous functions of t and we have

$$d_k(t) = (\dot{y}, \varphi_k)_H = [(y, \varphi_k)_H]^{\cdot} = \left[\sum_n \frac{1}{\varkappa_n^2} (y, u_n)_D (u_n, \varphi_k)_D\right]^{\cdot} =$$

$$= \frac{1}{\varkappa_k} [(y, u_k)_D]^{\cdot} = \frac{1}{\varkappa_k} \dot{c}_k(t).$$

Let $\gamma_n(t)$ be a function which is zero at the points $t = 0$ and $t = T$ and possesses a continuous derivative, but is otherwise arbitrary. Then

$$\eta(x, t) = \gamma_n(t) u_n(x)$$

is an admissible function satisfying conditions (28); inserting it in equation (27), we obtain

$$\int_0^T dt \left(\frac{\dot{c}_n \dot{\gamma}_n}{\varkappa_n^2} - c_n \gamma_n \right) = 0.$$

Denoting by $C_n(t)$ a primitive function of $c_n(t)$, we obtain by integrating by parts:

$$\int_0^T \left(\frac{\ddot{C}_n}{\varkappa_n^2} + C_n \right) \dot{\gamma}_n \, dt = 0.$$

Since $\dot{\gamma}_n$ can be an arbitrary function whose integral over $(0, T)$ is zero, it follows from this that the continuous function $\ddot{C}_n + \varkappa_n^2 C_n$ is constant, hence

$$c_n(t) = a_n \cos \varkappa_n t + b_n \sin \varkappa_n t.$$

A function $y(x, t)$ which corresponds to a movement of the string is therefore necessarily of the form

$$y(x, t) = \sum_n (a_n \cos \varkappa_n t + b_n \sin \varkappa_n t) u_n(x),$$

the coefficients being subject to the condition that the series

$$\sum_n a_n^2 \quad \text{and} \quad \sum_n b_n^2$$

converge; in fact, we have

$$\sum_n a_n^2 = \sum_n c_n^2(0) = \|y(x, 0)\|_D^2, \ \sum_n b_n^2 = \sum_n \frac{\dot{c}_n^2(0)}{\varkappa_n^2} = \|\dot{y}(x, 0)\|_H^2.$$

This condition is also sufficient, that is, it assures that equation (27) is verified, whatever be the admissible function $\eta(x, t)$ satisfying (28). Making use of the developments

$$\eta(x, t) = \sum_k \gamma_k(t) u_k(x), \qquad \dot{\eta}(x, t) = \sum_k \delta_k(t) \varphi_k(x),$$

where

$$\delta_k = \frac{\dot{\gamma}_k}{\varkappa_k} \quad \text{and} \quad \gamma_k(0) = \gamma_k(T) = 0,$$

the equation (27) to be proved takes on the form

$$\int_0^T dt \sum_k \left[\frac{\dot{c}_k \dot{\gamma}_k}{\varkappa_k^2} - c_k \gamma_k \right] = 0,$$

where

$$c_k(t) = a_k \cos \varkappa_k t + b_k \sin \varkappa_k t.$$

Since the integral of each term is equal to 0, there remains to show only that the series can be integrated term by term. But, this follows from the fact that its partial sums are majorized by

$$\left\{ \sum_k (a_k^2 + b_k^2) \cdot \sum_k \frac{\dot{\gamma}_k^2}{\varkappa_k^2} \right\}^{\frac{1}{2}} + \{ \sum_k (a_k^2 + b_k^2) \cdot \sum_k \gamma_k^2 \}^{\frac{1}{2}} = \{ \sum_k (a_k^2 + b_k^2) \}^{\frac{1}{2}} (\|\dot{\eta}\|_H + \|\eta\|_D),$$

which is a bounded function of t.

Hence we have the following result:

THEOREM. *The possible movements of the string are described by the functions which can be represented in the form*

(*) $$y(x, t) = \sum_k (a_k \cos \varkappa_k t + b_k \sin \varkappa_k t) u_k(x),$$

where the coefficients are subjected to the single condition that $\sum (a_k^2 + b_k^2)$ *converge. Under this condition, the series converges in the metric of the space D and, a fortiori, it converges absolutely and uniformly.*

Given two arbitrary functions belonging to the spaces D and H respectively:

$$f(x) = \sum_k a_k u_k(x), \quad g(x) = \sum_k b_k \varphi_k(x),$$

there therefore exists a movement $y(x, t)$ of the string satisfying the *initial conditions*

$$y(x, 0) = f(x), \quad \dot{y}(x, 0) = g(x),$$

namely, the function (*) formed by starting with the coefficients a_k, b_k of $f(x)$ and $g(x)$.

Therefore all the movements of the string result from the superposition of particular purely harmonic vibrations $(a_k \cos \varkappa_k t + b_k \sin \varkappa_k t)u_k(x)$, which we call *characteristic vibrations*.

The transformation A of the space D, associated with the bilinear form $(u, v)_H$ by means of the formula $(u, v)_H = (Au, v)_D$, can be easily calculated in the explicit form:

$$Au(x) = -\int_0^x dz\, [\int_0^z u(s)dm(s) - \int_0^1 dt \int_0^t u(r)dm(r)].$$

The functions $u_k(x)$ are characteristic functions of A,

$$Au_k(x) = \frac{1}{\varkappa_k^2}\, u_k(x).$$

We can write this equation in the obvious differential form

$$du_k'(x) + \varkappa_k^2 u_k(x)dm(x) = 0,$$

or, also, when (mx) possesses a density function $\varrho(x)$ of which it is the integral, in the form

$$u_k''(x) + \varkappa_k^2 \varrho(x)u_k(x) = 0.$$

In particular, if $\varrho(x) \equiv 1$, that is, if the string is *homogeneous*, we have the differential equation

$$u'' + \varkappa^2 u = 0;$$

keeping in mind the conditions

$$u(0) = 0, \quad u(1) = 0,$$

we obtain

$$\varkappa_k = k\pi \quad \text{and} \quad \varphi_k(x) \equiv \varkappa_k u_k(x) = \sqrt{2} \sin \pi k x \quad (k = 1, ,2 \ldots).$$

We can show, even in the general case, that the characteristic values $\mu_k = \dfrac{1}{\varkappa_k^2}$ are *simple*, that is, that the characteristic vibrations corresponding to the same "characteristic frequency" $\varkappa_k$ are all of the same "form" $u_k(x)$ and differ only in their phases and amplitudes. Furthermore, the number of zeros of $u_k(x)$ in the interior of (0, 1) equals $n - 1$, and between two consecutive zeros of $u_{n-1}(x)$ there is always a zero of $u_n(x)$ (Sturm's oscillation theorem).

101. The Space of Almost Periodic Functions

Among the applications of the theory of completely continuous symmetric transformations to various problems in analysis, one of the most interesting is that which relates to the theory of almost periodic functions.

According to the original definition, given by HARALD BOHR, *a continuous function $f(x)$ $(-\infty < x < \infty)$ is said to be almost periodic if for every $\varepsilon > 0$ it is possible to find a quantity $l > 0$ such that every interval of the x-axis of length l contains at least one number τ such that*

$$|f(x+\tau) - f(x)| < \varepsilon \quad \textit{for} \quad -\infty < x < \infty.$$

Later BOCHNER found the following equivalent definition: *The continuous function $f(x)$ is almost periodic if from every infinite sequence $\{f(x + \xi_n)\}$ of functions obtained from $f(x)$ by means of translations of the x-axis one can select a uniformly convergent subsequence.*

An introduction to the theory of almost periodic (abbreviated: a.p.) functions, and also the proof of the equivalence of the two definitions, is found in the book by FAVARD [*], among others.

The theory begins with the following theorems, the proofs of which are more or less simple:

a) Every continuous a. p. function is bounded and uniformly continuous on the x-axis.

b) Continuous periodic functions are also a. p.

c) The sums and products of a. p. functions and the limits of uniformly convergent sequences of a. p. functions are also a. p.

d) For every continuous a. p. function $f(x)$, the integral means

$$\frac{1}{2T} \int_{-T}^{T} f(x+t)dt$$

tend, as $T \to \infty$, to a limit independent of x, uniformly over the entire x-axis. This limit, which we denote by

$$\underset{t}{M}\{f(t)\},$$

is a linear functional defined for continuous a. p. functions; it has the same value for the functions $f(a \pm t)$ whatever be a, and we have in particular $\underset{t}{M}\{1\} = 1$. If $f(t)$ is real and non-negative, but is not identically zero, $\underset{t}{M}\{f(t)\}$ is positive.

e) If f and g are a. p., the composite function

$$f \times g = \underset{t}{M}\{f(x-t)g(t)\}$$

is also a. p., and we have

$$f \times g = g \times f, \quad f \times (g \times h) = (f \times g) \times h.$$

We denote the set of continuous a. p. functions by $\mathfrak{P}$. Setting

$$(f, g) = \underset{t}{M}\{f(t)\overline{g(t)}\} \text{ and } \|f\| = (f, f)^{\frac{1}{2}},$$

$\mathfrak{P}$ becomes an incomplete Hilbert space. We could complete it by adding certain ideal elements, but since these elements do not have an obvious representation as functions we prefer to use the incomplete space.

The dimension of $\mathfrak{P}$ equals the power of the continuum. In fact, it can not be greater, since the power of the set of *all* continuous functions is that of the continuum. On the other hand, it can not be smaller, because it contains an orthonormal system having the power of the continuum, for example, the system of functions

$$e_\nu(x) = e^{i\nu x},$$

where ν runs through all real numbers; in fact

$$\|e_\nu\|^2 = \underset{t}{M}\{|e_\nu|^2\} = \underset{t}{M}\{1\} = 1,$$

$$(e_\nu, e_\mu) = \underset{t}{M}\{e^{i(\nu-\mu)t}\} = \lim_{T\to\infty} \frac{1}{2T} \frac{e^{i(\nu-\mu)T} - e^{-i(\nu-\mu)T}}{i(\nu-\mu)} = 0 \text{ for } \nu \neq \mu.$$

We assign the "generalized Fourier coefficients"

$$c_\nu = (f, e_\nu)$$

to the continuous a. p. function $f(x)$. Whatever be the real numbers μ_k $(k = 1, 2, \ldots, r)$, we have the Bessel inequality:

$$\sum_{k=1}^{r} |c_{\mu_k}|^2 \leq \|f\|^2,$$

from which it follows that the number of values of ν for which $|c_\nu| \geq p$, where p is a given positive quantity, can not exceed $\frac{1}{p^2}\|f\|^2$. Consequently, $c_\nu = 0$ for all real values of ν with the exception of at most a finite or denumerable number of values $\nu_1, \nu_2, \nu_3, \ldots$. The series

$$\sum_k c_{\nu_k} e_{\nu_k}(x)$$

is called the "generalized Fourier series" of $f(x)$.

The fundamental problem of the theory is to know if, conversely, this series determines, in its turn, the function $f(x)$. The answer is in the affirmative:

THEOREM. *The generalized Fourier series associated with the continuous almost periodic function $f(x)$ converges in the metric of the space $\mathfrak{P}$ to the function $f(x)$.*

By virtue of "Bessel's identity"

$$\|f - \sum_{k=1}^{n} c_{\nu_k} e_{\nu_k}\|^2 = \|f\|^2 - \sum_{k=1}^{n} |c_{\nu_k}|^2,$$

everything reduces to proving "Parseval's formula":

$$\|f\|^2 = \sum_{k=1}^{\infty} |c_{\nu_k}|^2. \tag{33}$$

There are several approaches which lead to this fundamental theorem of BOHR; we shall follow one which starts from the theory which we have just discussed, and which was discovered by H. WEYL and, in its definitive form, by RELLICH.[10]

102. Proof of the Fundamental Theorem on Almost Periodic Functions

Before taking up the proof, we introduce the following notation:

$$f^*(x) = \overline{f(-x)}.$$

It is clear that $f^*(x)$ is, simultaneously with $f(x)$, a continuous a. p. function and that

$$(f^*)^* = f, \quad (f \times g)^* = g^* \times f^*, \quad f \times g^*(0) = (f, g),$$

and

$$(f^*, g^*) = \overline{(f, g)}. \tag{34}$$

We shall say that f is *symmetric* when $f = f^*$. In particular, $e_\nu(x) = e^{i\nu x}$ is such a function. The scalar product of two symmetric elements of the space $\mathfrak{P}$ is, by (34), real-valued. Moreover, every element $f(x)$ of $\mathfrak{P}$ admits a decomposition of the form $f = f_1 + if_2$ with f_1 and f_2 symmetric; we have only to set

$$f_1 = \tfrac{1}{2}(f + f^*) \text{ and } f_2 = \frac{1}{2i}(f - f^*).$$

Taking into consideration that the scalar products (f_1, f_2), (f_1, e_ν) and (f_2, e_ν)

[10] Cf. WEYL [5]; WINTNER [*]; RELLICH [1].

are real-valued, we easily verify the relations

$$\|f\|^2 = \|f_1\|^2 + \|f_2\|^2 \text{ and } |(f, e_\nu)|^2 = |(f_1, e_\nu)|^2 + |(f_2, e_\nu)|^2.$$

It follows that *it suffices to prove equation* (33) *for symmetric functions.*

Let $a(x)$ be a symmetric continuous a. p. function. By means of the formula

$$Ah = a \times h,$$

it generates a transformation A of the space $\mathfrak{P}$ into itself. A is obviously additive and homogeneous; furthermore, it is *bounded*, since we have

$$|Ah(x)|^2 = |\underset{t}{M}\{a(x-t)h(t)\}|^2 \leq \underset{t}{M}\{|a(x-t)|^2\}\underset{t}{M}\{|h(t)|^2\};$$

hence

$$|Ah(x)| \leq \|a\|\,\|h\| \tag{35}$$

and, a fortiori,

$$\|Ah\| \leq \|a\|\,\|h\|.$$

The transformation A is *symmetric*:

$$(Ag, h) = (a \times g) \times h^*(0) = h^* \times (a \times g)\,(0) = (h^* \times a) \times g(0) =$$
$$= (a \times h^*) \times g(0) = g \times (a \times h)^*(0) = (g, Ah).$$

Finally, A is *completely continuous.* We shall show, in fact, that from every bounded infinite set $\{h\}$ of elements of $\mathfrak{P}$, $\|h\| \leq C$, we can select a sequence $\{h_n\}$ for which the sequence $\{Ah_n\}$ is uniformly convergent; the limit is necessarily continuous and a. p. itself, hence it is an element of $\mathfrak{P}$; the sequence $\{Ah_n\}$ tends to this element, a fortiori, in the sense of the metric of the space $\mathfrak{P}$.

Let $\{r_m\}$ be a sequence of real numbers everywhere dense on the real axis, and set $a_m(x) = a(x - r_m)$. We use the diagonal process to select from $\{h\}$ a sequence $\{h_n\}$ such that for every m $\{(h_n, a_m)\}$ converges as $n \to \infty$. We shall show the sequence $\{Ah_n\}$ converges uniformly on the x-axis.

If this were not the case, there would be an $\varepsilon_0 > 0$, two sequences of integers tending to ∞, $\{n_k\}$ and $\{m_k\}$, and a sequence of real numbers $\{x_k\}$, such that

$$|Ah_{n_k}(x_k) - Ah_{m_k}(x_k)| > \varepsilon_0,$$

that is,

$$|\underset{t}{M}\{a(x_k - t)\,[h_{n_k}(t) - h_{m_k}(t)]\}| > \varepsilon_0, \tag{36}$$

for $k = 1, 2, \ldots$. By virtue of the uniform continuity of $a(x)$ we can obviously assume that the x_k are selected from the sequence $\{r_m\}$. By the second definition of a. p. functions, we can also assume that the sequence $\{a_{x_k}(x) = a(x - x_k)\}$

is uniformly convergent. We then choose the integer j so that

$$|a_{x_k}(x) - a_{x_j}(x)| < \frac{\varepsilon_0}{4C}$$

for $k \geq j$, and, fixing j, we choose the integer N such that

$$|(h_n - h_{n'}, a_{x_j})| < \frac{\varepsilon_0}{2}$$

for $n, n' \geq N$. We then have, when $k \geq j$ and $n_k, m_k \geq N$,

$$|(h_{n_k} - h_{m_k}, a_{x_k})| \leq |(h_{n_k}, a_{x_k} - a_{x_j})| + |(h_{n_k} - h_{m_k}, a_{x_j})| + |(h_{m_k}, a_{x_j} - a_{x_k})| \leq$$

$$\leq C\frac{\varepsilon_0}{4C} + \frac{\varepsilon_0}{2} + C\frac{\varepsilon_0}{4C} = \varepsilon_0,$$

in contradiction with (36). This proves our assertion.

The transformation A is thus linear, symmetric, and completely continuous. Therefore there exists, by Sec. 93, an orthonormal sequence of characteristic elements $\{\varphi_k\}$ of A, corresponding to the characteristic values μ_k, such that for all elements g of $\mathfrak{P}$:

(37) $$Ag = \sum_k \mu_k(g, \varphi_k)\varphi_k$$

in the sense of the convergence in the metric of $\mathfrak{P}$.

Making use of inequality (35), we see that

$$|\sum_{k=m}^{n} \mu_k(g, \varphi_k)\varphi_k(x)| = |A\sum_{k=m}^{n} (g, \varphi_k)\varphi_k(x)| \leq \|a\| \, \|\sum_{k=m}^{n} (g, \varphi_k)\varphi_k\| =$$

$$= \|a\| \, [\sum_{k=m}^{n} |(g, \varphi_k)|^2]^{\frac{1}{2}} \to 0$$

as $m, n \to \infty$. Series (37) therefore converges uniformly; its sum "in the mean" $Ag(x)$ is necessarily also its sum in the ordinary sense.

Setting in particular $g = a$ and $x = 0$, it follows that

$$Aa(0) = \sum_k \mu_k(a, \varphi_k)\varphi_k(0);$$

since

$$Aa(0) = a \times a(0) = (a, a) = \|a\|^2$$

and

$$\mu_k\varphi_k(0) = a \times \varphi_k(0) = \varphi_k \times a(0) = (\varphi_k, a),$$

this yields

(38) $$\|a\|^2 = \sum_k |(a, \varphi_k)|^2.$$

There remains to show only that the functions $\varphi_k(x)$ are of the form

$e^{i\nu x}$, or at least that they can be replaced by functions of this form. In fact, we had some freedom in the choice of the system $\{\varphi_k\}$: to a characteristic value μ of multiplicity p ($p \geq 1$) we were able to associate p *arbitrary* characteristic functions forming an orthonormal system in the characteristic subspace E_μ of dimension p corresponding to μ. We shall now select these functions in a definite manner, as follows.

We note first that if $\varphi(x)$ is an element of E_μ, the same is true for $\varphi(x+\xi)$; in fact,

$$A[\varphi(x+\xi)] = \underset{t}{M}\{a(x-t)\varphi(t+\xi)\} = \underset{t}{M}\{a(x+\xi-\tau)\varphi(\tau)\} =$$
$$= (A\varphi)\,(x+\xi) = \mu\varphi(x+\xi).$$

Thus to every real ξ there corresponds a transformation

$$U_\xi\varphi(x) = \varphi(x+\xi)$$

of E_μ into itself, it is obviously linear and we have

$$U_0 = I,\ U_\xi U_\eta = U_{\xi+\eta}, \tag{39}$$

$$U_\xi\varphi \to U_\eta\varphi \text{ for } \xi\to\eta, \tag{40}$$

$$(U_\xi\varphi, U_\xi\psi) = (\varphi,\psi). \tag{41}$$

It results from these conditions, as we shall show in the next section, that there exists in E_μ an orthonormal system $\psi_1(x), \psi_2(x), \ldots, \psi_p(x)$ and real numbers $\omega_1, \omega_2, \ldots, \omega_p$ such that

$$U_\xi[\psi_k(x)] = e^{i\omega_k\xi}\psi_k(x) \qquad (k = 1, 2, \ldots, p).$$

In particular, we have

$$\psi_k(\xi) = (U_\xi\psi_k)\,(0) = e^{i\omega_k\xi}\psi_k(0),$$

hence

$$\psi_k(x) = a_k e^{i\omega_k x}.$$

Since the ψ_k form an orthonormal system, the ω_k are distinct and $|a_k| = 1$. The functions $\psi_k(x)$ can therefore be replaced by the functions $e_{\omega_k}(x)$ ($k = 1, 2, \ldots, p$), which also form an orthonormal system in E_μ.

Summarizing, we can assume that the system $\{\varphi_k\}$ of characteristic elements of the transformation A is of the form $\{e_{\nu_k}\}$, where the "frequencies" $\nu_1, \nu_2, \ldots$ are distinct. Hence we have, by (38),

$$\|a\|^2 = \sum_k |c_{\nu_k}|^2,$$

which completes the proof.

103. Isometric Transformations of a Finite Dimensional Space

Here is the proof of the auxiliary theorem we just used concerning the family $\{U_\xi\}$ $(-\infty < \xi < \infty)$ of linear transformations of a complex space E of finite dimension p satisfying conditions (39)–(41).

Condition (41) means that the transformation U_ξ leaves scalar products and norms invariant — that it is therefore *isometric*. This condition can obviously be stated in the form: $U_\xi^* U_\xi = I$. Since, by (39), U_ξ possesses an inverse $U_\xi^{-1} = U_{-\xi}$, we therefore have

$$U_\xi^* = U_\xi^{-1}.$$

A transformation possessing this property is said to be *unitary*. It is obvious that in a finite dimensional space every isometric transformation has an inverse and consequently is also unitary.

It follows from condition (39) that the transformations belonging to the family $\{U_\xi\}$ are all mutually permutable:

$$U_\xi U_\eta = U_\eta U_\xi.$$

Now we have the

THEOREM. *A family $\{U\}$ of permutable unitary transformations of a complex space of finite dimension p always has a complete orthonormal system of common characteristic elements.*

The case of dimension $p = 1$ being clear, we assume our proposition has been established for spaces of dimension less than p. Setting aside the obvious case where all the transformations belonging to the family $\{U\}$ are numerical multiples of the identity transformation I, let U_0 be one of the given transformations which is not a numerical multiple of I. Let ϱ be a characteristic value of U_0 and let E_ϱ be the corresponding characteristic subspace. The dimension of E is then less than p.

All the transformations U belonging to the given family $\{U\}$ transform the subspace E_ϱ into itself; in fact, $U_0\varphi = \varrho\varphi$ implies

$$U_0(U\varphi) = UU_0\varphi = U(\varrho\varphi) = \varrho(U\varphi).$$

Since U is isometric, it can not diminish the number of dimensions; consequently the image of E_ϱ by the transformation U coincides with E_ϱ.

Let ψ be an element orthogonal to E_ϱ. By the isometry of U, the element $U\psi$ will be orthogonal to UE_ϱ, that is, to E_ϱ. The subspace F_ϱ, orthogonal and complementary to E_ϱ, is therefore transformed by U into itself.

Since the subspaces E_ϱ and F_ϱ are of dimensions less than p, the theorem holds, by hypothesis, in each of them, and consequently also in all of E; this completes the proof of the theorem.

We return to the family $\{U_\xi\}$; let $\psi_1, \psi_2, \ldots, \psi_p$ be a complete orthonormal system of common characteristic elements; we have

$$U_\xi \psi_k = \varrho_k(\xi)\psi_k \qquad (k = 1, 2, \ldots, p).$$

From conditions (39)–(41) it follows that

$$\varrho_k(0) = 1, \qquad \varrho_k(\xi)\varrho_k(\eta) = \varrho_k(\xi + \eta), \qquad \varrho_k(\xi) \to \varrho_k(\eta) \text{ for } \xi \to \eta,$$

and

$$|\varrho_k(\xi)| = 1.$$

But it is well known that the only complex-valued functions possessing these properties are the functions $e^{i\omega\xi}$. Hence

$$\varrho_k(\xi) = e^{i\omega_k\xi} \qquad (k = 1, 2, \ldots, p),$$

which was to be proved.

CHAPTER VII

BOUNDED SYMMETRIC, UNITARY, AND NORMAL TRANSFORMATIONS OF HILBERT SPACE

SYMMETRIC TRANSFORMATIONS

104. Some Fundamental Properties

In Sec. 92 we defined *symmetric* transformations in a Hilbert space $\mathfrak{H}$ and established their basic properties; then we studied, in particular, completely continuous symmetric transformations, and succeeded in developing them into series of characteristic elements. The problem arises of determining whether we can find an analogous development, say a *spectral decomposition*, even for a symmetric transformation which is not completely continuous. This problem was already posed and solved in the affirmative by HILBERT. Today we know several simpler methods of treating the problem; we shall discuss two of them (Secs. 107 and 108).

In the present section we shall establish some fundamental properties of symmetric transformations which will be of constant use in the sequel.

We begin by observing the obvious facts that real numerical multiples, sums, and limits (uniform, strong, or even weak) of symmetric transformations are also symmetric, whereas products are symmetric only when the factors are *mutually permutable* transformations.

It will be convenient to introduce a notation for permutability: we write

$$T \smile S$$

if T and S are two *permutable* transformations, $TS = ST$. Moreover, if $\{S\}$ is a family of transformations, we write

$$T \smile \{S\}$$

if the transformation T is permutable with all the transformations S belonging to $\{S\}$.

We define an *order relation* among the symmetric transformations by writing

$$A \geqq B \text{ or } B \leqq A$$

when

$$(Af, f) \geq (Bf, f)$$

for all elements f of $\mathfrak{H}$.

Linear transformations A such that $(Af, f) \geq 0$, are said to be *positive* (a notion which was introduced in Sec. 95). Every positive transformation of *complex* Hilbert space is symmetric; in fact, we have seen in Sec. 92 that every linear transformation A for which the quadratic form (Af, f) assumes only real values is symmetric.

For every positive symmetric transformation A and for two arbitrary elements f and g, we have *the generalized Schwarz inequality*

$$|(Af, g)|^2 \leq (Af, f)\,(Ag, g), \tag{1}$$

which is proved exactly as the Schwarz inequality proper (Sec. 83) by starting with the fact that for every real value of λ and for $h_\lambda = f + \lambda(Af, g)g$ we have

$$0 \leq (Ah_\lambda, h_\lambda) = (Af, f) + 2\lambda\,|(Af, g)|^2 + \lambda^2\,|(Af, g)|^2\,(Ag, g).$$

The *greatest lower* and *least upper bounds* of the symmetric transformation A are defined as the largest real number m and the smallest real number M for which

$$m(f, f) \leq (Af, f) \leq M(f, f),$$

that is, for which

$$mI \leq A \leq MI,$$

where I denotes the identity transformation. In other words, m is the greatest lower bound and M is the least upper bound of the quadratic form (Af, f) when f varies under the condition $\|f\| = 1$. But under this condition the least upper bound of $|(Af, f)|$ is equal to the norm of A (see Sec. 92), and consequently

$$\|A\| = \max\,\{|m|, |M|\}.$$

It follows, among other things, that *the relations $A \geq B$, $A \leq B$ can not hold simultaneously unless $A = B$*; in fact, these relations imply that $(Cf, f) \equiv 0$ for $C = A - B$; hence $m_C = M_C = 0$, $\|C\| = \max\,\{|m_C|, |M_C|\} = 0$, $C = O$.

It is clear that the order relation we have just introduced is *transitive*. We also see that $A \geq B$ implies

$$A + C \geq B + C \text{ and } cA \geq cB,$$

for every symmetric transformation C and for every number $c > 0$.

We see already the analogies which exist between the order relations of symmetric transformations on the one hand and those of real numbers on the other. But there is an essential difference: it is that two real numbers are always comparable, whereas there are pairs of symmetric transformations neither of which is equal to or greater than the other; the set of symmetric transformations is therefore only *partially ordered.*

The fact that a bounded monotonic sequence of real numbers is convergent has its analogue for transformations:

THEOREM.[1] *Every bounded monotonic sequence of symmetric transformations A_n converges (in the strong sense) to a symmetric transformation A.*

It obviously suffices to consider the case where

$$O \leqq A_1 \leqq A_2 \leqq \ldots \leqq I.$$

For $m < n$, we have $A_{mn} = A_n - A_m \geqq O$; using the generalized Schwarz inequality (1) we have for every f:

$$\|A_{mn}f\|^4 = (A_{mn}f, A_{mn}f)^2 \leqq (A_{mn}f, f)\,(A_{mn}^2 f, A_{mn}f).$$

Inasmuch as

$$O \leqq A_{mn} \leqq I, \text{ and hence } \|A_{mn}\| \leqq 1,$$

it follows that

$$\|A_n f - A_m f\|^4 \leqq [(A_n f, f) - (A_m f, f)]\,\|f\|^2.$$

But the numerical sequence $\{(A_n f, f)\}$ is bounded and nondecreasing, hence convergent, which implies, by the inequality obtained, that the sequence of elements $A_n f$ is also convergent. The transformation A defined by $Af = \lim A_n f$ is obviously symmetric and the theorem is proved.

The *square* of a symmetric transformation is always a positive transformation:

$$(A^2 f, f) = (Af, Af) \geqq 0.$$

We consider the converse problem, which is to find, for a given positive symmetric transformation A, a *square root,* that is, a symmetric transformation X such that $X^2 = A$.

Obviously we can assume that $A \leqq I$. Set $A = I - B$ (where $O \leqq B \leqq I$) and $X = I - Y$; the equation to be solved then takes on the form

$$Y = \tfrac{1}{2}(B + Y^2). \tag{2}$$

Let us try the method of successive approximations. We set

$$Y_0 = O,\ \ Y_1 = \tfrac{1}{2}B,$$

and in general

$$Y_{n+1} = \tfrac{1}{2}(B + Y_n^2) \qquad (n \geqq 0), \tag{3}$$

[1] In Sz.-Nagy [*] (p. 15), the theorem is proved under the additional condition that the A_n are permutable. The theorem in all its generality is due to Vigier [**1**].

and shall show that the sequence $\{Y_n\}$ converges to a solution of (2). We show first that Y_n is a polynomial in B with non-negative real coefficients and that the same is true of $Y_n - Y_{n-1}$. These propositions, obviously true for $n = 1$, are proved by induction on n by means of the recurrence formulas (3) and the formula

$$Y_{n+1} - Y_n = \tfrac{1}{2}(B + Y_n^2) - \tfrac{1}{2}(B + Y_{n-1}^2) = \tfrac{1}{2}(Y_n^2 - Y_{n-1}^2) =$$
$$= \tfrac{1}{2}(Y_n + Y_{n-1})(Y_n - Y_{n-1}).^2$$

Since $B \geq O$ implies $B^n \geq O$ for $n = 2, 3, \ldots,$[3] we therefore have $Y_n \geq O$ and $Y_n - Y_{n-1} \geq O$. Finally, $\|Y_n\| \leq 1$ for all n. This is true for $n = 0$ and is proved by induction with the recurrence formula (3).

Hence we can apply the theorem which we have just established for bounded monotonic sequences, from which it follows that the sequence $\{Y_n\}$ is convergent. Its limit Y necessarily satisfies equation (2), which is the limiting case of equation (3). We also have $\|Y\| \leq 1$, and consequently $Y \leq I$, $X = I - Y \geq O$.

We have thus constructed a solution of the equation $X^2 = A$; this solution X is symmetric and positive; furthermore, it is the limit of a sequence of polynomials in A.

We shall show this "positive square root" is *unique*, that is, if X' is a positive symmetric transformation such that $X'^2 = A$, then necessarily $X' = X$. In fact, since

$$X'A = AX' = X'^3,$$

X' is permutable with A and with polynomials in A as well as with their limits, and in particular, with X. Denote by Z and Z' the positive symmetric square roots of X and X' respectively obtained by the above procedure, but starting with X and X' instead of with A. Let f be an arbitrary element of the space $\mathfrak{H}$ and set $g = (X - X')f$. We have:

$$\|Zg\|^2 + \|Z'g\|^2 = (Z^2g, g) + (Z'^2g, g) = (Xg, g) + (X'g, g) =$$
$$= ((X + X')(X - X')f, g) = ((X^2 - X'^2)f, g) = ((A - A)f, g) = 0,$$

which yields: $Zg = Z'g = 0$; hence $Xg = ZZg = 0$, $X'g = Z'Z'g = 0$. It follows that

$$\|(X - X')f\|^2 = ((X - X')^2f, f) = ((X - X')g, f) = 0,$$

and hence $(X - X')f = 0$. Since this holds for all elements f, we have $X' = X$.

[2] Since the Y_n are polynomials in B, we have $Y_n \smile Y_m$.
[3] In fact, $(B^{2m}f, f) = \| B^m f \|^2 \geq 0$, $(B^{2m+1}f, f) = (BB^m f, B^m f) \geq 0$.

Hence we have the

THEOREM.[4] *Every positive symmetric transformation A possesses a positive symmetric square root, and only one, which we denote by $A^{\frac{1}{2}}$. It can be represented as the limit (in the strong sense) of a sequence of polynomials in A, and consequently is permutable with all transformations which are permutable with A.*

Here is a first application of this theorem. Let A and B be two permutable positive symmetric transformations. Since A is then also permutable with $B^{\frac{1}{2}}$, we have, for an arbitrary element f,

$$(ABf, f) = (AB^{\frac{1}{2}}B^{\frac{1}{2}}f, f) = (B^{\frac{1}{2}}AB^{\frac{1}{2}}f, f) = (AB^{\frac{1}{2}}f, B^{\frac{1}{2}}f) \geq 0.$$

This proves the following proposition:

The product of two permutable, positive, symmetric transformations is also a positive symmetric transformation.[5]

From this we derive the more general proposition:

The inequality $A \geq B$ remains valid if we multiply the two members by the same positive symmetric transformation C which is permutable with A and B:

$$AC = CA \geq CB = BC.$$

Now consider the problem of the *inversion* of a symmetric transformation A. When A^{-1} exists, it is also symmetric, since

$$(A^{-1}f, g) = (A^{-1}f, AA^{-1}g) = (AA^{-1}f, A^{-1}g) = (f, A^{-1}g).$$

It exists in particular if $\|I - A\| < 1$, and it can then be represented by the Neumann series

$$A^{-1} = [I - (I - A)]^{-1} = I + (I - A) + (I - A)^2 + \dots,$$

which converges in norm. This representation is also valid for a non-symmetric transformation. For A symmetric, the condition $\|I - A\| < 1$ is equivalent to the following:

$$0 < m \leq M < 2, \tag{4}$$

where m and M denote the greatest lower and least upper bounds of A.

If we assume only that $m > 0$, we can always find a positive quantity c such that the bounds m', M' of the transformation $A' = cA$ satisfy condition (4); we have only to take $c < \dfrac{2}{M}$. Consequently *we have the development*[6]

$$A^{-1} = (c^{-1}A')^{-1} = c(A')^{-1} = cI + c(I - cA) + c(I - cA)^2 + \dots \tag{5}$$

[4] The construction of the square root given in the text is due to VISSER [**1**]; the uniqueness proof follows Sz.-NAGY [**3**]. Another construction is found in WECKEN [**1**].

[5] Another proof was given by F. RIESZ [**13**] (pp. 33—34).

[6] Cf. HILB [**1**].

Since the terms of this series are positive transformations, the same is true of the sum A^{-1}. Multiplying the inequalities

$$m \cdot I \leq A \leq M \cdot I$$

by A^{-1}, we obtain

$$m \cdot A^{-1} \leq I \leq M \cdot A^{-1};$$

hence

$$M^{-1} \cdot I \leq A^{-1} \leq m^{-1} \cdot I.$$

We can also make use of these results in the problem of finding the inverse of a *non-symmetric* linear transformation T. Assume that the transformations T^*T and TT^*, which are symmetric, have positive lower bounds. This, as we have just seen, assures the existence of $(T^*T)^{-1}$ and $(TT^*)^{-1}$. It follows from the relations

$$(T^*T)^{-1}T^*T = I, \quad TT^*(TT^*)^{-1} = I,$$

that T has a left inverse $(T^*T)^{-1}T^*$ and a right inverse $T^*(TT^*)^{-1}$, which are then necessarily equal, that is,

$$T^{-1} = (T^*T)^{-1}T^* = T^*(TT^*)^{-1}$$

(see Sec. 67).

Moreover, the condition that T^*T and TT^* have positive lower bounds is also *necessary* for the existence of T^{-1}. In fact, if T^*T does not have a positive greatest lower bound, there exists a sequence $\{f_n\}$ such that

$$\|f_n\| = 1 \text{ and } \|Tf_n\|^2 = (Tf_n, Tf_n) = (T^*Tf_n, f_n) \to 0.$$

If T^{-1} existed, it would follow that

$$\|f_n\| = \|T^{-1}Tf_n\| \leq \|T^{-1}\| \cdot \|Tf_n\| \to 0,$$

which would contradict $\| f_n \| = 1$. Hence T has no inverse. Similarly, if TT^* does not have a positive greatest lower bound T^* does not have an inverse, and hence T can not have an inverse either.

Hence, we have the result:

THEOREM. *In order that the linear transformation T of Hilbert space possess an inverse, it is necessary and sufficient that the symmetric transformations T^*T and TT^* have positive greatest lower bounds.*

105. Projections

The simplest symmetric transformations are the orthogonal projections onto subspaces of the Hilbert space $\mathfrak{H}$.

Let $\mathfrak{M}$ and $\mathfrak{N}$ be two orthogonal and complementary subspaces of $\mathfrak{H}$ (see Sec. 34). Every element f of $\mathfrak{H}$ can then be uniquely decomposed into $f = g + h$, where g belongs to $\mathfrak{M}$ and h to $\mathfrak{N}$. The transformation P defined by $Pf = g$ is obviously linear and satisfies $P^2 = P$; it is called the orthogonal projection (or more briefly: the projection) onto the subspace $\mathfrak{M}$. The projection onto $\mathfrak{N}$ is then equal to $I - P$.

The transformation P is *symmetric*. In fact, if $f = g + h$ and $f' = g' + h'$ are decompositions of the elements f and f' with respect to the subspaces $\mathfrak{M}$ and $\mathfrak{N}$, we have $(g, h') = 0$ and $(h, g') = 0$, hence

$$(Pf, f') = (g, f') = (g, f') - (g, h') = (g, g') = (g, g') + (h, g') =$$
$$= (f, g') = (f, Pf').$$

Conversely, *every symmetric transformation P such that $P^2 = P$, is an (orthogonal) projection*, namely onto the subspace $\mathfrak{M}$ consisting of the elements of the form $h = Pf$, or equivalently, those for which $(I - P)h = 0$. In fact, every element f of $\mathfrak{H}$ admits the decomposition $f = Pf + (f - Pf)$, where Pf belongs to $\mathfrak{M}$ and $f - Pf$ is orthogonal to $\mathfrak{M}$:

$$(f - Pf, Pg) = (Pf - P^2f, g) = (0, g) = 0$$

for all g.

We have the frequently used relation $(Pf, f) = \|Pf\|^2$, valid for all elements f; in fact,

$$(Pf, f) = (P^2f, f) = (Pf, Pf) = \|Pf\|^2.$$

It follows in particular that $O \leq P \leq I$; we have $P = O$ if $\mathfrak{M}$ consists of the single element 0, and $P = I$ if $\mathfrak{M}$ coincides with $\mathfrak{H}$.

Let P_1 and P_2 be projections onto the subspaces $\mathfrak{M}_1$ and $\mathfrak{M}_2$. We verify the following statements without difficulty:

a) When $P_1 \smile P_2$, P_1P_2 is the projection onto the subspace $\mathfrak{M}_1 \cap \mathfrak{M}_2$ — the intersection of the subspaces $\mathfrak{M}_1$ and $\mathfrak{M}_2$.

b) When $P_1P_2 = O$, we also have $P_2P_1 = (P_1P_2)^* = O^* = O$, and the subspaces $\mathfrak{M}_1$ and $\mathfrak{M}_2$ are orthogonal; in this case we say that the projections P_1 and P_2 are *orthogonal* to one another. $P_1 + P_2$ is then also a projection, namely the projection onto the subspace $\mathfrak{M}_1 + \mathfrak{M}_2$.[7]

c) When P_1 and P_2 are permutable without necessarily being mutually orthogonal, $P_1 + P_2 - P_1P_2$ is always a projection, namely that onto the vector sum $\mathfrak{M}_1 + \mathfrak{M}_2$, which is in this case also a subspace of $\mathfrak{H}$. In fact, since P_1 and $(I - P_1)P_2$ are then mutually orthogonal projections, their sum is the projection onto

$$\mathfrak{M}_1 + (\mathfrak{H} - \mathfrak{M}_1) \cap \mathfrak{M}_2 = \mathfrak{M}_1 + \mathfrak{M}_2.^{8}$$

[7] It is verified without difficulty that the vector sum of two orthogonal subspaces is itself a *closed* linear set, and hence a subspace.

[8] It is clear that the vector sum in the first member is included in that of the second member. Therefore, in order to prove that they coincide, there remains only to observe that if $f_1 \in \mathfrak{M}_1$, $f_2 \in \mathfrak{M}_2$, we have $f_1 + f_2 = g_1 + g_2$, where

$$g_1 = f_1 + P_1f_2 \in \mathfrak{M}_1, \quad g_2 = f_2 - P_1f_2 \in (\mathfrak{H} - \mathfrak{M}_1) \cap \mathfrak{M}_2.$$

d) When $P_1P_2 = P_2$, we have $P_2P_1 = (P_1P_2)^* = P_2^* = P_2$, hence $\mathfrak{M}_1 \cap \mathfrak{M}_2 = \mathfrak{M}_2$, that is, $\mathfrak{M}_2 \subseteq \mathfrak{M}_1$. In this case, $P_1 - P_2$ is also a projection, namely, that onto the subspace $\mathfrak{M}_1 \ominus \mathfrak{M}_2$ consisting of the elements of $\mathfrak{M}_1$ which are orthogonal to $\mathfrak{M}_2$.

The relation $P_1P_2 = P_2$ implies therefore that $P_1 - P_2 \geq O$, $P_1 \geq P_2$. Conversely, $P_1 \geq P_2$ implies that $I - P_1 \leq I - P_2$, and consequently,

$$\|(I - P_1)P_2f\|^2 = ((I - P_1)P_2f, P_2f) \leq ((I - P_2)P_2f, P_2f) = 0;$$

therefore $(I - P_1)P_2 = O$, $P_1P_2 = P_2$. Hence *for two projections the relations* $P_1P_2 = P_2$ *and* $P_1 \geq P_2$ *are equivalent.*

It follows from the theorem on bounded monotonic sequences of symmetric transformations (Sec. 104) that every monotonic sequence of projections is convergent. A series whose terms are pairwise orthogonal projections has partial sums which form an increasing sequence of projections, hence it is always convergent; the subspace corresponding to the sum is that one which is spanned by the orthogonal subspaces corresponding to the terms.

We can define the *convergence* of a sequence of subspaces by the (strong) convergence of the sequence of corresponding projections. It can happen that a sequence of subspaces of infinite dimension tends, according to this definition, to a subspace of finite dimension. Such is the case, for example, for the sequence $\{\mathfrak{M}_n\}$ of subspaces of coordinate Hilbert space, if $\mathfrak{M}_n$ consists of the "vectors" whose first n components are zero. On the other hand, it can occur that a sequence of subspaces of finite dimension tends to a subspace of infinite dimension; we have only to take as an example the orthogonal complements of the subspaces $\mathfrak{M}_n$ which were just defined.

Such irregularities are excluded if we deal with *uniform convergence* of subspaces, defined by the uniform convergence of their projections. In fact, for a uniformly convergent sequence $\{\mathfrak{M}_n\}$ the dimension of $\mathfrak{M}_n$ is eventually constant. This follows from the following more precise proposition:

THEOREM.[9] *If the projections P and Q onto the subspaces $\mathfrak{P}$ and $\mathfrak{Q}$ satisfy the condition*

$$\|P - Q\| < 1,$$

then $\mathfrak{P}$ can be mapped linearly and isometrically onto $\mathfrak{Q}$.

It follows from our hypothesis that $\|P(Q - P)P\| < 1$, and that consequently the symmetric transformation

$$A = I + P(Q - P)P$$

has a positive greatest lower bound. Hence A^{-1} and $A^{-\frac{1}{2}} = (A^{-1})^{\frac{1}{2}}$ exist; they too are symmetric and positive. Consider the transformations

$$U = QA^{-\frac{1}{2}}P \quad \text{and} \quad U^* = PA^{-\frac{1}{2}}Q.$$

[9] Sz.-Nagy [*] (p. 58) and [4] (p. 350).

Since we obviously have $P \smile A$, we also have $P \smile A^{-\frac{1}{2}}$, and since furthermore

$$PQQP = PQP = P + P(Q - P)P = PA,$$

it follows that

$$U^*U = PA^{-\frac{1}{2}}QQA^{-\frac{1}{2}}P = A^{-\frac{1}{2}}PAA^{-\frac{1}{2}} = PA^{-\frac{1}{2}}AA^{-\frac{1}{2}} = P.$$

This implies that for elements of $\mathfrak{P}$ U is *isometric*; in fact, if f and g belong to $\mathfrak{P}$, we have

$$(Uf, Ug) = (U^*Uf, g) = (Pf, g) = (f, g).$$

Since $\mathfrak{P}$ is a closed set, its image under U is closed; it is therefore a subspace $\mathfrak{Q}'$ which is obviously contained in $\mathfrak{Q}$. Since U is zero for the elements orthogonal to $\mathfrak{P}$, $\mathfrak{Q}'$ is also the image under U of the entire space $\mathfrak{H}$.

Let h be an element orthogonal to $\mathfrak{Q}'$, that is, such that $(h, Uf) = 0$ for all elements f of $\mathfrak{H}$. Then $U^*h = 0$; hence

$$PQh = A^{\frac{1}{2}}A^{-\frac{1}{2}}PQh = A^{\frac{1}{2}}PA^{-\frac{1}{2}}Qh = A^{\frac{1}{2}}U^*h = 0,$$

and, consequently, $(Q - P)Qh = Qh$. In view of the hypothesis $\|Q - P\| < 1$, this equation is possible only if $Qh = 0$, that is, only if h is also orthogonal to $\mathfrak{Q}$. It follows that $\mathfrak{Q}' = \mathfrak{Q}$, which completes the proof.

We shall make use of this theorem later, in the theory of perturbations (Chapt. IX). It will be useful for us to note that the transformation U satisfies the relations

$$U^*U = P, \quad UU^* = Q,$$

the second of which follows from the fact that UU^* leaves the elements of $\mathfrak{Q}$ invariant:

$$UU^*(Uf) = U(U^*U)f = UPf = Uf,$$

and that U^*, hence also UU^*, is zero for all the elements h orthogonal to $\mathfrak{Q}$. From the second of these relations it follows that for elements of $\mathfrak{Q}$, U^* is the inverse of U.

106. Functions of a Bounded Symmetric Transformation

When A is a symmetric transformation, we assign to the polynomial with real coefficients

$$p(\lambda) = a_0 + a_1\lambda + a_2\lambda^2 + \ldots + a_n\lambda^n$$

the symmetric transformation

$$p(A) = a_0I + a_1A + a_2A^2 + \ldots + a_nA^n.$$

This correspondence is obviously *homogeneous, additive*, and *multiplicative*, that is, such that the transformations

$$cp(A), p(A) + q(A), p(A)q(A)$$

correspond to

$$cp(\lambda), p(\lambda) + q(\lambda), p(\lambda)q(\lambda)$$

respectively.

Moreover, this correspondence is of *positive type*, that is, if

$$p(\lambda) \geq 0 \text{ for } m \leq \lambda \leq M,$$

where m and M denote the greatest lower and least upper bounds of A, we also have

$$p(A) \geq O.$$

In order to see this, we decompose $p(\lambda)$ in the form

$$p(\lambda) = c \prod_i (\lambda - \alpha_i) \prod_j (\beta_j - \lambda) \prod_k [\lambda - \gamma_k)^2 + \delta_k^2],$$

where $c \geq 0$, $\alpha_i \leq m$, $\beta_j \leq M$, and where the quadratic factors correspond to complex conjugate zeros and to real zeros between m and M, the latter necessarily being of even multiplicity. Replacing λ by A, all the factors will be positive transformations, and since they are also permutable, the same will be true of the product $p(A)$.

More generally, the inequality

$$p(\lambda) \geq q(\lambda) \text{ for } m \leq \lambda \leq M$$

implies that

$$p(A) \geq q(A);$$

we have only to consider the difference $p(\lambda) - q(\lambda) \geq 0$. The correspondence is therefore *monotonic*.

The problem arises of *extending* this correspondence to larger classes of functions in such a way that the properties listed above are preserved. In order to do this we can imitate a procedure we used in Chapter II when we extended the notion of integral, defined first for certain simple functions (for step functions, or for continuous functions) to more general classes of functions. There we considered monotonic sequences of simple functions convergent almost everywhere. Now we shall content ourselves with considering monotonic sequences which converge *everywhere* (it will be a little more convenient to consider decreasing rather than increasing sequences).[10]

More precisely, let us consider first the class C_1 of non-negative real-valued functions defined in the interval $m \leq \lambda \leq M$ which are continuous or at least upper semi-continuous. To a function $u(\lambda)$ of this class we can attach a sequence of polynomials $p_n(\lambda)$ which decreases in this interval to $u(\lambda)$. The sequence of transformations $p_n(A)$ is then also decreasing and bounded below by O, hence it is convergent. The limit is a symmetric transformation which we assign to the function $u(\lambda)$ and denote by $u(A)$.

[10] This method of extending the correspondence between functions and transformations is due to F. RIESZ [*] (Chapt. V) and [**15**].

This definition is unique, that is, $u(A)$ does not depend on the particular choice of the sequence $\{p_n(\lambda)\}$: if $\{q_n(\lambda)\}$ is another sequence of the same type, $\lim p_n(A) = \lim q_n(A)$. In fact, whatever be the integer r, we have for s sufficiently large

$$p_s(\lambda) \leq q_r(\lambda) + \frac{1}{r}, \quad q_s(\lambda) \leq p_r(\lambda) + \frac{1}{r}$$

in the entire interval $m \leq \lambda \leq M$, which is a consequence of the well-known Dini theorem on monotonic sequences of continuous functions, or also of the Borel covering theorem. We then have

$$p_s(A) \leq q_r(A) + \frac{1}{r} I, \quad q_s(A) \leq p_r(A) + \frac{1}{r} I,$$

and taking the limit (first for $s \to \infty$, then $r \to \infty$), it follows that

$$\lim p_s(A) \leq \lim q_r(A), \quad \lim q_s(A) \leq \lim p_r(A);$$

hence

$$\lim p_n(A) = \lim q_n(A),$$

which was to be proved.

It follows by the same reasoning, or rather from half of it, that if

$$u_1(\lambda) \geq u_2(\lambda) \qquad (m \leq \lambda \leq M),$$

then

$$u_1(A) \geq u_2(A).$$

Hence the correspondence thus extended is monotonic, and furthermore it is positively homogeneous, additive, and multiplicative. In fact, if the sequences $\{p_n(\lambda)\}$ and $\{q_n(\lambda)\}$ decrease to $u(\lambda)$ and $v(\lambda)$, the sequences $\{cp_n(\lambda)\}$ with $c > 0$, $\{p_n(\lambda) + q_n(\lambda)\}$, and $\{p_n(\lambda)q_n(\lambda)\}$ decrease to $cu(\lambda)$, $u(\lambda) + v(\lambda)$, and $u(\lambda)v(\lambda)$, respectively.

Now we consider the class C_2 of functions admitting a decomposition into the difference of two functions of class C_1. We assign to the function $w(\lambda) = u(\lambda) - v(\lambda)$ the transformation $w(A) = u(A) - v(A)$. This definition is unique, since

$$u_1(\lambda) - v_1(\lambda) = u_2(\lambda) - v_2(\lambda)$$

implies

$$u_1(A) - v_1(A) = u_2(A) - v_2(A),$$

as becomes clear if we write the two equations in the form $u_1 + v_2 = v_1 + u_2$ and use additivity for the class C_1.

For the class C_2 the correspondence is homogeneous, additive, and multiplicative — a consequence of the corresponding properties of the class C_1

and the following decompositions:

$$c(u-v) = cu - cv \text{ (for } c > 0), \; c(u-v) = (-c)v - (-c)u \text{ (for } c < 0),$$

$$(u_1 - v_1) + (u_2 - v_2) = (u_1 + u_2) - (v_1 + v_2),$$

$$(u_1 - v_1)(u_2 - v_2) = (u_1u_2 + v_1v_2) - (u_1v_2 + u_2v_1).$$

Finally, the correspondence is also monotonic,

$$u_1(\lambda) - v_1(\lambda) \geq u_2(\lambda) - v_2(\lambda)$$

implying

$$u_1(A) - v_1(A) \geq u_2(A) - v_2(A);$$

this becomes evident if we write these inequalities in the form $u_1 + v_2 \geq$ $\geq v_1 + u_2$ and use monotonicity for the class C_1.

107. Spectral Decomposition of a Bounded Symmetric Transformation

Among the "functions" of the symmetric transformation A which we have just defined there are projections, namely those which correspond to the functions $e(\lambda)$ taking on only the values 0 and 1, because we then have $[e(\lambda)]^2 = e(\lambda)$, hence also $[e(A)]^2 = e(A)$.

Let us consider in particular the function $e_\mu(\lambda)$, depending on the real parameter μ, which is 1 for $\lambda \leq \mu$ and 0 for $\lambda > \mu$. This function obviously belongs to the class C_1; hence there corresponds to it a transformation $e_\mu(A)$ which is a projection and which we denote by E_μ. Since $e_\mu(\lambda)e_\nu(\lambda) = e_\mu(\lambda)$ for $\mu < \nu$, we have $E_\mu E_\nu = E_\nu E_\mu = E_\mu$, hence $E_\mu \leq E_\nu$; and since on the segment $m \leq \lambda \leq M$, $e_\mu(\lambda) \equiv 0$ if $\mu < m$ and $e_\mu(\lambda) \equiv 1$ if $\mu \geq M$, we have $E_\mu = 0$ if $\mu < m$ and $E_\mu = I$ if $\mu \geq M$.

Finally, as a function of μ, E_μ is continuous from the right. In order to see this, we fix μ and construct a sequence of polynomials $p_n(\lambda)$ which decrease in $[m, M]$ to $e_\mu(\lambda)$, and in addition satisfy

$$p_n(\lambda) \geq e_{\mu+\frac{1}{n}}(\lambda).$$

Then we have

$$p_n(A) \geq E_{\mu+\frac{1}{n}} \geq E_\mu.$$

Since $p_n(A) \to E_\mu$, it follows that $E_{\mu+\frac{1}{n}} \to E_\mu$ for $n \to \infty$, and since E_μ is a monotone function of μ, this implies that

$$E_{\mu+\varepsilon} \to E_\mu \text{ for } 0 < \varepsilon \to 0.$$

For $\mu < \nu$ we obviously have

$$\mu[e_\nu(\lambda) - e_\mu(\lambda)] \leq \lambda[e_\nu(\lambda) - e_\mu(\lambda)] \leq \nu[e_\nu(\lambda) - e_\mu(\lambda)],$$

from which it follows that

(6) $$\mu(E_\nu - E_\mu) \leq A(E_\nu - E_\mu) \leq \nu(E_\nu - E_\mu).$$

We consider a sequence of points $\mu_0, \mu_1, \ldots, \mu_n$ such that

$$\mu_0 < m < \mu_1 < \mu_2 < \ldots < \mu_{n-1} < M \leq \mu_n.$$

We write (6) with $\mu = \mu_{k-1}$, $\nu = \mu_k$ $(k = 1, 2, \ldots, n)$ and take the sum; it follows that

$$\sum_{k=1}^{n} \mu_{k-1}(E_{\mu_k} - E_{\mu_{k-1}}) \leq A \sum_{k=1}^{n} (E_{\mu_k} - E_{\mu_{k-1}}) \leq \sum_{k=1}^{n} \mu_k(E_{\mu_k} - E_{\mu_{k-1}}).$$

The middle member is equal to $A(E_{\mu_n} - E_{\mu_0}) = A(I - O) = A$, and if $\max(\mu_k - \mu_{k-1}) \leq \varepsilon$, the difference of the third and first members is less than εI. It follows that if λ_k is any point between μ_{k-1} and μ_k, we have

$$\|A - \sum_{k=1}^{n} \lambda_k(E_{\mu_k} - E_{\mu_{k-1}})\| \leq \varepsilon.$$

If we increase indefinitely the number n of decomposition intervals (μ_{k-1}, μ_k), in such a way that their maximum length tends to 0, the sums

$$\sum_k \lambda_k(E_{\mu_k} - E_{\mu_{k-1}})$$

will therefore tend to A in norm. Since E_λ is constant for $\lambda \geq M$ and for $\lambda < m$, we can express this result by writing, in analogy with ordinary Stieltjes integrals,

$$A = \int_{-\infty}^{\infty} \lambda dE_\lambda = \int_{m-0}^{M} \lambda dE_\lambda.^{11}$$

Moreover, for every integer $r > 0$ we have

$$A^r = \int_{m-0}^{M} \lambda^r dE_\lambda,$$

since

$$[\sum_k \lambda_k(E_{\mu_k} - E_{\mu_{k-1}})]^r = \sum_k \lambda_k^r(E_{\mu_k} - E_{\mu_{k-1}});$$

this follows from the fact that, as a consequence of the relation $E_\nu E_\mu = E_{\min\{\nu,\mu\}}$, the differences $E_{\mu_k} - E_{\mu_{k-1}}$ are pairwise orthogonal projections. Inasmuch as this relation remains valid when $r = 0$, we have for every polynomial $p(\lambda)$:

$$p(A) = \int_{m-0}^{M} p(\lambda) dE_\lambda.$$

[11] It is a question here of an integral from m to M, but with respect to the function E_λ which has been modified at the point m by replacing E_m by E_{m-0}, that is, by O. Moreover, if the condition of continuity from the right, which is of little importance, is omitted we can set $E_m = O$ by definition.

The formal (or "algebraic") correspondence from which we started thus takes on an "analytic" form.

From this, we can pass immediately to an arbitrary *continuous* function $u(\lambda)$ in $[m, M]$. In fact, given an arbitrary $\varepsilon > 0$, we can find a polynomial $p(\lambda)$ such that

$$-\frac{\varepsilon}{3} \leq u(\lambda) - p(\lambda) \leq \frac{\varepsilon}{3}$$

in $[m, M]$; we then also have

$$-\frac{\varepsilon}{3} I \leq u(A) - p(A) \leq \frac{\varepsilon}{3} I,$$

hence

$$\|u(A) - p(A)\| \leq \frac{\varepsilon}{3}.$$

On the other hand, for every decomposition of the λ-axis we shall have for the sum

$$S_u = \sum_k u(\lambda_k)\,(E_{\mu_k} - E_{\mu_{k-1}}),$$

and for the analogous one S_p with $p(\lambda_k)$ instead of $u(\lambda_k)$,

$$-\frac{\varepsilon}{3} I \leq S_u - S_p \leq \frac{\varepsilon}{3} I,$$

hence

$$\|S_u - S_p\| \leq \frac{\varepsilon}{3}.$$

When the decomposition is sufficiently fine that

$$\|p(A) - S_p\| \leq \frac{\varepsilon}{3},$$

we shall have

$$\|u(A) - S_u\| \leq \|u(A) - p(A)\| + \|p(A) - S_p\| + \|S_p - S_u\| \leq \varepsilon.$$

Thus for every continuous function $u(\lambda)$,

$$u(A) = \int_{m-0}^{M} u(\lambda)\, dE_\lambda$$

in the sense of convergence in the norm of sums of Stieltjes type. It follows that for every pair of elements f, g:

$$(u(A)f, g) = \int_{m-0}^{M} u(\lambda)\, d(E_\lambda f, g) \tag{7}$$

in the ordinary Stieltjes sense.

By the Lebesgue theorem on term by term integration of bounded convergent sequences, formula (7) even extends to all functions $u(\lambda)$ belonging to the classes C_1 and C_2 — functions for which the correspondence $u(\lambda) \to u(A)$ has been established. Integral (7) must then be taken, of course, in the Stieltjes-Lebesgue sense.

Moreover, formula (7) furnishes us with a means of extending the notion of a function of the transformation A beyond the class C_2; in fact, we can define by (7) the transformation $u(A)$ for every function $u(\lambda)$ which is bounded and summable with respect to the function (of bounded variation) $(E_\lambda f, g)$ for any two elements f and g. We can even admit unbounded summable functions $u(\lambda)$, but then we can no longer expect $u(A)$ to be a bounded transformation. We shall return to these questions in Chapters VIII and IX.

We summarize the essential points:

THEOREM.[12] *To every symmetric transformation A in Hilbert space, with greatest lower and least upper bounds equal to m and M, we can assign a "spectral family" on the interval $[m, M]$ — that is, a family of projections $\{E_\lambda\}$ depending on the real parameter λ such that*

a) $E_\lambda \leq E_\mu$, *or equivalently*, $E_\lambda E_\mu = E_\lambda$ *for* $\lambda \leq \mu$,
b) $E_{\lambda+0} = E_\lambda$,
c) $E_\lambda = 0$ *for* $\lambda < m$ *and* $E_\lambda = I$ *for* $\lambda \geq M$ —

[12] This fundamental theorem is due to HILBERT [*] (Note 4). The proof given in the text is due to F. RIESZ, [*] and [**13**]. Other proofs are found in HELLINGER [**2**], RIESZ [**5**], VON NEUMANN [**1**], STONE [*] (Chapt. V); LENGYEL–STONE [**1**]; WINTNER [*] (Chapt. V); SZ.-NAGY [*] (Chapt. IV); LORCH [**6**]. The first proof given by RIESZ ([**5**]; cf. also EBERLEIN [**1**]) reduces the problem to that of the moments

$$\int_m^M \lambda^r \, d\alpha_f(\lambda) = (A^r f, f) \qquad (r = 0, 1, 2, \ldots),$$

where f is an arbitrary given element of $\mathfrak{H}$. Since the correspondence between polynomials in λ and in A is of positive type (cf. Sec. 106), the problem of moments is also of positive type in the sense of Sec. 53; consequently it possesses a nondecreasing solution $\alpha_f(\lambda)$. This solution will be uniquely determined if we agree to normalize it suitably, for example, in such a way that in the interior of (m, M) it is continuous from the right and that it is zero at the point m. Next it is proved that there exists a nondecreasing family $\{E_\lambda\}$ of projections such that for every f, $\alpha_f(\lambda) = (E_\lambda f, f)$; then it is easy to pass to the formulas

$$(A^r f, g) = \int_m^M \lambda^r \, d(E_\lambda f, g) \qquad (r = 0, 1, 2, \ldots)$$

for arbitrary f and g. We thus obtain the decomposition (8), but with the difference that instead of convergence *in the norm* of the sums of the Stieltjes-Riemann type, we have established only their *weak* convergence. If the functions $\alpha(\lambda)$ are normalized in the manner indicated, the family $\{E_\lambda\}$ will even be continuous from the right except perhaps at the point $\lambda = m$, where we shall have $E_m = O$.

A very similar reasoning will be applied later, in Sec. 138, when we prove Stone's theorem by the method of Bochner.

in such a way that we have

(8)
$$A = \int_{m-0}^{M} \lambda dE_\lambda.$$

Moreover, these properties uniquely determine the family $\{E_\lambda\}$. For every fixed value of the parameter, E_λ is the limit (in the strong sense) of a sequence of polynomials in A.

The statement concerning uniqueness is yet to be proved. Since $\{E_\lambda\}$ is a spectral family on the interval $[m, M]$ and satisfies (8), we conclude as above that it also satisfies (7), and in particular satisfies (7) for all continuous functions $u(\lambda)$. But since the first member of (7) is defined in a manner which does not depend on $\{E_\lambda\}$, it follows from our theorems in Sec. 51 that for every pair of elements f, g the function $(E_\lambda f, g)$ is determined up to an additive constant by the relation (7) at its points of continuity and at $m - 0$ and M. Since the function is continuous from the right and has the fixed value (f, g) at the point M, it is therefore uniquely determined everywhere.

The representation of A in the form of integral (8), or rather the corresponding representation of the element Af:

(9)
$$Af = \int_{m-0}^{M} \lambda dE_\lambda f,$$

is a generalization of the representation

$$Af = \sum_k \mu_k(f, \varphi_k)\varphi_k$$

which we obtained in Sec. 93 for completely continuous transformations A. In fact, this latter series can also be written in the form of a Stieltjes integral of type (9), we have only to define E_λ by

$$E_\lambda f = \begin{cases} \sum\limits_{\mu_k \leqq \lambda} (f, \varphi_k)\varphi_k & \text{for} \quad \lambda < 0, \\ f - \sum\limits_{\mu_k > \lambda} (f, \varphi_k)\varphi_k & \text{for} \quad \lambda \geq 0. \end{cases}$$

In this case E_λ is, as a function of λ, constant between two consecutive characteristic values of A, equal to 0 when λ is less than all the characteristic values, and equal to I when λ is greater than the latter. The jump of E_λ when λ passes through a characteristic value μ, namely $E(\mu) = E_\lambda - E_{\lambda-0}$, is the projection onto the characteristic subspace corresponding to μ:

$$E(\mu)f = \sum_{\mu_k=\mu} (f, \varphi_k)\varphi_k \quad \text{for} \quad \mu \neq 0,$$

$$E(0)f = f - \sum_k (f, \varphi_k)\varphi_k.$$

In the case of a transformation A which is not completely continuous,

E_λ can increase in a continuous manner. Consider as an example the space L^2 (0, 1) and assign to every function $f(x)$ in L^2 the function $x \cdot f(x)$. This transformation A is obviously linear and symmetric, and its greatest lower and least upper bounds are 0 and 1. For every polynomial $p(\lambda)$, and therefore for every continuous or semi-continuous function $u(\lambda)$, we have

$$u(A)f(x) = u(x) \cdot f(x).$$

In particular, we have

$$E_\mu f(x) = e_\mu(x) \cdot f(x)$$

where $e_\mu(x)$ is the function defined at the beginning of this section. In this case, formula (9) reduces, therefore, to the obvious formula:

$$x = \int_0^1 \lambda de_\lambda(x) \qquad (0 \leq x \leq 1).$$

108. Positive and Negative Parts of a Symmetric Transformation. Another Proof of the Spectral Decomposition

Let A be a symmetric transformation. We consider the transformations which correspond, in the sense of Sec. 106, to the functions $|\lambda|$, λ^+, λ^-, which we shall denote by $|A|$, A^+, A^-. Since the functions under consideration are non-negative, and since furthermore,

$$|\lambda|^2 = \lambda^2, \ \lambda^+ = \tfrac{1}{2}(|\lambda| + \lambda), \ \lambda^- = \tfrac{1}{2}(|\lambda| - \lambda),$$

we also have $|A| \geq 0$, $A^+ \geq 0$, $A^- \geq 0$ and

$$|A|^2 = A^2, \ A^+ = \tfrac{1}{2}(|A| + A), \ A^- = \tfrac{1}{2}(|A| - A). \tag{10}$$

$|A|$ is therefore a positive square root of the positive symmetric transformation A^2. But we have already established the existence and uniqueness of this square root in Sec. 104. This permits us to *define* the transformation $|A|$ *directly*, without recourse to the results of Sec. 106, by setting

$$|A| = (A^2)^{\frac{1}{2}}.$$

The transformations A^+ and A^- will then be defined by means of formulas (10); since $(A^2)^{\frac{1}{2}}$ is the limit of a sequence of polynomials in A^2, the transformations $|A|$, A^+ and A^- will be permutable with A and among themselves, as well as with all the transformations which are permutable with A. Since $|A| \smile A$, we shall have

$$A^+A^- = \tfrac{1}{2}(|A| + A)\,(|A| - A) = \tfrac{1}{4}(|A|^2 - A^2) = O.$$

Let us denote by $\mathfrak{L}$ the set of all elements f of the space $\mathfrak{H}$ for which

$$A^+f = 0;$$

this is evidently a subspace of $\mathfrak{H}$. Let E be the orthogonal projection onto $\mathfrak{L}$; then

$$A^+E_m = O \quad \text{and} \quad EA^+ = (A^+E)^* = O^* = O.$$

On the other hand, the relation $A^+A^- = O$ implies that every element of the form A^-g belongs to $\mathfrak{L}$, from which it follows that

$$EA^- = A^- \quad \text{and} \quad A^-E = (EA^-)^* = (A^-)^* = A^-.$$

Hence we have

$$EA^+ = A^+E = O \quad \text{and} \quad EA^- = A^-E = A^-. \tag{11}$$

Since $A = A^+ - A^-$, it follows that

$$EA = AE = -A^- \quad \text{and} \quad (I - E)A = A(I - E) = A^+. \tag{12}$$

Since, by (11), E is permutable with A^+ and A^-, hence also with $|A| = = A^+ + A^-$, and since

$$E \geq O,\ I - E \geq O, \quad \text{and} \quad |A| \geq O,$$

we have

$$A^- = EA^+ + EA^- = E|A| \geq O,\ A^+ = |A| - A^- = |A| - E|A| = (I - E)|A| \geq O.$$

With the positiveness of A^+ and A^- thus established, the inequalities

$$|A| = A^+ + A^- \geq \pm A^+ \mp A^- = \pm A,\ A^+ \geq A^+ - A^- = A,$$
$$A^- \geq A^- - A^+ = -A$$

follow immediately.

The projection E is also permutable with every transformation T which permutes with A. In fact, since T then also permutes with A^+, we have

$$A^+TE = TA^+E = TO = O;$$

this means that every element of the form TEf belongs to the subspace $\mathfrak{L}$, hence $TE = ETE$, and consequently $ET = (TE)^* = (ETE)^* = ETE$, which proves that $ET = TE$.

We shall show that $|A|$ *is the smallest symmetric transformation which permutes with* A *and majorizes both* A *and* $-A$.

Since the inequalities $|A| \geq \pm A$ are already established, we have only to show that the relations $T \smile A$ and $T \geq \pm A$ imply that $T \geq |A|$. Now since the projection E is permutable with A and with T, the inequalities

$$T \geq A, \quad T \geq -A$$

remain valid if we multiply the first by $I - E \geq O$ and the second by $E \geq O$, which yields:

$$T(I - E) \geq A(I - E) = A^+, \quad TE \geq -AE = A^-.$$

It follows, by addition, that

$$T \geqq A^+ + A^- = |A|,$$

which completes the proof of our proposition.

It follows from this that A^+ *is the smallest positive symmetric transformation which permutes with* A *and majorizes* A, *and that* A^- *is the smallest positive symmetric transformation which permutes with* A *and majorizes* $-A$.

All these propositions are included in the following:

If A *and* B *are two permutable symmetric transformations, the transformation* $\frac{1}{2}(A + B + |A - B|)$ *is the smallest symmetric transformation which both majorizes* A *and* B *and permutes with them.*

In fact, this transformation is larger than

$$\tfrac{1}{2}(A + B + (A - B)) = A \text{ and } \tfrac{1}{2}(A + B - (A - B)) = B;$$

on the other hand, if C is a symmetric transformation majorizing A and B and permutable with them, we have

$$C - \tfrac{1}{2}(A + B) \geqq \begin{cases} A - \tfrac{1}{2}(A + B) = \tfrac{1}{2}(A - B), \\ B - \tfrac{1}{2}(A + B) = -\tfrac{1}{2}(A - B), \end{cases}$$

and thus

$$C - \tfrac{1}{2}(A + B) \geqq \tfrac{1}{2}|A - B|, \quad C \geqq \tfrac{1}{2}(A + B + |A - B|),$$

q. e. d.

The transformation $\frac{1}{2}(A + B + |A - B|)$ therefore possesses the properties of an *upper envelope*; we can denote it by sup $\{A, B\}$. The transformation $\frac{1}{2}(A + B - |A - B|)$ possesses, in an analogous manner, the properties of a *lower envelope*; we can denote it by inf $\{A, B\}$.

*

We now show how we can use these considerations as the base for a new proof of the spectral decomposition (9).[13]

In addition to the given symmetric transformation A, let us consider the transformations $A_\lambda = A - \lambda I$, where λ is a real parameter. For $\lambda < \mu$ we obviously have $A_\lambda^+ \geqq A_\lambda \geqq A_\mu$ and $A_\lambda^+ \geqq O$, hence also $A_\lambda^+ \geqq A_\mu^+$ (the conditions for permutability are satisfied, since A_λ^+ is the limit of polynomials in A_λ, hence also of polynomials in A_μ). For $\lambda < m$ we have $A_\lambda \geqq O$, therefore $A_\lambda^+ = A_\lambda \geqq (m - \lambda)I$. For $\lambda \geqq M$ we have $A_\lambda \leqq O$, therefore $A_\lambda^+ = O$.

It follows from these relations that the subspace $\mathfrak{L}_\lambda$ of elements for which A_λ^+ vanishes increases with λ, consists of the single element 0 for $\lambda < m$, and coincides with the entire space for $\lambda \geqq M$. In fact, to show the first of these

[13] Cf. Sz.-Nagy [*] (Chapt. IV, Sec. 1).

properties, note that $A_\lambda^+ \geq A_\mu^+$ for $\lambda < \mu$, and hence $A_\mu^+ A_\lambda^+ \geq (A_\mu^+)^2$ and

$$(A_\mu^+ A_\lambda^+ f, f) \geq (A_\mu^{+2} f, f) = (A_\mu^+ f, A_\mu^+ f) = \|A_\mu^+ f\|^2$$

for all f. Therefore when $A_\lambda^+ f = 0$ we also have $A_\mu^+ f = 0$.

Let E_λ be the projection upon $\mathfrak{L}_\lambda$. It also increases with λ, and is equal to O for $\lambda < m$ and to I for $\lambda \geq M$. For $\lambda < \mu$ we set $E_{\lambda\mu} = E_\mu - E_\lambda$. Since E_λ is increasing we have

$$E_\mu E_{\lambda\mu} = (I - E_\lambda) E_{\lambda\mu} = E_{\lambda\mu}. \tag{13}$$

Since the transformations A_λ^+, A_μ^+, and $E_{\lambda\mu}$ are positive and permutable, we have, by (11) and (12),

$$(\mu I - A) E_{\lambda\mu} = - A_\mu E_{\lambda\mu} = - A_\mu E_\mu E_{\lambda\mu} = A_\mu^- E_{\lambda\mu} \geq O,$$
$$(A - \lambda I) E_{\lambda\mu} = A_\lambda E_{\lambda\mu} = A_\lambda (I - E_\lambda) E_{\lambda\mu} = A_\lambda^+ E_{\lambda\mu} \geq O.$$

These inequalities can also be written in the form

$$\lambda E_{\lambda\mu} \leq A E_{\lambda\mu} \leq \mu E_{\lambda\mu}. \tag{14}$$

These are inequalities (6) of the preceding section; we have seen that they imply the formula

$$A = \int_{m-0}^{M} \lambda dE_\lambda.$$

In order to show that the family $\{E_\lambda\}$ which we have just defined is a spectral family, there remains only to verify that E_λ as a function of λ is continuous from the right, that is,

$$P_\lambda = \lim_{\mu \to \lambda + 0} E_{\lambda\mu} = O.$$

But it follows from (14) that $\lambda P_\lambda \leq A P_\lambda \leq \lambda P_\lambda$, hence $A P_\lambda = \lambda P_\lambda$, $A_\lambda P_\lambda = O$, and by (12), $A_\lambda^+ P_\lambda = (I - E_\lambda) A_\lambda P_\lambda = O$. This means that $P_\lambda f$ belongs to $\mathfrak{L}_\lambda$ for every f, hence that $P_\lambda = E_\lambda P_\lambda$. Comparing this result with the equation $(I - E_\lambda) P_\lambda = P_\lambda$, the limit case of (13), we obtain $P_\lambda = O$, q. e. d.

UNITARY AND NORMAL TRANSFORMATIONS

109. Unitary Transformations

The linear transformation U of the Hilbert space $\mathfrak{H}$ is said to be *isometric* (cf. Sec. 103) if it leaves scalar products invariant,

$$(Uf, Ug) = (f, g),$$

or equivalently, if

$$U^* U = I.$$

If the image of the space $\mathfrak{H}$ under the transformation U coincides with $\mathfrak{H}$, U is said to be *unitary*. Since in this case the equation $Ug = f$ has a solution g for arbitrary given f, we have

$$UU^*f = U(U^*Ug) = Ug = f, \text{ hence } UU^* = I.$$

The two equations

$$U^*U = I \text{ and } UU^* = I,$$

or the equivalent equation

$$U^* = U^{-1},$$

are obviously characteristic for unitary transformations.

We have already remarked in Sec. 103 that in a finite-dimensional space every isometric transformation is unitary. By contrast, in an infinite-dimensional space there are isometric transformations which are not unitary, for example the transformation

$$U(x_1, x_2, \ldots) = (0, x_1, x_2, \ldots)$$

of Hilbert coordinate space.

For unitary transformations there is a decomposition theorem analogous to the one we obtained for symmetric transformations:

THEOREM.[14] *Every unitary transformation U has a spectral decomposition*

$$U = \int_{-0}^{2\pi} e^{i\varphi}\, dE_\varphi, \tag{15}$$

where $\{E_\varphi\}$ is a spectral family over the segment $0 \leq \varphi \leq 2\pi$. We can require that E_φ be continuous at the point $\varphi = 0$, that is, $E_0 = 0$; $\{E_\varphi\}$ will then be determined uniquely by U. Moreover, E_φ is the limit of a sequence of polynomials in U and U^{-1}.

This theorem can be deduced from the one on symmetric transformations, or from the theorem on trigonometric moments (Sec. 53), but we prefer to sketch a direct proof, analogous to the one in Sec. 107 for symmetric transformations.

To begin, we assign to the trigonometric polynomial

$$p(e^{i\varphi}) = \sum_{-n}^{n} c_k e^{ik\varphi}$$

the transformation

$$p(U) = \sum_{-n}^{n} c_k U^k;$$

here we admit *complex* coefficients c_k. The correspondence is obviously *linear*

[14] Cf. WINTNER [**1**]; VON NEUMANN [**1**] (p. 281); FRIEDRICHS [**4**]; WECKEN [**1**]; STONE [*] (p. 302).

(homogeneous and additive), *multiplicative*, and such that the transformation corresponding to the conjugate polynomial

$$\overline{p(e^{i\varphi})} = \sum_{-n}^{n} \bar{c}_k e^{-ik\varphi}$$

is the adjoint of the one corresponding to $p(e^{i\varphi})$. If $p(e^{i\varphi})$ is real-valued, $p(U)$ is symmetric. The correspondence is also of *positive type*, that is, if

$$p(e^{i\varphi}) \geq 0, \quad \text{then} \quad p(U) \geq 0.$$

To see this, we can use the lemma by L. FEJÉR and F. RIESZ which asserts that every positive trigonometric polynomial can be represented by the square of the absolute value of another trigonometric polynomial (see Sec. 53). Now if

$$p(e^{i\varphi}) = |q(e^{i\varphi})|^2 = \overline{q(e^{i\varphi})}q(e^{i\varphi}),$$

we have

$$p(U) = q(U)^* q(U),$$

hence

$$(p(U)f, f) = (q(U)f, q(U)f) \geq 0, \quad p(U) \geq O.$$

The correspondence established in this manner for trigonometric polynomials extends to more general functions of period 2π, namely, first to functions which are limits of decreasing sequences of positive trigonometric polynomials, then to linear combinations of these functions (with real or complex coefficients); the method is analogous to that we followed in Sec. 106 for symmetric transformations. The correspondence thus extended continues to be linear, multiplicative, and of positive type, and the transformations which correspond to two conjugate functions are adjoint to one another.

The class of these functions includes, in particular, the functions $e_\psi(\varphi)$ which depend on the real parameter $\psi (0 \leq \psi \leq 2\pi)$ and are defined as follows: $e_0(\varphi) \equiv 0$, $e_{2\pi}(\varphi) \equiv 1$, and for $0 < \psi < 2\pi$,

$$e_\psi(\varphi) = \begin{cases} 1 \text{ when } 2k\pi < \varphi \leq 2k\pi + \psi \\ 0 \text{ when } 2k\pi + \psi < \varphi \leq 2(k+1)\pi \end{cases} \quad (k = 0, \pm 1, \pm 2, \ldots).$$

Since these functions are equal to their squares, the corresponding transformations E_ψ will be projections. We shall have, in particular, $E_0 = O$ and $E_{2\pi} = I$, and since $e_\psi(\varphi) \leq e_\chi(\varphi)$ for $\psi \leq \chi$, we shall also have $E_\psi \leq E_\chi$.

Moreover, E_ψ is a function of ψ which is continuous on the right. To see this we consider first, for $0 \leq \psi < 2\pi$, the functions $e'_\psi(\varphi) = e_\psi(\varphi) + e'_0(\varphi)$, where $e'_0(\varphi)$ is equal to 1 at the points $\varphi = 2k\pi$ and zero elsewhere. These functions are upper semi-continuous; therefore we can construct, for each fixed ψ, a sequence of trigonometric polynomials $p_n(e^{i\varphi})$ which decreases to $e'_\psi(\varphi)$, and has the additional property that for n sufficiently large

$$P_n(e^{i\varphi}) \geq e'_{\psi+\frac{1}{n}}(\varphi).$$

This implies for the corresponding transformations that $E'_{\psi+\frac{1}{n}} \to E'_\psi$, hence also that $E_{\psi+\frac{1}{n}} \to E_\psi$ $(n \to \infty)$, and more generally, that $\lim\limits_{\chi\to\psi+0} E_\chi = E_\psi$.

The transformations E_ψ therefore form a spectral family over the segment $[0, 2\pi]$; moreover $E_0 = O$. By its construction, E_ψ is the limit of polynomials in U and in $U^* = U^{-1}$.

In order to establish relation (15), we consider a decomposition of the segment $[0, 2\pi]$ by means of the points

$$0 = \psi_0 < \psi_1 < \ldots < \psi_n = 2\pi$$

such that $\max(\psi_k - \psi_{k-1}) \leq \varepsilon$. We choose an arbitrary point φ_k in each of the intervals $[\psi_{k-1}, \psi_k]$. For $\psi_{h-1} < \varphi \leq \psi_h$ we have

$$|e^{i\varphi} - \sum_{k=1}^{n} e^{i\varphi_k}[e_{\psi_k}(\varphi) - e_{\psi_{k-1}}(\varphi)]| = |e^{i\varphi} - e^{i\varphi_h}| \leq |\varphi - \varphi_h| \leq \varepsilon,$$

and an analogous result for $\varphi = 0$. Hence for every value of φ,

$$0 \leq \overline{[e^{i\varphi} - \sum_{k=1}^{n} e^{i\varphi_k}(e_{\psi_k}(\varphi) - e_{\psi_{k-1}}(\varphi))]}\,[e^{i\varphi} - \sum_{k=1}^{n} e^{i\varphi_k}(e_{\psi_k}(\varphi) - e_{\psi_{k-1}}(\varphi))] \leq \varepsilon^2.$$

Passing to the corresponding transformations, it follows that

$$O \leq [U - \sum_{k=1}^{n} e^{i\varphi_k}(E_{\psi_k} - E_{\psi_{k-1}})]^*\,[U - \sum_{k=1}^{n} e^{i\varphi_k}(E_{\psi_k} - E_{\psi_{k-1}})] \leq \varepsilon^2 I,$$

hence that

$$\|U - \sum_{k=1}^{n} e^{i\varphi_k}(E_{\psi_k} - E_{\psi_{k-1}})\| \leq \varepsilon,$$

which proves (15).

Inasmuch as the projections $E_{\psi_k} - E_{\psi_{k-1}}$ are pairwise orthogonal, we also have, for every integer $r \geq 0$,

$$\sum e^{ir\varphi_k}(E_{\psi_k} - E_{\psi_{k-1}}) = [\sum e^{i\varphi_k}(E_{\psi_k} - E_{\psi_{k-1}})]^r \Rightarrow U^r,$$

$$\sum e^{-ir\varphi_k}(E_{\psi_k} - E_{\psi_{k-1}}) = [\sum e^{-i\varphi_k}(E_{\psi_k} - E_{\psi_{k-1}})]^r \Rightarrow (U^*)^r = U^{-r};$$

hence

$$\int_0^{2\pi} e^{in\varphi} dE_\varphi = U^n \qquad (n = 0, \pm 1, \pm 2, \ldots).$$

It follows that for every trigonometric polynomial and even for every continuous function $u(e^{i\varphi})$,

$$u(U) = \int_0^{2\pi} u(e^{i\varphi}) dE_\varphi,$$

in the sense of convergence in norm of the sums of Stieltjes type. The same formula, interpreted in the sense of weak convergence, is also valid for the other functions $u(e^{i\varphi})$ for which the correspondence is established, that is, we have

$$(u(U)f, g) = \int_0^{2\pi} u(e^{i\varphi}) d(E_\varphi f, g), \tag{16}$$

where the integral is taken in the Stieltjes-Lebesgue sense.

This formula, being a consequence of (15), is valid for an arbitrary spectral family $\{F_\varphi\}$ over $[0, 2\pi]$ for which (15) holds. Taking in particular $u(e^{i\varphi}) = e_\psi(\varphi)$, we obtain

$$(E_\psi f, g) = \int_0^{2\pi} e_\psi(\varphi) d(F_\varphi f, g) = \int_0^{\psi} d(F_\varphi f, g) = ((F_\psi - F_0)f, g).$$

When we have in addition $F_0 = O$, it follows that $E_\psi = F_\psi$. This proves the *uniqueness* of the spectral family corresponding to U.

110. Normal Transformations. Factorizations

Symmetric and unitary transformations are particular types of *normal* transformations, that is, linear transformations N which are permutable with their adjoints: $N^*N = NN^*$.

Every normal transformation N can be written in the form

$$N = X + iY \tag{17}$$

where X and Y are permutable symmetric transformations: we have only to set

$$X = \tfrac{1}{2}(N + N^*), \quad Y = \frac{1}{2i}(N - N^*);$$

it is clear that $\|X\| \leq \|N\|$, $\|Y\| \leq \|N\|$.

Another type of decomposition, which is less immediate, is the following:

$$N = RU = UR, \tag{18}$$

where R is a *positive* symmetric transformation and U is a *unitary* transformation.

Just as the decomposition (17) is the analogue of the decomposition of a complex number into its real and imaginary parts: $z = x + iy$, the decomposition (18) is the analogue of the decomposition of z into the product of its modulus and a factor of unit modulus: $z = re^{i\varphi}$.

In order to obtain the decomposition (18), we take for R the positive square root of the positive transformation $N^*N = NN^*$; since R is the limit of a sequence of polynomials in N^*N, it is permutable with N and with N^*.

We have for every element f:

$$\|Rf\|^2 = (Rf, Rf) = (R^2f, f) = \begin{cases}(N^*Nf, f) = (Nf, Nf) = \|Nf\|^2, \\ (NN^*f, f) = (N^*f, N^*f) = \|N^*f\|^2,\end{cases}$$

hence

$$\|Nf\| = \|N^*f\| = \|Rf\|. \tag{19}$$

We denote by $\mathfrak{L}$ the subspace of $\mathfrak{H}$ consisting of elements of the form Rf and of their limits, and let $\mathfrak{M}$ be its orthogonal complement. $\mathfrak{M}$ obviously consists of the elements h for which $Rh = 0$, or equivalently, by (19), those for which $Nh = 0$ or $N^*h = 0$. But the set of elements h such that $N^*h = 0$ is the orthogonal complement of the subspace $\mathfrak{L}'$ which consists of the elements of the form Nf and of their limits. Consequently $\mathfrak{L} = \mathfrak{L}'$.

We assign to each element of the form $g = Rf$ the element $Ug = Nf$; the latter is uniquely determined, because if $Rf = Rf'$ we have $R(f - f') = 0$, hence by (19), $N(f - f') = 0$ and $Nf = Nf'$. This correspondence obviously is homogeneous, additive, and moreover isometric: $\|Ug\| = \|g\|$. We can extend it by continuity to all elements of $\mathfrak{L}$ and we thus obtain an isometric transformation U of $\mathfrak{L}$ into itself; it will even be unitary, because the elements of the form Nf and their limits fill the subspace entirely. We can extend the transformation U to the entire space $\mathfrak{H}$ in such a way that it remains unitary: we have only to define U in the complementary subspace $\mathfrak{M}$ by an arbitrary unitary transformation of $\mathfrak{M}$ into itself (for example by the identity transformation) and then define it in the entire space $\mathfrak{H} = \mathfrak{L} + \mathfrak{M}$ by linearity.

The equation $Nf = URf$ is verified for elements of the form Rf by the very definition of the transformation U. As for the equation $Nf = RUf$, it is obvious for elements f of $\mathfrak{M}$ and it also holds for elements g of $\mathfrak{L}$, since these elements are of the form

$$g = \lim_{n\to\infty} Rf_n$$

and we thus have

$$Ng = \lim_{n\to\infty} NRf_n = \lim_{n\to\infty} RNf_n = R \lim_{n\to\infty} Nf_n = R \lim_{n\to\infty} URf_n = $$
$$= RU \lim_{n\to\infty} Rf_n = RUg;$$

the equation is therefore true for all elements of $\mathfrak{H}$. This completes the proof of (18).

We observe further that the transformation R is obviously permutable with every linear transformation A which is permutable with N and N^*. The same is true of U, if it is defined to be the identity in the subspace $\mathfrak{M}$. In fact, we have on the one hand

$$AURf = ANf = NAf = URAf = UARf,$$

hence

$$AUg = UAg$$

for all elements g of the form Rf, and consequently for all elements of $\mathfrak{L}$. On the other hand, for g belonging to $\mathfrak{M}$, Ag also belongs to $\mathfrak{M}$, since $RAg = ARg = 0$; therefore $AUg = Ag = UAg$.

We remark that part of this argument is also applicable to the case of an arbitrary linear transformation T instead of a normal transformation N. We can form the positive symmetric transformation $R = (T^*T)^{\frac{1}{2}}$ and we have further $\|Tf\| = \|Rf\|$ for all elements f, from which it follows as above that the transformation U, defined for elements of the form $g = Rf$ by $Ug = Tf$, is homogeneous, additive, and isometric and that it can be extended by continuity to the entire subspace $\mathfrak{L}$ with the preservation of these properties. However, the elements Ug will not in general belong to $\mathfrak{L}$ and we shall not be able to extend U to a unitary transformation of $\mathfrak{H}$. But by setting $Ug = 0$ for elements g of the orthogonal complement $\mathfrak{M}$, we extend U to a *partially isometric* transformation of the space $\mathfrak{H}$; this is a linear transformation of $\mathfrak{H}$ which is isometric for the elements of a certain subspace of $\mathfrak{H}$ and is zero for the elements of the orthogonal complement of the subspace.

We summarize:

THEOREM. *Every linear transformation T of Hilbert space can be written in the form UR where R is a positive symmetric transformation and U is a partially isometric transformation. When T is normal, $T = N$, U can be chosen to be unitary and such that U and R permute with one another and with all linear transformations which permute with N and N^*.*

111. The Spectral Decomposition of Normal Transformations. Functions of Several Transformations

From each of the decompositions (17) and (18) a *spectral decomposition* of the normal transformation N can be obtained in the following manner:

Let $\{E_\lambda^X\}$ and $\{E_\lambda^Y\}$ be spectral families of the symmetric transformations X and Y over the segment $-\|N\| \leq \lambda \leq \|N\|$. For every fixed value of x and of y, E_x^X and E_y^Y are limits of polynomials in X and in Y respectively, and consequently the limits of polynomials in N and N^*. It follows, in particular, that $E_x^X \smile E_y^Y$. We have

$$N = X + iY = \int_{-\infty}^{\infty} x dE_x^X \cdot \int_{-\infty}^{\infty} dE_y^Y + i \int_{-\infty}^{\infty} dE_x^X \cdot \int_{-\infty}^{\infty} y dE_y^Y = \tag{20}$$

$$= \int_{-\infty}^{\infty} \int_{-\infty}^{\infty} (x + iy) dE_x^X dE_y^Y$$

in the sense that the sum

$$\sum_{h,k} z_{hk}(E^X_{x_h} - E^X_{x_{h-1}})(E^Y_{y_k} - E^Y_{y_{k-1}}),$$

corresponding to a decomposition of the complex plane into rectangles

$$\delta_{hk} = [x_{h-1} < x \leq x_h, y_{k-1} < y \leq y_k]$$

and arbitrary points $z_{hk} = x_{hk} + iy_{hk}$ of δ_{hk}, converges in norm to the transformation N as the decomposition becomes arbitrarily fine. Since $E^X_x \smile E^Y_y$, the products

$$E(\delta_{hk}) = (E^X_{x_h} - E^X_{x_{h-1}})(E^Y_{y_k} - E^Y_{y_{k-1}})$$

are also projections; moreover they are pairwise orthogonal, and consequently they define a decomposition of the entire space into the vector sum of mutually orthogonal subspaces.

We have, by analogy with (20):

(20a) $$N^* = X - iY = \int_{-\infty}^{\infty}\int_{-\infty}^{\infty}(x - iy)dE^X_x dE^Y_y.$$

More generally, it follows from the relations

$$X^rY^s = \int_{-\infty}^{\infty} x^r dE^X_x \cdot \int_{-\infty}^{\infty} y^s dE^Y_y = \int_{-\infty}^{\infty}\int_{-\infty}^{\infty} x^r y^s dE^X_x dE^Y_y \quad (r, s = 0, 1, 2, \ldots)$$

that

(21) $$p(X, Y) = \int_{-\infty}^{\infty}\int_{-\infty}^{\infty} p(x, y)dE^X_x dE^Y_y$$

for every polynomial

$$p(x, y) = \sum_{r,s} c_{rs} x^r y^s$$

and for the corresponding transformation

$$p(X, Y) = \sum_{r,s} c_{rs} X^r Y^s.$$

Equivalently, we have

(22) $$q(N, N^*) = \int_{-\infty}^{\infty}\int_{-\infty}^{\infty} q(z, \bar{z})dE^X_x dE^Y_y$$

for every polynomial

$$q(z, \bar{z}) = \sum_{r,s} d_{rs} z^r \bar{z}^s$$

in $z = x + iy$ and $\bar{z} = x - iy$, and for the corresponding transformation

$$q(N, N^*) = \sum_{r,s} d_{rs} N^r N^{*s};$$

formula (22) obviously includes (20) and (20a).

The projection $E(\delta)$, which is a function of the variable rectangle δ, is *additive* and *multiplicative* in the sense that for two disjoint rectangles δ_1, δ_2 whose union is a rectangle we have

$$E(\delta_1) + E(\delta_2) = E(\delta_1 \cup \delta_2),$$

and for two arbitrary rectangles δ_1, δ_2 we have

$$E(\delta_1)E(\delta_2) = E(\delta_1 \cap \delta_2)$$

where we set the second member equal to O if the set $\delta_1 \cap \delta_2$ is empty. If δ includes the closed rectangle

$$\varDelta = [m_X \leq x \leq M_X,\ m_Y \leq y \leq M_Y]$$

where m_X, M_X, m_Y, M_Y denote the greatest lower and least upper bounds of X and Y, we have $E(\delta) = I$; consequently $E(\delta) = O$ if the rectangle δ lies entirely in the exterior of $\varDelta$. In all these statements, which are easy to verify, we see a certain advantage in half-open rectangles

$$\delta = [x_1 < x \leq x_2, y_1 < y \leq y_2],$$

since the intersection of two rectangles of this type is either a rectangle of the same type or empty.

Moreover, it is easy to extend the definition of $E(\delta)$ to rectangles of other types. For example, for an open rectangle $\delta = [x_1 < x < x_2,\ y_1 < y < y_2]$ we set

$$E(\delta) = (E^X_{x_2-0} - E^X_{x_1})\,(E^Y_{y_2-0} - E^Y_{y_1}),$$

and for a closed rectangle $\delta = [x_1 \leq x \leq x_2,\ y_1 \leq y \leq y_2]$ we set

$$E(\delta) = (E^X_{x_2} - E^X_{x_1-0})\,(E^Y_{y_2} - E^Y_{y_1-0}).$$

The additive and multiplicative properties remain preserved and the function $E(\delta)$ will even be denumerably additive, or equivalently, it will be *continuous* in the following sense: if $\delta_1 \subset \delta_2 \subset \delta_3 \subset \ldots$, then

$$\lim_n E(\delta_n) = E(\bigcup_n \delta_n),$$

and if $\delta_1 \supset \delta_2 \supset \delta_3 \supset \ldots$, then

$$\lim_n E(\delta_n) = E(\bigcap_n \delta_n).$$

In formulas (20)—(22) the integrals are taken with respect to this additive and multiplicative rectangle function $E(\delta)$, a fact which we can conveniently express by writing $E(dxdy)$ instead of $dE^X_x dE^Y_y$.

We can then state our results as follows:

THEOREM.[15] *To every normal transformation N there corresponds a family $\{E(\delta)\}$ of projections such that $E(\delta)$ is an additive and multiplicative rectangle*

[15] Cf. WINTNER [*] (p. 281); VON NEUMANN [2].

function,

$$N = \int_{-\infty}^{\infty}\int_{-\infty}^{\infty} zE(dxdy), \qquad N^* = \int_{-\infty}^{\infty}\int_{-\infty}^{\infty} \bar{z}E(dxdy),$$

and more generally,

$$q(N, N^*) = \int_{-\infty}^{\infty}\int_{-\infty}^{\infty} q(z, \bar{z})E(dxdy),$$

where $q(z, \bar{z})$ *is an arbitrary polynomial in* $z = x + iy$ *and* $\bar{z} = x - iy$; *for every fixed rectangle* δ, $E(\delta)$ *is the limit of a sequence of polynomials in* N *and* N^*.

The domain of integration can be restricted to the rectangle Δ, or rather to the disc $x^2 + y^2 = z\bar{z} \leq \|N\|^2$, since for every rectangle δ lying in the exterior of this disc and at a distance $\varepsilon > 0$ from it we have $E(\delta) = 0$. In fact, for every element of the form $g = E(\delta)f$ we have:

$$\|N\|^2 \|g\|^2 \geq \|Ng\|^2 = (N^*Ng, g) = \int_{-\infty}^{\infty}\int_{-\infty}^{\infty} \bar{z}z(E(dxdy)g, g) \geq$$

$$\geq (\|N\| + \varepsilon)^2 \int_{-\infty}^{\infty}\int_{-\infty}^{\infty} (E(dxdy)g, g) = (\|N\| + \varepsilon)^2 \|g\|^2,$$

as follows from the fact that, inasmuch as $E(\delta')g = E(\delta')E(\delta)f = 0$ for every δ' in the exterior of δ, the integration can be restricted to the domain δ, and that in this domain $|z| \geq \|N\| + \varepsilon$. But the inequality obtained obviously is possible only if $\|g\| = 0$, $g = 0$.

In the case where there exists a quantity $m > 0$ such that $\|Nf\| \geq m \|f\|$ for all f, we see by means of an analogous argument that $E(\delta) = 0$ for every rectangle δ lying entirely in the disc $x^2 + y^2 = \bar{z}z \leq m^2$.

In particular, if the transformation N is *unitary*, $N = U$, we have $\|Uf\| = \|f\|$ for every f, and consequently we have $E(\delta) = 0$ for every rectangle δ which lies either entirely in the exterior or entirely in the interior of the unit circle $x^2 + y^2 = \bar{z}z = 1$. We can therefore say that all the "spectral mass" is concentrated on the unit circle. Let $\delta_1, \delta_2, \ldots, \delta_n$ be (open, half-open, or closed) rectangles, covering the arc $0 < \psi \leq \varphi$ of this circle in a simple manner and having no other points in common with the circle. The sum $\sum_1^n E(\delta_k)$ then depends only on the arc in question, that is, it depends only on φ; denoting it by E_φ^U, it is easy to see that we have found the spectral family $\{E_\varphi^U\}$ of the unitary transformation U, that is,

$$U = \int_0^{2\pi} e^{i\varphi} dE_\varphi^U.$$

On the other hand, when we start with the factorization (18) of the given normal transformation N and make use of the spectral families $\{E_r^R\}$,

$\{E^U_\varphi\}$ of the positive symmetric transformation R and of the unitary transformation U, we arrive at the following spectral representation of N:

$$N = RU = \int_0^\infty r dE^R_r \cdot \int_0^{2\pi} e^{i\varphi} dE^U_\varphi = \int_0^\infty \int_0^{2\pi} re^{i\varphi} dE^R_r dE^U_\varphi = \int_0^\infty \int_0^{2\pi} z\widehat{E}(drd\varphi),$$

where $z = re^{i\varphi}$ and $\widehat{E}(\widehat{\delta})$ denotes the additive and multiplicative function of the "circular rectangle"

$$\widehat{\delta} = [r_1 < r \leq r_2,\ \varphi_1 < \varphi \leq \varphi_2],$$

defined by

$$\widehat{E}(\widehat{\delta}) = (E^R_{r_2} - E^R_{r_1})\,(E^U_{\varphi_2} - E^U_{\varphi_1}).$$

We can pass from one of the integral formulas to the other by substituting the polar coordinates r, φ for the rectangular coordinates x, y, or conversely.

Starting with formula (21), valid for polynomials, we can define the transformations $u(X, Y)$ corresponding to more general functions $u(x, y)$, in particular, to all continuous functions: the correspondence continues to be homogeneous, additive, and multiplicative.

Passage to the case of n variables is also possible. Given n permutable symmetric transformations

$$X_k = \int_{-\infty}^{\infty} \lambda dE^{(k)}_\lambda \qquad (k = 1, 2, \ldots, n),$$

we assign to each "n-dimensional rectangle"

$$\delta = [a_k < x_k \leq b_k;\ k = 1, 2, \ldots, n]$$

the transformation

$$E(\delta) = \prod_{k=1}^{n} (E^{(k)}_{b_k} - E^{(k)}_{a_k});$$

for a rectangle to which we have added other of its "faces," or for an open rectangle, the definition of $E(\delta)$ is modified in an obvious manner by replacing certain of the $E^{(k)}_\lambda$ by $E^{(k)}_{\lambda-0}$. We thus arrive at a family of projections such that $E(\delta)$ is an additive (and even denumerably additive) and multiplicative rectangle function, $E(\Theta) = O$ for the empty set Θ, and $E(\Delta) = I$ for the rectangle

$$\Delta = [m_{X_k} \leq x_k \leq M_{X_k};\quad k = 1, 2, \ldots, n].$$

We immediately see that

$$p(X_1, X_2, \ldots, X_n) = \int \underset{\Delta}{\ldots} \int p(x_1, x_2, \ldots, x_n) E(dx_1 dx_2 \ldots dx_n)$$

for every polynomial $p(x_1, x_2, \ldots, x_n)$, the integral being defined as the limit in the norm of the sums of the Stieltjes-Riemann type. The same formula permits us to assign transformations to other more general functions, in particular, to all functions continuous in Δ, and the correspondence will be homogeneous, additive, and multiplicative.

It is even possible to extend this method to the case of an *infinite number of permutable symmetric transformations*

$$X_k = \int_{-\infty}^{\infty} \lambda dE_\lambda^{(k)} \qquad (k = 1, 2, \ldots).$$

In the corresponding space Q^∞, consisting of the points $x = (x_1, x_2, \ldots)$ with an infinite number of coordinates, the role of a rectangle is played by a set consisting of all the points a finite number of whose coordinates belong to given linear intervals, with the others arbitrary. The rectangle function $E(\delta)$ is then defined as the product of the projections $E_{b_k}^{(k)} - E_{a_k}^{(k)}$ (or of the analogous ones with $b_k - 0$ or with $a_k - 0$ instead of b_k or of a_k) corresponding to the intervals in question. The additive and multiplicative properties are verified without difficulty, and we define the transformation $u(X_1, X_2, \ldots)$ corresponding to the function $u(x_1, x_2, \ldots)$ as the integral of u with respect to this additive rectangle function, at least for the functions which are continuous in the closed domain

$$\Delta = [m_{X_k} \leq x_k \leq M_{X_k};\ k = 1, 2, \ldots].$$

Continuity at a point x^* means we can assign to every $\varepsilon > 0$ a rectangle containing the point x^* in its interior and such that for every other point x of this rectangle we have $|u(x) - u(x^*)| < \varepsilon$. In other words, the function is continuous at the point $x^* = (x_1^*, x_2^*, \ldots)$, if its value does not change noticeably when a finite number of coordinates are varied a little, and the others are varied arbitrarily. The Borel covering theorem can be extended to the space Q^∞, or rather to the closed domain Δ and to its coverings by open rectangles, and it follows that if the function $u(x)$ is continuous at every point of Δ, then there exists for every given $\varepsilon > 0$ a finite system of rectangles covering Δ such that, in the portion of Δ contained in any one rectangle, the oscillation of the function is less than ε. The existence of the integral of $u(x)$ with respect to $E(\delta)$, that is, of the limit (in norm) of the sums of the Riemann-Stieltjes type, can be proved as in the case of ordinary integrals; the sequence of decompositions must be constructed so that the induced decompositions of an arbitrary fixed "edge" of Δ become infinitely fine.

UNITARY TRANSFORMATIONS OF THE SPACE L^2

112. A Theorem of Bochner

Let us consider the space L^2 of functions defined in the interval (a, b), where $-\infty \leq a < b \leq \infty$; the unitary transformations of this space can be characterized "analytically" in the following manner:

THEOREM.[16] *To every unitary transformation $g = Uf$ of the space L^2 we can attach two functions, $K(\xi, x)$ and $H(\xi, x)$, which are defined on the square $(a < \xi < b;\ a < x < b)$ and for every fixed value of ξ belong to the space L^2,*

[16] BOCHNER [**2**].

and which satisfy

$$(23)\qquad \int_0^\xi g(x)dx = \int_a^b \overline{K(\xi, x)}f(x)dx \quad \textit{and} \quad \int_0^\xi f(x)dx = \int_a^b \overline{H(\xi, x)}g(x)dx;$$

moreover, these functions satisfy the equations

$$\left.\begin{aligned}&\text{(a)}\quad \int_a^b \overline{K(\xi, x)}K(\eta, x)dx\\ &\text{(b)}\quad \int_a^b \overline{H(\xi, x)}H(\eta, x)dx\end{aligned}\right\} = \min\{|\xi|, |\eta|\} \text{ or } 0, \text{ depending on whether } \xi\eta \geq 0 \text{ or } \leq 0,$$

$$\text{(c)}\quad \int_0^\eta K(\xi, x)dx = \int_0^\xi \overline{H(\eta, x)}dx.$$

Conversely, every pair of functions satisfying these conditions generates by formulas (23) *a unitary transformation of the space L^2 and its inverse.*

In order to see this, let U be a given unitary transformation of L^2 and define the functions K and H by

$$H(\xi, x) = Ue_\xi(x) \text{ and } K(\xi, x) = U^{-1}e_\xi(x),$$

where $e_\xi(x)$ denotes the function which is equal to sgn ξ for all x lying between 0 and ξ and which is zero outside this interval. Since U and U^{-1} are isometric, we have for $g = Uf$:

$$(g, e_\xi) = (Uf, e_\xi) = (f, U^{-1}e_\xi), \ (f, e_\xi) = (U^{-1}g, e_\xi) = (g, Ue_\xi);$$

these are precisely formulas (23). Choosing, in particular, first

$$f = U^{-1}e_\eta \text{ and } g = Uf = e_\eta,$$

then

$$f = e_\eta \text{ and } g = Uf = Ue_\eta,$$

formulas (23) reduce to equations (a)—(c).

Let us consider now the converse proposition. With the functions $K(\xi, x)$ and $H(\xi, x)$ given, we begin by defining two transformations, U and V, first for the functions $e_\xi(x)$ by setting

$$Ue_\xi(x) = H(\xi, x), \quad Ve_\xi(x) = K(\xi, x).$$

Equations (a)—(c) assure us that

$$(24)\qquad (Ve_\xi, Ve_\eta) = (e_\xi, e_\eta), \quad (Ue_\xi, Ue_\eta) = (e_\xi, e_\eta), \quad (Ve_\xi, e_\eta) = (e_\xi, Ue_\eta).$$

Let $f(x)$ be a step function. It can be expressed as a linear combination of the functions $e_\xi(x)$ in a unique manner, and we can then define Uf and Vf by the same linear combinations of the corresponding Ue_ξ and Ve_ξ. It follows from relations (24) that we have for every pair of step functions f and g

$$(Vf, Vg) = (f, g), \quad (Uf, Ug) = (f, g), \quad (Vf, g) = (f, Ug).$$

This means that U and V are isometric transformations which are adjoint to one another, at least in their present domain — the linear set formed by the step functions. But since this set is everywhere dense in L^2, the transformations U and V can be extended to the entire space by continuity, and they remain isometric and adjoint, that is, such that

$$U^*U = I, \quad V^*V = I, \quad \text{and} \quad U = V^*.$$

We see by these relations that U has the left inverse U^* and the right inverse V; therefore U^{-1} exists and equals U^* and V (cf. Sec. 67).

The transformation U is therefore unitary, and since the functions K and H attached to it by the theorem are obviously equal to the functions with which we started, U and U^{-1} are represented analytically by formulas (23); this completes the proof of the theorem.

113. Fourier-Plancherel and Watson Transformations

Let us consider in particular the functions

$$K(\xi, x) = \frac{\overline{\chi(\xi x)}}{x} \quad \text{and} \quad H(\xi, x) = \frac{\chi(\xi x)}{x},$$

where the function χ is so chosen that $\chi(x)/x$ belongs to $L^2(a, b)$ and that for every ξ and η in (a, b)

$$\int_a^b \frac{\overline{\chi(\xi x)}\chi(\eta x)}{x^2} dx = \begin{cases} \min\{|\xi|, |\eta|\} & \text{if } \xi\eta \geq 0, \\ 0 & \text{if } \xi\eta \leq 0. \end{cases}$$

Since all the hypotheses of the converse theorem we just proved are then verified (condition (c) automatically), the formulas

$$g(x) = \frac{d}{dx}\int_a^b \frac{\chi(xy)}{y} f(y)dy \quad \text{and} \quad f(x) = \frac{d}{dx}\int_a^b \frac{\overline{\chi(xy)}}{x} g(y)dy$$

define a unitary transformation of the space $L^2(a, b)$ into itself, and its inverse.

These are the "general transforms" of WATSON.[17] Specifically, they generalize the FOURIER-PLANCHEREL transformation which corresponds to

[17] WATSON [**1**], TITCHMARSH [**1**], PLANCHEREL [**2**].

the interval $(-\infty, \infty)$ and to the function

$$\chi(x) = \frac{1}{\sqrt{2\pi}} \frac{e^{-ix}-1}{-i}.$$

In fact, the function $(e^{-ix}-1)/x$ belongs to $L^2(-\infty, \infty)$ and

$$\frac{1}{2\pi}\int_{-\infty}^{\infty} \frac{(e^{i\xi x}-1)(e^{i\eta x}-1)}{x^2}\,dx = \frac{1}{2\pi}\int_{-\infty}^{\infty} \frac{\cos(\xi-\eta)x - \cos\xi x - \cos\eta x + 1}{x^2}\,dx =$$

$$= \frac{1}{2\pi}\{|\xi|+|\eta|-|\xi-\eta|\}\int_{-\infty}^{\infty}\frac{\sin^2 u}{u^2}\,du = \tfrac{1}{2}\{|\xi|+|\eta|-|\xi-\eta|\} =$$

$$= \begin{cases} \min\{|\xi|, |\eta|\}, & \text{if } \xi\eta \geq 0. \\ 0 & \text{if } \xi\eta \leq 0. \end{cases}$$

Therefore we have the

THEOREM.[18] *The formulas*

$$g(x) = \frac{1}{\sqrt{2\pi}}\frac{d}{dx}\int_{-\infty}^{\infty}\frac{e^{-ixy}-1}{-iy}f(y)dy, \quad f(x) = \frac{1}{\sqrt{2\pi}}\frac{d}{dx}\int_{-\infty}^{\infty}\frac{e^{ixy}-1}{iy}g(y)dy$$

define a unitary transformation of the space $L^2(-\infty, \infty)$ and its inverse. These formulas can also be written in the classical form

$$g(x) = \frac{1}{\sqrt{2\pi}}\int_{-\infty}^{\infty} e^{-ixy}f(y)dy, \quad f(x) = \frac{1}{\sqrt{2\pi}}\int_{-\infty}^{\infty} e^{ixy}g(y)dy,$$

if we agree to denote here by the integral from $-\infty$ to ∞ the limit in the mean (with respect to the variable x) of the integral from $-\omega$ to ω when $\omega \to \infty$.

In order to prove this last statement we consider the function $f_\omega(x)$ which equals $f(x)$ in the finite interval $-\omega \leq x \leq \omega$ and which is zero elsewhere, and we denote its transform by $g_\omega(x)$. We have

$$g_\omega(x) = \frac{1}{\sqrt{2\pi}}\lim_{h\to 0}\frac{1}{h}\int_{-\omega}^{\omega}\frac{e^{-i(x+h)y}-e^{-ixy}}{-iy}f(y)dy =$$

$$= \frac{1}{\sqrt{2\pi}}\lim_{h\to 0}\int_{-\omega}^{\omega}\frac{\sin\frac{hy}{2}}{\frac{hy}{2}}e^{-i\frac{hy}{2}}e^{-ixy}f(y)dy.$$

The function under the integral sign is majorized in absolute value by the function $|f(x)|$, which is summable in the finite interval $(-\omega, \omega)$; therefore by

[18] PLANCHEREL [**1**].

Lebesgue's theorem we can interchange the order of integration and the passage to the limit, which yields

$$g_\omega(x) = \frac{1}{\sqrt{2\pi}} \int_{-\omega}^{\omega} e^{-ixy} f(y)\,dy.$$

But, $f_\omega(x)$ obviously converges to $f(x)$ in the mean when $\omega \to \infty$. Since the transformation is isometric, this implies that $g_\omega(x)$ also converges in the mean to the transform of $f(x)$, that is, to $g(x)$, which proves the first formula. The second, relative to the inverse transformation, is established in an analogous manner.

Denoting the Fourier-Plancherel transformation by U, we see that

$$Uf(x) = U^{-1}f(-x),$$

hence that

$$U^2 f(x) = f(-x),\ U^4 f(x) = f(x).$$

A simple calculation, based only on the relations

$$U^* = U^{-1},\ U^{k+4} = U^k U^4 = U^k \quad (k = 0, \pm 1, \ldots),$$

verifies that the transformations

$$P_0 = \tfrac{1}{4}(I + U + U^2 + U^3),\ P_1 = \tfrac{1}{4}(I - iU - U^2 + iU^3),$$
$$P_2 = \tfrac{1}{4}(I - U + U^2 - U^3),\ P_3 = \tfrac{1}{4}(I + iU - U^2 - iU^3)$$

are pairwise orthogonal projections whose sum is I; moreover,

$$UP_k = i^k P_k \qquad (k = 0, 1, 2, 3).$$

Therefore the transformation U has for characteristic values

$$1, i, -1, -i,$$

and every element f of L^2 can be decomposed into the sum of four orthogonal elements, which are characteristic elements corresponding to these characteristic values. The "spectral decomposition" of U is therefore of the simple form

$$U = U(P_0 + P_1 + P_2 + P_3) = P_0 + iP_1 - P_2 - iP_3,$$

that is,

$$U = \int_0^{2\pi} e^{i\lambda}\,dE_\lambda,$$

where E_λ is equal to

$$O, \qquad P_1, \qquad P_1 + P_2, \qquad P_1 + P_2 + P_3, \qquad I,$$

according as

$$0 \leq \lambda < \frac{\pi}{2},\ \frac{\pi}{2} \leq \lambda < \pi,\ \pi \leq \lambda < \frac{3\pi}{2},\ \frac{3\pi}{2} \leq \lambda < 2\pi,\ \lambda = 2\pi.$$

Chapter VIII

UNBOUNDED LINEAR TRANSFORMATIONS OF HILBERT SPACE

GENERALIZATION OF THE CONCEPT OF LINEAR TRANSFORMATION

114. A Theorem of Hellinger and Toeplitz. Extension of the Concept of Linear Transformation

Up to now we have always considered linear transformations of a Hilbert space $\mathfrak{H}$ which by definition were defined for all elements of $\mathfrak{H}$ and which were additive, homogeneous, and bounded. But there are examples imposed by analysis and by mathematical physics of transformations which are additive and homogeneous, but which are not defined for all the elements of $\mathfrak{H}$ and which are not bounded. In the space L^2, such a transformation is the transformation which assigns to every absolutely continuous function its derivative provided the latter also belongs to L^2; this transformation plays a fundamental role in wave mechanics. It is clear that this transformation is defined only for elements of a linear set which is everywhere dense in the space L^2 without exhausting it. It is not bounded; in fact, the function $e^{2\pi inx}$, as an element of the space $L^2(0,1)$, has unit norm, while its derivative $2\pi ine^{2\pi inx}$ has norm $2\pi n$, a quantity which increases indefinitely with n.

We should point out, at least for the symmetric transformations, the connection which exists between the properties of being defined everywhere and of being bounded. We have, namely, the following theorem, due to Hellinger and Toeplitz:[1]

THEOREM. *Every transformation A which is defined for all elements of Hilbert space and which is additive, homogeneous, and symmetric, that is, satisfies $(Af, g) = (f, Ag)$, is necessarily bounded.*

If not, there would be a sequence of elements g_n such that

$$\|g_n\| = 1 \text{ and } \|Ag_n\| \to \infty.$$

[1] Hellinger–Toeplitz [**1**] (in particular pages 321–327); see also Stone [*] (Theorem 2.23).

We consider the sequence of functionals operating on the variable element f:

$$L_n(f) = (Af, g_n) = (f, Ag_n);$$

these are obviously linear, since L_n has $\|Ag_n\|$ for bound. Moreover, we have

$$|L_n(f)| \leq \|Af\| \, \|g_n\| = \|Af\| \qquad (n = 1, 2, \ldots),$$

that is, for each fixed element f the values of $L_n(f)$ constitute a bounded numerical sequence. But we know that a sequence of linear functionals can not be bounded for every element f of the space $\mathfrak{H}$ unless the sequence of norms is also bounded (see Sec. 31). Hence there exists a constant C such that

$$|L_n(f)| \leq C\|f\|$$

for $n = 1, 2, \ldots$ and for all the elements f of $\mathfrak{H}$. In particular, we set $f = Ag_n$; it follows that

$$\|Ag_n\|^2 = (Ag_n, Ag_n) = L_n(Ag_n) \leq C\|Ag_n\|,$$

in contradiction to the hypothesis that

$$\|Ag_n\| \to \infty.$$

Hence the theorem is proved.

We shall see soon that this theorem can also be extended to certain non-symmetric transformations.

After this digression let us return to our problem, which is to so extend the notion of linear transformation that it includes transformations which are not bounded and which are not even defined for all the elements of the Hilbert space $\mathfrak{H}$; an example is the differentiation transformation.

Let us agree upon the following more general *definition.*

A linear transformation of the Hilbert space $\mathfrak{H}$ is a function T which assigns to the elements f of a certain linear subset $\mathfrak{D}_T$ of $\mathfrak{H}$ elements Tf of $\mathfrak{H}$ in an additive and homogeneous manner; that is, it satisfies

$$1^0 \quad T(f_1 + f_2) = Tf_1 + Tf_2, \qquad 2^0 \quad T(cf) = cTf.$$

The linear set $\mathfrak{D}_T$ is said to be the *domain of definition,* or more briefly the *domain,* of the linear transformation T; it contains at least the element 0 and we always have $T0 = 0$.

Two linear transformations are regarded as equal only if they both have the same domain and always assume the same values. If the domain of the transformation T' includes that of the transformation T, and if in $\mathfrak{D}_T$ the two transformations coincide, we say that T' is an *extension* of T; this is written

$$T' \supseteq T \text{ or } T \subseteq T'.$$

When the linear transformation T is *bounded,* that is, when there exists a constant C such that

$$\|Tf\| \leq C\|f\|$$

for all elements f of $\mathfrak{D}_T$, it can be extended to the closure $[\mathfrak{D}_T]$ of $\mathfrak{D}_T$ by continuity. If the linear set $\mathfrak{D}_T$ is not dense in $\mathfrak{H}$, that is, if its closure $[\mathfrak{D}_T]$ is a proper subspace of $\mathfrak{H}$, we can even extend the transformation beyond $[\mathfrak{D}_T]$, for example by setting $Tf = 0$ in the orthogonal complement and then defining T in the entire space $\mathfrak{H}$ by linearity. The linear transformation thus extended will obviously have the same bound C.

Hence there is no loss of generality if we assume that the bounded linear transformations considered are already defined everywhere in $\mathfrak{H}$. When we speak in the sequel of *bounded linear transformations* we shall assume, unless otherwise specified, that they are *defined everywhere* in $\mathfrak{H}$, that is, that they are linear transformations in the sense admitted up to now.

Sums and products of linear transformations are defined, at least from the formal point of view, just as before:

$$(T_1 + T_2)f = T_1f + T_2f, \quad (cT)f = cTf, \quad (T_1T_2)f = T_1(T_2f).$$

But there are complications here resulting from the fact that the domains of definition do not coincide in general with the entire space. *The domain of $T_1 + T_2$ is therefore only the common part of the domains of T_1 and of T_2, whereas the domain of T_1T_2 consists of those elements f of the domain of T_2 for which T_2f belongs to the domain of T_1.* It can even occur, according to this definition, that the sum or the product of two linear transformations is defined only for the single element 0. Moreover, this can even occur for the square of a linear transformation whose domain is everywhere dense in $\mathfrak{H}$.

The concept of an *inverse* can also be extended in an obvious manner. In fact, we can define the inverse T^{-1} of every *one-to-one* linear transformation, that is, every transformation not taking on the same value g at two different points f, by setting

$$T^{-1}g = f \quad \text{if} \quad g = Tf.$$

The domain of definition of T^{-1} therefore coincides with the "range" of T, which is a linear set; T^{-1} is obviously a linear transformation also. In this general sense it is possible that a linear transformation, even a bounded one, should have an inverse which is not defined everywhere. The linear transformation

$$T(x_1, x_2, x_3, \ldots) = (0, x_1, x_2, \ldots)$$

of Hilbert coordinate space is as an example. Here T^{-1} is defined only for vectors whose first component is zero. It is clear that for a linear transformation T the condition that T be one-to-one is equivalent to the condition: Tf is zero only for $f = 0$.

Let us pass to the generalization of the concept of *limit* of a sequence of transformations. If $\{T_n\}$ is an *arbitrary* sequence of linear transformations, we define its limit

$$T = \lim_{n\to\infty} T_n$$

by the formula

$$Tf = \lim_{n\to\infty} T_n f$$

which must be interpreted in the following manner: Tf is defined only for the elements f which belong to the domains of all the transformations T_n, at least starting with some index n_0 which can depend on f, and for which the sequence $\{T_n f\}$ converges. For other f, the limit transformation will not be defined. We easily verify the following rules of calculation:

a) $T_1 + T_2 = T_2 + T_1$;

b) $(T_1 + T_2) + T_3 = T_1 + (T_2 + T_3)$;

c) $OT \subseteq O$;

d) $(T_1 T_2)T_3 = T_1(T_2 T_3)$;

e) $(T_1 + T_2)T_3 = T_1 T_3 + T_2 T_3$;

f) $T_1(T_2 + T_3) \supseteq T_1 T_2 + T_1 T_3$ (the equality sign being valid, for example, if T_1 is defined everywhere);

g) $(\lim_{n\to\infty} T_n)T = \lim_{n\to\infty} (T_n T)$;

finally, when T_1^{-1} and T_2^{-1} exist,

h) $(T_1 T_2)^{-1} = T_2^{-1} T_1^{-1}$.

115. Adjoint Transformations

For a bounded linear transformation T, we defined the *adjoint transformation* T^* by the equation

$$(Tf, g) = (f, T^*g),$$

assumed valid for all elements f and g of $\mathfrak{H}$. If we wish to use the same formula in the general case also, it is necessary to assume that $\mathfrak{D}_T$ is everywhere dense in $\mathfrak{H}$, since in the contrary case the element T^*g would not be uniquely determined by the values which (f, T^*g) takes on when f runs through $\mathfrak{D}_T$.

We are thus led to the following *definition*:

Let T be a linear transformation whose domain $\mathfrak{D}_T$ is dense in $\mathfrak{H}$. Let g be an element of $\mathfrak{H}$ to which there corresponds an element g^* of $\mathfrak{H}$ such that

$$(Tf, g) = (f, g^*)$$

for all elements f of $\mathfrak{D}_T$; g^* is then uniquely determined by g, and setting

$$g^* = T^*g$$

we define a transformation T^* which is the adjoint of the linear transformation T.

It is clear that this transformation T^* is also linear; in fact, if it is defined for g_1 and for g_2 we have

$$(Tf, c_1g_1 + c_2g_2) = \bar{c}_1(Tf, g_1) + \bar{c}_2(Tf, g_2) = \bar{c}_1(f, T^*g_1) + \bar{c}_2(f, T^*g_2) =$$
$$= (f, c_1T^*g_1 + c_2T^*g_2)$$

for all f in $\mathfrak{D}_T$; hence $c_1g_1 + c_2g_2$ also belongs to the domain of T^* and we have

$$T^*(c_1g_1 + c_2g_2) = c_1T^*g_1 + c_2T^*g_2.$$

If the transformation T is bounded, then as we have already seen in Sec. 84 its adjoint T^* is everywhere defined and bounded, and has the same norm as T.

In the general case, the domain of T^* does not coincide with the entire space $\mathfrak{H}$; in fact, one can not be sure that this domain contains elements other than 0. Whatever its domain of definition, the transformation T^* possesses the following property:

If $\{g_n\}$ is a sequence of elements of $\mathfrak{D}_{T^*}$ such that

$$g_n \to g \quad \textit{and} \quad T^*g_n \to h,$$

the limit element g also belongs to $\mathfrak{D}_{T^*}$, and we have $T^*g = h$.

This is an immediate consequence of the fact that the scalar product is a continuous function of its factors; in fact,

$$(Tf, g) = \lim_n (Tf, g_n) = \lim_n (f, T^*g_n) = (f, h).$$

We express this property of the adjoint transformation by saying that it is *closed*. In general, *a linear transformation S is called closed if it has the property that for every sequence $\{g_n\}$ of elements of $\mathfrak{D}_S$ such that*

$$g_n \to g \quad \textit{and} \quad Sg_n \to h,$$

the limit element g also belongs to $\mathfrak{D}_S$ and $Sg = h$.

It is obvious that every *continuous* transformation, hence every *bounded* linear transformation, is also *closed*, but the converse is not true in general.

The following relations are obvious:

a) $(cT)^* = \bar{c}T^*$ (for $c \neq 0$),

b) $(T_1 + T_2)^* \supseteq T_1^* + T_2^*$,

c) $T_1 T_2)^* \supseteq T_2^* T_1^*$;

of course, b) and c) have meaning only if $T_1 + T_2$ or T_1T_2 have domains everywhere dense in $\mathfrak{H}$.

It is more difficult to point out the cases where the *equality sign* is valid. Let us show that, in particular, *this is the case if the transformation T_1 is bounded.* It suffices to prove that in this case we also have the converse relations

b′) $(T_1 + T_2)^* \subseteq T_1^* + T_2^*$,

c′) $(T_1 T_2)^* \subseteq T_2^* T_1^*$.

We begin by applying relation b) to the sum $(T_1 + T_2) + (-T_1)$, which is equal to T_2 since T_1 is defined everywhere. We obtain

$$T_2^* \supseteq (T_1 + T_2)^* - T_1^*,$$

or, adding T_1^* to the two members,

$$T_1^* + T_2^* \supseteq (T_1 + T_2)^* - T_1^* + T_1^* = (T_1 + T_2)^*;$$

here we used the fact that T_1^* is also defined everywhere. This proves b′). To prove c′), we consider an element f from the domain of $(T_1 T_2)^*$. For every element g of the domain of T_2, we have

$$(T_2 g, T_1^* f) = (T_1 T_2 g, f) = (g, (T_1 T_2)^* f);$$

hence $T_1^* f$ also belongs to the domain of T_2^* and we have $T_2^* T_1^* f = (T_1 T_2)^* f$. This proves c′).

We observe further the obvious fact:

d) $T_1 \subseteq T_2$ implies $T_1^* \supseteq T_2^*$.

116. Permutability. Reduction

Let us consider now the question of *permutability*. It would be natural to define permutability by the equation $T_1 T_2 = T_2 T_1$, but then a bounded transformation B would not even be permutable with its inverse B^{-1} if the latter existed without being defined everywhere. In fact, we have in this case $B^{-1}B = I$ and $BB^{-1} \neq I$, since BB^{-1} is defined only for the elements of the domain of B^{-1}. In order to include this case, we agree to make the following definition.

If B is a bounded linear transformation and T is a linear transformation of general type, we say that B is permutable with T, and we write $B \smile T$, if

$$BT \subseteq TB.$$

This definition is justified by the fact that the following rules of calculation are valid:

e) $B \smile T_1$, $B \smile T_2$ *imply* $B \smile (T_1 + T_2)$ *and* $B \smile T_1 T_2$,

f) $B_1 \smile T$, $B_2 \smile T$ *imply* $(B_1 + B_2) \smile T$ *and* $B_1 B_2 \smile T$,

g) *if* T^{-1} *exists,* $B \smile T$ *implies* $B \smile T^{-1}$,

h) $B \smile T_n$ $(n = 1, 2, \ldots)$ *implies* $B \smile \lim T_n$,

i) $B_n \smile T$ $(n = 1, 2, \ldots)$ *implies* $\lim B_n \smile T$, *if* $\lim B_n$ *is bounded and if* T *is closed.*

j) *if* T^* *exists,* $B \smile T$ *implies* $B^* \smile T^*$.

The propositions e) and f) are clear; g) is proved by observing that for every element f of the domain of T^{-1},

$$Bf = BTT^{-1}f = TBT^{-1}f,$$

and that consequently Bf also belongs to the domain of T^{-1} and

$$T^{-1}Bf = BT^{-1}f.$$

Proposition *h*) follows from the fact that B is continuous; in fact, if f is an element of the domain of $\lim T_n$,

$$B \cdot \lim T_n f = \lim BT_n f = \lim T_n Bf = (\lim T_n)Bf.$$

In order to verify i), we have only to observe that for every element f of $\mathfrak{D}_T$,

$$B_n f \to (\lim B_n)f, \quad TB_n f = B_n Tf \to (\lim B_n)Tf;$$

inasmuch as T is closed, it follows that $(\lim B_n)f$ belongs to the domain of T and that $T(\lim B_n)f = (\lim B_n)Tf$. Finally, j) is proved as follows: since f is an element of $\mathfrak{D}_{T^*}$,

$$(Tg, B^*f) = (BTg, f) = (TBg, f) = (Bg, T^*f) = (g, B^*T^*f)$$

for all elements g of $\mathfrak{D}_T$, which shows that B^*f also belongs to $\mathfrak{D}_T{}^*$ and that $T^*B^*f = B^*f^*T$.

*

The generalization of the concept of permutability which we have just given justifies itself in particular when it is a question of the permutability of a *projection* P with an arbitrary linear transformation T.

Let $P \smile T$. The relation $PT \subseteq TP$ implies that

$$PTP = (PT)P \subseteq (TP)P = TP,$$

and since PTP and TP have the same domain of definition,

$$PTP = TP. \tag{1}$$

Since the complementary projection $Q = I - P$ is also permutable with T, we have, similarly,

$$QTQ = TQ. \tag{2}$$

Moreover,

$$T = (P + Q)T = PT + QT \subseteq TP + TQ = T(P + Q) = T,$$

and since the two extreme members are equal,

$$T = TP + TQ. \tag{3}$$

Let us consider the orthogonal and complementary subspaces

$$\mathfrak{P} = P\mathfrak{H}, \quad \mathfrak{Q} = Q\mathfrak{H}.$$

Equations (1) and (2) express the fact that if we restrict the definition of the transformation T to elements of its domain which are contained in $\mathfrak{P}$ or in $\mathfrak{Q}$, we obtain linear transformations T_P, T_Q which also have their values in $\mathfrak{P}$ and in $\mathfrak{Q}$ respectively, and which consequently can be considered as linear transformations in $\mathfrak{P}$ and in $\mathfrak{Q}$. Equation (3) expresses that T can be reconstructed from its "parts" in $\mathfrak{P}$ and $\mathfrak{Q}$, that is, from T_P and T_Q, the domain of T consisting precisely of those elements whose projections onto $\mathfrak{P}$ and $\mathfrak{Q}$ belong to the domains of T_P and T_Q, respectively.

We express these facts by saying that the subspaces $\mathfrak{P}$ and $\mathfrak{Q}$ *reduce* the transformation T.

*

Let us introduce a notation which will be useful in the sequel. We shall write

$$T \smile\!\smile \{T_\alpha\}$$

to express the fact that every bounded symmetric transformation which is permutable with all the transformations T_α is also permutable with the transformation T.

117. The Graph of a Transformation

Ordinary functions $y = f(x)$ are usually represented by their curves or graphs, that is, by the sets of points

$$\{x, f(x)\}$$

of the $\{x, y\}$ plane.

If we wish to obtain an analogous "graphical" representation of a transformation in the Hilbert space $\mathfrak{H}$, it is necessary first to form what corresponds to the plane: this will be the "Cartesian product" $\boldsymbol{H} = \mathfrak{H} \times \mathfrak{H}$ formed by all the pairs

$$\{f, g\}$$

of elements of $\mathfrak{H}$; $\boldsymbol{H}$ is also a Hilbert space if we define the fundamental operations in it by the formulas

$$c\{f, g\} = \{cf, cg\},$$

$$\{f_1, g_1\} + \{f_2, g_2\} = \{f_1 + f_2, g_1 + g_2\},$$

$$(\{f_1, g_1\}, \{f_2, g_2\}) = (f_1, f_2) + (g_1, g_2).$$

The transformation T of $\mathfrak{H}$ will then be represented in $\boldsymbol{H}$ by the set of elements

$$\{f, Tf\},$$

where f runs through all the elements of $\mathfrak{D}_T$. We call this set the *graph* of the transformation T and denote it by $\boldsymbol{G}_T$.[2]

It is obvious that the relations

$$T_1 = T_2 \quad \text{and} \quad T_1 \supseteq T_2,$$

are respectively equivalent to the relations

$$\boldsymbol{G}_{T_1} = \boldsymbol{G}_{T_2} \quad \text{and} \quad \boldsymbol{G}_{T_1} \supseteq \boldsymbol{G}_{T_2};$$

of course, for the graphs the symbol $\supseteq$ indicates that $\boldsymbol{G}_{T_2}$ is a subset of $\boldsymbol{G}_{T_1}$.

It is also obvious that the graph $\boldsymbol{G}_T$ of a linear transformation T is a linear set; in order that this set be also closed, and hence a subspace of $\boldsymbol{H}$, it is necessary and sufficient that the transformation T be closed.

We consider, in $\boldsymbol{H}$, the following transformations:

$$\boldsymbol{U}\{f, g\} = \{g, f\}, \quad \boldsymbol{V}\{f, g\} = \{g, -f\};$$

they are obviously unitary and satisfy

$$\boldsymbol{UV} = -\boldsymbol{VU}, \quad \boldsymbol{U}^2 = -\boldsymbol{V}^2 = \boldsymbol{I}$$

(where $\boldsymbol{I}$ denotes the identity transformation of $\boldsymbol{H}$).

With this notation, the equation

$$(Tf, g) = (f, g^*),$$

which defines the adjoint transformation $T^*g = g^*$, can be written in the form

$$(\boldsymbol{V}\{f, Tf\}, \{g, g^*\}) = 0.$$

This expresses the fact that $\boldsymbol{G}_{T^*}$ consists of those elements of $\boldsymbol{H}$ which are orthogonal to $\boldsymbol{VG}_T$. $\boldsymbol{G}_{T^*}$ is therefore a subspace of $\boldsymbol{H}$, namely the orthogonal complement of the subspace $[\boldsymbol{VG}_T]$, the closure of the set $\boldsymbol{VG}_T$. Since we obviously have

$$[\boldsymbol{VG}_T] = \boldsymbol{V}[\boldsymbol{G}_T],$$

we can write:

$$\boldsymbol{G}_{T^*} = \boldsymbol{H} \ominus \boldsymbol{V}[\boldsymbol{G}_T].$$

This relation offers a very handy means for the study of transformations and their adjoints.

For example, the proposition:

"*If T is a linear transformation, the existence of T^{-1}, T^*, and $(T^{-1})^*$ implies the existence of $(T^*)^{-1}$, and $(T^*)^{-1} = (T^{-1})^*$*" can be proved by means of graphs

[2] This idea, as well as the facts we shall prove in sections 117 and 118, are due to von Neumann [**4**].

in the following manner. We observe first that the graph of T^{-1} is obtained from that of T by interchanging the two "coordinates," that is,

$$\boldsymbol{G}_{T^{-1}} = \boldsymbol{U}\boldsymbol{G}_T.$$

Hence we have

$$\boldsymbol{G}_{(T^{-1})^*} = \boldsymbol{H} \ominus \boldsymbol{V}[\boldsymbol{G}_{T^{-1}}] = \boldsymbol{H} \ominus \boldsymbol{V}\boldsymbol{U}[\boldsymbol{G}_T] = \boldsymbol{U}(\boldsymbol{U}\boldsymbol{H} \ominus \boldsymbol{V}[\boldsymbol{G}_T]) =$$
$$= \boldsymbol{U}(\boldsymbol{H} \ominus \boldsymbol{V}[\boldsymbol{G}_T]) = \boldsymbol{U}\boldsymbol{G}_{T^*},$$

which proves that $(T^*)^{-1}$ exists and is equal to $(T^{-1})^*$.

The following is another important proposition:

THEOREM. *If the linear transformation T is closed and if its domain is dense in $\mathfrak{H}$, the domain of T^* is also dense in $\mathfrak{H}$, hence $T^{**} = (T^*)^*$ exists; moreover, $T^{**} = T$.*

Let us suppose, to the contrary, that $\mathfrak{D}_{T^*}$ is not dense in $\mathfrak{H}$, and that consequently there exists an element $h \neq 0$ orthogonal to $\mathfrak{D}_{T^*}$. The element

$$\{0, h\}$$

of $\boldsymbol{H}$ will then be orthogonal to all elements of the form

$$\{T^*g, \ -g\},$$

where g runs through $\mathfrak{D}_{T^*}$, hence it will be orthogonal to $\boldsymbol{V}\boldsymbol{G}_{T^*}$. We know that the orthogonal complement of $\boldsymbol{G}_{T^*}$ is equal to $\boldsymbol{V}[\boldsymbol{G}_T]$; therefore, since $\boldsymbol{V}$ is isometric, the orthogonal complement of $\boldsymbol{V}\boldsymbol{G}_{T^*}$ is equal to

$$\boldsymbol{V}^2[\boldsymbol{G}_T] = -\boldsymbol{I}[\boldsymbol{G}_T] = [\boldsymbol{G}_T].$$

But we have

$$[\boldsymbol{G}_T] = \boldsymbol{G}_T,$$

since T is a *closed* linear transformation. Consequently $\{0, h\}$ belongs to $\boldsymbol{G}_T$, which implies that $h = T0 = 0$, contrary to the hypothesis that $h \neq 0$.

This contradiction proves that $\mathfrak{D}_{T^*}$ is dense in $\mathfrak{H}$, and that consequently the transformation T^{**} also exists. But since $G_{T^{**}}$ is the orthogonal complement of $\boldsymbol{V}\boldsymbol{G}_{T^*}$, we necessarily have

$$\boldsymbol{G}_{T^{**}} = \boldsymbol{G}_T,$$

hence

$$T^{**} = T, \qquad \text{q. e. d.}$$

It is easy to complete this argument (we leave it to the reader) so that it proves the following more general proposition:

THEOREM. *If T is a linear transformation with domain dense in $\mathfrak{H}$, a necessary and sufficient condition that the domain of T^* should be dense in $\mathfrak{H}$ is that T possess a closed linear extension. The transformation T^{**} is then the smallest* *closed linear extension of T, that is, every closed linear extension of T is also an extension of T^{**}.*[3]

We consider now a linear transformation T which is defined *everywhere* in $\mathfrak{H}$. We shall show that its adjoint T^* is *bounded* (in $\mathfrak{D}_{T^*}$).

Let us assume the contrary, that is, that there exists a sequence $\{g_n\}$ of elements of $\mathfrak{D}_{T^*}$ such that

$$\|g_n\| = 1 \quad \text{and} \quad \|T^*g_n\| \to \infty.$$

The functionals

$$L_n(f) = (Tf, g_n) = (f, T^*g_n) \qquad (n = 1, 2, \ldots)$$

are obviously linear, L_n having the norm $\|T^*g_n\|$. Moreover, the sequence $\{L_n(f)\}$ is bounded for each fixed element f:

$$|L_n(f)| \leq \|Tf\| \, \|g_n\| = \|Tf\|.$$

Consequently, these functionals have a common bound C (see Sec. 31):

$$|L_n(f)| \leq C\|f\| \qquad (n = 1, 2, \ldots).$$

Setting

$$f = T^*g_n \qquad (n = 1, 2, \ldots),$$

we arrive at a contradiction.

Therefore T^* is bounded in $\mathfrak{D}_{T^*}$, that is, there exists a constant M such that

$$\|T^*f\| \leq M\|f\|$$

for every element f of $\mathfrak{D}_{T^*}$.

Let f^* be the limit of a sequence $\{f_n\}$ of elements of $\mathfrak{D}_{T^*}$. Since

$$\|T^*(f_n - f_m)\| \leq M\|f_n - f_m\| \to 0 \quad (\text{for } m, n \to \infty),$$

the sequence $\{T^*f_n\}$ is also convergent. Inasmuch as the transformation T^* is closed, this implies that f also belongs to $\mathfrak{D}_{T^*}$. Therefore $\mathfrak{D}_{T^*}$ is a closed linear set — a subspace of $\mathfrak{H}$.

If the transformation T is also *closed*, $\mathfrak{D}_{T^*}$ is dense in $\mathfrak{H}$, and—being closed—coincides with $\mathfrak{H}$; that is, in this case T^* is a transformation which is defined and bounded everywhere. The same is then true of $T^{**} = T$.

Hence we have arrived at the following result:

THEOREM. *Every closed linear transformation which is defined everywhere is necessarily bounded.*

This theorem includes the theorem of Hellinger and Toeplitz (Sec. 114); in fact, the symmetry relation $(Tf, g) = (f, Tg)$, together with the fact that T is defined everywhere, implies that T is equal to T^*, hence that it is closed.

[3] Of course, the extensions here are not necessarily proper.

This theorem is, in its turn, a particular case of a general theorem of BANACH[4] which asserts the same fact for transformations in a Banach space or even in spaces of a still more general type.

118. The Transformations $B = (I + T^*T)^{-1}$ and $C = T(I + T^*T)^{-1}$

If the linear transformation T is bounded, it is clear that the transformation B appearing in the title of this section is also bounded, symmetric, and such that $O \leq B \leq I$; $C = TB$ is then bounded too. If T is a linear transformation with dense domain, we know that T^*, and consequently also T^*T, exist, but we know nothing of their domains of definition. However, it will be possible for us to prove the rather surprising fact expressed by the following theorem:

THEOREM. *If the linear transformation T is closed and if its domain is dense in $\mathfrak{H}$, the transformations*

$$B = (I + T^*T)^{-1}, \quad C = T(I + T^*T)^{-1}$$

are defined everywhere and bounded,

$$\|B\| \leq 1, \quad \|C\| \leq 1;$$

moreover, B is symmetric and positive.

In order to prove this theorem, we shall again make use of the graph of T, which in this case is a *closed* linear set.

Let h be an arbitrary element of $\mathfrak{H}$. Since $\boldsymbol{G}_T$ and $\boldsymbol{VG}_{T^*}$ are complementary orthogonal subspaces of $\boldsymbol{H}$ (see the preceding section), we can decompose the element $\{h, 0\}$ of $\boldsymbol{H}$ into the sum of an element of $\boldsymbol{G}_T$ and an element of $\boldsymbol{VG}_{T^*}$, and this in only one way:

$$\{h, 0\} = \{f, Tf\} + \{T^*g, -g\}. \tag{4}$$

This means, passing to the components, that the system of equations

$$h = f + T^*g, \quad 0 = Tf - g$$

has a unique solution f in $\mathfrak{D}_T$ and g in $\mathfrak{D}_{T^*}$. Writing

$$f = Bh, \quad g = Ch,$$

we define two transformations of $\mathfrak{H}$ into itself which are obviously linear. The system of equations can then be written in the form

$$I = B + T^*C, \quad O = TB - C,$$

from which is obtained

$$C = TB, \quad I = B + T^*TB = (I + T^*T)B. \tag{5}$$

[4] BANACH [1] and [*] (p. 41, theorem 7).

Since the two terms in the second member of (4) are orthogonal, we have
$\|h\|^2 = \|\{h, 0\}\|^2 = \|\{f, Tf\}\|^2 + \|\{T^*g, -g\}\|^2 = \|f\|^2 + \|Tf\|^2 + \|T^*g\|^2 + \|g\|^2$,
from which

$$\|Bh\|^2 + \|Ch\|^2 = \|f\|^2 + \|g\|^2 \leq \|h\|^2;$$

therefore

$$\|B\| \leq 1, \quad \|C\| \leq 1.$$

For any element u in the domain of T^*T, we have

$$((I + T^*T)u, u) = (u, u) + (Tu, Tu) \geq (u, u),$$

hence

$$(I + T^*T)u = 0 \text{ implies that } u = 0.$$

This assures that the inverse transformation $(I + T^*T)^{-1}$ exists. According to the second equation (5), it is defined everywhere and equal to B:

$$B = (I + T^*T)^{-1}.$$

The transformation B is symmetric and positive; in fact,

$$(Bu, v) = (Bu, (I + T^*T)Bv) = (Bu, Bv) + (Bu, T^*TBv) =$$
$$= (Bu, Bv) + (T^*TBu, Bv) = ((I + T^*T)Bu, Bv) = (u, Bv)$$

and

$$(Bu, u) = (Bu, (I + T^*T)Bu) = (Bu, Bu) + (TBu, TBu) \geq 0.$$

This completes the proof of the theorem.

SELF-ADJOINT TRANSFORMATIONS. SPECTRAL DECOMPOSITION

119. Symmetric and Self-adjoint Transformations. Definitions and Examples

In the case of bounded linear transformations, symmetric transformations were characterized by the relation

$$(Tf, g) = (f, Tg).$$

The same relation can also serve as the definition for unbounded symmetric transformations, if, of course, we require that it hold for all pairs of elements f, g belonging to the domain $\mathfrak{D}_T$ of T. We shall also assume that $\mathfrak{D}_T$ is dense in $\mathfrak{H}$, and consequently, that T^* exists. The definition given is then obviously equivalent to the following:

A transformation T is said to be symmetric if it is linear, has a domain dense in $\mathfrak{H}$, and satisfies $T \subseteq T^*$.

If T is symmetric, T^{**} is also symmetric; in fact, $T \subseteq T^*$ implies that $T^* \supseteq T^{**}$, hence

$$T^{**} \subseteq T^* = (T^*)^{**} = (T^{**})^*.$$

Therefore every symmetric transformation has a closed and symmetric linear extension, namely T^{**}.

*

Let us consider, as an example, the transformation

$$Tf(x) = if'(x)$$

of the space $L^2(0, 1)$, defined for all absolutely continuous functions $f(x)$ which are zero at the points 0 and 1 and for which the derivative $f'(x)$ (which exists almost everywhere) belongs to L^2. This transformation is obviously linear; it is also symmetric, since

$$(Tf, g) - (f, Tg) = i\int_0^1 (f'\bar{g} + f\bar{g}')dx = i[f\bar{g}]_0^1 = 0.$$

Moreover, T is closed. In fact, if a sequence $\{f_n(x)\}$ of functions in the domain of T is such that

$$\left.\begin{aligned} f_n(x) &\to f(x) \\ if_n'(x) &\to ih(x) \end{aligned}\right\} \text{in the mean}$$

we have

$$if_n(x) = i\int_0^x f_n'(\xi)d\xi \to i\int_0^x h(\xi)d\xi \quad (= 0 \text{ for } x = 0 \text{ and for } x = 1)$$

in the sense of ordinary convergence, from which it follows that

$$f(x) = \int_0^x h(\xi)d\xi$$

almost everywhere. Since the limit in the mean is determined only up to a set of measure zero, we can choose $f(x)$ in such a way that this equation is verified everywhere. We then see that $f(x)$ also belongs to $\mathfrak{D}_T$ and that $Tf(x) = ih(x)$.

Let us calculate T^*.

It is necessary to find all the pairs g, g^* of elements of L^2 such that

$$(if', g) = (f, g^*)$$

for all elements f of $\mathfrak{D}_T$. Denoting by g^{**} an indefinite integral of g^*, we obtain by integrating by parts:

$$i\int_0^1 f'(\bar{g} + \overline{ig^{**}})dx = 0. \tag{6}$$

In order that this equation be satisfied, it obviously suffices that the function

$$h(x) = g(x) + ig^{**}(x)$$

be constant almost everywhere. This is also necessary, as we see by choosing $f(x)$ in such a way that almost everywhere

$$f'(x) = h(x) - c,$$

where

$$c = \int_0^1 h(x)dx;$$

in fact, we then have

$$\int_0^1 |h(x) - c|^2 dx = \int_0^1 [h(x) - c]\overline{h(x)}dx - \bar{c}\int_0^1 h(x)dx + \bar{c}c =$$

$$= \int_0^1 f'(x)\overline{h(x)}dx = 0,$$

hence

$$h(x) = c$$

almost everywhere. Modifying $g(x)$ on a set of measure zero if necessary, we therefore have at all points

$$g(x) + ig^{**}(x) = c,$$

and at almost all points

$$g'(x) + ig^{*}(x) = 0.$$

The domain of T^* therefore consists of all absolutely continuous functions $g(x)$ whose derivatives belong to L^2, but which are not subjected to any condition at the extremities 0 and 1, and we have

$$T^*g(x) = ig'(x).$$

The transformation T which we have just considered therefore provides an example of a closed symmetric transformation for which T^* is a *proper* extension of T.

If in the example considered we replace the limit conditions

$$f(0) = f(1) = 0$$

by the less stringent condition

$$f(1) = Cf(0),$$

where C is a given number, we obtain a transformation T_C which is again an extension of T. We see without difficulty that T_C is also linear and closed.

When $|C| = 1$, T_C is also symmetric:

$$(T_C f, g) - (f, T_C g) = i[f\bar{g}]_0^1 = i[Cf(0)\overline{Cg(0)} - f(0)\overline{g(0)}] = 0.$$

Let us calculate T_C^*.

First, we consider the case $C = 1$. Let g be an element of $\mathfrak{D}_{T_1^*}$, and let $g^* = T^*g$. The function $f(x) \equiv 1$ belongs to $\mathfrak{D}_{T_1}$; substituting it for f in the relation

$$(if', g) = (f, g^*),$$

we obtain

$$\int_0^1 g^*(x)dx = 0;$$

the indefinite integral $g^{**}(x)$ of $g^*(x)$ therefore satisfies the condition $g^{**}(1) = g^{**}(0)$. Integrating by parts, we arrive again at relation (6). It follows that, by modifying $g(x)$ on a set of measure zero if necessary, $g(x) + ig^{**}(x)$ will be constant, which implies

$$g(1) = g(0) \text{ and } g^*(x) = ig'(x).$$

Therefore g also belongs to $\mathfrak{D}_{T_1}$ and we have $T_1^*g = T_1 g$. Consequently,

$$T_1^* = T_1.$$

The case $C \neq 1$, $|C| = 1$, can be studied in an analogous manner. Among the indefinite integrals of $g^*(x)$ there is one, $g^{**}(x)$, for which

$$g^{**}(1) = Cg^{**}(0).$$

For this integral, (6) holds. But there exists a function $f(x)$ in $\mathfrak{D}_{T_C}$ for which almost everywhere

$$f'(x) = g(x) + ig^{**}(x);$$

inserting the function $f(x)$ in (6) we obtain that $g(x) + ig^{**}(x)$ is zero almost everywhere. Modifying $g(x)$ on a set of measure zero if necessary, we have

$$g(x) = -ig^{**}(x);$$

hence $g(x)$ also belongs to $\mathfrak{D}_{T_C}$ and we have

$$g^*(x) = ig'(x).$$

Consequently, we also have in this case

$$T_C^* = T_C.$$

*

Let us agree upon the following definition:

A linear transformation which is equal to its adjoint will be called self-adjoint.

Every bounded symmetric transformation is self-adjoint, but this is not always so for an unbounded symmetric transformation, even if it is closed, as we just saw from the example

$$Tf = if' \qquad (f(0) = f(1) = 0).$$

It can occur that a symmetric transformation which is not self-adjoint possesses self-adjoint extensions, for example the transformation T in question which has the extensions

$$T_C f = if' \qquad (f(1) = Cf(0), |C| = 1),$$

but this is not always the case. There are symmetric transformations which are *maximal*, that is, which have no proper symmetric extensions, and which are not self-adjoint. *Every self-adjoint transformation A is maximal symmetric*, since if

$$A \subseteq T, \quad T \subseteq T^*,$$

we have

$$A = A^* \supseteq T^* \supseteq T \supseteq A,$$

hence

$$A = T$$

With *bounded* linear transformations there is a simple means of constructing symmetric transformations: it is to take an arbitrary linear transformation T and to form T^*T.

It is a remarkable fact that *the same construction furnishes symmetric transformations—indeed self-adjoint extensions—even in the unbounded case, at least if the linear transformation T is closed and has a domain dense in $\mathfrak{H}$.*

We know, in fact, that, in this case the transformation

$$B = (I + T^*T)^{-1}$$

exists, is defined everywhere, bounded, and symmetric, hence also self-adjoint. Inasmuch as

$$T^*T = B^{-1} - I,$$

there remains to prove only the two following propositions:

a) *The inverse of a self-adjoint transformation A, when it exists, is also self-adjoint.*

b) *When A is a self-adjoint transformation, $A + cI$ is also self-adjoint, whatever the real number c.*

Proposition b) is clear. As to a), we have only to show that $(A^{-1})^*$ exists, that is, that the domain of A^{-1} is dense in $\mathfrak{H}$, because then

$$(A^{-1})^* = (A^*)^{-1} = A^{-1}.$$

Since the domain of A^{-1} consists of those elements of the form Af, if it were not dense in $\mathfrak{H}$ there would be an element $h \neq 0$ such that

$$(Af, h) = 0$$

for all elements f of $\mathfrak{D}_A$, hence h would belong to the domain of $A^* = A$ and we would have $Ah = 0$. Since A^{-1} exists, it would follow that $h = 0$, contrary to the hypothesis.

We consider, *as an example*, the transformation

$$Tf(x) = if'(x)$$

studied above with the limit conditions

$$f(0) = f(1) = 0.$$

We have seen that this transformation is symmetric without being self-adjoint; we have, namely,

$$T^*g(x) = ig'(x),$$

but for all $g(x)$, without any limit conditions. Consequently

$$T^*Tf(x) = -f''(x)$$

with the limit conditions $f(0) = f(1) = 0$. According to what we have just proved, this transformation is self-adjoint.[5]

120. Spectral Decomposition of a Self-adjoint Transformation

The question at once arises of whether or not there exists a spectral decomposition of every symmetric transformation analogous to that existing in the case of bounded transformations.

Important works, especially those of CARLEMAN [*] dealing with integral equations with "singular" symmetric kernels, have made it clear that it is impossible to arrive in the general case at a complete analogy. ERHARD SCHMIDT was the first to observe that if we wish to obtain an analogous spectral decomposition it is necessary to restrict ourselves to *self-adjoint*[6] transformations. The theorem of the spectral decomposition of a general self-adjoint transformation, proved in several ways by VON NEUMANN,[7] STONE,[8] RIESZ,[9] and others[10], has become the point of departure of the new theory of linear transformations in Hilbert space.

[5] Concerning differential operators, cf. in particular WEYL [**1**]; COURANT [**2**]; FRIEDRICHS [**1**], [**2**], [**3**]; STONE [*] (Chapters V, X), KREIN [**1**] (second part).

[6] Cf. the remark by VON NEUMANN, [**1**] (p. 62, footnote 23).

[7] VON NEUMANN [**1**]. We shall reproduce this proof in Sec. 121.

[8] STONE [*] (Chapt. V); this proof is partially based on a method invented and applied by CARLEMAN [*] in his theory of integral equations with "singular" symmetric kernel.

[9] RIESZ [**13**].

[10] Cf. KOOPMAN–DOOB [**1**]; RIESZ–LORCH [**1**]; LENGYEL [**1**]; COOPER [**1**]; LORCH [**6**].

The proof due to RIESZ and LORCH, which we shall reproduce in this section, depends largely on the following lemma.

LEMMA. *Let*

$$\mathfrak{L}_1, \mathfrak{L}_2, \ldots, \mathfrak{L}_i, \ldots$$

be a sequence of subspaces of the Hilbert space $\mathfrak{H}$ *which are pairwise orthogonal and span the entire space* $\mathfrak{H}$. *If* f *is an arbitrary element of* $\mathfrak{H}$, *we denote its projection on* $\mathfrak{L}_i$ *by* f_i. *Let*

$$A_1, A_2, \ldots, A_i, \ldots$$

be a given sequence of linear transformations with the property that A_i *reduces in* $\mathfrak{L}_i$ *to a bounded self-adjoint transformation of* $\mathfrak{L}_i$ *into itself. Then there is one and only one self-adjoint transformation* A *of* $\mathfrak{H}$, *in general not bounded, which reduces in each* $\mathfrak{L}_i$ *to* A_i $(i = 1, 2. \ldots)$. *Its domain consists of the elements* f *for which the series*

$$\sum_{i=1}^{\infty} \|A_i f_i\|^2$$

converges, and for these f

$$Af = \sum_{i=1}^{\infty} A_i f_i.$$

We observe first that the transformation A thus defined is linear; its domain $\mathfrak{D}_A$ is dense in $\mathfrak{H}$, since it contains all elements of the form $\sum_{i=1}^{n} f_i$. Moreover, A is symmetric:

$$(Af, g) = \sum (A_i f_i, g_i) = \sum (f_i, A_i g_i) = (f, Ag)$$

for all elements f and g of $\mathfrak{D}_A$.

Let g be an element of $\mathfrak{D}_{A^*}$; then we have

$$(Af, g) = (f, A^*g)$$

for all elements f of $\mathfrak{D}_A$; hence

$$\sum_{i=1}^{\infty} (A_i f_i, g_i) = \sum_{i=1}^{\infty} (f_i, (A^*g)_i).$$

If in particular we choose for f an arbitrary element of $\mathfrak{L}_j$, we shall have $f_i = 0$ for $i \neq j$; hence

$$(A_j f_j, g_j) = (f_j, (A^*g)_j).$$

Since A_j was assumed to be self-adjoint in $\mathfrak{L}_j$, we deduce from this that

$$(A^*g)_j = A_j g_j.$$

It follows that

$$\sum_{j=1}^{\infty} \|A_j g_j\|^2 = \sum_{j=1}^{\infty} \|(A^*g)_j\|^2 = \|A^*g\|^2;$$

hence g also belongs to $\mathfrak{D}_A$ and we have

$$Ag = \sum_{j=1}^{\infty} A_j g_j = \sum_{j=1}^{\infty} (A^*g)_j = A^*g.$$

This proves that

$$A^* \subseteq A,$$

and since A is symmetric,

$$A^* = A.$$

It remains to prove the uniqueness. Let A' be an arbitrary self-adjoint transformation which reduces to A_i in each $\mathfrak{L}_i$. Since A' is closed, it is necessarily defined for all elements f for which the series

$$\sum_{i=1}^{\infty} A'f_i$$

converges, and the sum of this series is then equal to $A'f$. Since $A'f_i = A_i f_i$, and since the convergence of a series of orthogonal elements is obviously equivalent to the convergence of the series of the squares of the norms, the set of these elements f coincides with $\mathfrak{D}_A$, and for these f we have

$$A'f = Af;$$

hence

$$A' \supseteq A.$$

But A is self-adjoint, and consequently maximal symmetric; hence we necessarily have

$$A' = A.$$

This completes the proof of the lemma.

It is clear that the self-adjoint transformation A will not be bounded unless the transformations A_i (in $\mathfrak{L}_i$) possess a common bound. Therefore our lemma furnishes a means of constructing unbounded self-adjoint transformations starting with bounded self-adjoint transformations. But it is rather the converse which is essential: that every unbounded self-adjoint transformation can be constructed in this manner starting with bounded self-adjoint transformations. This is what we shall prove and this will lead us to the general theorem on spectral decomposition.

*

Let us consider a self-adjoint transformation A and let

$$B = (I + A^2)^{-1} \text{ and } C = AB = A(I + A^2)^{-1}$$

be the corresponding transformations which we have already studied in section 118.

Since B is bounded and symmetric, and such that

$$O \leq B \leq I,$$

there corresponds a spectral family $\{F_\lambda\}$ over the closed interval $[0, 1]$ such that

$$B = \int_{-0}^{1} \lambda dF_\lambda.$$

The fact that B^{-1} exists implies that $\lambda = 0$ is a point of continuity of F, that is, that $F_0 = O$. In fact, since $F_\lambda F_0$ is equal to O for $\lambda < 0$, and to F_0 for $\lambda \geq 0$, we have

$$BF_0 = \int_{-0}^{1} \lambda dF_\lambda F_0 = O, \quad F_0 = B^{-1}BF_0 = O.$$

Consider the projections

$$P_n = F_{\frac{1}{n}} - F_{\frac{1}{n+1}} \qquad (n = 1, 2, \ldots);$$

they are pairwise orthogonal and we have

$$\sum_n P_n = F_1 - F_0 = I - O = I.$$

Hence the corresponding subspaces $\mathfrak{L}_n$ are pairwise orthogonal and span the entire space $\mathfrak{H}$.

We shall show that A transforms each subspace $\mathfrak{L}_n$ into itself and that, in $\mathfrak{L}_n$, A is a *bounded* self-adjoint transformation, or equivalently, that $P_n \smile A$ and that the product AP_n is a bounded transformation which is defined everywhere.

Multiplying the two members of the obvious equality $A(I + A^2) = (I + A^2)A$ on the left and right by B, and taking into consideration the fact that $(I + A^2)B = I$ and $B(I + A^2) \subseteq I$, we obtain first that

$$BA \subseteq AB.$$

It follows that

$$BC = BAB \subseteq ABB = CB,$$

and since BC is defined everywhere,

$$BC = CB.$$

Then C is also permutable with the polynomials in B as well as with their limits, in particular, with F_λ, P_n, and with the transformation $s_n(B)$

corresponding to the function

$$s_n(\lambda) = \begin{cases} \frac{1}{\lambda} & \text{for } \frac{1}{n+1} < \lambda \leq \frac{1}{n} \\ 0 & \text{elsewhere.} \end{cases}$$

Since the product of the functions λ and $s_n(\lambda)$ equals 1 for $\frac{1}{n+1} < \lambda \leq \frac{1}{n}$ and is zero elsewhere, we have

$$Bs_n(B) = s_n(B)B = P_n$$

and

$$AP_n = ABs_n(B) = Cs_n(B),$$

$$P_nA = s_n(B)BA \subseteq s_n(B)AB = s_n(B)C.$$

Since C and $s_n(B)$ are bounded, symmetric, and permutable, this proves our assertion that $P_n \smile A$ and that the transformation AP_n is defined everywhere and bounded.

We have thus obtained a decomposition of the space $\mathfrak{H}$ into the vector sum of the orthogonal subspaces $\mathfrak{L}_n$ each of which reduces A to a *bounded* self-adjoint transformation.

Let us denote the spectral family of A, considered as a transformation in $\mathfrak{L}_n$, by $\{E_{\lambda,n}\}$; it is a spectral family over some finite segment of the λ-axis determined by the bounds of A in $\mathfrak{L}_n$. According to the lemma, there exists a self-adjoint transformation E_λ of $\mathfrak{H}$ which reduces in each $\mathfrak{L}_n$ to $E_{\lambda,n}$. It is easy to see that E_λ is also a projection, and that moreover it possesses the following properties:

a) $E_\lambda \leq E_\mu$ for $\lambda < \mu$,
b) $E_{\lambda+0} = E_\lambda$,
c) $E_\lambda \to O$ for $\lambda \to -\infty$ and $E_\lambda \to I$ for $\lambda \to \infty$.

It is therefore a spectral family over the entire line $(-\infty, \infty)$, or more briefly a *spectral family*.

If a bounded linear transformation T is permutable with A, it is also permutable with $B = (I + A^2)^{-1}$ and with the polynomials in B, as well as with their limits; in particular, with $Bs_n(B) = P_n$. This means that T is reduced by each subspace $\mathfrak{L}_n$. But in $\mathfrak{L}_n$, $E_{\lambda,n}$ is the limit of a sequence of polynomials in A and consequently T is permutable with $E_{\lambda,n}$ there. Hence T is permutable in the entire space $\mathfrak{H}$ with

$$\sum_n E_{\lambda,n}P_n = E_\lambda.$$

Thus we have

(7) $$E_\lambda \smile\smile A.$$

Our principal goal is to establish the formula

$$A = \int_{-\infty}^{\infty} \lambda dE_\lambda, \tag{8}$$

but since neither the domain of integration nor the function under the integral sign is bounded, it is first necessary to make precise the meaning of an integral of this type. Reserving the definition and study of the integral of general functions for the following chapter, we content ourselves here with defining the integral appearing in the second member of (8), which integral we shall denote briefly by J. This definition will be valid for an arbitrary spectral family (this means that in the definition we shall only make use of properties a)—c) of the family of projections E_λ). Let us consider the projections

$$E_m - E_{m-1} \qquad (m = 0, \pm 1, \pm 2, \ldots)$$

and the corresponding subspaces $\mathfrak{M}_n$; the latter are obviously pairwise orthogonal and span the entire space $\mathfrak{H}$. Set

$$J_m = \int_{m-1}^{m} \lambda dE_\lambda;$$

this is a bounded self-adjoint transformation which transforms $\mathfrak{M}_m$ into itself. Making use of the lemma, *we define the integral J as the uniquely determined self-adjoint transformation in $\mathfrak{H}$ which reduces to the transformation J_m in each subspace $\mathfrak{M}_m$.*

Setting $f_m = (E_m - E_{m-1})f$, the domain of definition of J therefore consists of the elements f for which the series

$$\sum_{-\infty}^{\infty} \|J_m f_m\|^2 = \sum_{-\infty}^{\infty} (J_m^2 f_m, f_m) = \sum_{-\infty}^{\infty} \int_{m-1}^{m} \lambda^2 d\, \|E_\lambda f_m\|^2$$

converges, or equivalently, since $E_\lambda f_m = E_\lambda f - E_{m-1} f$ in the interval $m - 1 \leq \lambda \leq m$, those for which the integral

$$\int_{-\infty}^{\infty} \lambda^2 d\, \|E_\lambda f\|^2 \tag{9}$$

converges; for these f,

$$Jf = \sum_{-\infty}^{\infty} J_m f_m = \sum_{-\infty}^{\infty} \int_{m-1}^{m} \lambda dE_\lambda f_m = \sum_{-\infty}^{\infty} \int_{m-1}^{m} \lambda dE_\lambda f.$$

It is clear that if f belongs to the domain of J, the same is true of $E_\mu f$ and we have

$$JE_\mu f = \sum_{-\infty}^{\infty} J_m (E_\mu f)_m = \sum_{-\infty}^{\infty} J_m E_\mu f_m = E_\mu \sum_{-\infty}^{\infty} J_m f_m = E_\mu J f;$$

hence

$$E_\mu \smile J. \tag{9a}$$

It is easy to see that if instead of starting with the sequence of integers $m = 0, \pm 1, \pm 2, \ldots$ we start with another sequence of real numbers which goes to infinity in both directions, we arrive at the same definition of the integral J.

This being the case, in order to establish formula (8)—that is, that the given self-adjoint transformation A is equal to the integral J formed starting with the spectral family of A—it suffices, by virtue of the lemma, to verify that the two self-adjoint transformations A and J coincide in each of the orthogonal subspaces $\mathfrak{L}_n$ ($n = 1, 2, \ldots$). But for an element f of $\mathfrak{L}_n$ we have, by definition,

$$E_\lambda f = E_{\lambda,n} f.$$

Since $\{E_{\lambda,n}\}$ is a spectral family over the finite interval $[a, b]$, $E_\lambda f$ is constant for $\lambda < a$ and for $\lambda \geq b$, and consequently the integral (9) converges; hence f belongs to the domain of J and we have

$$Jf = \sum_{-\infty}^{\infty} \int_{m-1}^{m} \lambda dE_\lambda f = \int_{a-0}^{b} \lambda dE_\lambda f = \int_{a-0}^{b} \lambda dE_{\lambda,n} f = Af,$$

by the definition of $\{E_{\lambda,n}\}$ as a spectral family corresponding to A in the subspace $\mathfrak{L}_n$. This completes the proof of the fundamental formula (8).

We now consider the question: to what extent is the spectral family $\{E_\lambda\}$ determined by formula (8)?

We saw (in section 107) that if A is *bounded*, $mI \leq A \leq MI$, it has only one spectral family over the finite segment $[m, M]$. But this last restriction is not essential, since if $\{E'_\lambda\}$ is an arbitrary spectral family such that equation (8) holds, we necessarily have

$$E'_\lambda = O \text{ for } \lambda < m \text{ and } E'_\lambda = I \text{ for } \lambda \geq M.$$

In fact, let $\varkappa < m$; for every element of the form $g = E'_\varkappa f$ we have

$$E'_\lambda g = E'_\lambda E'_\varkappa f = E'_\varkappa f \text{ for } \lambda \geq \varkappa \text{ and } E'_\varkappa g = g,$$

hence

$$(Ag, g) = \int_{-\infty}^{\infty} \lambda d(E'_\lambda g, g) = \int_{-\infty}^{\varkappa} \lambda d(E'_\lambda g, g) \leq \varkappa \int_{-\infty}^{\varkappa} d(E'_\lambda g, g) = \varkappa(E'_\varkappa g, g) = \varkappa(g, g);$$

inasmuch as

$$(Ag, g) \geq m(g, g),$$

this is possible only if $g = 0$. Hence $E'_\varkappa = O$. The assertion concerning the values $\lambda \geq M$ is proved in an analogous manner.

If the transformation A is *unbounded*, we prove the uniqueness of the spectral family by reduction to the bounded case. Let $\{E'_\lambda\}$ be an arbitrary spectral family such that the integral $\int_{-\infty}^{\infty} \lambda dE'_\lambda$ is also equal to A; then we have

$E'_\lambda \smile A$ (see (9a)), and since $E_\mu \smile\smile A$ (see (7)), we also have

$$E'_\lambda \smile E_\mu.$$

The projection $Q_n = E_n - E_{n-1}$ is therefore permutable not only with all the projections E_λ, but also with all the projections E'_λ. Considered as transformations of the subspace $\mathfrak{D}_n = Q_n \mathfrak{H}$ into itself, the projections E_λ and E'_λ therefore form two spectral families such that the integral of λ with respect to either is equal to the same *bounded* symmetric transformation AQ_n. Therefore these two spectral families coincide in $\mathfrak{D}_n$, and since these subspaces are pairwise orthogonal and span the entire space $\mathfrak{H}$, it follows from the lemma that they coincide everywhere.

We have thus proved the fundamental theorem on spectral decomposition:

THEOREM. *Every self-adjoint transformation A has the representation*

$$A = \int_{-\infty}^{\infty} \lambda dE_\lambda,$$

where $\{E_\lambda\}$ is a spectral family which is uniquely determined by the transformation A; E_λ is permutable with A, as well as with all the bounded transformations which permute with A.

121. Von Neumann's Method. Cayley Transforms

The transformations B and C, which play an essential role in the above discussions, are obviously the symmetric components of the normal transformation

$$C + iB = (A + iI)(I + A^2)^{-1} = (A - iI)^{-1}.$$

This transformation and its adjoint

$$C - iB = (A + iI)^{-1},$$

or more generally the transformations

$$R_z = (A - zI)^{-1},$$

where z is a real or complex parameter, also play an essential role in other proofs of the theorem.

The existence of

$$R_{\pm i} = (A \mp iI)^{-1}$$

can be proved directly from the relations

(10) $$\|(A \mp iI)f\|^2 = (Af, Af) \mp i(f, Af) \pm i(Af, f) + (f, f) = \|Af\|^2 + \|f\|^2.$$

In fact, they show that neither of the equations $(A - iI)f = 0$, $(A + iI)f = 0$

is possible unless $f = 0$, which suffices for the existence of the inverses. Furthermore, we see that

$$\|(A \mp iI)f\| \geq \|f\|,$$

which implies that

(11) $$\|g\| \geq \|R_{\pm i}g\|$$

for all elements g in the domain of R_i, respectively R_{-i}.

Now these domains coincide with the entire space; this will follow from the fact that these domains are a) closed, and b) everywhere dense in $\mathfrak{H}$. Proposition a) follows from the fact that the transformations R_i and R_{-i} are continuous (consequence of (11)) and closed (since A and $A \mp iI$ are closed). Proposition b) is proved, for example for R_i, in the following manner. If the domain of R_i, which is a linear set, were not everywhere dense in $\mathfrak{H}$, there would be an element $h \neq 0$ orthogonal to the domain of R_i, that is, to all elements of the form $(A - iI)f$. But it then follows from the equation

$$((A - iI)f, h) = 0$$

that h is in the domain of $(A - iI)^* = A + iI$ and that $(A + iI)h = 0$. Hence $h = 0$, which contradicts the hypothesis that $h \neq 0$.

The transformations $R_{\pm i}$ are therefore defined everywhere and bounded. The same is true for $R_z = R_{x+iy}$ when $y \neq 0$, since

$$(A - (x + iy)I)^{-1} = \frac{1}{y}\left(\frac{A - xI}{y} - iI\right)^{-1}.$$

Of course, R_z can exist and be bounded even for certain real values of the parameter z.

We return to relations (10). They show that

$$\|(A - iI)f\| = \|(A + iI)f\|,$$

that is,

$$\|(A - iI)\,(A + iI)^{-1}g\| = \|g\|.$$

The transformation

$$V = (A - iI)\,(A + iI)^{-1},$$

called the *Cayley transform of A*, is, therefore, *isometric*. It is defined for elements of the form

(12) $$g = (A + iI)f$$

by

(13) $$Vg = (A - iI)f,$$

where f runs through $\mathfrak{D}_A$. Then g and Vg each run through the entire space $\mathfrak{H}$; hence V is also *unitary*.

It is easy to recover A starting with V. It follows from (12) and (13), by addition and subtraction, that

$$(I+V)g = 2Af, \quad (I-V)g = 2if,$$

from which we see that $(I-V)g = 0$ implies that $f = 0$ and consequently, by (12), that $g = 0$ also; hence $(I-V)^{-1}$ exists and

$$2Af = (I+V)(I-V)^{-1}2if,$$

that is,

(14) $$A = i(I+V)(I-V)^{-1}.$$

Let

$$V = \int_0^{2\pi} e^{i\varphi} dF_\varphi \qquad (F_0 = O, \quad F_{2\pi} = I)$$

be the spectral decomposition of the unitary transformation V (Sec. 109). Using relation (14), we can deduce the spectral decomposition of A from that of V in the following manner.

We begin by observing that F_φ is a continuous function of φ not only at the point $\varphi = 0$, but also at the point $\varphi = 2\pi$. For if not, V would have the characteristic value 1; hence $(I-V)^{-1}$ would not exist, contradicting (14).

Let us decompose the interval $(0, 2\pi)$ by means of an infinite number of points having the two endpoints for limit points, say, by means of the points φ_m for which

$$-\cot\frac{\varphi_m}{2} = m \qquad (m = 0, \pm 1, \pm 2, \ldots).$$

The projections

$$P_m = F_{\varphi_m} - F_{\varphi_{m-1}}$$

are then pairwise orthogonal and

$$\sum_{-\infty}^{\infty} P_m = \lim_{\varphi\to 2\pi} F_\varphi - \lim_{\varphi\to 0} F_\varphi = I - O = I.$$

The projection P_m, being permutable with V, is also permutable with A; the subspace $\mathfrak{L}_m$ corresponding to P_m therefore reduces the transformations V and A. Since the function $(1-e^{i\varphi})^{-1}$ is bounded in the interval $\varphi_{m-1} \leq \varphi \leq$ $\leq \varphi_m$ we have, for f in $\mathfrak{L}_m$:

$$Af = AP_m f = i(I+V)(I-V)^{-1}P_m f = \int_{\varphi_{m-1}}^{\varphi_m} i(1+e^{i\varphi})(1-e^{i\varphi})^{-1} dF_\varphi f =$$

$$= \int_{\varphi_{m-1}}^{\varphi_m} \left(-\cot\frac{\varphi}{2}\right) dF_\varphi f.$$

or

$$Af = \int_{m-1}^{m} \lambda dE_\lambda f,$$

where we have set

$$E_\lambda = F_{-2\,\mathrm{arccot}\,\lambda};$$

$\{E_\lambda\}$ obviously is a spectral family over $(-\infty, \infty)$.

Now in the preceding section, we defined the integral

$$\int_{-\infty}^{\infty} \lambda dE_\lambda$$

as the self-adjoint transformation (which exists and is uniquely determined in view of the lemma), which reduces in each of the subspaces $\mathfrak{L}_m = (F_{\varphi_m} - F_{\varphi_{m-1}})\mathfrak{H} = (E_m - E_{m-1})\mathfrak{H}$ to the bounded self-adjoint transformation

$$\int_{m-1}^{m} \lambda dE_\lambda \qquad (m = 0, \pm 1, \pm 2, \ldots).$$

We have thus arrived anew at the formula

$$A = \int_{-\infty}^{\infty} \lambda dE_\lambda.$$

It is in this manner that J. VON NEUMANN [1] first proved the spectral decomposition of an unbounded self-adjoint transformation.

122. Semi-bounded Self-adjoint Transformations

A symmetric transformation S is said to be *lower semi-bounded* if there exists a real quantity c such that

$$(Sf, f) \geq c(f, f)$$

for all f in $\mathfrak{D}_S$; it is said to be *upper semi-bounded* if the opposite inequality is valid. If, in particular,

$$(Sf, f) \geq 0,$$

we shall say, following the definition set down for bounded transformations, that S is *positive.*

Since every semi-bounded symmetric transformation is obtained from a positive transformation T by one or the other of the formulas

$$S = T + cI, \quad S = -T + cI,$$

it suffices to consider positive transformations in the sequel.

For a *positive self-adjoint* transformation A, the spectral decomposition can be deduced very simply from that for a bounded self-adjoint transformation.[11] This is done with the aid of a linear transformation of the semi-axis $\lambda \geq 0$ into a finite segment of the μ-axis, for example, the transformation

$$\mu = \frac{\lambda - 1}{\lambda + 1},$$

which carries the semi-axis $\lambda \geq 0$ into the segment $-1 \leq \mu \leq 1$. This is the analogue of the linear transformation

$$\mu = \frac{\lambda - i}{\lambda + i},$$

which maps the circumference of the unit circle in the plane of complex numbers onto the entire λ-axis—the transformation which led to the idea of the "Cayley transform." Since we now deal only with the semi-axis $\lambda \geq 0$, it is not necessary to use imaginary numbers in order to transform it into a bounded curve.

Hence we form the transformation

$$B = (A - I)\,(A + I)^{-1}$$

instead of the Cayley transform. Since $((A + I)f, f) \geq (f, f)$, the transformation

$$C = (A + I)^{-1}$$

exists and $(g, Cg) \geq (Cg, Cg)$ for all g in $\mathfrak{D}_C$. It follows that

$$(Cg, g) \geq 0 \text{ and } \|Cg\| \leq \|g\|.$$

Since C is the inverse of a self-adjoint transformation, it is also self-adjoint (see Sec. 119, proposition a)), and since it is bounded in its domain $\mathfrak{D}_C$, this domain necessarily coincides with the entire space $\mathfrak{H}$. Thus the transformation

$$I - 2C = (A + I)C - 2C = (A - I)C = B$$

is also self-adjoint and bounded, and since $O \leq C \leq I$, we have $\|B\| \leq 1$.

Let

$$B = \int_{-1-0}^{1} \mu dF_\mu$$

be the spectral decomposition of B. Since the transformation $I - B = 2C$ possesses an inverse (namely, $\frac{1}{2}(A + I)$), the value 1 is not a characteristic value of B, hence F_μ is a continuous function of μ at the point $\mu = 1$, that is, $F_{1-0} = F_1 = I$. Consequently we have

$$A = (I + B)\,(I - B)^{-1} = \int_{-1-0}^{1} \frac{1 + \mu}{1 - \mu}\, dF_\mu = \int_{-0}^{\infty} \lambda dE_\lambda, \tag{15}$$

[11] Friedrichs [**1**] (Note 1).

where

$$E_\lambda = F_\mu \text{ for } \mu = \frac{\lambda - 1}{\lambda + 1};$$

$\{E_\lambda\}$ is obviously a spectral family over the semi-axis $\lambda \geq 0$. For a rigorous proof of (15), we can use a decomposition of the segment $-1 \leq \mu < 1$ by means of an infinite number of points which tend to 1, and then reason as in sections 120 and 121. Since $E_\lambda = F_\mu$ is the limit of polynomials in B, it obviously is permutable with A and with all the bounded transformations which permute with A.

EXTENSIONS OF SYMMETRIC TRANSFORMATIONS

123. Cayley Transforms. Deficiency Indices

Since, as we have just seen, it is the self-adjoint transformations which have spectral decompositions, it is important to know whether or not a given symmetric transformation possesses a self-adjoint extension. More generally, the problem arises of characterizing all the symmetric extensions of a given symmetric transformation S.

Cayley transforms have been used in the study of this problem since their introduction by J. VON NEUMANN [**1**]. The Cayley transform of a symmetric transformation S is defined just as for a self-adjoint transformation (see Sec. 121), namely, by

$$V = (S - iI)(S + iI)^{-1};$$

just as there, we show that V is *isometric* and that we can recover S from V by means of the formula

$$S = i(I + V)(I - V)^{-1}.$$

By using relations (10), (12), (13), (p. 319), written for S instead of A, it is easy to see that if S is closed then V is also closed, and conversely. Since every symmetric transformation S has the closed extension S^{**} (its closure), we shall consider only *closed* symmetric transformations.

We know that if S is *self-adjoint*, its Cayley transform V is *unitary*; we shall show that the converse is also true. Suppose that V is unitary; let g be an element of $\mathfrak{D}_{S^*}$ and set $g^* = S^*g$. Then

$$(Sf, g) = (f, g^*)$$

for all elements f of $\mathfrak{D}_S$, and since these elements f are of the form $f =$

$= (I - V)h$, where h runs through $\mathfrak{D}_V = \mathfrak{H}$, we have

$$(i(I + V)h, g) = ((I - V)h, g^*),$$

or

$$i(h, g) + i(Vh, g) = (h, g^*) - (Vh, g^*),$$

for all elements h of $\mathfrak{H}$. Since V is unitary (hence defined everywhere and isometric), we can replace (h, g) by (Vh, Vg) and (h, g^*) by (Vh, Vg^*) and obtain

$$(Vh, - iVg - ig - Vg^* + g^*) = 0.$$

The values Vh of the unitary transformation V exhaust the space $\mathfrak{H}$; this implies that

$$- iVg - ig - Vg^* + g^* = 0,$$

from which we deduce that

$$g = (I - V)\frac{g - ig^*}{2}, \quad g^* = i(I + V)\frac{g - ig^*}{2}.$$

Consequently g also belongs to the domain of S and $Sg = g^*$. This proves that $S^* = S$; S is therefore a self-adjoint transformation.

In the case of an arbitrary closed symmetric transformation S, the domain of definition $\mathfrak{D}_V$ and the set of values $\mathfrak{W}_V = V\mathfrak{D}_V$ do not in general coincide with the entire space $\mathfrak{H}$; but since V is isometric and closed, $\mathfrak{D}_V$ and $\mathfrak{W}_V$ are closed sets, that is, subspaces of $\mathfrak{H}$, one or the other of which may coincide with $\mathfrak{H}$. The orthogonal complements $\mathfrak{H} - \mathfrak{D}_V$ and $\mathfrak{H} - \mathfrak{W}_V$ are called the *deficiency subspaces*, and their dimensions the *deficiency indices* of the symmetric transformation S (or also of the isometric transformation V). Let us recall that $\mathfrak{D}_V$ is the set of values of $S + iI$ and that $\mathfrak{W}_V$ is the set of values of $S - iI$.

It follows from what we have just proved that a closed symmetric transformation is *self-adjoint* if and only if its deficiency indices are $(0, 0)$.

We now pass to the problem of extension. It is clear that if S' is an extension of S (we suppose that both S and S' are symmetric and closed), the Cayley transform V' of S' will be an extension of the Cayley transform V of S. $\mathfrak{D}_V$ will be a subspace of $\mathfrak{D}_{V'}$, and its orthogonal complement $\mathfrak{D}_{V'} - \mathfrak{D}_V$ will be carried by the isometric transformation V' into the orthogonal complement of $\mathfrak{W}_V$ with respect to $\mathfrak{W}_{V'}$. It follows that when we pass from S to S', *the deficiency indices diminish by the same* (finite or infinite) *number*.

We now show that, conversely, every isometric extension U of the Cayley transform V of S determines a symmetric extension S' of S whose Cayley transform V' equals U.

First we observe that $(I - U)^{-1}$ exists, that is, that $(I - U)h = 0$

implies $h = 0$. In fact, if $(I - U)h = 0$, then for every element of the form $f = (I - U)g$:

$$(h, f) = (h, g) - (h, Ug) = (Uh, Ug) - (h, Ug) = -((I - U)h, Ug) = 0;$$

hence h is orthogonal to the set of values of $I - U$, and a fortiori, h is orthogonal to the set of values of $I - V$, and therefore to the domain of definition of S. Since this domain is dense in $\mathfrak{H}$, we necessarily have $h = 0$.

Now let us form the transformation

$$S' = i(I + U)(I - U)^{-1},$$

which obviously is an extension of S. S' is symmetric; in fact, if f and g are elements of $\mathfrak{D}_{S'}$, they are of the form

$$f = (I - U)\varphi, \quad g = (I - U)\psi,$$

and we have

$$S'f = i(I + U)\varphi, \quad S'g = i(I + U)\psi;$$

hence, inasmuch as $(\varphi, \psi) = (U\varphi, U\psi)$,

$$(S'f, g) = (i(I + U)\varphi, (I - U)\psi) = i[(U\varphi, \psi) - (\varphi, U\psi)] =$$
$$= ((I - U)\varphi, i(I + U)\psi) = (f, S'g).$$

Finally, the relation $f = (I - U)\varphi$ implies that

$$S'f = i(I + U)\varphi, \ (S' + iI)f = 2i\varphi, \ (S' - iI)f = 2iU\varphi,$$

from which we see that the domain of the Cayley transform V' consists of elements of the form $2i\varphi$, where φ runs through $\mathfrak{D}_U$, and that

$$V'(2i\varphi) = 2iU\varphi = U(2i\varphi).$$

This proves that $V' = U$, which was to be shown.

We note that if U is an arbitrary isometric transformation for which the set of values of $I - U$ is dense in $\mathfrak{H}$, the same reasoning proves that $S' = i(I + U)(I - U)^{-1}$ is a symmetric transformation whose Cayley transform equals U.

Thus *the problem of finding all the (closed) symmetric extensions of the closed symmetric transformation S reduces to the problem of finding all the isometric extensions of its Cayley transform V*; this problem is obviously much simpler than the original problem.

In fact, in order to extend V we have only to map the deficiency subspace $\mathfrak{H} - \mathfrak{D}_V$, or a subspace of the latter, isometrically into the deficiency subspace $\mathfrak{H} - \mathfrak{B}_V$; this is accomplished, for example, with the aid of two orthonormal systems taken in $\mathfrak{H} - \mathfrak{D}_V$ and in $\mathfrak{H} - \mathfrak{B}_V$. It is thus possible to exhaust the deficiency subspace with the smaller dimension; the corresponding symmetric transformation S' will then be a maximal extension of S. If the two deficiency subspaces are of the same dimension, they can be exhausted simultaneously, and we obtain a unitary extension of V, and consequently a self-adjoint extension of S.

Let us summarize the essential points in the

THEOREM. *In order that the closed symmetric transformation S be maximal, it is necessary and sufficient that one or the other of its deficiency indices be equal to* 0; *in order that it admit a self-adjoint transformation as an extension, it is necessary and sufficient that its deficiency indices be equal; finally, in order that it itself be self-adjoint, it is necessary and sufficient that its two deficiency indices be equal to* 0.

Let us give an example in the Hilbert space $\mathfrak{H}$ whose dimension is denumerably infinite, of an isometric non-unitary transformation V_0: let $\{g_n\}$ be a complete orthonormal sequence in $\mathfrak{H}$ and set

$$V_0 \sum_1^\infty c_k g_k = \sum_1^\infty c_k g_{k+1}.$$

Then the domain of V_0 is the entire space, while the set of values $V_0 f$ has the orthogonal complement of dimension 1 consisting of the elements of the form cg_1. It is easily shown that the set of values of $I - V_0$ is dense in $\mathfrak{H}$. Hence V_0 is the Cayley transform of the symmetric transformation

$$S_0 = i(I + V_0)(I - V_0)^{-1};$$

S_0 is defined by

$$S_0 \sum_1^\infty c_k g_k = ic_1 g_1 + i(2c_1 + c_2)g_2 + i(2c_1 + 2c_2 + c_3)g_3 + \dots$$

for all elements $f = \Sigma c_k g_k$ for which $(I - V_0)^{-1}f$ has a meaning, that is, for which

$$|c_1|^2 + |c_1 + c_2|^2 + |c_1 + c_2 + c_3|^2 + \dots$$

converges (for these f we have, in particular, $\sum_1^\infty c_k = 0$).

Therefore the transformation S_0 has (0,1) for deficiency indices; it is called the *elementary symmetric transformation.*

It can be shown that every symmetric transformation S of a Hilbert space $\mathfrak{H}$ of arbitrary dimension with the deficiency indices $(0, m)$ (where m is an arbitrary finite or infinite cardinal number) is composed of m elementary symmetric transformations plus possibly a self-adjoint transformation, in the following sense: there are m mutually orthogonal subspaces $\mathfrak{M}_a$ with denumerably infinite dimension in $\mathfrak{H}$, each of which reduces S to an elementary symmetric transformation, such that, in the subspace $\mathfrak{N}$ of elements orthogonal to all the $\mathfrak{M}_a$ (a subspace which may consist of the single element 0), S reduces to a self-adjoint transformation.

The problem of maximal symmetric transformations whose deficiency indices are $(m, 0)$ presents nothing new, since, in general, if S has (m, n) for indices, $-S$ will have (n, m) for indices. This follows from the fact that the Cayley transform of $-S$ is obviously equal to the inverse of that of S.

Let us remark that the real symmetric transformations of the space L^2 always have equal deficiency indices, hence they are either self-adjoint or possess self-adjoint extensions. The transformation S is said to be *real* if its domain contains with a function $f(x)$ its conjugate $\overline{f(x)}$, and if in addition $S\overline{f(x)} = \overline{Sf(x)}$.

Our proposition is verified as follows. The domain $\mathfrak{D}_V$ and the range $\mathfrak{W}_V$ of the Cayley transform of S consist of the functions

$$u(x) = Sf(x) = if(x) \text{ and } v(x) = Sg(x) - ig(x),$$

respectively, where f and g run through the domain of S. Setting $g(x) = \overline{f(x)}$ we have $v(x) = \overline{u(x)}$, hence $\mathfrak{W}_V$ consists of the conjugates of the functions belonging to $\mathfrak{D}_V$. Since the conjugates of two orthogonal functions are also orthogonal, $\mathfrak{H} - \mathfrak{W}_V$ consists of the conjugates of functions belonging to $\mathfrak{H} - \mathfrak{D}_V$. Consequently the dimension of $\mathfrak{H} - \mathfrak{W}_V$ equals that of $\mathfrak{H} - \mathfrak{D}_V$, that is, the two deficiency indices are equal.

Furthermore, it is possible to define "real" transformations in an abstract Hilbert space $\mathfrak{H}$, if one first introduces an operation J which corresponds to "conjugation." J must be a transformation of $\mathfrak{H}$ into itself such that

$$J(f+g) = Jf + Jg, \quad J(cf) = \bar{c}Jf, \quad J^2 = I,$$

$$(Jf, Jg) = \overline{(f, g)} = (g, f).$$

The transformation T is said to be "real with respect to the conjugation J" if its domain contains Jf whenever it contains f and if $TJf = JTf$, that is, $J \smile T$. The above proposition for real symmetric transformations remains valid in the abstract case.[12]

124. Semi-bounded Symmetric Transformations. The Method of Friedrichs

Among the differential operators of mathematical physics, one frequently encounters operators which, under the given boundary conditions, give rise to semi-bounded symmetric transformations of a Hilbert space. This is the case, for example, for the operator

$$Au(x) = -[p(x)u'(x)]' + q(x)u(x),$$

where

$$p(x) \geq 0, \quad q(x) \geq q_0 \quad (0 \leq x \leq 1),$$

if it is considered as a linear transformation in the space $L^2(0, 1)$, defined

[12] Cf. VON NEUMANN [1] (p. 101); STONE [*] (Chapt. IX, Sec. 2).

for functions $u(x)$ which are twice continuously differentiable and satisfy the conditions

$$u(0) = 0, \quad u'(1) + hu(1) = 0,$$

where h is a constant $\geqq 0$; the functions $p(x)$, $p'(x)$, $q(x)$ are assumed to be continuous. We can immediately verify that

$$(Au, u) = hp(1)\,|u(1)|^2 + \int_0^1 p(x)\,|u'(x)|^2 dx + \int_0^1 q(x)\,|u(x)|^2 dx \geqq$$

$$\geqq \int_0^1 q_0\,|u(x)|^2 dx = q_0(u, u).$$

Another important example is furnished, in the space of functions $u(x, y, z)$ which are square-summable in ordinary (x, y, z) space, by the Schrödinger operator for the hydrogen atom

$$Au = -\,(u_{xx} + u_{yy} + u_{zz}) - \frac{c}{r}\,u \qquad (r = \sqrt{x^2 + y^2 + z^2});$$

we define this operator first for sufficiently regular functions, for example, for functions u which are zero when $r \leqq r_0$ and when $r \geqq R_0$ (r_0 and R_0 being positive quantities which vary with the function considered), and which are continuous together with their partial derivatives up to and including the second order. It is easy to see that A is then symmetric, and we obtain by a calculation that[13]

$$(Au, u) = \iiint \left(|u_x|^2 + |u_y|^2 + |u_z|^2 - \frac{c}{r}\,|u|^2\right) dx\,dy\,dz \geqq -\,2c^2(u, u),$$

where the integral extends over the entire (x, y, z) space.

We could multiply the number of examples drawn from quantum theory and other branches of mathematical physics. It is important to know whether these transformations can be extended to *self-adjoint* transformations, that is, to transformations for which the spectral decomposition is possible. The answer to this question is in the affirmative:

THEOREM. *Every semi-bounded symmetric transformation S can be extended to a semi-bounded self-adjoint transformation A in such a way that A has the same (greatest lower or least upper) bound as S.*

This theorem was stated by J. VON NEUMANN,[14] but he proved only the less precise proposition that the bound of A differs from that of S by as little as one wishes. The first proofs of the complete theorem are due to STONE

[13] Cf. COURANT–HILBERT [*] (p. 387) and RELLICH [**3**] (Note 4, p. 380).
[14] VON NEUMANN [**1**] (p. 103); cf. also WINTNER [*] (Sec. 111).

and FRIEDRICHS.[15] STONE used an approximation of the given symmetric transformation S by bounded symmetric transformations (even of finite rank) S_n which tend to S in the sense that

$$\lim S_n \subseteq S \subseteq (\lim S_n)^{**}.$$

The method of FRIEDRICHS, which we are going to reproduce, can be characterized as an extension by "closure."

It suffices to prove the theorem for positive transformations, or even for transformations bounded below by 1.

Hence let S be a symmetric transformation such that

$$(Sf, f) \geq (f, f)$$

for all elements f of $\mathfrak{D}_S$. We introduce in $\mathfrak{D}_S$ the new scalar product

$$[f, g] = (Sf, g) = (f, Sg)$$

and the new norm

$$[[f]] = (Sf, f)^{\frac{1}{2}};$$

we obviously have

$$[[f]] \geq \|f\|. \tag{16}$$

With this metric, the set $\mathfrak{D}_S$ becomes a Hilbert space which, in general, is not complete. We can always *complete* it by the adjunction of certain ideal elements, by assigning to each Cauchy sequence $\{f_n\}$, that is, each sequence such that

$$[[f_n - f_m]] \to 0 \quad \text{for} \quad m, n \to \infty,$$

an ideal limit element, provided that $\{f_n\}$ has no limit element in $\mathfrak{D}_S$; to two equivalent Cauchy sequences $\{f_n\}$ and $\{f'_n\}$, that is, sequences such that

$$[[f_n - f'_n]] \to 0 \quad \text{for} \quad n \to \infty,$$

we assign the same ideal limit element. If $\{f_n\}$ and $\{g_n\}$ are two Cauchy sequences with the limits f and g, where f and g can be elements of $\mathfrak{D}_S$ or ideal elements, the sequence of scalar products

$$[f_n, g_n]$$

is also convergent and its limit depends only on f and g, that is, it remains invariant when we replace the Cauchy sequences in question by others which are equivalent. If f and g belong to $\mathfrak{D}_S$, then

$$\lim\, [f_n, g_n] = [f, g];$$

if f or g or both are ideal, this equation can serve as the *definition* of $[f, g]$.

[15] STONE [*] (p. 387); FRIEDRICHS [**1**] (Note 1); cf. also FREUDENTHAL [**1**] and EBERLEIN [**2**].

It is easy to see that after the adjunction of these elements we obtain a complete space, that is, a space in which every Cauchy sequence is convergent. We denote this space by $\mathfrak{H}_0$.

This is the usual procedure for completing an incomplete Hilbert space, but in the case we are considering we can say more: namely, that the ideal elements can be identified with certain elements of the space $\mathfrak{H}$ and that in this way $\mathfrak{H}_0$ can be considered to be a linear subspace of $\mathfrak{H}$:

$$\mathfrak{D}_S \subseteq \mathfrak{H}_0 \subseteq \mathfrak{H}.$$

In fact, it follows from inequality (16) that if a sequence of elements of $\mathfrak{D}_S$ is a Cauchy sequence in the new metric, it is also a Cauchy sequence in the original metric, and that two Cauchy sequences equivalent in the new metric are also equivalent in the original metric, so that they converge in the orginal metric to a well-determined element of the space $\mathfrak{H}$. We can thus assign to each element g^* of $\mathfrak{H}_0$ (ideal or not) a definite element g of $\mathfrak{H}$, the correspondence obviously being linear and such that $g^* = g$ for the elements g^* of $\mathfrak{D}_S$. Since the equation

$$[f, g^*] = (Sf, g),$$

where g denotes the element of $\mathfrak{H}$ which we have just made correspond to the element g^* of $\mathfrak{H}_0$, is valid by definition if f and g^* both belong to $\mathfrak{D}_S$, by continuity it remains valid in the case where f is an element of $\mathfrak{D}_S$ and g^* is an arbitrary element of $\mathfrak{H}_0$. It follows from this equation that two different elements of $\mathfrak{H}_0$ can not be represented by the same element g of $\mathfrak{H}$, or equivalently, that $g = 0$ implies $g^* = 0$. In fact, $g = 0$ implies that $[f, g^*] = 0$ for all the elements of $\mathfrak{D}_S$; but $\mathfrak{D}_S$ is dense in $\mathfrak{H}_0$ (in the new metric), hence $g^* = 0$. It is therefore legitimate to *identify* the elements of $\mathfrak{H}_0$ with the corresponding elements of $\mathfrak{H}$: that is what we wanted to prove.

Inequality (16) carries over by continuity to all elements f of $\mathfrak{H}_0$.

Let us now fix an arbitrary element h of $\mathfrak{H}$ and consider the functional

$$L_h(f) = (f, h).$$

If f is in $\mathfrak{H}_0$, then

$$|L_h(f)| \leq \|f\|\,\|h\| \leq [[f]]\,\|h\|.$$

Considered in the Hilbert space $\mathfrak{H}_0$, $L_h(f)$ is therefore a linear functional whose norm does not exceed $\|h\|$. Consequently (see Sec. 30) there exists an element g in $\mathfrak{H}_0$ for which

$$L_h(f) = [f, g]$$

and whose norm $[[g]]$ equals that of the functional, hence

$$\|g\| \leq [[g]] \leq \|h\|. \tag{17}$$

Moreover, this element g is uniquely determined by the functional, and hence by h, so that if we write

$$g = Bh$$

we define in a unique manner a transformation B whose domain is the entire space $\mathfrak{H}$ and whose values are contained in $\mathfrak{H}_0$. This transformation is obviously linear, and by definition

$$(f, h) = [f, Bh] \tag{18}$$

for f in $\mathfrak{H}_0$, h in $\mathfrak{H}$; by (17),

$$\|Bh\| \leq \|h\|.$$

Setting $f = Bh'$ in (18), where h' is again an arbitrary element of $\mathfrak{H}$, we obtain

$$(Bh', h) = [Bh', Bh] = \overline{[Bh, Bh']} = \overline{(Bh, h')} = (h', Bh),$$

and in particular, for $h = h'$:

$$(Bh, h) = [Bh, Bh] \geq (Bh, Bh) \geq 0, \tag{19}$$

hence B is a positive and symmetric transformation in the space $\mathfrak{H}$.

We shall show that B has an inverse. In fact, by (18) $Bh = 0$ implies that $(f, h) = 0$ for all elements f of $\mathfrak{H}_0$, and since $\mathfrak{H}_0$ is dense in $\mathfrak{H}$, this implies that $h = 0$.

Since B is bounded and symmetric, and therefore self-adjoint, its inverse

$$A = B^{-1}$$

will also be a *self-adjoint* transformation (see Sec. 119, proposition a)). A is semi-bounded below by 1; in fact, if $g = Bh$ is an element of the domain of A, then by virtue of (19)

$$(g, Ag) \geq (g, g).$$

Moreover, by (18)

$$(f, Ag) = [f, g] \tag{20}$$

for f in $\mathfrak{H}_0$ and g in $\mathfrak{D}_A$. Furthermore, the domain $\mathfrak{D}_A$, which is a linear subset of $\mathfrak{H}_0$, is dense in $\mathfrak{H}_0$ in the sense of the new metric as well as the old. In fact, in the contrary case there would be an element $f_0 \neq 0$ in $\mathfrak{H}_0$ such that

$$(f_0, Ag) = [f_0, g] = 0$$

for all elements g of $\mathfrak{D}_A$. But the range of A exhausts the space $\mathfrak{H}$, and consequently $f_0 = 0$, in contradiction with the hypothesis that $f_0 \neq 0$.

There remains to prove that the self-adjoint transformation A is an extension of the given symmetric transformation S.

Let f and g be two arbitrary elements of $\mathfrak{D}_S$. We have on the one hand, by (18),

$$(f, Sg) = [f, BSg],$$

and on the other hand, by the definition of the new metric,

$$(f, Sg) = [f, g].$$

Since $\mathfrak{D}_S$ is dense in $\mathfrak{H}_0$ (in the sense of the new metric), the equation

$$[f, BSg] = [f, g]$$

is possible for all elements f of $\mathfrak{D}_S$ only if

$$BSg = g.$$

This shows that g belongs to the domain of A and that

$$Ag = Sg.$$

Hence $A \supseteq S$. This completes the proof of the theorem.

It is possible for S to have still other self-adjoint transformations as extensions. But among these transformations there is only one — the transformation A we have just constructed — whose domain is contained in $\mathfrak{H}_0$. Moreover, if S' is an arbitrary symmetric extension of S whose domain is contained in $\mathfrak{H}_0$, then $A \supseteq S'$.

In fact, let f' be an element of $\mathfrak{D}_{S'}$ and let f be an arbitrary element of $\mathfrak{D}_S$; by (18) we have

$$[f, BS'f'] = (f, S'f') = (S'f, f') = (Sf, f'),$$

and since the relation $(Sf, g) = [f, g]$, which is first valid for f and g in $\mathfrak{D}_S$, extends by continuity to f in $\mathfrak{D}_S$ and g in $\mathfrak{H}_0$, we have

$$[f, BS'f'] = [f, f'].$$

It follows that $BS'f' = f'$, and that consequently A is defined for f' and

$$Af' = ABS'f' = S'f',$$

which proves that $A \supseteq S'$.

*

In conclusion, we consider our problem under a slightly different aspect. Instead of starting with a semi-bounded symmetric transformation S and with the form $[f, g]$ which corresponds to it, we consider an arbitrary form $[f, g]$ which possesses analogous properties, that is, we suppose $[f, g]$ to be a real-valued function defined for the elements f, g of a linear set $\mathfrak{D} \subseteq \mathfrak{H}$ such that

$$(21)\qquad \begin{cases} [f_1 + f_2, g] = [f_1, g] + [f_2, g], \\ [cf, g] = c[f, g], \\ [f, g] = \overline{[g, f]} \end{cases}$$

$$(22)\qquad [[f]]^2 = [f, f] \geq (f, f) = \|f\|^2.$$

Briefly, let $[f, g]$ be a symmetric bilinear form, defined in $\mathfrak{D}$ and semi-bounded below by 1.

The same reasoning as above leads us to the

THEOREM.[16] *The form $[f, g]$ can be extended by continuity to the linear set $\mathfrak{H}_0$ consisting of the elements g of $\mathfrak{H}$ for which it is possible to find a sequence $\{f_n\}$ of elements of $\mathfrak{D}$ such that*

$$\|g - f_n\| \to 0 \text{ and } [[f_n - f_m]] \to 0 \text{ for } m, n \to \infty.$$

There exists a self-adjoint transformation A, and only one, whose domain of definition $\mathfrak{D}_A$ is contained in $\mathfrak{H}_0$ and for which

$$[f, g] = (f, Ag)$$

for all elements f of $\mathfrak{H}_0$ and g of $\mathfrak{D}_A$. This transformation is semi-bounded below by 1.

In fact, if A is constructed as above, $A = B^{-1}$, we have (20), and if S is an arbitrary symmetric transformation, whose domain is contained in $\mathfrak{H}_0$ and for which

$$[f, g] = (f, Sg)$$

for f in $\mathfrak{H}_0$ and g in $\mathfrak{D}_S$, then by (18)

$$[f, BSg] = (f, Sg) = [f, g],$$

and consequently

$$BSg = g;$$

hence

$$Ag = Sg,$$

that is,

$$A \supseteq S.$$

The case of an arbitrary semi-bounded symmetric bilinear form, that is, one for which (21) is valid, and instead of (22) one or the other of the inequalities

$$[f, f] \geq c(f, f), \quad [f, f] \leq c(f, f),$$

can be reduced to the preceding by replacing $[f, g]$ by the bilinear form

$$(1 \mp c)\,(f, g) \pm [f, g].$$

[16] FRIEDRICHS [**1**] (Note 1, pp. 478–479).

125. Krein's Method

By the preceding method we have obtained a well-defined extension of the semi-bounded symmetric transformation S to a semi-bounded self-adjoint transformation A, but we have not gathered any information about other possible extensions. The following method due to M. KREIN [**1**], which is the "real" analogue of the method of J. VON NEUMANN for transformations which are not semi-bounded (Sec. 121), covers all the extensions in question.

Let S be a symmetric transformation,

$$S \geq cI,$$

and let us seek all its extensions to self-adjoint transformations A such that

$$A \geq \gamma I$$

where $\gamma \leq c$. Replacing S by $S - \gamma I$, the problem reduces to the following:

Given a positive symmetric transformation S, find all its extensions to positive self-adjoint transformations A.

In order to solve this problem, let us form, just as in Sec. 122, the transformation

$$B = (S - I)(S + I)^{-1};$$

we see, precisely as in the above-mentioned section, that B is symmetric and bounded, that is,

$$(Bf, g) = (f, Bg) \quad \text{and} \quad \|Bf\| \leq \|f\|$$

for all elements f, g of the domain of B. We denote this domain by $\mathfrak{L}$; in this case, neither does it in general coincide with the entire space $\mathfrak{H}$ nor is it even dense in $\mathfrak{H}$. When the transformation S is *closed*, which we can suppose without loss of generality, B is also closed, and since it is also bounded and hence continuous, its domain $\mathfrak{L}$ is necessarily a closed linear set, that is, a subspace of $\mathfrak{H}$ (which perhaps coincides with $\mathfrak{H}$).

It is clear that if S' is a positive symmetric transformation such that

$$S' \supseteq S,$$

then

$$B' \supseteq B$$

for the corresponding transformations.

Conversely, if $\bar{B}$ is a linear transformation defined in a subspace $\mathfrak{L}' \supseteq \mathfrak{L}$ which is symmetric, bounded by 1, and such that $\bar{B} \supseteq B$, there corresponds to it a positive symmetric transformation $S' \supseteq S$, namely,

$$S' = (I + \bar{B})(I - \bar{B})^{-1},$$

and

$$B' = \bar{B}.$$

In fact, all this is immediately verified as soon as we show that $(I - \bar{B})^{-1}$ exists, that is, that $(I - \bar{B})f = 0$ implies $f = 0$. But $(I - \bar{B})f = 0$ implies that

$$(f, (I - B)g) = (f, (I - \bar{B})g) = ((I - \bar{B})f, g) = 0$$

for all elements g of the domain $\mathfrak{L}$ of B, that is, that f is orthogonal to the range of the transformation

$$I - B = I - (S - I)(S + I)^{-1} = [(S + I) - (S - I)](S + I)^{-1} = 2(S + I)^{-1},$$

hence to the domain of $S + I$. Since the latter is dense in $\mathfrak{H}$, we necessarily have $f = 0$.

We know that if S' is a *self-adjoint* transformation, then B' is defined *everywhere* (see Sec. 122). Conversely, if B' is defined everywhere it is self-adjoint, and then

$$S' = (I + B')(I - B')^{-1} = -I + 2(I - B')^{-1}$$

is also self-adjoint (see propositions a) and b), Sec. 119).

Thus the problem we are dealing with reduces to the following:

To construct all the extensions of the transformation

$$B = (S - I)(S + I)^{-1}$$

which are defined everywhere and which are symmetric and of norm ≤ 1.

The *existence* of such extensions can be proved not only for the transformation B in question but also *for an arbitrary transformation B which is defined in a proper subspace $\mathfrak{L}$ of $\mathfrak{H}$ and which is symmetric and has bound* 1 *there.*

Here is how we prove it.

Let f be an arbitrary element of $\mathfrak{L}$ and denote by B_0f and B_1f the orthogonal projections of Bf onto $\mathfrak{L}$ and onto its orthogonal complement $\mathfrak{M}$. We thus define two linear transformations B_0 and B_1 having $\mathfrak{L}$ for domain and such that in $\mathfrak{L}$,

$$B = B_0 + B_1.$$

We shall extend B_0, and then B_1, to the entire space $\mathfrak{H}$. First consider B_0. If f and g belong to $\mathfrak{L}$, then obviously

$$(f, B_0 g) = (Bf, g). \tag{23}$$

If g does not belong to $\mathfrak{L}$, the same equation can serve as the *definition* of $B_0 g$ as an element of $\mathfrak{L}$. In fact, the second member is, for fixed g and for f varying in $\mathfrak{L}$, a linear functional $F_g(f)$, such that

$$|F_g(f)| \leq \|Bf\|\,\|g\| \leq \|f\|\,\|g\|,$$

that is,

$$\|F_g\| \leq \|g\|.$$

Hence there exists a well-defined element of $\mathfrak{L}$, which we shall denote by $B_0 g$, which satisfies equation (23) for every f in $\mathfrak{L}$; we have

$$\|B_0 g\| = \|F_g\| \leq \|g\|. \tag{24}$$

The transformation B_0 thus extended to the entire space $\mathfrak{H}$ obviously remains linear.

Now we shall extend B_1 so that it takes on its values in $\mathfrak{M}$ and that the inequality

$$\|B_0 g\|^2 + \|B_1 g\|^2 \leq \|g\|^2, \tag{25}$$

which is valid in $\mathfrak{L}$ by the hypothesis $\|Bg\| \leq \|g\|$, is preserved after the extension.

In this connection we consider the symmetric bilinear form

$$[f, g] = (f, g) - (B_0 f, B_0 g).$$

By (24) we have

$$[f, f] = \|f\|^2 - \|B_0 f\|^2 \geq 0.$$

Setting

$$[[f]] = [f, f]^{\frac{1}{2}} \geq 0,$$

inequality (25) can be written in the following form:

$$\|B_1 g\| \leq [[g]]. \tag{26}$$

$[f, g]$ and $[[f]]$ possess all the properties of a scalar product and of a norm, except that $[[f]]$ can very well become zero for certain $f \neq 0$. But this does not keep the inequalities of Schwarz and Minkowski

$$|[f, g]| \leq [[f]]\,[[g]] \text{ and } [[f + g]] \leq [[f]] + [[g]]$$

from being valid; see the proof of these inequalities in Sec. 83. This shows, in particular, that the elements f for which $[[f]] = 0$ form a linear set $\mathfrak{O}$; furthermore, this set is also closed, since B_0 is continuous. We introduce in $\mathfrak{H}$ a new sort of inequality by writing

$$f_1 \equiv f_2 \text{ if } [[f_1 - f_2]] = 0;$$

this amounts to passing to the quotient space

$$\mathfrak{H}/\mathfrak{O}.$$

We thus obtain a new linear space with $[f, g]$ as the scalar product and $[[f]]$ as norm—a space which in general will not be complete. However, we can always complete it by the addition of certain ideal elements (see Sec. 124), and we thus obtain a new Hilbert space which we shall denote by $\bar{\mathfrak{H}}$.

We observe that if g_1 and g_2 are two elements of $\mathfrak{L}$ such that

$$g_1 \equiv g_2,$$

then by (26) (setting $g = g_1 - g_2$ there):

$$B_1 g_1 = B_1 g_2.$$

This permits us to consider B_1 as a linear transformation of a linear subset $\bar{\mathfrak{L}}$ of $\bar{\mathfrak{H}}$ into the subspace $\mathfrak{M}$ of $\mathfrak{H}$; $\bar{\mathfrak{L}}$ consists of the elements of $\bar{\mathfrak{H}}$ which arise from $\mathfrak{L}$. By (26) this transformation can be extended by continuity to the subspace $[\bar{\mathfrak{L}}]$ of $\bar{\mathfrak{H}}$ which is the closure of $\bar{\mathfrak{L}}$ in the sense of the metric of $\bar{\mathfrak{H}}$. The values of B_1 will always belong to the subspace $\mathfrak{M}$ of $\mathfrak{H}$, and inequality (26) remains valid.

Denoting by P the linear transformation of $\bar{\mathfrak{H}}$ which assigns to every element u of $\bar{\mathfrak{H}}$ its orthogonal projection onto $[\bar{\mathfrak{L}}]$ (in the sense of the metric of $\bar{\mathfrak{H}}$), let us extend B_1 to the entire space $\bar{\mathfrak{H}}$ by setting

$$B_1 u = B_1 P u;$$

B_1 will then be a linear transformation which carries the entire space $\bar{\mathfrak{H}}$ into the subspace $\mathfrak{M}$ of $\mathfrak{H}$; moreover, we shall have

$$\|B_1 u\| = \|B_1 P u\| \leq [[Pu]] \leq [[u]]. \tag{27}$$

The transformation B_1 thus extended to the entire space $\bar{\mathfrak{H}}$ generates a linear transformation of the original space $\mathfrak{H}$; in fact, if f is an element of $\mathfrak{H}$ we have only to set

$$B_1 f = B_1 u,$$

where u denotes the element which represents f in $\bar{\mathfrak{H}}$. We then have, by (27),

$$\|B_1 f\|^2 \leq [[f]]^2 = \|f\|^2 - \|B_0 f\|^2;$$

inequality (25) is therefore preserved.

Since $B_0 f$ always belongs to $\mathfrak{L}$ and $B_1 f$ to $\mathfrak{M}$, we have $(B_0 f, B_1 f) = 0$; hence

$$\|B_0 f + B_1 f\|^2 = \|B_0 f\|^2 + \|B_1 f\|^2 \leq \|f\|^2.$$

The sum of the two extended linear transformations,

$$C = B_0 + B_1,$$

is therefore a linear transformation which is defined everywhere, extends the given transformation B, and satisfies

$$\|C\| \leq 1.$$

But in general C is not symmetric. However C^* is also an extension of B, since for all f in $\mathfrak{L}$ and all g in $\mathfrak{H}$,

$$(C^* f, g) = (f, Cg) = (f, B_0 g + B_1 g) = (f, B_0 g) = (Bf, g),$$

and hence

$$C^* f = Bf.$$

The symmetric transformation

$$\tilde{B} = \tfrac{1}{2}(C + C^*)$$

is thus an extension of B, and since $\|C^*\| = \|C\| \leq 1$,

$$\|\tilde{B}\| \leq 1.$$

Therefore our problem concerning the existence of an extension of the required type is solved, and in such a way that we also have a new proof of the existence of an extension of the positive symmetric transformation S to a positive self-adjoint transformation A.

But in general there are several solutions. KREIN has shown that among the extensions $\tilde{B}$ of B which are everywhere defined, symmetric, and of norm ≤ 1 there are two extremal transformations, $\tilde{B}_{\min}$ and $\tilde{B}_{\max}$ ($\tilde{B}_{\min} \leq \tilde{B}_{\max}$); that is,

$$\tilde{B}_{\min} \leq \tilde{B} \leq \tilde{B}_{\max}$$

for all the other transformations $\tilde{B}$ in question. For the proof of this proposition as well as for a profound study of the structure of the various extensions we must refer the reader to KREIN's paper [1].

Further, one can prove by the same method the following theorem of CALKIN,[17] as was observed by KREIN.

THEOREM. *Every symmetric transformation S with the property that*

$$\|Sf\| \geq \|f\|$$

for all elements f of the domain of S can be extended to a self-adjoint transformation satisfying the same inequality.

In the proof we can assume that S is closed; its inverse $B = S^{-1}$ then has a certain subspace $\mathfrak{L}$ of $\mathfrak{H}$ for domain, and we have

$$(Bf, g) = (f, Bg), \quad \|Bf\| \leq \|f\|.$$

By what we have just seen, B can be extended to a transformation $\tilde{B}$ which is everywhere defined, symmetric, and of norm ≤ 1; $A = \tilde{B}^{-1}$ is then, when it exists, a self-adjoint transformation of the required type. But the existence of this inverse is assured by the fact that if $\tilde{B}f = 0$, then f is orthogonal to all elements of the form $\tilde{B}g$, and in particular to all elements of the form Bg; that is, f is orthogonal to the domain of S, and since the latter is dense in $\mathfrak{H}$, we have $f = 0$.

Moreover, this theorem extends immediately to symmetric transformations whose "*spectrum has a gap containing a finite interval* (a, b)," that is, for which

$$\left\| \left(S - \frac{a+b}{2} \right) f \right\| \geq \frac{b-a}{2} \|f\|.$$

[17] CALKIN [**1**]. Cf. also KRASNOSELSKY [**1**] where deficiency indices for an arbitrary closed linear transformation are defined.

CHAPTER IX

SELF-ADJOINT TRANSFORMATIONS FUNCTIONAL CALCULUS, SPECTRUM, PERTURBATIONS

FUNCTIONAL CALCULUS

126. Bounded Functions

If the spectral family $\{E_\lambda\}$ of the bounded or unbounded self-adjoint transformation A is known, we can make use of the formula

$$(u(A)f, g) = \int_{-\infty}^{\infty} u(\lambda)d(E_\lambda f, g) \tag{1}$$

(cf. Sec. 107) to define "functions" of A.[1]

Let us assume at first that the function $u(\lambda)$ is real-valued, bounded, and summable over the entire λ-axis, in the Stieltjes-Lebesgue sense, with respect to all the functions of bounded variation $(E_\lambda f, g)$; or equivalently, by virtue of the relation

$$(E_\lambda f, g) = \left\| E_\lambda \frac{f+g}{2} \right\|^2 - \left\| E_\lambda \frac{f-g}{2} \right\|^2 + i \left\| E_\lambda \frac{f+ig}{2} \right\|^2 - i \left\| E_\lambda \frac{f-ig}{2} \right\|^2,$$

that it is summable with respect to all the nondecreasing real functions

$$(E_\lambda f, f) = \|E_\lambda f\|^2,$$

where f runs through all the elements of the space $\mathfrak{H}$.

For fixed f, the second member of (1) is the complex conjugate of an additive and homogeneous functional $L_f(g)$; let us show that the latter is also bounded. Denoting by M an upper bound of $|u(\lambda)|$, we obviously have

$$|L_f(g)| \leq M \times \text{total variation of } (E_\lambda f, g).$$

[1] This definition of a "function" of a self-adjoint transformation is due to v. NEUMANN [**4**] and to STONE [*] (Chapt. VI). See also LORCH [**1**], [**2**], who does not use the bilinear form appearing in the second member of (1) in his definition of $u(A)$, but instead constructs a notion of "measure" whose values are projections (or, what amounts to the same thing, subspaces) and defines $u(A)$ as the Stieltjes-Lebesgue type integral with respect to this "measure." Moreover, the discussion by LORCH [**2**] also includes the case of more general spaces, namely, reflexive Banach spaces (cf. Sec. 87).

This total variation does not exceed $\|f\| \cdot \|g\|$. It suffices to see that the same inequality holds for the variation corresponding to an arbitrary subdivision of the λ-axis by a finite number of points,

$$-\infty = \lambda_0 < \lambda_1 < \ldots < \lambda_n = \infty.$$

Since the differences $E(\delta_k) = E_{\lambda_k} - E_{\lambda_{k-1}}$ $(k = 1, 2, \ldots, n)$ are mutually orthogonal projections with sum equal to I,

$$\sum_1^n |(E_{\lambda_k} f, g) - (E_{\lambda_{k-1}} f, g)| = \sum_1^n |(E(\delta_k)f, g)| = \sum_1^n |(E(\delta_k)f, E(\delta_k)g)| \leq$$

$$\leq \sum_1^n \|E(\delta_k)f\| \, \|E(\delta_k)g\| \leq \{\sum_1^n \|E(\delta_k)f\|^2 \sum_1^n \|E(\delta_k)g\|^2\}^{\frac{1}{2}} = \|f\| \cdot \|g\|;$$

here we have first used Schwarz's inequality and then that of Cauchy:

$$\sum a_k b_k \leq \{\sum a_k^2 \cdot \sum b_k^2\}^{\frac{1}{2}}.$$

Therefore

$$|L_f(g)| = M\|f\| \cdot \|g\|,$$

that is, $L_f(g)$ is a linear functional such that

$$\|L_f\| \leq M\|f\|.$$

Consequently there exists (cf. Sec. 30) a uniquely determined element f^* such that

$$L_f(g) = (g, f^*).$$

The second member of (1) is therefore equal to (f^*, g). We are thus led *to define*: $u(A)f = f^*$; this transformation $u(A)$ is obviously linear, and since

$$\|f^*\|^2 = (f^*, f^*) = L_f(f^*) \leq M\|f\| \cdot \|f^*\|,$$

we have $\|f^*\| \leq M\|f\|$; hence $\|u(A)\| \leq M$.

It follows directly from the definition that if $u(\lambda)$ is equal to a constant c, then $u(A) = cI$, and that the correspondence between the functions $u(\lambda)$ and the transformations $u(A)$ is *homogeneous* and *additive*:

$$(cu)\,(A) = cu(A), \quad (u_1 + u_2)\,(A) = u_1(A) + u_2(A).$$

Furthermore, it is *multiplicative*. In fact, if

$$u(\lambda) = u_1(\lambda)u_2(\lambda),$$

then, inasmuch as $(E_\mu f, E_\lambda g) = (E_\lambda f, g)$ for $\mu \geq \lambda$,

$$(u_1(A)u_2(A)f, g) = \int_{-\infty}^{\infty} u_1(\lambda) d(E_\lambda u_2(A)f, g) = \int_{-\infty}^{\infty} u_1(\lambda) d(u_2(A)f, E_\lambda g) =$$

$$= \int_{-\infty}^{\infty} u_1(\lambda) d_\lambda \int_{-\infty}^{\infty} u_2(\mu) d_\mu (E_\mu f, E_\lambda g) = \int_{-\infty}^{\infty} u_1(\lambda) d_\lambda \int_{-\infty}^{\lambda} u_2(\mu) d_\mu (E_\mu f, g) =$$

$$= \int_{-\infty}^{\infty} u_1(\lambda) u_2(\lambda) d_\lambda (E_\lambda f, g) = (u(A)f, g);$$

therefore,

$$u(A) = u_1(A)u_2(A).$$

This also shows that all the functions of A which can be obtained in this manner are mutually *permutable*.

Moreover, the transformation $u(A)$ is *symmetric*:

$$(u(A)f, g) = \int_{-\infty}^{\infty} u(\lambda)d(E_\lambda f, g) = \overline{\int_{-\infty}^{\infty} u(\lambda)d(E_\lambda g, f)} = \overline{(u(A)g, f)} = (f, u(A)g).$$

If $u(\lambda) \geq 0$, we have also

$$(u(A)f, f) = \int_{-\infty}^{\infty} u(\lambda)d(E_\lambda f, f) \geq 0,$$

that is, the correspondence is *of positive type*.

It follows from the multiplicative property that

$$\|u(A)f\|^2 = ([u(A)]^2f, f) = \int_{-\infty}^{\infty} [u(\lambda)]^2d(E_\lambda f, f). \tag{2}$$

We deduce from this that the correspondence is also *continuous* in the sense that if $\{u_n(\lambda)\}$ is a uniformly bounded sequence of functions which tends to $u(\lambda)$, then the sequence $\{u_n(A)\}$ tends to $u(A)$. We have only to write (2) with $u_n - u$ in place of u and apply Lebesgue's theorem on term-by-term integration. Furthermore, it obviously suffices to require the convergence of $\{u_n(\lambda)\}$ almost everywhere with respect to all the functions $(E_\lambda f, f) = \|E_\lambda f\|^2$, or briefly, *almost everywhere with respect to* $\{E_\lambda\}$. It also follows from (2) that if the functions $u_n(\lambda)$ converge uniformly in $-\infty < \lambda < \infty$, then the transformations $u_n(A)$ also converge in norm.

Further, the transformations $u(A)$ possess the important property of *permuting with all the bounded transformations B which permute with A*, in particular with all the bounded symmetric transformations which permute with A; hence $u(A) \smile\smile A$.

In fact, since B then permutes with E_λ (cf. Sec. 120),

$$(u(A)Bf, g) = \int_{-\infty}^{\infty} u(\lambda)d(E_\lambda Bf, g) = \int_{-\infty}^{\infty} u(\lambda)d(E_\lambda f, B^*g) =$$
$$= (u(A)f, B^*g) = (Bu(A)f, g).$$

We shall see shortly that this property is characteristic of "functions" of A. But first let us remove the hypothesis that the functions $u(\lambda)$ are bounded.

127. Unbounded Functions. Definitions

We shall extend the correspondence between functions and transformations to all functions which are *measurable, finite, and defined almost every-*

where with respect to the spectral family $\{E_\lambda\}$, that is, almost everywhere with respect to all the functions $(E_\lambda f, f) = \|E_\lambda f\|^2$.

Let $u(\lambda)$ be one of these functions and assume further that it is real-valued. Denote by $e_k(\lambda)$ the characteristic function of the set where

$$k - 1 \leq u(\lambda) < k \qquad (k = 0, \pm 1, \pm 2, \ldots).$$

Since the functions $e_k(\lambda)$ and $u_k(\lambda) = u(\lambda)e_k(\lambda)$ are bounded and measurable, and hence also summable with respect to $\{E_\lambda\}$, the transformations $e_k(A)$ and $u_k(A)$ are already defined; moreover, since $e_k^2(\lambda) = e_k(\lambda)$, $e_k(\lambda)e_h(\lambda) = 0$ $(k \neq h)$, and

$$\sum_{-\infty}^{\infty} e_k(\lambda) = 1,$$

we also have $[e_k(A)]^2 = e_k(A)$, $e_k(A)e_h(A) = 0$ $(k \neq h)$, and

$$\sum_{-\infty}^{\infty} e_k(A) = I,$$

that is, the $e_k(A)$ are pairwise orthogonal projections which have the sum I. Let $\mathfrak{L}_k$ be the subspace corresponding to the projection $e_k(A)$; since $u_k(A)$ permutes with $e_k(A)$, it reduces, in $\mathfrak{L}_k$, to a bounded symmetric transformation.

This being the case, we *define* the transformation $u(A)$ to be the self-adjoint transformation of $\mathfrak{H}$ which is uniquely determined and which reduces in each $\mathfrak{L}_k$ to $u_k(A)$ $(k = 0, \pm 1, \pm 2, \ldots)$; see the lemma, Sec. 120).

By virtue of this definition, the *domain* of $u(A)$ consists of those elements $f = \sum_{-\infty}^{\infty} f_k$ (where f_k is the component of f in $\mathfrak{L}_k$, $f_k = e_k(A)f$), for which the series

$$\sum_{-\infty}^{\infty} \|u_k(A)f_k\|^2 = \sum_{-\infty}^{\infty} \|u_k(A)f\|^2 = \sum_{k=-\infty}^{\infty} \int_{-\infty}^{\infty} u_k^2(\lambda)d(E_\lambda f, f)$$

converges, or, what amounts to the same thing in view of the equation

$$\sum_{k=-\infty}^{\infty} u_k^2(\lambda) = u^2(\lambda)$$

and by virtue of the theorems of Lebesgue and Beppo Levi, those elements for which the integral

$$\int_{-\infty}^{\infty} u^2(\lambda)d(E_\lambda f, f)$$

converges, the value of this integral being then equal to $\|u(A)f\|^2$. For these f we have, always by virtue of the definition,

$$u(A)f = \sum_{k=-\infty}^{\infty} u_k(A)f_k = \sum_{k=-\infty}^{\infty} u_k(A)f.$$

hence also

$$(u(A)f, g) = \sum_k \int_{-\infty}^{\infty} u_k(\lambda) d(E_\lambda f, g) = \int_{-\infty}^{\infty} u(\lambda) d(E_\lambda f, g),$$

for every element g of $\mathfrak{H}$, whether or not it belongs to $\mathfrak{D}_{u(A)}$. In the last step we could have made use of Lebesgue's theorem on term-by-term integration, since the function $u(\lambda)$ is summable with respect to the indefinite total variation $\varrho(\lambda)$ of the function $(E_\lambda f, g)$. In fact, we have already seen in the preceding section that

$$\varrho(\infty) - \varrho(-\infty) \leq \|f\| \cdot \|g\|,$$

and it is shown in an analogous manner that

$$\varrho(b) - \varrho(a) \leq \|E(\delta)f\| \cdot \|E(\delta)g\|,$$

where $E(\delta) = E_b - E_a$; therefore if $\gamma(\lambda)$ is a step function which assumes the constant values c_k in the intervals $\delta_k = (a_k < \lambda \leq b_k)$, by Cauchy's inequality

$$\sum_k |c_k| \, [\varrho(b_k) - \varrho(a_k)] \leq \sum_k |c_k| \cdot \|E(\delta_k)f\| \cdot \|E(\delta_k)g\| \leq$$

$$\leq \{\sum_k c_k^2 \, \|E(\delta_k)f\|^2 \cdot \sum_k \|E(\delta_k)g\|^2\}^{\frac{1}{2}} = \{\sum_k c_k^2[\|E_{b_k}f\|^2 - \|E_{a_k}f\|^2\}^{\frac{1}{2}} \cdot \|g\|,$$

or

$$\int_{-\infty}^{\infty} |\gamma(\lambda) d\varrho(\lambda) \leq \int_{-\infty}^{\infty} \gamma^2(\lambda) d \, \|E_\lambda f\|^2 \cdot \|g\|,$$

and this inequality then carries over to all functions which are square-summable with respect to $\|E_\lambda f\|^2$, in particular to the function $u(\lambda)$ which we have in mind.

It is easy to show that the relations $u_k(A) \smile\smile A$ imply the relation

$$u(A) \smile\smile A.$$

Up to now we have considered only real-valued functions. For a *complex-valued* function

$$w(\lambda) = u(\lambda) + iv(\lambda),$$

where $u(\lambda)$, $v(\lambda)$ denote the real and imaginary parts, we define $w(A)$ by

$$w(A) = u(A) + iv(A).$$

The domain of $w(A)$ is therefore equal to the intersection of the domains of $u(A)$ and of $v(A)$; it consists of those elements f for which

(3) $$\int_{-\infty}^{\infty} u^2(\lambda) d(E_\lambda f, f) + \int_{-\infty}^{\infty} v^2(\lambda) d(E_\lambda f, f) < \infty,$$

and for these f and for arbitrary g in $\mathfrak{H}$ we have

$$(w(A)f, g) = \int_{-\infty}^{\infty} w(\lambda) d(E_\lambda f, g). \tag{4}$$

Since $u^2(\lambda) + v^2(\lambda) = |w(\lambda)|^2$, the domain of $w(A)$ coincides with the domain of the self-adjoint transformation $|w|(A)$ which corresponds to the non-negative function $|w(\lambda)|$, and consequently this domain is everywhere dense in $\mathfrak{H}$. On the other hand,

$$\begin{aligned} \|w(A)f\|^2 &= \|u(A)f + iv(A)f\|^2 = \\ &= \|u(A)f\|^2 + \|v(A)f\|^2 + i[(v(A)f, U(A)f) - (u(A)f, v(A)f)] = \\ &= \|u(A)f\|^2 + \|v(A)f\|^2 + 2 \operatorname{Im} (v(A)f, U(A)f) = \|u(A)f\|^2 + \|v(A)f\|^2, \end{aligned}$$

since $(v(A)f, u(A)f)$, being equal to

$$\int_{-\infty}^{\infty} v(\lambda) d(E_\lambda f, u(A)f) = \int_{-\infty}^{\infty} v(\lambda) d_\lambda \int_{-\infty}^{\infty} u(\mu) d(E_\lambda f, E_\mu f) = \int_{-\infty}^{\infty} v(\lambda) u(\lambda) d(E_\lambda f, f),$$

is real-valued. Therefore,

$$\|w(A)f\|^2 = \int_{-\infty}^{\infty} [u^2(\lambda) + v^2(\lambda)]\, d(E_\lambda f, f) = \int_{-\infty}^{\infty} |w(\lambda)|^2\, d(E_\lambda f, f). \tag{5}$$

128. Unbounded Functions. Rules of Calculation

It is clear that $w(A) \smile\smile A$ and that the correspondence between functions and transformations, thus extended, remains *homogeneous* and *of positive type*. It also remains *additive* and *multiplicative* in the generalized sense which we shall now state.

Instead of *additivity* in the ordinary sense, we have the relation

$$(w_1 + w_2)(A) \supseteq w_1(A) + w_2(A) \tag{6}$$

which follows, on the one hand from the fact that the sum of two functions which are square-summable with respect to $(E_\lambda f, f)$ is likewise square-summable, and on the other hand from formula (4). By the same reasoning, we see that the domain of the second member of (6) is equal to the intersection of the domain of the first member with the domain of $w_1(A)$ (or, just as well, with the domain of $w_2(A)$) and that consequently the equality sign is valid in (6) if and only if the domain of the first member is contained in the domain of $w_1(A)$ *or* of $w_2(A)$, which happens in particular if one or the other of the functions $w_1(\lambda)$, $w_2(\lambda)$ is bounded almost everywhere with respect to $\{E_\lambda\}$.

Instead of ordinary *multiplicativity* we have the relation

$$(w_1 w_2)(A) \supseteq w_1(A) w_2(A). \tag{7}$$

To establish this relation, consider an element f of the domain of $w_2(A)$. By virtue of (5),

$$\|E_\lambda w_2(A)f\|^2 = \|w_2(A)E_\lambda f\|^2 = \int_{-\infty}^{\infty} |w_2(\mu)|^2 d_\mu \|E_\mu E_\lambda f\|^2 = \int_{-\infty}^{\infty} |w_2(\mu)|^2 \, d\|E_\mu f\|^2,$$

and therefore

$$\int_{-\infty}^{\infty} |w_1(\lambda)|^2 \, d\|E_\lambda w_2(A)f\|^2 = \int_{-\infty}^{\infty} |w_1(\lambda)w_2(\lambda)|^2 \, d\|E_\lambda f\|^2$$

in the extended sense that if one of the two members is infinite, the same is true of the other. This means that if f is an element of the domain of $w_2(A)$, it either belongs to the domains of both $w_1(A)w_2(A)$ and $(w_1w_2)(A)$, or to neither of them. In other words, the domain of the product $w_1(A)w_2(A)$ is equal to the intersection of the domains of $w_2(A)$ and of $(w_1w_2)(A)$. For an element f of this domain and for every element g of $\mathfrak{H}$, we have

$$(w_1(A)w_2(A)f, g) = \int_{-\infty}^{\infty} w_1(\lambda)w_2(\lambda)d(E_\lambda f, g),$$

which we obtain, using formula (4), by means of a calculation which is already familiar (cf. Sec. 126). It follows that

$$w_1(A)w_2(A)f = (w_1w_2)(A)f,$$

which proves (7).

We have even proved more, namely, that *$w_1(A)w_2(A)$ is the "restriction" of $(w_1w_2)(A)$ to the intersection of the domain of the latter transformation with the domain of $w_2(A)$.* In order that the two members of (7) be equal, it is therefore necessary and sufficient that the domain of $w_2(A)$ contain that of $(w_1w_2)(A)$.

In particular, equality holds in (7) if $w_2(\lambda) = w_1^n(\lambda)$, where n is a positive integer, because the convergence of the integral of $|w_1^{n+1}(\lambda)|^2$ with respect to $(E_\lambda f, f)$ implies the convergence of the integral of $|w_1^n(\lambda)|^2$ also. Applying this fact successively for $n = 1, 2, \ldots$, we obtain, in general:

(8) $$(w^n)(A) = [w(A)]^n \qquad (n = 1, 2, \ldots).$$

In particular,

$$A^n = \int_{-\infty}^{\infty} \lambda^n dE_\lambda,$$

from which it follows, among other things, that *all the iterates $A^2, A^3, \ldots$ of a self-adjoint transformation A are also self-adjoint.*

Relation (8) is also valid for negative integral exponents; it suffices to establish this fact for $n = -1$. Suppose the function $w(\lambda)$ is different from 0 almost everywhere with respect to $\{E_\lambda\}$. The function $w^{-1}(\lambda) = \dfrac{1}{w(\lambda)}$ then belongs to our class and the relation

$$w^{-1}(\lambda)w(\lambda) = w(\lambda)w^{-1}(\lambda) = 1,$$

which is valid almost everywhere with respect to $\{E_\lambda\}$, implies that $w^{-1}(A)w(A)$ is the restriction of the identity transformation I to the domain of $w(A)$ and that $w(A)w^{-1}(A)$ is the restriction of I to the domain of $w^{-1}(A)$, which means that the transformation $w(A)$ possesses an inverse and that this inverse is equal to $w^{-1}(A)$.

Conversely, the existence of $[w(A)]^{-1}$ implies the existence of $w^{-1}(A)$. In fact, the set of zeros of $w(\lambda)$ is then of measure zero with respect to all the functions $(E_\lambda f, f)$. To see this, denote by $e(\lambda)$ the characteristic function of the set of zeros; then $w(\lambda)e(\lambda) = 0$, consequently $w(A)e(A) = 0$ and

$$e(A) = [w(A)]^{-1}\, w(A)e(A) = 0,$$

and therefore

$$\int_{-\infty}^{\infty} e^2(\lambda)d(E_\lambda f, f) = \|e(A)f\|^2 = 0.$$

We summarize: *The relation*

$$[w(A)]^{-1} = w^{-1}(A)$$

holds whenever one or the other of the two members has a meaning.

In particular,

$$(A - zI)^{-1} = \int_{-\infty}^{\infty} \frac{1}{\lambda - z}\, dE_\lambda$$

if one or the other of the two members has a meaning.

*

Under what conditions is the transformation $(A - zI)^{-1}$, or more generally, the transformation $w(A)$ *bounded*? A sufficient condition follows from formula (5): it is that the function $w(\lambda)$ be bounded almost everywhere with respect to $\{E_\lambda\}$. In fact, if $|w(\lambda)| \leq M$ almost everywhere with respect to $\{E_\lambda\}$, then

$$\|w(A)f\|^2 = \int_{-\infty}^{\infty} |w(\lambda)|^2\, d(E_\lambda f, f) \leq M^2 \int_{-\infty}^{\infty} d(E_\lambda f, f) = M^2\, \|f\|^2,$$

hence

$$\|w(A)\| \leq M. \tag{9}$$

But this condition is also necessary. In fact, (9) implies, by virtue of (8), that

$$\|w^n(A)\| = \|[w(A)]^n\| \leq \|w(A)\|^n \leq M^n;$$

hence, if we denote by e the set of points where $|w(\lambda)| > M$,

$$\int_e \left|\frac{w(\lambda)}{M}\right|^{2n} d(E_\lambda f, f) \leq \int_{\infty}^{\infty} \left|\frac{w(\lambda)}{M}\right|^{2n} d(E_\lambda f, f) = \frac{1}{M^{2n}}\, \|w^n(A)f\|^2 \leq \|f\|^2;$$

inasmuch as $|w(\lambda)/M|^{2n}$ increases to infinity in e, it follows that e is necessarily of measure zero with respect to each function $(E_\lambda f, f)$.

This yields the following proposition.

A necessary and sufficient condition that the transformation $w(A)$ be bounded is that the function $w(\lambda)$ be bounded almost everywhere with respect to $\{E_\lambda\}$; under this condition,

$$\|w(A)\| = \text{true max } |w(\lambda)| \qquad (\textit{with respect to } \{E_\lambda\}). \tag{10}$$

Let us pass to the study of *adjoint* transformations. When the function $w(\lambda)$ is bounded, then obviously

$$[w(A)]^* = [u(A) + iv(A)]^* = u(A) - iv(A) = \bar{w}(A),$$

where $\bar{w}(A)$ denotes the transformation which corresponds to the conjugate function $\overline{w(\lambda)}$. In the general case, we decompose the function $w(\lambda)$ in the form

$$w(\lambda) = r(\lambda)u(\lambda)$$

where

$$r(\lambda) = |w(\lambda)| \geq 0 \text{ and } |u(\lambda)| = 1.$$

Since the transformations $w(A)$ and $r(A)$ have the same domain,

$$w(A) = u(A)r(A);$$

since $u(A)$ is bounded and $r(A)$ is self-adjoint, it follows that

$$[w(A)]^* = [u(A)r(A)]^* = [r(A)]^* [u(A)]^* = r(A)\bar{u}(A) = \bar{w}(A).$$

Hence, we have the relations $[w(A)]^* = \bar{w}(A)$ or $w(A) = [\bar{w}(A)]^*$, from which it follows, in particular, that every transformation $w(A)$ is closed.

We also have

$$[w(A)]^* w(A) = \bar{w}(A)w(A) \subseteq (\bar{w}w)(A) \supseteq w(A)\bar{w}(A) = w(A)\, [w(A)]^*;$$

and since all these transformations are self-adjoint, the two extreme members by Sec. 119, and the middle term since the function $\bar{w}w = w\bar{w} = |w|^2$ is real-valued, we necessarily have

$$[w(A)]^* w(A) = (|w|^2)(A) = w(A)\, [w(A)]^*.$$

We called those bounded linear transformations T for which

$$T^*T = TT^* \tag{11}$$

normal. We retain the same nomenclature for unbounded, but *closed*, linear transformations which satisfy equation (11). Obviously, all self-adjoint transformations are normal.

The formula we have just established makes clear that *all the transformations $w(A)$ are normal.*

In particular, $w(A)$ is a *self-adjoint* transformation if the function $w(\lambda)$ is real-valued, and it is a *unitary* transformation if $|w(\lambda)| = 1$ almost everywhere with respect to $\{E_\lambda\}$. The latter proposition follows from the fact that, under this hypothesis, $\overline{w(\lambda)}w(\lambda) = 1$, and hence $[w(A)]^* w(A) = w(A)[w(A)]^* = I$.

Let us now consider the problem of *composite* functions. First, we must determine the spectral family of the self-adjoint transformation

$$B = u(A)$$

corresponding to the real-valued function $u(\lambda)$. We shall show that it is the family $\{F_\mu = e_\mu(A)\}$, where $e_\mu(\lambda)$ denotes the characteristic function of the set in which the function $u(\lambda)$ does not exceed the value μ. Since the characteristic properties of a spectral family can be established without difficulty, there only remains to show that

$$\int_{-\infty}^{\infty} \mu^2 d(e_\mu(A)f, f) = \int_{-\infty}^{\infty} u^2(\lambda) d(E_\lambda f, f) \tag{12}$$

for all elements f of the domain of B and that, for these f,

$$\int_{-\infty}^{\infty} \mu d(e_\mu(A)f, f) = \int_{-\infty}^{\infty} u(\lambda) d(E_\lambda f, f). \tag{13}$$

In fact, these relations express the fact that the domain of B is contained in that of

$$B' = \int_{-\infty}^{\infty} \mu de_\mu(A)$$

and that, for every element f of the domain of B, we have $(B'f, f) = (Bf, f)$. It also follows that $(B'f_1, f_2) = (Bf_1, f_2)$ for two arbitrary elements f_1, f_2 of the domain of B, and that, consequently, $B' \supseteq B$. Since B and B' are self-adjoint, and hence maximal symmetric, this is possible only if $B' = B$. Now, since

$$(e_\mu(A)f, f) = \int_{-\infty}^{\infty} e_\mu(\lambda) d(E_\lambda f, f),$$

the relations (12), (13) to be verified follow from the general theorem on Stieltjes-Lebesgue integrals which was proved at the end of Sec. 58.

Let us now consider a function $v(\lambda)$ for which

$$v(B) = v(u(A))$$

has a meaning. By virtue of the theorem we have just recalled, the composite function $w(\lambda) = v(u(\lambda))$ is measurable with respect to $\{E_\lambda\}$ and we have

$$\int_{-\infty}^{\infty} |v(\mu)|^2 d(F_\mu f, f) = \int_{-\infty}^{\infty} |w(\lambda)|^2 d(E_\lambda f, f)$$

and

$$\int_{-\infty}^{\infty} v(\mu)d(F_\mu f, f) = \int_{-\infty}^{\infty} w(\lambda)d(E_\lambda f, f).$$

This yields the following proposition:

Let $u(\lambda)$ be a real-valued function which is measurable and finite almost everywhere with respect to the spectral family $\{E_\lambda\}$ of the self-adjoint transformation A, and let $v(\mu)$ be a real or complex-valued function which is measurable and finite almost everywhere with respect to the spectral family $\{F_\mu\}$ of the self-adjoint transformation $B = u(A)$. The composite function $w(\lambda) = v(u(\lambda))$ is then measurable and finite almost everywhere with respect to $\{E_\lambda\}$, and

$$w(A) = v(B) = v(u(A)). \tag{14}$$

129. Characteristic Properties of Functions of a Self-adjoint Transformation

We have seen that every function $w(A)$ of the self-adjoint transformation A is a closed linear transformation, with domain dense in $\mathfrak{H}$, and such that

$$w(A) \smile\smile A.$$

Now, at least in the case where the space $\mathfrak{H}$ is separable, these properties characterize the functions of A. Namely, we have the

THEOREM.[2] *Every closed linear transformation T with domain dense in the (separable) space $\mathfrak{H}$ and such that*

$$T \smile\smile A,$$

is a (real or complex) function of A.

Consequently, if we wish to retain these very natural properties we can not extend the notion of "function" of A.

We may assume that A is *bounded*; in fact, the case of an unbounded transformation A can always be reduced to the bounded case by (14), for example by considering instead of A the bounded transformation $A' =$ $=$ arc tan A; if $T = w(A')$, then $T = v(A)$, where $v(\lambda) = w(\text{arc tan } \lambda)$.

[2] One finds this condition, although it is not explicitly formulated, by combining Theorem 6 of VON NEUMANN's paper [**3**] with Theorem 5 of his paper [**5**]. It was formulated explicitly, and proved using a process of differentiation, by F. RIESZ [**16**]. In each of these proofs, the discussion is restricted to the case of bounded A and T. The method of F. RIESZ was extended to the case of unbounded A and T by MIMURA [**1**]. The above method, which is a simplified form of that of RIESZ and MIMURA, is due to SZ.-NAGY [*] (pp. 63–65). Cf. further NAKANO [**2**], [**3**].

We first prove the

LEMMA. *To each element f_0 of the domain of T can be associated a function $F(\lambda)$, finite and measurable (B), such that $F(A)f_0 = Tf_0$.*

In order to show this, consider the subspace $\mathfrak{L} = \mathfrak{L}(f_0)$ determined by the elements

$$f_0, Af_0, A^2f_0, \ldots, A^nf_0, \ldots.$$

Clearly, this subspace is transformed by A into itself. Denoting the corresponding projection by L, we have $LAL = AL$, consequently $LA = (AL)^* = (LAL)^* = LAL$, and hence $L \smile A$. It follows from the hypothesis that $L \smile T$, and in particular, that $Tf_0 = TLf_0 = LTf_0$; that is, that Tf_0 is an element of the subspace $\mathfrak{L}$. Therefore there is a sequence of polynomials $\{P_m(\lambda)\}$ such that

$$P_m(A)f_0 \to Tf_0.$$

Since

$$\|[P_n(A) - P_m(A)]f_0\|^2 = \int_{-\infty}^{\infty} |P_n(\lambda) - P_m(\lambda)|^2 \, d\|E_\lambda f_0\|^2 \to 0$$

for $m, n \to \infty$, the sequence $\{P_n(\lambda)\}$ satisfies the Cauchy criterion in the space L^2 of functions which are square-summable with respect to $\|E_\lambda f_0\|^2$. Thus by the Riesz-Fischer theorem there exists an element $F(\lambda)$ of L^2 such that

$$\int_{-\infty}^{\infty} |P_m(\lambda) - F(\lambda)|^2 \, d\|E_\lambda f_0\|^2 \to 0; \tag{15}$$

since the values of $F(\lambda)$ are determined only up to a set of measure zero (with respect to the nondecreasing function $\|E_\lambda f_0\|^2$), we can assume that the function $F(\lambda)$ is measurable (B) [see Sec. 43 where we proved that every measurable function coincides almost everywhere with a function measurable (B), a fact which remains valid for a Stieltjes-Lebesgue measure]. The function $F(\lambda)$ is thus measurable with respect to any nondecreasing function, and consequently the transformation $F(A)$ exists. Since moreover $F(\lambda)$ belongs to L^2, that is,

$$\int_{-\infty}^{\infty} |F(\lambda)|^2 \, d\|E_\lambda f_0\|^2 < \infty,$$

f_0 belongs to the domain of $F(A)$. Since integral (15) is equal to

$$\|[P_m(A) - F(A)]f_0\|^2,$$

$$P_m(A)f_0 \to F(A)f_0 \text{ when } m \to \infty,$$

and therefore

$$F(A)f_0 = Tf_0, \qquad \text{q.e.d.}$$

This lemma established, our theorem is proved in the following manner.

Since the space $\mathfrak{H}$ is assumed separable, we can select from $\mathfrak{D}_T$ a sequence of elements $g_1, g_2, \ldots$ which is everywhere dense in $\mathfrak{D}_T$, and therefore also in $\mathfrak{H}$. We set

$$(16) \qquad f_1 = g_1,\ f_2 = g_2 - L_1 g_2,\ \ldots,\ f_n = g_n - \sum_{k=1}^{n-1} L_k g_n,\ \ldots,$$

where L_k denotes the projection onto the subspace $\mathfrak{L}_k = \mathfrak{L}(f_k)$. Since $L_k \smile T$ (see the proof of the lemma), $L_k g_n$ also belongs to $\mathfrak{D}_T$. We show that the projections L_k are pairwise orthogonal:

$$L_i L_k = 0 \text{ for } i \neq k.$$

Assume that this has been established for $i, k < n$. Then we have, for $i < n$:

$$L_i f_n = L_i g_n - \sum_{k=1}^{n-1} L_i L_k g_n = L_i g_n - L_i g_n = 0,$$

hence also

$$L_i A^j f_n = A^j L_i f_n = 0.$$

The elements $A^j f_n$ $(j = 0, 1, \ldots)$, and consequently the entire subspace $\mathfrak{L}_n$, are therefore annihilated by the projection L_i, and we have $L_i L_n = L_n L_i = 0$. The orthogonality of the projections is thus proved by induction.

The sum P of the projections L_k is equal to I, that is, the corresponding subspaces determine the entire space $\mathfrak{H}$. Since the sequence $\{g_n\}$ is dense in $\mathfrak{H}$, it suffices to show that $Pg_n = g_n$ for all n. For, according to (16),

$$Pg_n = Pf_n + \sum_{k=1}^{n-1} PL_k g_n = f_n + \sum_{k=1}^{n-1} L_k g_n = g_n,$$

since

$$PL_k = L_k \quad\text{and}\quad Pf_n = PL_n f_n = L_n f_n = f_n.$$

Now we choose a sequence $\{c_n\}$ of positive numbers such that the series

$$\sum c_n f_n \quad\text{and}\quad \sum c_n T f_n$$

converge; let, for example,

$$c_n = [2(1 + \|f_n\| + \|Tf_n\|)]^{-n}.$$

By the hypothesis that T is closed, $f_0 = \sum c_n f_n$ necessarily belongs to the domain of T.

Therefore, by the lemma, there is a function $F(\lambda)$, which is measurable (B) and such that $F(A)f_0 = Tf_0$. Since

$$F(A) \smile\smile A \quad\text{and}\quad T \smile\smile A,$$

$F(A)$ and T are defined for all elements of the form Bf_0, where B is an arbitrary bounded symmetric transformation which permutes with A, and we have

$$F(A)Bf_0 = TBf_0.$$

In particular, we set

$$B = \frac{1}{c_m} P_n A^k L_m,$$

where $P_n = e_n(A)$, and $e_n(\lambda)$ denotes the characteristic function of the set where $|F(\lambda)| \leq n$. Since

$$L_m f_0 = \sum_{n=1}^{\infty} c_n L_m f_n = c_m f_m,$$

it follows that

$$F(A)P_n A^k f_m = TP_n A^k f_m.$$

The equation

$$F(A)P_n h = TP_n h$$

is then also valid for linear combinations h of the elements $A^k f_m$. Now for fixed m the latter are dense in the subspace $\mathfrak{L}_m$, and letting m vary, we obtain a set dense in $\mathfrak{H}$. Therefore for every element h_0 of $\mathfrak{H}$ there exists a sequence $\{h_i\}$ of these linear combinations which tends to h_0. Since the transformation $F(A)P_n$ is bounded, and hence continuous,

$$TP_n h_i = F(A)P_n h_i \to F(A)P_n h_0;$$

since also $P_n h_i \to P_n h_0$ and since T is closed, $P_n h_0$ belongs to the domain of T and

$$TP_n h_0 = \lim_{i\to\infty} TP_n h_i = F(A)P_n h_0.$$

Hence

$$TP_n = F(A)P_n.$$

Let g be an element of the domain of $F(A)$ and set $g_n = P_n g$; we obviously have $g_n \to g$. Since

$$Tg_n = TP_n g = F(A)P_n g = P_n F(A)g \to F(A)g,$$

and since T is closed, g necessarily belongs to the domain of T and we have $Tg = F(A)g$. This means that $T \supseteq F(A)$.

Since we not only have $P_n \smile F(A)$, but also $P_n \smile T$ (because $P_n \smile A$), and since $F(A)$, being a function of A, is also closed, we can interchange the rôle of T and of $F(A)$ in this reasoning; we thus obtain $F(A) \supseteq T$.

Therefore $T = F(A)$, q.e.d.

We remark that in a non-separable space $\mathfrak{H}$ the theorem is not necessarily true.[3]

[3] This remark is due to NAKANO [**3**]; cf. also WECKEN [**2**] and SZ.-NAGY [*] (p. 65).

130. Finite or Denumerable Sets of Permutable Self-adjoint Transformations

We have seen in Sec. 111 that if X and Y are two bounded permutable symmetric transformations, there exists a family $\{E(\delta)\}$ of projections, where $E(\delta)$ is an additive, multiplicative, and continuous function of a (closed, half-open, or open) rectangle δ which varies in the (x, y) plane and has sides parallel to the axes, with the property that

$$p(X, Y) = \int_{-\infty}^{\infty} \int_{-\infty}^{\infty} p(x, y) E(dxdy)$$

for every polynomial in the variables x, y and for the corresponding polynomial of the transformations X, Y; moreover, the domain of integration can be reduced to the rectangle

$$\Delta = [m_X \leqq x \leqq M_X,\ m_Y \leqq y \leqq M_Y],$$

since $E(\Delta) = I$ and $E(\delta) = 0$ for every rectangle δ which is exterior to Δ. The same formula allowed us to extend the correspondence between functions and transformations to more general functions, namely to all functions $u(x, y)$ which are continuous in Δ, the integral being always defined as the limit in norm of sums of the Riemann-Stieltjes type.

Now, starting with the formula for scalar products

$$(u(X, Y)f, g) = \iint_{\Delta} u(x, y)\ (E(dxdy)f, g),$$

the correspondence can be extended, just as in the case of a single variable (Secs. 126—128), to all functions $u(x, y)$ which are measurable, finite, and defined almost everywhere with respect to each of the additive, and of bounded variation rectangle functions $(E(\delta)f, g)$, or, what amounts to the same thing, with respect to each additive and non-negative rectangle function $(E(\delta)f, f)$. The correspondence possesses the same properties as in the case of a single variable; in particular,

$$u(X, Y) \smile\smile \{X, Y\},$$

and it can be shown, just as in Sec. 129, that this property is also characteristic of the "functions" thus defined of X and Y, at least in the case of a separable Hilbert space.

This functional calculus in X and Y furnishes at the same time a functional calculus for the *normal* transformation $N = X + iY$ (cf. Sec. 111); in particular we have

$$(N - \zeta I)^{-1} = \iint_{\Delta} \frac{1}{z - \zeta} E(dxdy) \qquad (z = x + iy) \tag{17}$$

provided one or the other of the two members exists.

The case where X and Y are arbitrary self-adjoint transformations which are permutable in the sense that their spectral families consist of permutable projections (this reduces to ordinary permutability if one or the other of the transformations is bounded) can be treated in the same manner.

On the other hand, everything that has just been said carries over to the case of more than two or even an infinity of permutable self-adjoint transformations.

The theorem on composite functions, proved at the end of Sec. 128, also generalizes to functions of several variables. We limit ourselves to the case of two variables. If the self-adjoint transformations X and Y are "functions" of the same self-adjoint transformation A, that is, if $X = x(A)$, $Y = y(A)$, and if $u(x, y)$ is a function such that $u(X, Y)$ exists, we shall have $u(X, Y) = v(A)$, where $v(A)$ is the transformation corresponding to the composite function $v(\lambda) = u(x(\lambda), y(\lambda))$.

This poses the following problem: *Given two permutable self-adjoint transformations X and Y, construct, if possible, a self-adjoint transformation A such that X and Y are "functions" of A.*

We shall show that this is always possible. The idea of the following construction is borrowed from the principle of transition which allowed us, in Sec. 39, to reduce the problem of the integration of functions of several variables to that of functions of a single variable.

Without loss of generality, we can suppose the transformations X and Y to be bounded, and, moreover, that the corresponding domain Δ is contained in the half-open unit square

$$Q^2 = [0 < x \leq 1,\ 0 < y \leq 1].$$

Consider, on the square, a network of successive divisions into squares $C^{(1)}$, $C^{(2)}, \ldots$, the elements of the division into squares $C^{(m)}$ being half-open squares of the same type as Q^2 itself and with sides of length 2^{-m}. We assign to this network a network on the linear interval

$$Q = [0 < \lambda \leq 1],$$

formed by successive decompositions $D^{(1)}, D^{(2)}, \ldots$, the elements of $D^{(m)}$ being half-open intervals of the same type as Q and of length 2^{-2m}. Let the correspondence between the two networks be established just as in Sec. 39;[4] let the elements of $C^{(m)}$ and of $D^{(m)}$ be numbered so that corresponding elements have the same indices:

$$\delta_1^{(m)}, \delta_2^{(m)}, \delta_3^{(m)}, \ldots; \qquad i_1^{(m)}, i_2^{(m)}, i_3^{(m)}, \ldots.$$

This done, let us, for fixed m and λ $(0 < \lambda \leq 1)$, denote by $E_\lambda^{(m)}$ the sum

[4] It makes no essential difference that we have used here squares of width 2^{-m} instead of width 3^{-m}, and intervals of length 2^{-2m} instead of intervals of length 3^{-2m}.

of the projections $E(\delta_n^{(m)})$, where $\delta_n^{(m)}$ runs through those elements of $C^{(m)}$ whose correspondents $i_n^{(m)}$ lie at least partially to the left of the point λ; $E_\lambda^{(m)}$ is also a projection, and we obviously have

$$E_\lambda^{(m)} \geq E_\lambda^{(m-1)} \text{ and } E_\lambda^{(m)} \leq E_\mu^{(m)} \text{ for } \lambda < \mu,$$

and furthermore

$$E_1^{(m)} = I.$$

Consequently,

$$E_\lambda = \lim_{m\to\infty} E_\lambda^{(m)}$$

is also a projection; moreover, $E_\lambda \leq E_\mu$ for $\lambda < \mu$ and $E_1 = I$. Setting $E_0 = E_{+0}$ and $E_\lambda = 0$ for $\lambda < 0$, $E_\lambda = I$ for $\lambda \geq 1$, we define a spectral family over the closed interval $0 \leq \lambda \leq 1$; it generates the self-adjoint transformation

$$A = \int_{-0}^{1} \lambda dE_\lambda.$$ [5]

Denote by $x_n^{(m)}, y_n^{(m)}$ the coordinates of the upper right vertex of the rectangle $\delta_n^{(m)}$, and let $\varphi_n^{(m)}(\lambda)$ be the characteristic function of the corresponding linear interval $i_n^{(m)}$. The functions

$$x_m(\lambda) = \sum_m x_n^{(m)} \varphi_n^{(m)}(\lambda) \geq 0,$$

where the sum is taken over all elements of the decomposition $D^{(m)}$, decrease to a limit $x(\lambda) \geq 0$ which is summable with respect to $\{E_\lambda\}$. Hence the transformation $x(A)$ exists and

$$x(A) = \int_{-0}^{1} x(\lambda)dE_\lambda = \lim_{m\to\infty} \int_{-0}^{1} x_m(\lambda)dE_\lambda = \lim_{m\to\infty} \sum_n x_n^{(m)} \int_{-0}^{1} \varphi_n^{(m)}(\lambda)dE_\lambda =$$

$$= \lim_{m\to\infty} \sum_n x_n^{(m)} E(\delta_n^{(m)}) = \iint_{Q^2} xE(dxdy) = X.$$

In an analogous manner, we construct a function $y(\lambda)$ for which

$$y(A) = Y.$$

Therefore X and Y are "functions" of the self-adjoint transformation A.

The same method also applies to the case of *more than two*, or even a *denumerable infinity*, of self-adjoint transformations which are mutually permutable. We have only to use a network on the unit cube of the corresponding dimension, and to apply it to a network suitably constructed on the interval $0 < \lambda \leq 1$. In the case of a denumerable infinity of transformations, the sequence of successive decompositions $C^{(1)}, C^{(2)}, \ldots$ into which the

[5] Actually we have not verified that E_λ is a continuous function from the right of λ. But this is not important, because by replacing E_λ if necessary with $E_{\lambda+0}$ we arrive at the same transformation A.

network decomposes on the cube Q^∞ must be such that $C^{(m)}$ decomposes only a finite number of the edges of the cube, and that, on each fixed edge, the decomposition becomes infinitely fine when $m \to \infty$; we construct $C^{(m)}$, for example, by decomposing each of the first m edges of Q^∞ into 2^m equal parts.[6]

Thus we have arrived at the following result.

THEOREM.[7] *Let $\{X_m\}$ be a finite or denumerable infinity of permutable self-adjoint transformations of the Hilbert space $\mathfrak{H}$. Then there exists a self-adjoint transformation A of the space $\mathfrak{H}$ of which all the X_m are "functions":*

$$X_m = x_m(A) \qquad (m = 1, 2, \ldots);$$

moreover,

$$A \smile\!\smile \{X_m\}.$$

The last assertion follows directly from the above construction; in fact, the spectral projections E_λ of A were obtained as the limits of sums whose terms are projections $E(\delta)$ which, in their turn, are obtained from spectral projections of X_m by a finite number of subtractions and multiplications.

131. Arbitrary Sets of Permutable Self-adjoint Transformations

The method of the preceding section is not applicable to a non-denumerable set of transformations because it is impossible to cover the cube of a non-denumerable number of dimensions with a network which can be mapped onto a linear network. Nevertheless, the above theorem remains valid in the general case, at least if we restrict ourselves to a *separable* space $\mathfrak{H}$. In this case, the set of all the bounded linear transformations of $\mathfrak{H}$ is also separable in a sense we shall make precise. Therefore we have only to apply the theorem to a dense denumerable subset of transformations in order to immediately deduce the representation of the entire given set.

The fact to be established is the following:

THEOREM.[8] *From every infinite set $\Sigma = \{T\}$ of bounded linear transformations of the separable Hilbert space $\mathfrak{H}$, a denumerable subset $\sigma = \{T_n\}$ can be selected such that every transformation $\mathfrak{H}$ belonging to Σ is the limit (in the strong sense) of a subsequence $\{T_{n_i}\}$ from σ, and that, at the same time, T^* is the limit of $\{T^*_{n_i}\}$.*

[6] Such networks are considered by JESSEN [**1**] in his theory of functions of an infinite number of variables. Cf. also SZ.-NAGY [**2**] (p. 221), [*] (p. 62) and NAKANO [**3**].

[7] Cf. VON NEUMANN [**2**] (Theorem 10), HAAR [**1**] (pp. 781–790), or SZ.-NAGY [*] (pp. 66–69).

[8] Cf. VON NEUMANN [**2**] (pp. 386–388); the above simplified proof is due to SZ.-NAGY [*] (pp. 12–13).

It suffices to prove the theorem in the case where the transformations T have a common bound,

$$\|T\| \leq C;$$

in fact, the general case reduces to this if we decompose the set Σ into its subsets Σ_N ($N = 1, 2, \ldots$) characterized by the inequalities $N - 1 \leq \|T\| < N$.

Consider all the sequences $\{f_k\}$ ($k = 1, 2, \ldots$) of elements of $\mathfrak{H}$ such that $\sum \|f_k\|^2$ converges. Setting, by definition,

$$c\{f_k\} = \{cf_k\}, \ \{f_k\} + \{g_k\} = \{f_k + g_k\},$$
$$(\{f_k\}, \{g_k\}) = \sum (f_k, g_k),$$

these sequences constitute a new Hilbert space, which we shall denote by $\mathfrak{H}^\omega$. Since $\mathfrak{H}$ is separable, there exists a sequence of elements $g_1, g_2, \ldots$, different from 0 and everywhere dense in $\mathfrak{H}$. It is easy to see that the sequences of the type

$$\{g_{n_1}, g_{n_2}, \ldots, g_{n_r}, 0, 0, 0, \ldots\}$$

form a denumerable set which is everywhere dense in $\mathfrak{H}^\omega$; therefore $\mathfrak{H}^\omega$ is also separable.

To each transformation T belonging to the set Σ we assign an element φ_T of $\mathfrak{H}^\omega$, namely the element

$$\varphi_T = \{c_1 Tg_1, c_1 T^* g_1, c_2 Tg_2, c_2 T^* g_2, \ldots, c_k Tg_k, c_k T^* g_k, \ldots\},$$

where $c_k = (2^k \|g_k\|)^{-1}$. The set of these elements φ_T, a subset of the separable space $\mathfrak{H}^\omega$, is itself separable.[9] Therefore it contains a denumerable dense subset $\{\varphi_{T_n}\}$ ($n = 1, 2, \ldots$). Consequently every element φ_T is the limit of a suitable subsequence $\{\varphi_{T_{n_i}}\}$. Passing to components, we see that

$$T_{n_i} g_k \to Tg_k \text{ and } T^*_{n_i} g_k \to T^* g_k$$

for $i \to \infty$, for any k.

Let f be an arbitrary element of $\mathfrak{H}$. We have

$$\|Tf - T_{n_i} f\| \leq \|Tf - Tg_k\| + \|Tg_k - T_{n_i} g_k\| + \|T_{n_i} g_k - T_{n_i} f\| \leq$$
$$\leq C\|f - g_k\| + \|Tg_k - T_{n_i} g_k\| + C\|g_k - f\|,$$

and hence

$$\limsup_{i \to \infty} \|Tf - T_{n_i} f\| \leq 2C\|f - g_k\|$$

for every k. Since the sequence $\{g_k\}$ is dense in $\mathfrak{H}$, this limit superior is necessarily equal to 0, and therefore

$$T_{n_i} f \to Tf.$$

[9] Cf. Sec. 33, p. 68, footnote[15].

In an analogous manner, we find that

$$T^*_{n_i} f \to T^* f.$$

The sequence $\{T_n\}$ therefore possesses the properties stated in our proposition.

With this established, we can extend the theorem of the preceding section to an arbitrary set $\{X\}$ of permutable self-adjoint transformations of a separable Hilbert space $\mathfrak{H}$. We can assume that all the X are bounded. Let $\{X_n\}$ be a denumerable subset of $\{X\}$, dense in the sense we have just made precise. According to the theorem already proved, there exists a bounded self-adjoint transformation A of which all the X_n are functions and such that $A \smile\smile \{X_n\}$. Since

$$X_n = x_n(A) \smile\smile A \qquad (n = 1, 2, \ldots),$$

we also have

$$X \smile\smile A$$

for all the transformations belonging to the given system. By the theorem of Sec. 129, this implies that X is also a function of A. Therefore, we have the

THEOREM.[10] *Given an arbitrary family $\{X\}$ of self-adjoint transformations of the separable Hilbert space $\mathfrak{H}$ which are mutually permutable, there exists a bounded self-adjoint transformation A such that $A \smile\smile \{X\}$ and such that all the given transformations are functions of A.*

THE SPECTRUM OF A SELF-ADJOINT TRANSFORMATION AND ITS PERTURBATIONS

132. The Spectrum of a Self-adjoint Transformation. Decomposition in Terms of the Point Spectrum and the Continuous Spectrum

We have seen in Sec. 128 that if the self-adjoint transformation A has the spectral decomposition

$$A = \int_{-\infty}^{\infty} \lambda dE_\lambda,$$

then

$$(A - zI)^{-1} = \int_{-\infty}^{\infty} \frac{1}{\lambda - z}\, dE_\lambda$$

[10] Cf. von Neumann [**2**] (Theorem 10), or Sz.-Nagy [*] (pp. 65–69).

for every real or complex value of z for which the function $\frac{1}{\lambda - z}$ is finite on the λ-axis almost everywhere with respect to $\{E_\lambda\}$. This is clearly the case for all the non-real values of z as well as for those real values $z = \lambda$ for which E_λ is a continuous function of the parameter λ. In addition, we have seen that for the transformation $(A - zI)^{-1}$ to be defined everywhere and bounded, it is necessary and sufficient that the function $\frac{1}{\lambda - z}$ be essentially bounded with respect to $\{E_\lambda\}$ on the λ-axis; this is the case for all non-real complex values of z and for those real values which are in the interior of an interval in which E_λ is constant: all these values z constitute the *resolvent set* of A; it is obviously an *open* set. The complementary set in the complex plane is called the *spectrum* of A; it is therefore a *closed* set on the real axis, consisting of the points of increase of E_λ as a function of λ.

Since the intervals of constancy of E_λ are finite in number or at most denumerable, the part of the λ-axis which does not belong to the spectrum is of measure zero with respect to $\{E_\lambda\}$. Consequently all the integrals with respect to $\{E_\lambda\}$ reduce to integrals on the spectrum; the transformation $w(A)$ corresponding to the function $w(\lambda)$ is bounded if and only if the latter is essentially bounded on the spectrum, and then the norm of $w(A)$ is equal to the true maximum of $|w(\lambda)|$ on the spectrum with respect to $\{E_\lambda\}$ (cf. formula (10), page 349).

The spectrum of A contains in particular the points μ of the real axis where E_λ has a jump, that is, the characteristic values of A. The jump

$$E(\mu) = E_{\mu+0} - E_{\mu-0} = E_\mu - E_{\mu-0}$$

is equal to the orthogonal projection on the *characteristic subspace* $\mathfrak{M}_\mu$ which consists of all solutions f of the equation

$$Af = \mu f,$$

that is, of all the *characteristic elements* corresponding to the characteristic value μ (and of the element 0). The characteristic values form the *point spectrum* of A. The characteristic subspaces corresponding to two different characteristic values are mutually orthogonal. If the space $\mathfrak{H}$ is separable, it obviously can contain only a finite or denumerable number of orthogonal elements different from 0; therefore in this case the point spectrum of the transformation A is a finite or denumerable set.

It might happen that the transformation A has no characteristic values; in this case we say that A has a *purely continuous* spectrum. The opposite case arises if the characteristic elements of A form a complete set in $\mathfrak{H}$; in this case we say that A has a *pure point* spectrum (actually, the spectrum of A then consists of the point spectrum plus its accumulation points).

We shall show that *the general case can be generated by the superposition of these two extreme cases.*

Let us consider first the case where there is only a finite or denumerable number of jumps of E_λ, say at the points $\mu_1, \mu_2, \ldots$, as is true whenever the space $\mathfrak{H}$ is separable. Since the jumps $E(\mu_k)$ are projections which are pairwise orthogonal, their sum

$$E = \sum_k E(\mu_k)$$

exists and is also a projection; E is obviously permutable with the E_λ and hence also with

$$A = \int_{-\infty}^{\infty} \lambda dE_\lambda.$$

Therefore the subspaces

$$\mathfrak{H}_p = E\mathfrak{H} \text{ and } \mathfrak{H}_c = (I - E)\mathfrak{H}$$

reduce the transformation A (cf. Sec. 116); the corresponding parts A_p and A_c of A are also self-adjoint transformations.[11] It follows from the definition that $\mathfrak{H}_p$ is determined by the characteristic subspaces $\mathfrak{M}_{\mu_k}$, and that consequently there is no characteristic element of A in $\mathfrak{H}_c$. A_p therefore has a pure point spectrum and A_c has a purely continuous spectrum; the spectrum of A_c is also called the *continuous spectrum* of A. According to this definition, the continuous spectrum of A can very well be superposed on the point spectrum and the set of accumulation points of the latter.[12]

In the case where there is a non-denumerable infinity of jumps $E(\mu)$, which can happen if the space $\mathfrak{H}$ is not separable, the sum

$$E = \sum_\mu E(\mu)$$

is defined as follows. For every fixed element f, there is only a finite number or a denumerable infinity of values of μ for which $E(\mu)f \neq 0$; this is a

[11] In fact, if $(A_p g, h) = (g, h^*)$ for two elements h, h^* of $\mathfrak{H}_p$ and for all elements g of the domain of A_p, then for all the elements f of the domain of A:

$$(Af, h) = (A_p Ef, h) = (Ef, h^*) = (f, Eh^*) = (f, h^*);$$

since A is self-adjoint, this is possible only if h belongs to the domain of A and if $h^* = Ah = A_p h$. This proves that A_p is a self-adjoint transformation of $\mathfrak{H}_p$. Similar reasoning applies to A_c.

[12] The definitions adopted by STONE [*] (pp. 128—129 and p. 164) and SZ.-NAGY [*] (p. 54) differ from those of the text in that the continuous spectrum is defined as the set of points λ of continuous growth of E_λ, that is, such that $E_{\lambda+\varepsilon} - E_{\lambda-\varepsilon} \neq 0$ for $\varepsilon > 0$ and $\lim_{\varepsilon\to 0} (E_{\lambda+\varepsilon} - E_{\lambda-\varepsilon}) = 0$; according to this definition a point can not belong simultaneously to the point spectrum and to the continuous spectrum. The definition of the text is that given originally by HILBERT [*] (Note 4).

consequence of the orthogonality of the characteristic subspaces $\mathfrak{M}_\mu$ and of Bessel's inequality. We define Ef as the sum of these non-zero components; because of the orthogonality, the arrangement of the terms is immaterial. It is easy to verify that the transformation E thus defined is a projection and that the corresponding subspace $\mathfrak{H}_p = E\mathfrak{H}$ is the smallest subspace containing all the characteristic subspaces $\mathfrak{M}_\mu$. Moreover, E is permutable with the E_λ, since for arbitrary fixed λ and f,

$$EE_\lambda f = \sum_\mu E(\mu)E_\lambda f = \sum_\mu E_\lambda E(\mu)f = E_\lambda \sum_\mu E(\mu)f = E_\lambda Ef,$$

where μ runs through the set—at most denumerable—of values for which $E(\mu)E_\lambda f \neq 0$. The argument is completed as above.

Let us add that analogous considerations apply in the case of an arbitrary *normal* transformation

$$N = \int_{-\infty}^{\infty}\int_{-\infty}^{\infty} zE(dxdy) \qquad (z = x + iy).^{13}$$

The spectrum of N is a closed set in the complex plane—the complement of the open resolvent set which is the union of these open rectangles δ for which $E(\delta) = 0$ [cf. formula (17) for $(N - \zeta I)^{-1}$, p. 355]. In particular, the *unitary* transformations have their spectrum lying on the unit circle $|z| = 1$ (cf. Sec. 111).

133. Limit Points of the Spectrum

Let us restrict ourselves to self-adjoint transformations. Points of the continuous spectrum, limit points of the point spectrum, and characteristic values of infinite multiplicity are said to be *limit points* of the spectrum. The point ∞, respectively $-\infty$, is included when the spectrum is not bounded above, respectively below.

If the finite real value μ is not a limit point of the spectrum, there is an interval $\delta = (a, b)$ containing μ in its interior which contains no point of the spectrum except possibly the point μ itself, which is then a characteristic value of finite multiplicity. Therefore, in this case the projection

$$E(\delta) = E_b - E_a$$

is *of finite rank*. (The rank of a projection P is defined to be equal to the dimension of the corresponding subspace $\mathfrak{L} = P\mathfrak{H}$.) Conversely, suppose that

[13] We have established this formula only for normal *bounded* transformations (Sec. 111), but it is also valid for normal *unbounded* transformations (that is, for linear transformations N with domain dense in $\mathfrak{H}$ and such that $NN^* = N^*N$, cf. Sec. 128). This can be proved, for example, by adapting the method of Sec. 120; for a detailed discussion the reader is referred to Sz.-Nagy [*] (pp. 48—50). Cf. further Nakano [**1**].

there exists an interval δ containing the point μ in its interior and such that $E(\delta)$ is of finite rank. Since for λ in this interval

$$E_\lambda - E_a \leq E_b - E_a = E(\delta),$$

the rank of E_λ can not increase when λ runs through the interval δ, except at a finite number of points and by finite jumps; consequently μ can not be a limit point of the spectrum.

From all this, it follows that *the finite real value μ is a limit point of the transformation A if and only if the projection $E(\delta) = E_b - E_a$ is of infinite rank for every interval (a, b) containing μ in its interior.*

This criterion has the inconvenience that it brings in the spectral family of the transformation A which, in general, one does not know *a priori*. We can avoid this as follows. Suppose that $E(\delta)$ is of infinite rank for any interval δ containing the point μ in its interior. Consider a sequence of nested intervals δ_n which contract to the point μ. Since each subspace $\mathfrak{L}_n = E(\delta_n)\mathfrak{H}$ is of infinite dimension, we can choose an orthonormal sequence $\{f_n\}$ such that f_n is an element of $\mathfrak{L}_n$ $(n = 1, 2, \ldots)$. Since

$$\|(A - \mu I)f_n\|^2 = \|(A - \mu I)E(\delta_n)f_n\|^2 = \int_{\delta_n} (\lambda - \mu)^2 d\|E_\lambda f_n\|^2 \leq |\delta_n|^2\|f_n\|^2 = |\delta_n|^2 \to 0,$$

and since every infinite orthonormal sequence converges weakly to 0 (by Bessel's inequality), the sequence $\{f_n\}$ satisfies the following three conditions:

$$\|f_n\| = 1, \; f_n \rightharpoonup 0, \text{ and } (A - \mu I)f_n \to 0.$$

We shall show that, conversely, if there exists a sequence of elements $\{f_n\}$ satisfying these conditions, then the point μ is necessarily a limit point of the spectrum. In fact, since $\delta = (a, b)$ is an arbitrary interval containing μ in its interior, we have for all elements f of the domain of A:

$$\|(A - \mu I)f\|^2 = \int_{-\infty}^{\infty} (\lambda - \mu)^2 d\|E_\lambda f\|^2 \geq (b - \mu)^2 \int_b^{\infty} d\|E_\lambda f\|^2 + (a - \mu)^2 \int_{-\infty}^{a} d\|E_\lambda f\|^2 =$$

$$= (b - \mu)^2 \|(I - E_b)f\|^2 + (a - \mu)^2 \|E_a f\|^2;$$

it follows that $(I - E_b)f_n \to 0$ and $E_a f_n \to 0$, therefore that $(I - E(\delta))f_n \to 0$ when $n \to \infty$; consequently,

$$\lim_n \|E(\delta)f_n\| = \lim_n \|f_n\| = 1.$$

The subspace $\mathfrak{L}(\delta) = E(\delta)\mathfrak{H}$ can not be of finite dimension because if it were the weak convergence of the sequence $\{f_n\}$ would imply its strong convergence, to the same limit 0, which would contradict the fact that $\|E(\delta)f_n\| \to 1$.

We have thus proved the following criterion, due to H. WEYL [4].

A necessary and sufficient condition that the finite real value μ be a limit point of the spectrum of the self-adjoint transformation A is that there exist a *sequence of elements f_n belonging to the domain of A and such that*

$$\|f_n\| = 1, \quad f_n \rightharpoonup 0, \quad (A - \mu I)f_n \to 0.$$

*

Completely continuous self-adjoint transformations can be characterized as those whose spectrum has 0 as the only possible limit point (as we have already observed at the end of Sec. 93). In this case we have the simple decomposition

$$Af = \sum_k \mu_k (f, \varphi_k)\varphi_k,$$

where μ_k runs through the characteristic values of A which are different from 0 (each as many times as its multiplicity indicates) and where $\{\varphi_k\}$ is an orthonormal system of characteristic elements.

If we not only have $\mu_k \to 0$, but also

$$\sum_k \mu_k^2 < \infty,$$

we say that the transformation A is *of Hilbert-Schmidt type*. This nomenclature is motivated by the fact that if the Hilbert space in which A operates is the functional space $L^2(a, b)$, then (almost everywhere)

$$Af(x) = \int_a^b K(x, y)f(y)dy, \tag{18}$$

where

$$K(x, y) = \sum_k \mu_k \varphi_k(x)\overline{\varphi_k(y)}$$

(the series converges in the mean) and

$$\int_a^b \int_a^b |K(x, y)|^2 \, dxdy = \sum_k \mu_k^2 < \infty,$$

that is, that A is generated by a kernel of the type considered by HILBERT and SCHMIDT (cf. Sec. 97).

We are thus led to seek a characterization of the spectrum of the transformations of L^2 which admit the representation (18), but with a symmetric kernel of a more general type, subject, for example, only to the condition that the integral

$$\int_a^b |K(x, y)|^2 dy$$

exist for almost all x (a kernel which is said to be of CARLEMAN[14] type). The class of transformations of this type is much more general than that we have

[14] Cf. CARLEMAN [*] and STONE [*] (Chapt. X, Sec. 1).

just considered; it includes unbounded transformations, but it does not include *all* the self-adjoint transformations, among others it does not include the identity transformation. The following characterization of these transformations, which we state here without proof, is due to VON NEUMANN [7]:

THEOREM. *Every self-adjoint transformation A of the space $L^2(a, b)$ which is generated by a kernel of Carleman type $K(x, y)$ in such a way that the representation* (18) *is valid for all the elements f of the domain of A and almost everywhere in x, has the point* 0 *for limit point (not necessarily unique) of its spectrum. Conversely, every self-adjoint transformation A of the space $L^2(a, b)$ whose spectrum has* 0 *for limit point is unitarily equivalent to a transformation of this type.*

We say the self-adjoint transformations A and A' are *unitarily equivalent* if there exists a unitary transformation U such that

$$A = U^{-1}A'U.$$

Denoting by $\{E_\lambda\}$, $\{E'_\lambda\}$ the corresponding spectral families, we then obviously have

$$E_\lambda = U^{-1}E'_\lambda U,$$

which shows that A and A' have the same point spectrum and the same continuous spectrum, with equal characteristic values having the same multiplicity. Conversely, if A and A' have the same pure point spectrum, including multiplicities, then A and A' are unitarily equivalent. In fact, the corresponding spectral families are then saltus functions:

$$E_\lambda = \sum_{\lambda \leqq \mu} E(\mu), \quad E'_\lambda = \sum_{\lambda \leqq \mu} E'(\mu),$$

the "jumps" $E(\mu)$, $E'(\mu)$ being projections onto the subspaces $\mathfrak{L}(\mu)$, $\mathfrak{L}'(\mu)$ of the same dimension. Hence we can map $\mathfrak{L}(\mu)$ linearly and isometrically onto $\mathfrak{L}'(\mu)$ for every value of μ, thus generating a unitary transformation U of the entire space $\mathfrak{H}$ such that

$$U^{-1}E'_\lambda U = E_\lambda, \quad U^{-1}A'U = A.$$

In the case where the continuous spectrum is not empty, the problem of equivalence is much more complicated. HELLINGER was the first to succeed in finding, in this general case, a complete system of "unitary invariants"; the theory was simplified later by HAHN.[15] These authors had in mind only the case of coordinate Hilbert space, that is, a *separable* space. The non-separable spaces present essential difficulties; we ask the reader to refer to the works cited in the footnote [16].

[15] Cf. HELLINGER [**1**] and HAHN [**1**]; a discussion which also covers the case of unbounded transformations is found in STONE [*] (Chapt. VII).

[16] WECKEN [**2**], NAKANO [**4**], [**5**]; PLESSNER and ROKHLIN [**1**]; HALMOS [*].

134. Perturbation of the Spectrum by the Addition of a Completely Continuous Transformation

We have seen in Sec. 95 how the characteristic values of a completely continuous symmetric transformation A vary when we replace A by $A' = A + B$, where B is a transformation of the same type. For a transformation A which is not completely continuous, the situation is more complicated. The following theorem is due to H. WEYL [**4**].

THEOREM. *If a completely continuous symmetric transformation B is added to a bounded symmetric transformation A, the set of limit points of the spectrum remains invariant.*

In fact, since $f_n \rightharpoonup 0$ implies that $Bf_n \to 0$ by virtue of the complete continuity of B (cf. Sec. 85), the hypotheses

$$\|f_n\| = 1, \quad f_n \rightharpoonup 0, \quad (A - \mu I)f_n \to 0$$

imply that

$$\|f_n\| = 1, \quad f_n \rightharpoonup 0, \quad (A' - \mu I)f_n \to 0,$$

and conversely, and we have only to apply Weyl's criterion for the limit points of the spectrum (Sec. 133).

Furthermore, since the treatment of the possible limit points ∞, $-\infty$ is clear, the theorem extends to unbounded transformations A also.

Therefore the addition of a completely continuous symmetric transformation does not change the limit points of the spectrum. But *by the addition of a completely continuous symmetric transformation we can change the continuous spectrum into a pure point spectrum, at least if the space $\mathfrak{H}$ is separable.* This was proved by H. WEYL [**4**]; J. VON NEUMANN [**7**] later gave a simpler proof and showed that *the added transformation can even be chosen in the Hilbert-Schmidt class, and moreover, that the corresponding sum $\sum_k \mu_k^2$ can be made arbitrarily small.*

Using these results, VON NEUMANN [**7**] proved the following converse of Weyl's theorem:

THEOREM. *If the spectra of the bounded symmetric transformations A and A' of the separable Hilbert space $\mathfrak{H}$ have the same limit points, then A' is unitarily equivalent to a transformation A'' such that $B = A - A''$ is completely continuous.*

In this theorem, no restrictions can be placed upon the class of transformations B; the exclusion of a single completely continuous transformation would make the theorem false.

135. Continuous Perturbations

By virtue of the above theorem of Weyl-Neumann, every self-adjoint transformation, even if it has purely continuous spectrum, is the uniform limit of self-adjoint transformations with pure point spectra.

Let us now study the variation of the spectral family during the passage to the limit. Let $\{A_n\}$ be a sequence of self-adjoint transformations all of which are bounded and which tend (strongly) to the transformation A. For definiteness, let $\|A_n\| \leq 1$. Then for every polynomial $p(\mu)$,

$$p(A_n) \to p(A).$$

Using Weierstrass's approximation theorem, we deduce from this the same behavior for all the continuous functions $u(\mu)$ defined on the closed interval $-1 \leq \mu \leq 1$. In particular, let $u(\mu) = e_0(\mu) . \mu$, where $e_0(\mu) = 1$ for $\mu \leq 0$ and $e_0(\mu) = 0$ for $\mu > 0$. We obtain

$$E_{n0} A_n \to E_0 A,$$

where $\{E_{n\lambda}\}$ and $\{E_\lambda\}$ denote the spectral families corresponding to A_n and to A, respectively. From this we deduce, by virtue of the inequality

$$\|E_{n0}Af - E_0 Af\| \leq \|E_{n0}(A - A_n)f\| + \|E_{n0}A_n f - E_0 Af\| \leq$$
$$\leq \|(A - A_n)f\| + \|E_{n0}A_n f - E_0 Af\|,$$

that

$$E_{n0}A \to E_0 A.$$

If 0 is not a characteristic value of A, then A^{-1} is a self-adjoint transformation, and for every element f of its domain,

$$E_{n0} f = E_{n0} AA^{-1}f \to E_0 AA^{-1}f = E_0 f.$$

Since this domain is dense in $\mathfrak{H}$ and inasmuch as the projections E_{n0} are uniformly bounded, it follows that

$$E_{n0} \to E_0.$$

By the same reasoning, we show that, in general,

$$E_{n\lambda} \to E_\lambda$$

for every value λ which does not belong to the point spectrum of A.

The same proposition holds also for a sequence of *not uniformly bounded* self-adjoint transformations A_n which converges to the self-adjoint transformation A in the extended sense that A is equal to the smallest closed extension of $\lim A_n$. Let λ be a value not belonging to the point spectrum of A; we can assume, without loss of generality, that $\lambda = 0$. First, we observe that the spectral projections E_0, E_{n0} corresponding to A and A_n, respectively, are identical to those corresponding to

$$C = A(I + A^2)^{-1} \text{ and } C_n = A_n(I + A_n^2)^{-1},$$

and that $\|C_n\| \leq 1$. Hence our proposition will be reduced to the case already established as soon as we show that $C_n \to C$.

Now, setting

$$B = (I + A^2)^{-1} \text{ and } B_n = (I + A_n^2)^{-1},$$

$\{Bh, Ch\}$ and $\{B_nh, C_nh\}$ are the projections of the element $\{h, 0\}$ of the doubled space $\boldsymbol{H}$ onto the graphs $\boldsymbol{G}_A, \boldsymbol{G}_{A_n}$ (see Sec. 118). Hence, we have only to show that the projection onto $\boldsymbol{G}_{A_n}$ tends to the projection onto $\boldsymbol{G}_A$; this is established without difficulty[17] by considering first the elements of the subspace $\boldsymbol{G}_A$, then those of the complementary orthogonal subspace $\boldsymbol{H} - \boldsymbol{G}_A$, which is equal to the transform $\boldsymbol{VG}_A$ of $\boldsymbol{G}_A$ (see Sec. 117).

We summarize:

THEOREM.[18] *Let $\{A_n\}$ be a sequence of self-adjoint transformations which tend to the self-adjoint transformation A in the extended sense that A is equal to the smallest closed extension of $\lim A_n$. Denoting the respective spectral families by $\{E_{n\lambda}\}$ and $\{E_\lambda\}$, we have*

$$E_\lambda = \lim_n E_{n\lambda}$$

at every point λ which does not belong to the point spectrum of A.

In particular, let $\mathfrak{H}$ be a separable space and let $\{\varphi_k\}$ be a complete orthonormal sequence in $\mathfrak{H}$. Let P_n be the projection onto the subspace determined by the first n of the φ_k, then

$$P_n \to I \text{ and } A_n = P_n A P_n \to A$$

for every bounded transformation A. Therefore the spectral family of a bounded symmetric transformation A is the limit of the spectral families of its approximations A_n, at least in a set dense on the λ-axis. This is essentially the way in which HILBERT was the first to obtain the spectral family of a bounded symmetric transformation. This method was adapted by STONE to obtain the spectral family of an unbounded self-adjoint transformation.[19]

We add to the above theorem that *the convergence $E_{n\lambda} \to E_\lambda$ is in general not uniform even if $A_n \Rightarrow A$.*

Consider, for example, the sequence of transformations

$$A_n f(x) = \left(x - \frac{1}{n}\right) f(x)$$

of the space $L^2(0, 1)$, which converges uniformly to the transformation

$$Af(x) = xf(x).$$

[17] Cf. SZ.-NAGY [*] (pp. 56–57).
[18] Cf. RELLICH [3] (Note 2), or SZ.-NAGY [*] (pp. 56–57).
[19] Cf. STONE [*] (pp. 165–167).

We have

$$E_\lambda f(x) = e_\lambda(x) f(x),$$

where $e_\lambda(x)$ is the characteristic function of the semi-axis $x \leq \lambda$, and

$$E_{n\lambda} f(x) = e_{\lambda + \frac{1}{n}}(x) f(x);$$

hence $E_{n\lambda} - E_\lambda$ is the projection onto the subspace of those functions which are annihilated in the exterior of $\left(\lambda, \lambda + \frac{1}{n}\right)$ and arbitrary in the interior. Therefore $E_{n\lambda} - E_\lambda \to 0$, but

$$\|E_{n\lambda} - E_\lambda\| = 1 \quad \text{for} \quad 0 \leq \lambda < 1; \quad n = 1, 2, \ldots .$$

*

The situation is simpler for *uniformly convergent* sequences $\{A_n\}$ and for the *isolated parts of the spectrum* of the limit transformation A. It even suffices to assume that the convergence is "relatively uniform" in the sense that the transformations A_n have the same domain $\mathfrak{D}_A$, and that

(19) $$\eta_n = \sup_f \frac{\|(A_n - A)f\|}{\|f\| + \|Af\|} \to 0 \quad \text{when} \quad n \to \infty.$$

We say that the interval $\delta = (\mu_1, \mu_2)$ of the λ-axis carries an isolated part of the spectrum of A if its extremities do not belong to the spectrum of A. This part of the spectrum can therefore be separated from the remainder by a closed rectifiable Jordan curve J which is entirely contained in the resolvent set of A; for definiteness, let J be the circle having the segment $[\mu_1, \mu_2]$ for diameter.

In the sequel, z will denote a variable point of J. Since

$$R_z = (A - zI)^{-1}$$

transforms the space $\mathfrak{H}$ into $\mathfrak{D}_A$, $(A_n - A)R_z$ is defined everywhere and it follows from (19) that

(20) $$\|(A_n - A)R_z f\| \leq \eta_n(\|R_z f\| + \|AR_z f\|) \leq c\eta_n \|f\|,$$

where c denotes the maximum of $\|R_z\| + \|AR_z\|$ on J. We have

$$c \leq \frac{1}{d} + \left(1 + \frac{\varrho}{d}\right),$$

where d is the minimum distance from J to the spectrum of A and ϱ is the maximum distance from J to the point 0.[20] From (20) it follows that

$$A_n - zI = (A - zI) + (A_n - A) = [I + (A_n - A)R_z]\,(A - zI)$$

also has a bounded inverse as soon as $c\eta_n < 1$, namely

$$R_{nz} = (A_n - zI)^{-1} = R_z[I - (A - A_n)R_z]^{-1} = R_z \sum_{\nu=0}^{\infty} [(A - A_n)R_z]^\nu, \tag{21}$$

and we have

$$\begin{aligned} \|R_{nz} - R_z\| &= \|R_z \sum_{\nu=1}^{\infty} [(A - A_n)R_z]^\nu\| \le \frac{1}{d} \sum_{\nu=1}^{\infty} (c\eta_n)^\nu = \\ &= \frac{1}{d}\,\frac{c\eta_n}{1 - c\eta_n} = \varkappa_n \to 0. \end{aligned} \tag{22}$$

Since the transformations R_z and R_{nz} are continuous functions of the parameter z,[21] the integrals

$$P(J) = -\frac{1}{2\pi i} \int_J R_z\,dz, \quad P_n(J) = -\frac{1}{2\pi i} \int_J R_{nz}\,dz$$

have an obvious meaning. Denoting the perimeter of J by $|J|$, we have

$$\|P_n(J) - P(J)\| \le \frac{1}{2\pi} \int_J \|R_{nz} - R_z\|\,|dz| \le \frac{|J|}{2\pi}\,\varkappa_n \to 0.$$

Now

$$P(J) = -\frac{1}{2\pi i} \int_J dz \int_{-\infty}^{\infty} \frac{1}{\lambda - z}\,dE_\lambda = \int_{-\infty}^{\infty} \left(-\frac{1}{2\pi i} \int_J \frac{1}{\lambda - z}\,dz \right) dE_\lambda,$$

and since the expression in parentheses is equal to 1 for λ interior to J and to 0 for λ exterior to J, we have $P(J) = E_{\mu_2} - E_{\mu_1} = E(\delta)$. For the same reason, $P_n(J) = E_n(\delta)$. Therefore

$$E_n(\delta) \Rightarrow E(\delta) \text{ when } n \to \infty.$$

Moreover, writing

$$A(\delta) = AE(\delta) \text{ and } A_n(\delta) = A_n E_n(\delta),$$

[20] In fact, $\|R_z\| = \max_\lambda \left| \frac{1}{\lambda - z} \right|$ and $\|AR_z\| = \max_\lambda \left| \frac{\lambda}{\lambda - z} \right| = \max_\lambda \left| 1 + \frac{z}{\lambda - z} \right|$, where λ runs through the spectrum of A.

[21] This follows from the relation $R_z - R_\zeta = (z - \zeta) R_z R_\zeta$, which is an immediate consequence of the identity $\frac{1}{\lambda - z} - \frac{1}{\lambda - \zeta} = \frac{z - \zeta}{(\lambda - z)(\lambda - \zeta)}$, and from the analogous relation for R_{nz}, keeping in mind that on J, $\|R_z\| \le \frac{1}{d}$ and, by (21), $\|R_{nz}\| \le \le \frac{1}{d}\,\frac{1}{1 - c\eta_n}$.

and inasmuch as $AR_z = (A - zI + zI)R_z = I + zR_z$, we have

$$A(\delta) = -\frac{1}{2\pi i}\int_J AR_z\,dz = -\frac{1}{2\pi i}\int_J (I + zR_z)dz,$$

$$A_n(\delta) = -\frac{1}{2\pi i}\int_J (I + zR_{nz})dz,$$

$$\|A_n(\delta) - A(\delta)\| = \left\| -\frac{1}{2\pi i}\int_J z(R_{nz} - R_z)dz \right\| \leq \frac{\varrho|J|}{2\pi}\,\varkappa_n \to 0.$$

We summarize:

THEOREM. *Let $\{A_n\}$ be a sequence of self-adjoint transformations all having the same domain $\mathfrak{D}$ and assume that $\{A_n\}$ tends "relatively uniformly" in $\mathfrak{D}$ to the self-adjoint transformation A, that is, in such a way that*

$$\sup \frac{\|(A_n - A)f\|}{\|f\| + \|Af\|} \to 0 \text{ when } n \to \infty$$

(this occurs, in particular, if the transformations A_n are bounded and tend uniformly to the transformation A). Denoting the corresponding spectral families by $\{E_{n\lambda}\}$ and $\{E_\lambda\}$, we have

$$E_n(\delta) \Rightarrow E(\delta) \text{ and } A_nE_n(\delta) \Rightarrow AE(\delta)$$

for every interval $\delta = (\mu_1, \mu_2)$ whose extremities do not belong to the spectrum of A.[22]

In particular, consider the case where the interval δ *contains only a single point λ_0 of the spectrum of A*, which is a characteristic value *of finite multiplicity m*. Since $E(\delta)$ is the projection onto the corresponding characteristic subspace, it is of rank m. As soon as

$$\|E_n(\delta) - E(\delta)\| < 1,$$

[22] The above proof follows the method of Sz.-Nagy [**4**]. The theorem was stated by F. Rellich [**3**] (Note 2); the proof which he gave, as well as that of Sz.-Nagy [*] (pp. 57–58), deduce the theorem from the following more general proposition; it is a proposition the proofs of which in the places cited, however, include only the case of bounded transformations A_n and A:

If the self-adjoint transformations A_n, all having the same domain, tend to the self-adjoint transformation A "relatively uniformly," and if $\{E_{n\lambda}\}$, $\{E_\lambda\}$ denote the corresponding spectral families, then $E_{n\lambda} \Rightarrow E_\lambda$ at every point λ which does not belong to the spectrum of the limit-transformation A.

In the general case, which includes unbounded transformations also, this theorem was proved only very recently, by E. Heinz [**1**]. His proof is based on the formula

$$E_n\lambda = \tfrac{1}{2}I - \frac{1}{2\pi i}\int_{\lambda - i\infty}^{\lambda + i\infty} R_{nz}\,dz$$

and requires a rather delicate analysis.

the projection $E_n(\delta)$ will therefore also be of rank m (cf. Sec. 105). This means that for n sufficiently large, the part of the spectrum of A_n contained in the interval δ consists of characteristic values of total multiplicity m. Since we could have chosen for δ an arbitrarily small interval about the point λ_0, the characteristic values of A_n in question converge to λ_0 as n goes to infinity.

136. Analytic Perturbations

It is clear that everything that was said in the preceding section about the sequences $\{A_n\}$ can be formulated as well for families $\{A(\varepsilon)\}$, which depend on a real parameter ε, such that $A(\varepsilon) \to A$ for $\varepsilon \to 0$. However, new problems arise when $A(\varepsilon)$ is a regular analytic function of ε in the neighborhood of 0; namely, we can then ask whether the spectrum also varies, in a sense we shall make precise, in a regular analytic manner as a function of ε. This problem is of great practical importance since it is at the foundation of the "calculus of perturbations" which is often used in wave mechanics.[23]

For definiteness, we assume that $A(\varepsilon)$ is represented by the entire series

$$A + \varepsilon A^{(1)} + \varepsilon^2 A^{(2)} + \dots \tag{23}$$

whose coefficients $A^{(k)}$ $(k = 1, 2, \dots)$ are symmetric transformations having the same domain $\mathfrak{D}$ as the "non-perturbed" *self-adjoint* transformation $A = A(0)$. Moreover, we assume that

$$\|A^{(k)}f\| \leq \frac{M}{r^{k-1}} (\|f\| + \|Af\|) \qquad (k = 1, 2, \dots), \tag{24}$$

which assures the convergence of (23) for $|\varepsilon| < r$; the sum $A(\varepsilon)$ is clearly a symmetric transformation.

Let $\delta = (\mu_1, \mu_2)$ be an interval supporting an isolated part of the spectrum of A. Fixing the curve J and the corresponding quantities d, ϱ, c as in the preceding section, we have, for z varying on J,

$$\|R_z\| \leq \frac{1}{d}$$

and, by (24),

$$\|A^{(k)} R_z\| \leq \frac{Mc}{r^{k-1}}.$$

[23] A rigorous mathematical theory of analytic perturbations was proposed for the first time by RELLICH [**3**] (Notes 1, 3, 4, 5). The simplified and generalized method we are going to present is due to SZ.-NAGY [**4**]. This method, slightly modified, applies, at least partially, even to *closed* transformations of general type, not necessarily self-adjoint; cf. SZ.-NAGY [**8**].

From this, we deduce that the coefficients of the series expansion

(25) $$R_z(\varepsilon) = [A(\varepsilon) - zI]^{-1} = R_z \sum_{\nu=0}^{\infty} [- \sum_{k=1}^{\infty} \varepsilon^k A^{(k)} R_z]^\nu,$$

which is analogous to (21), are majorized in norm by those of the numerical series

$$\frac{1}{d} \sum_{\nu=0}^{\infty} \left[\sum_{k=1}^{\infty} \varepsilon^k \frac{Mc}{r^{k-1}} \right]^\nu.$$

Now this series converges for

$$|\varepsilon| < \frac{r}{1 + Mcr},$$

and for these values of ε it can be rearranged into the entire series

$$\frac{1}{d} + \frac{Mc}{d} \sum_{n=1}^{\infty} \varepsilon^n \left(\frac{1}{r} + Mc \right)^{n-1}$$

For these values of ε, our calculation (25) is therefore justified; furthermore, the last member can also be rearranged into an entire series

$$R_z(\varepsilon) = R_z + \varepsilon R_z^{(1)} + \varepsilon^2 R_z^{(2)} + \ldots,$$

and we have

$$\|R_z^{(k)}\| \leq \frac{Mc}{d} \left(\frac{1}{r} + Mc \right)^{k-1} \qquad (k = 1, 2, \ldots).$$

In particular, $R_{\mu_1}(\varepsilon) = [A(\varepsilon) - \mu_1 I]^{-1}$ is a bounded transformation, and since it is symmetric, its inverse $A(\varepsilon) - \mu_1 I$, and thus also $A(\varepsilon)$, are *self-adjoint*. Denoting by $\{E_\lambda\}$ the spectral family of $A = A(0)$ and by $\{E_\lambda(\varepsilon)\}$ that of $A(\varepsilon)$, and setting

$$E(\delta) = E_{\mu_2} - E_{\mu_1}, \quad E(\delta; \varepsilon) = E_{\mu_2}(\varepsilon) - E_{\mu_1}(\varepsilon),$$

and

$$A(\delta; \varepsilon) = A(\varepsilon) E(\delta; \varepsilon),$$

we have

$$E(\delta; \varepsilon) = - \frac{1}{2\pi i} \int_J R_z(\varepsilon) dz = E(\delta) + \varepsilon E^{(1)} + \varepsilon^2 E^{(2)} + \ldots$$ [24]

and

$$A(\delta; \varepsilon) = - \frac{1}{2\pi i} \int_J (I + z R_z(\varepsilon)) dz = A E(\delta) + \varepsilon B^{(1)} + \varepsilon^2 B^{(2)} + \ldots,$$

[24] HEINZ [**1**] has just proved that for every point λ not belonging to the spectrum of A, $E_\lambda(\varepsilon)$ will be, for $|\varepsilon|$ sufficiently small, an entire series in ε. This proposition, whose proof is delicate, obviously implies that our $E(\delta; \varepsilon)$ is an entire series in ε.

where

$$E^{(k)} = -\frac{1}{2\pi i}\int_J R_z^{(k)}\,dz, \quad B^{(k)} = -\frac{1}{2\pi i}\int_J zR_z^{(k)}\,dz \quad (k = 1, 2, \ldots).$$

The series for $E(\delta;\varepsilon)$ and for $A(\delta;\varepsilon)$ converge in norm; their coefficients can be evaluated with the aid of the inequalities just obtained for the $R_z^{(k)}$.

*

Suppose, in particular, that δ contains *a single point* $\lambda^{(0)}$ of the spectrum of A and that this is *a simple characteristic value*; let $\varphi^{(0)}$ be a corresponding normed characteristic element. For z sufficiently small, in particular, as soon as

$$\|E(\delta;\varepsilon) - E(\delta)\| < 1,$$

$E(\delta;\varepsilon)$ is the projection onto a subspace of dimension 1; therefore δ contains a single point $\lambda(\varepsilon)$ of the spectrum of $A(\varepsilon)$, which is a simple characteristic value. $\psi(\varepsilon) = E(\delta, \varepsilon)\varphi^{(0)}$ is then a corresponding characteristic element,

$$A(\varepsilon)\psi(\varepsilon) = \lambda(\varepsilon)\psi(\varepsilon).$$

From this it follows that

$$\lambda(\varepsilon) = \frac{(A(\delta;\varepsilon)\varphi^{(0)}, \varphi^{(0)})}{(E(\delta;\varepsilon)\varphi^{(0)}, \varphi^{(0)})} = \frac{\lambda^{(0)} + \varepsilon(B^{(1)}\varphi^{(0)}, \varphi^{(0)}) + \varepsilon^2(B^{(2)}\varphi^{(0)}, \varphi^{(0)}) + \ldots}{1 + \varepsilon\,(E^{(1)}\varphi^{(0)}, \varphi^{(0)}) + \varepsilon^2(E^{(2)}\varphi^{(0)}, \varphi^{(0)}) + \ldots},$$

and hence

(26) $$\lambda(\varepsilon) = \lambda^0 + \varepsilon\lambda^{(1)} + \varepsilon^2\lambda^{(2)} + \ldots$$

for sufficiently small $|\varepsilon|$. Since

$$\|\psi(\varepsilon)\|^2 = (E(\delta;\varepsilon)\varphi^{(0)}, \varphi^{(0)})$$

is a regular function at the point $\varepsilon = 0$ and such that $\|\psi(0)\|^2 = 1$,

$$\varphi(\varepsilon) = \frac{\psi(\varepsilon)}{\|\psi(\varepsilon)\|}$$

will also be a regular function at the same point:

(27) $$\varphi(\varepsilon) = \varphi^{(0)} + \varepsilon\varphi^{(1)} + \varepsilon^2\varphi^{(2)} + \ldots,$$

the coefficients being elements of the space $\mathfrak{H}$. $\varphi(\varepsilon)$ is a *normalized* characteristic element; moreover,

$$(\varphi(\varepsilon), \varphi^{(0)}) = \frac{(E(\delta;\varepsilon)\varphi^{(0)}, \varphi^{(0)})}{\|\psi(\varepsilon)\|} \geq 0,$$

which means that $\varphi(\varepsilon)$ is also normalized with respect to its "phase."

The *existence* of the developments (26) and (27) being established, we calculate the coefficients $\lambda^{(k)}$, $\varphi^{(k)}$ more and more closely with the aid of the system of equations which are obtained by comparing the coefficients of the developments of the two members of the equations

$$A(\varepsilon)\varphi(\varepsilon) = \lambda(\varepsilon)\varphi(\varepsilon), \quad (\varphi(\varepsilon), \varphi(\varepsilon)) = 1,$$

using the fact that $(\varphi(\varepsilon), \varphi^{(0)}) \geq 0$ implies that the quantities $(\varphi^{(k)}, \varphi^{(0)})$ are real. In this manner we obtain, in particular, that

$$\lambda^{(1)} = (A^{(1)}\varphi^{(0)}, \varphi^{(0)}).$$

*

Let us now pass to the more general case where the only point of the spectrum of A which is contained in δ is a characteristic value of finite multiplicity m. We shall prove the following theorem:

THEOREM. *If the interval δ contains only a single point $\lambda^{(0)}$ of the spectrum of the transformation A, which is a characteristic value of finite multiplicity m, then for $|\varepsilon|$ sufficiently small there exist m real values*

$$\lambda_i(\varepsilon) = \lambda^{(0)} + \varepsilon\lambda_i^{(0)} + \varepsilon^2\lambda_i^{(2)} + \ldots \qquad (i = 1, 2, \ldots, m),$$

and m elements of $\mathfrak{H}$

$$\varphi_i(\varepsilon) = \varphi_i^{(0)} + \varepsilon\varphi_i^{(1)} + \varepsilon^2\varphi_i^{(2)} + \ldots \qquad (i = 1, 2, \ldots, m),$$

such that the spectrum of A in δ consists of the characteristic values $\lambda_i(\varepsilon)$, each counted as many times as its multiplicity indicates, and that the $\varphi_i(\varepsilon)$ form an orthonormal system of corresponding characteristic elements.

Since the case $m = 1$ has already been established, we can argue by induction with respect to m. We shall show that if the proposition is true for multiplicities less than m, it is also true for the multiplicity m.

We write, for brevity, $E(\delta) = P$ and $E(\delta; \varepsilon) = P(\varepsilon)$; we call the corresponding subspaces $\mathfrak{M}$ and $\mathfrak{M}(\varepsilon)$. As soon as $\|P(\varepsilon) - P\| < 1$, these subspaces have the same dimension (equal to m) and, as we have seen in Sec. 105, the transformation

$$U(\varepsilon) = P(\varepsilon)\,[I + P(P(\varepsilon) - P)P]^{-\frac{1}{2}}P$$

maps $\mathfrak{M}$ onto $\mathfrak{M}(\varepsilon)$ isometrically, the inverse transformation from $\mathfrak{M}(\varepsilon)$ to $\mathfrak{M}$ being furnished by $U^*(\varepsilon) = [U(\varepsilon)]^*$. As transformations of the entire space $\mathfrak{H}$, $U(\varepsilon)$ and $U^*(\varepsilon)$ are "partially isometric" (cf. Sec. 110); we have the relations

$$U^*(\varepsilon)U(\varepsilon) = P, \;\; U(\varepsilon)U^*(\varepsilon) = P(\varepsilon).$$

For small values of ε, $U(\varepsilon)$ and $U^*(\varepsilon)$ depend on ε in a regular analytic manner,

$$U(\varepsilon) = P + \varepsilon U^{(1)} + \varepsilon^2 U^{(2)} + \ldots, \quad U^*(\varepsilon) = P + \varepsilon U^{(1)*} + \varepsilon^2 U^{(2)*} + \ldots;$$

to show this, we develop $[I + P(P(\varepsilon) - P)P]^{-\frac{1}{2}}$ by the binomial formula, introduce the entire series of $P(\varepsilon)$ and rearrange according to powers of ε, then justify all the formal calculations by constructing convergent majorizing series.

Let us now consider the bounded self-adjoint transformation

$$C(\varepsilon) = U^*(\varepsilon)A(\delta; \varepsilon)U(\varepsilon) = U^*(\varepsilon)A(\varepsilon)P(\varepsilon)U(\varepsilon) = U^*(\varepsilon)A(\varepsilon)U(\varepsilon);$$

it is also a regular function of ε:

$$C(\varepsilon) = C^{(0)} + \varepsilon C^{(1)} + \varepsilon^2 C^{(2)} + \dots;$$

$C(\varepsilon)$ transforms $\mathfrak{M}$ into itself and annihilates all the elements orthogonal to $\mathfrak{M}$. The same is also true, consequently, for the coefficients $C^{(k)}$. We have in particular

$$C^{(0)} = U^*(0)A(0)U(0) = PAP = \lambda^{(0)}P.$$

It might happen that the other coefficients $C^{(k)}$ are also numerical multiples of P,

$$C^{(k)} = \lambda^{(k)}P \qquad (k = 0, 1, \dots);$$

$C(\varepsilon)$ will then be itself a numerical multiple of P; we shall have $C(\varepsilon) = \lambda(\varepsilon)P$, with

$$\lambda(\varepsilon) = \lambda^{(0)} + \varepsilon\lambda^{(1)} + \varepsilon^2\lambda^{(2)} + \dots.$$

Since, in general,

$$(28) \qquad U(\varepsilon)C(\varepsilon) = U(\varepsilon)U^*(\varepsilon)A(\varepsilon)U(\varepsilon) = P(\varepsilon)A(\varepsilon)U(\varepsilon) = A(\varepsilon)U(\varepsilon)$$

we shall have, in this case,

$$A(\varepsilon)U(\varepsilon) = \lambda(\varepsilon)U(\varepsilon)P = \lambda(\varepsilon)U(\varepsilon).$$

Let us choose, arbitrarily, a complete orthonormal system $\{\varphi_i^{(0)}\}$ in $\mathfrak{M}$ ($i = 1, 2, \dots, m$); by virtue of the isometric property of $U(\varepsilon)$, the elements $\varphi_i(\varepsilon) = U(\varepsilon)\varphi_i^{(0)}$ then form a complete orthonormal system in $\mathfrak{M}(\varepsilon)$, and it follows from (28) that these elements are all characteristic elements of $A(\varepsilon)$ belonging to the same characteristic value $\lambda(\varepsilon)$. Therefore in this case the characteristic value $\lambda^{(0)}$ of multiplicity m does not split with the perturbation; it is only displaced to $\lambda(\varepsilon)$. The characteristic value $\lambda(\varepsilon)$ and the corresponding characteristic elements $\varphi_i(\varepsilon)$ are obviously regular analytic functions of ε.

Now consider the case where there are coefficients $C^{(k)}$ which are not multiples of P; let $C^{(s)}$ be the first of them, $s \geq 1$. For $k < s$ we now have $C^{(k)} = \lambda^{(k)}P$. Set

$$D(\varepsilon) = C^{(s)} + \varepsilon C^{(s+1)} + \dots.$$

$D(0) = C^{(s)}$, considered as a self-adjoint transformation of the subspace $\mathfrak{M}$ of dimension m into itself, is not a multiple of the identity; consequently, it has at least two distinct characteristic values. Denote its distinct characteristic values by $\varkappa_1, \varkappa_2, \ldots, \varkappa_n$; let their multiplicities be respectively equal to $m_1, m_2, \ldots, m_n$ $(m_1 + m_2 + \ldots + m_n = m)$. Since each characteristic value $\varkappa$ is obviously isolated and of multiplicity less than m, by the hypothesis we can apply the theorem to each of them. From this follows the existence of m real values $\varkappa_i(\varepsilon)$ and of m orthonormal elements $\psi_i(\varepsilon)$ of $\mathfrak{M}$ which are regular functions of ε in a neighborhood of $\varepsilon = 0$ and such that

$$\varkappa_i^{(0)} = \varkappa_1 \text{ for } i \leq m_1, \quad \varkappa_i^{(0)} = \varkappa_2 \text{ for } m_1 < i \leq m_1 + m_2, \text{ etc.,}$$

and that

$$D(\varepsilon)\psi_i(\varepsilon) = \varkappa_i(\varepsilon)\psi_i(\varepsilon) \qquad (i = 1, 2, \ldots, m).$$

It follows that

$$C(\varepsilon)\psi_i(\varepsilon) = [\sum_{k=0}^{s-1} \varepsilon^k \lambda^{(k)} P + \varepsilon^s D(\varepsilon)]\psi_i(\varepsilon) = \sum_{k=0}^{s-1} \varepsilon^k \lambda^{(k)} \psi_i(\varepsilon) + \varepsilon^s \varkappa_i(\varepsilon)\psi_i(\varepsilon),$$

and therefore, setting

$$\lambda_i(\varepsilon) = \lambda^{(0)} + \varepsilon\lambda^{(1)} + \ldots + \varepsilon^{s-1}\lambda^{(s-1)} + \varepsilon^s \varkappa_i(\varepsilon),$$

we have

$$C(\varepsilon)\psi_i(\varepsilon) = \lambda_i(\varepsilon)\psi_i(\varepsilon);$$

consequently, by (28),

$$A(\varepsilon)U(\varepsilon)\psi_i(\varepsilon) = \lambda_i(\varepsilon)U(\varepsilon)\psi_i(\varepsilon).$$

The values $\lambda_i(\varepsilon)$ and the elements $\varphi_i(\varepsilon) = U(\varepsilon)\psi_i(\varepsilon)$ therefore satisfy the theorem; this completes the proof.

*

It is clear that a characteristic value of *infinite* multiplicity can very well transform into a continuous spectrum. For example, the transformation

$$A(\varepsilon)f(x) = \varepsilon x f(x)$$

of the space $L^2(-1, 1)$ has for $\varepsilon = 0$ the isolated characteristic value 0, and for $\varepsilon > 0$ the continuous spectrum fills the closed interval $-\varepsilon \leq \lambda \leq \varepsilon$.

On the other hand, we shall observe that this reasoning applies also to transformations which depend on *several perturbation parameters* in the form of an entire series, at least if it is a question of the perturbation of a *simple* isolated characteristic value. Even for a multiple characteristic value, our proof carries over to the case of several parameters up to the point where we constructed the transformation $C(\varepsilon)$, but at this point we run into an obstacle.

In the case of two parameters, we have, namely,

$$C(\varepsilon, \eta) = C^{(0,0)} + \varepsilon C^{(1,0)} + \eta C^{(0,1)} + \varepsilon^2 C^{(2,0)} + \varepsilon\eta C^{(1,1)} + \eta^2 C^{(0,2)} + \ldots,$$

but when there are coefficients which are not numerical multiples of the projection P we can not form, in general, the analogue of the transformation $D(\varepsilon)$. In this case the theorem is even false, as we see from the following example: the transformation of two-dimensional space which has the matrix

$$\varepsilon \begin{pmatrix} 2 & 1 \\ 1 & 0 \end{pmatrix} + \eta \begin{pmatrix} 0 & 1 \\ 1 & 2 \end{pmatrix}$$

has for characteristic values

$$\varepsilon + \eta \pm \sqrt{2}\sqrt{\varepsilon^2 + \eta^2};$$

as functions of ε, η, these are not regular at the origin $\varepsilon = \eta = 0$.

CHAPTER X

GROUPS AND SEMI-GROUPS OF TRANSFORMATIONS

UNITARY TRANSFORMATIONS

137. Stone's Theorem

Let U be a unitary transformation in Hilbert space, and let

$$U = \int_0^1 e^{2\pi i\varphi} dE_\varphi$$

be its spectral decomposition (with $E_0 = 0$; cf. Sec. 109). Then we also have

$$U^n = \int_0^1 e^{2\pi i n\varphi} dE_\varphi$$

for every positive or negative n. This suggests defining U^t for every real value of t by the integral

$$U^t = \int_0^1 e^{2\pi i t\varphi} dE_\varphi. \tag{1}$$

We thus obtain a family $\{U^t\}$ of unitary transformations such that

$$U^o = I, \quad U^t U^s = U^{t+s}$$

and

$$U^s \Rightarrow U^t \text{ when } s \to t.$$

Let us consider the converse problem. We assume given a family $\{T_t\}$ $(-\infty < t < \infty)$ of bounded linear transformations in Hilbert space possessing the properties

$$T_0 = I, \ T_t T_s = T_{t+s};$$

we shall call it a *one-parameter group of transformations*. We assume, moreover, that T_t depends continuously on the parameter t, that is, that if $s \to t$, T_s tends to T_t in one sense or another: uniformly, strongly, or weakly. Furthermore, weak continuity means that for every f and g, $(T_t f, g)$ is a continuous real-valued function of t. The problem is to determine whether every continuous one-parameter group of *unitary* transformations $\{U_t\}$ admits an integral representation analogous to (1).

We owe to M. H. STONE the theorem which asserts that, even if we assume only *weak* continuity, this representation is always possible, under the condition that we do not restrict ourselves to spectral families on the finite segment [0, 1]. Moreover, here weak continuity implies strong continuity, by virtue of the theorem proved in Sec. 29, according to which

$$f_n \rightharpoonup f \text{ and } \|f_n\| \to \|f\| \text{ imply } f_n \to f.$$

First, we consider the particular case where U_t is a *periodic* function of t, say of period 1, that is, $U_1 = U_0 = I$, from which it follows that

$$U_{t+1} = U_t U_1 = U_t I = U_t.$$

We shall show that in this case U_t admits a sort of Fourier development

$$U_t = \sum_{n=-\infty}^{\infty} e^{2\pi i n t} P_n \tag{2}$$

whose coefficients P_n are mutually orthogonal projections with sum I which are calculated by means of the familiar formulas

$$P_n = \int_a^{a+1} e^{-2\pi i n t} U_t \, dt, \tag{3}$$

understood in the sense that

$$(P_n f, g) = \int_a^{a+1} e^{-2\pi i n t} (U_t f, g) dt. \tag{4}$$

Since the second member of equation (4) is clearly a bounded bilinear form, this equation uniquely determines the bounded linear transformation P_n (cf. Sec. 84). P_n is *symmetric* since, inasmuch as $U_t^* = U_{-t}$, we have

$$P_n^* = \int_0^1 e^{2\pi i n t} U_{-t} \, dt = -\int_0^{-1} e^{-2\pi i n \tau} U_\tau \, d\tau = \int_{-1}^{0} = P_n.$$

Furthermore, we easily see that

$$U_s P_n = \int_0^1 e^{-2\pi i n t} U_{s+t} dt = e^{2\pi i n s} \int_s^{s+1} e^{-2\pi i n \tau} U_\tau \, d\tau = e^{2\pi i n s} P_n, \tag{5}$$

and that consequently

$$P_m P_n = \int_0^1 e^{-2\pi i m t} U_t P_n dt = \int_0^1 e^{-2\pi i m t} e^{2\pi i n t} P_n dt = \begin{cases} P_n \text{ for } m = n, \\ 0 \text{ for } m \neq n. \end{cases}$$

This proves that the P_n are mutually orthogonal projections.

There remains to prove that the sum

$$P = \sum_{-\infty}^{\infty} P_n$$

equals I and that the development (2) is valid. In any case, P and $Q = I - P$

are projections, and Q is clearly orthogonal to all the projections P_n, hence we have

$$\int_0^1 e^{-2\pi int}(U_tQf, g)dt = (P_nQf, g) = 0 \quad (n = 0, \pm 1, \pm 2, \ldots)$$

for all elements f and g. The *continuous* real-valued function (U_tQf, g) therefore has all its Fourier coefficients equal to 0, which implies it vanishes identically. It follows that

$$U_tQ = O,\ Q = U_{-t}U_tQ = O,\ P = I - Q = I.$$

Finally, development (2) follows very simply from the equation $I = \sum_{-\infty}^{\infty} P_n$ if we multiply the two members by U_t and use relations (5).

Development (2) can also be put in integral form:

$$U_t = \int_{-\infty}^{\infty} e^{i\lambda t}dE_\lambda,$$

where $\{E_\lambda\}$ is the spectral family defined by

$$E_\lambda = \sum_{n \leqq \frac{\lambda}{2\pi}} P_n.$$

It is clear from their definition (3) or (4) that the projections P_n possess the property

$$P_n \smile\smile \{U_t\}.$$

Let us pass now to the general case in which $U_1 \neq I$. We set

$$V_t = U_tU_1^{-t},$$

the powers U_1^t being defined in the sense (1), that is,

$$U_1^t = \int_0^1 e^{2\pi it\varphi}dF_\varphi.$$

U_t, being permutable with U_1, and therefore also with F_φ, is permutable with U_1^s, from which follows the property

$$V_sV_t = V_{s+t}.$$

Furthermore, $U_s \rightharpoonup U_t$ and $U_1^{-s} \Rightarrow U_1^{-t}$ imply that

$$V_s \rightharpoonup V_t$$

(cf. Sec. 84). Finally, we have $V_1 = I$. But this assures, by what precedes, that V_t admits the development

$$V_t = \sum_{-\infty}^{\infty} e^{2\pi int}P_n,$$

from which it finally follows that

$$U_t = \sum_{n=-\infty}^{\infty} e^{2\pi int} P_n U_1^t = \sum_{n=-\infty}^{\infty} e^{2\pi int} P_n \int_0^1 e^{2\pi it\varphi} dF_\varphi = \sum_{n=-\infty}^{\infty} \int_n^{n+1} e^{2\pi i\mu t} dP_n F_{\mu-n}$$

or

$$U_t = \int_{-\infty}^{\infty} e^{i\lambda t} dE_\lambda,$$

the spectral family $\{E_\lambda\}$ being defined by the formula

$$E_{2\pi\mu} = \sum_{n<[\mu]} P_n + P_{[\mu]} F_{\mu-[\mu]},$$

where $[\mu]$ denotes the greatest integer $\leq \mu$. The relations

$$F_\varphi \smile\smile U_1, \; P_n \smile\smile \{V_t\}, \; V_s \smile\smile \{U_t\}$$

imply that

$$E_\lambda \smile\smile \{U_t\}.$$

Hence we have arrived at

STONE'S THEOREM.[1] *Every one-parameter group $\{U_t\}$ $(-\infty < t < \infty)$ of unitary transformations for which $(U_t f, g)$ is a continuous function of t, for all elements f and g, admits the spectral representation*

$$U_t = \int_{-\infty}^{\infty} e^{i\lambda t} dE_\lambda, \tag{6}$$

where $\{E_\lambda\}$ is a spectral family such that $E_\lambda \smile\smile \{U_t\}$.

Furthermore, E_λ is uniquely determined. In fact, it follows from (6) that if

$$\pi_n(\lambda) = \sum_k a_{nk} e^{i\lambda t_k} \tag{7}$$

is a bounded sequence of generalized trigonometric polynomials tending to the characteristic function of the half-line $\lambda \leq \mu$,[2] then the sequence of

[1] Cf. STONE [**1**] (Note 3) and [**2**]; VON NEUMANN [**5**]; BOCHNER [**1**]; RIESZ [**14**]; SZ.-NAGY [**1**]; YOSIDA [**2**]. The above proof is due to SZ.-NAGY.

[2] The existence of such a sequence can be proved in the following manner. We take first a sequence $\{f_n(\lambda)\}$ of periodic functions (of period which increases indefinitely with n), which are continuous, uniformly bounded, and tending toward the required limit; let for example $f_n(\lambda)$ be the function of period $2n$, defined in the interval $\mu - n \leq \lambda \leq \mu + n$ as follows:

$$f(\mu - n) = 0, \; f\left(\mu - n + \frac{1}{n}\right) = 1, \; f(\mu) = 1, \; f\left(\mu + \frac{1}{n}\right) = 0, \; f(\mu + n) = 0,$$

and $f_n(\lambda)$ is linear between two indicated consecutive points.

We then approximate $f_n(\lambda)$ uniformly to within $\frac{1}{n}$ by a trigonometric polynomial of the same period (Weierstrass's approximation theorem). The sequence of trigonometric polynomials obtained clearly satisfies our requirements.

corresponding transformations

$$\Pi_n = \sum_k a_{nk} U_{t_k}$$

tends to E_λ. This again brings out that $E_\lambda \smile\smile \{U_t\}$.

We add that *in the case where the space* $\mathfrak{H}$ *is separable, the theorem holds even under the weaker assumption that the functions* $(U_t f, g)$ *of* t *be measurable in the Lebesgue sense for any* f *and* g.

In fact, in the periodic case ($U_1 = I$) the projections P_n can be defined by the same integrals (4), which must now be taken in the Lebesgue sense. Their properties are established in the same way, except for one, that concerning the sum ΣP_n. Since the function $(U_t Qf, g)$ is not assumed continuous, we can conclude from the fact that its Fourier coefficients are zero only that $(U_t Qf, g) = 0$ *almost everywhere*. But if f is fixed and if g runs through a sequence which is everywhere dense in $\mathfrak{H}$, the sum of exceptional sets will also be of measure zero, hence there will certainly be values of t such that $(U_t Qf, g) = 0$ for all these g, consequently $U_t Qf = 0$, $Qf = U_t^{-1} U_t Qf = 0$. Since this holds for all f, it follows that $Q = 0$.

The general case ($U_1 \neq I$) reduces to the periodic case as above. The only thing to be proved is that $(V_t f, g)$ is also a measurable function of t. But this follows from the development

$$(V_t f, g) = (U_t f, U_1^t g) = \sum_n (U_t f, \varphi_n)(\varphi_n, U_1^t g),$$

where $\{\varphi_n\}$ is a complete orthonormal sequence, since the sum of a series with measurable terms is also measurable.

We return to formula (6); it can be written in the form $U_t = e^{itA}$, where A denotes the *self-adjoint* transformation

$$A = \int_{-\infty}^{\infty} \lambda dE_\lambda.$$

This transformation A, or rather the transformation iA, plays the role of an "infinitesimal" tranformation in the sense of the theory of continuous groups: namely, we have the relation

$$iA = \lim_{h\to 0} \frac{U_h - 1}{h}.$$

In fact, if f is an element of the domain of A,

$$\left\| \left[\frac{1}{h}(U_h - I) - iA \right] f \right\|^2 = \int_{-\infty}^{\infty} \left| \frac{1}{h}(e^{i\lambda h} - 1) - i\lambda \right|^2 d\|E_\lambda f\|^2 \to 0$$

when $h \to 0$, because the function under the integral sign tends to 0 and is majorized by a function summable with respect to $\|E_\lambda f\|^2$, namely, by a numerical multiple of λ^2. Conversely, if f is an element for which

$$\lim_{h\to 0} \frac{1}{h} (U_h - I)f$$

exists, then

$$\lim_{h\to\infty} \left\| \frac{1}{h} (U_h - I)f \right\|^2 = \lim_{h\to 0} \int_{-\infty}^{\infty} \left| \frac{1}{h} (e^{i\lambda h} - 1) \right|^2 d\|E_\lambda f\|^2$$

also exists, and since the function under the integral sign has the limit λ^2, it follows from Fatou's theorem (Sec. 20) that λ^2 is summable with respect to $\|E_\lambda f\|^2$, hence that f belongs to the domain of A also. This completes the proof of our proposition.

Therefore Stone's theorem can also be expressed in the following form:

Every continuous one-parameter group of unitary transformations $\{U_t\}$ is generated by an infinitesimal transformation iA, where A is a self-adjoint transformation which is, in general, not bounded:

$$U_t = e^{itA}, \quad iA = \lim_{h\to 0} \frac{1}{h} (U_h - I).$$

If the space is separable, continuity follows from measurability.

138. Another Proof, Based on a Theorem of Bochner

The above proof of Stone's theorem, due to one of the authors and making use of Fourier series, was preceded by proofs which made use (in varying degrees) of the theory of Fourier integrals, cf. footnote [1], page 383. BOCHNER based his proof on the following theorem, the analogue for Fourier integrals of the theorem of trigonometric moments (Sec. 53), which we state here without proof:

BOCHNER'S THEOREM.[3] *In order that the function $p(t)$ $(-\infty < t < \infty)$ admit the representation*

$$p(t) = \int_{-\infty}^{\infty} e^{i\lambda t} dV(\lambda) \tag{8}$$

with a nondecreasing and bounded real function $V(\lambda)$, it is necessary and sufficient that $p(t)$ be continuous and of positive type, that is, that

$$\sum_{\mu,\nu=1}^{m} p(t_\mu - t_\nu)\varrho_\mu\bar{\varrho}_\nu \geq 0,$$

whatever be the positive integer m, the real numbers $t_1, t_2, \ldots, t_m$, and the complex numbers $\varrho_1, \varrho_2, \ldots, \varrho_m$.

[3] BOCHNER [*] (§ 20).

The fact which connects the Stone and Bochner theorems is that the function

$$p_f(t) = (U_t f, f)$$

is of positive type; in fact, we have

$$\sum_{\mu, \nu=1}^{m} (U_{t_\mu - t_\nu} f, f) \varrho_\mu \bar{\varrho}_\nu = (\sum_{\mu=1}^{m} \varrho_\mu U_{t_\mu} f, \sum_{=1}^{m} \varrho_\nu U_{t_\nu} f) \geq 0.$$

Hence, by the Bochner theorem,

$$p_f(t) = \int_{-\infty}^{\infty} e^{i\lambda t} d V_f(\lambda),$$

and since

$$4(U_t f, g) = p_{f+g}(t) - p_{f-g}(t) + i p_{f+ig}(t) - i p_{f-ig}(t),$$

we also have

$$(9) \qquad (U_t f, g) = \int_{-\infty}^{\infty} e^{i\lambda t} dV(f, g; \lambda),$$

where $V(f, g; \lambda)$ is a function of bounded variation. We can normalize the function by requiring that it tend to 0 when λ tends to $-\infty$ and that it be continuous from the right. It is then determined uniquely; in fact, if $\{\pi_n(\lambda)\}$ is the sequence of trigonometric polynomials (7) and if $\{\Pi_n\}$ is the corresponding sequence of transformations, then

$$(\Pi_n f, g) \to V(f, g; \mu).$$

Since the first member of (9) is a bilinear form in f and g, by the uniqueness established the same is true of $V(f, g; \lambda)$; this bilinear form is bounded, since we have on the one hand

$$|V(f, g; \lambda)|^2 \leq V(f, f; \lambda) V(g, g; \lambda),$$

the Schwarz inequality for the quadratic form

$$V(f, f; \lambda) = V_f(\lambda) \geq 0,$$

and on the other hand, since $V_f(\lambda)$ is a nondecreasing function of λ,

$$V_f(\lambda) \leq V_f(\infty) = \int_{-\infty}^{\infty} dV_f(\lambda) = p_f(0) = \|f\|^2.$$

This assures the existence of a bounded linear transformation E_λ such that

$$V(f, g; \lambda) = (E_\lambda f, g)$$

(see Sec. 84). Let us show that $\{E_\lambda\}$ is a spectral family. We again consider sequence (7) tending to the characteristic function of the half-line $\lambda \leq \mu$. It follows immediately from the construction of E_μ that $\Pi_n \rightharpoonup E_\mu$. In order to show that actually $\Pi_n \to E_\mu$, it suffices to show that $\Pi_n - \Pi_m \to O$ as $m, n \to \infty$. This is a consequence of the following relation between an arbitrary trigonometric polynomial $\pi(\lambda) = \Sigma c_k e^{i\lambda t_k}$ and the corresponding transformation $\Pi = \Sigma c_k U_{t_k}$:

$$\|\Pi f\|^2 = (\Pi^*\Pi f, f) = ((\sum_k \bar{c}_k U_{t_k}^*) (\sum_h c_h U_{t_h})f, f) = \sum_k \sum_h \bar{c}_k c_h (U_{t_h - t_k} f, f) =$$

$$= \sum_k \sum_h \bar{c}_k c_h \int_{-\infty}^{\infty} e^{i(t_h - t_k)\lambda} d(E_\lambda f, f) = \int_{-\infty}^{\infty} |\pi(\lambda)|^2 d(E_\lambda f, f).$$

Since the sequences

$$\{\pi_n(\lambda)\}, \quad \{\overline{\pi_n(\lambda)}\}, \quad \text{and} \quad \{[\pi_n(\lambda)]^2\}$$

all have the same limit, we have not only $\Pi_n \to E_\mu$, but also

$$\Pi_n^* \to E_\mu \text{ and } \Pi_n^2 \to E_\mu,$$

from which it follows that

$$E_\mu = E_\mu^* = E_\mu^2,$$

that is, that E_μ is an orthogonal projection. The other properties of a spectral family can be established immediately.

We have thus arrived again at Stone's theorem, at least in the case where each $(U_t f, g)$ is a *continuous* function of t.

If in the case of a separable space $\mathfrak{H}$ we assume only the *measurability* of $(U_t f, g)$ as a function of t, we can use a more precise form of Bochner's theorem asserting that *if the function of positive type $p(t)$ is measurable, it admits the representation* (8) *for almost all values of t.*[4] The presence of the exceptional set of measure zero is eliminated when we are dealing with the functions $p_f(t) = (U_t f, f)$, by virtue of the separability of the space $\mathfrak{H}$ and the group property.

But *continuity* can also be deduced from *measurability* in a more direct manner. We form the transformations

$$T_x = \int_0^x U_s ds$$

with the aid of the numerical integrals

$$(T_x f, g) = \int_0^x (U_s f, g) ds.$$

Since

$$(U_t T_x h, g) = (T_x h, U_{-t} g) = \int_0^x (U_s h, U_{-t} g) ds = \int_0^x (U_{t+s} h, g) ds = \int_t^{t+x} (U_\tau h, g) d\tau,$$

[4] F. Riesz [**14**]; cf. also Hopf [*] (p. 21).

$U_t f$ is a weakly (and hence also strongly) continuous function in t for all elements of the form $f = T_x h$. Now these elements are dense in $\mathfrak{H}$. If they were not, there would be a $g \neq 0$ such that

$$(T_x h, g) = \int_0^x (U_s h, g) ds = 0$$

for all elements h of $\mathfrak{H}$ and all values of x. It would follow that $(U_s h, g) = 0$ for every value of s with the possible exception of a set of measure zero (depending on h). Letting h run through a denumerable set everywhere dense in $\mathfrak{H}$ and forming the union E of the exceptional sets of s, which is also of measure zero, we would have $(h, U_s^* g) = (U_s h, g) = 0$ for all these h and for all values of s not belonging to E, hence at least for one value s_0. We would then have $U_{s_0}^* g = 0$, $g = U_{s_0} U_{s_0}^* g = 0$, in contradiction with the hypothesis that $g \neq 0$. The elements $T_x h$ are therefore certainly dense in $\mathfrak{H}$. But then $U_t f$ is a continuous function of t for all elements f without exception, since if $\{f_n\}$ is a sequence of elements of the form $T_x h$ tending to f, $\{U_t f_n\}$ will be a sequence of continuous functions in t tending to $U_t f$ uniformly in t,

$$\|U_t f - U_t f_n\| = \|f - f_n\| \to 0,$$

which implies that $U_t f$ is also a continuous function of t.

139. Some Applications of Stone's Theorem

Stone's theorem has found its foremost application in statistical mechanics, namely, in "ergodic" theory.

Let Φ be the "phase space" of a holonomic mechanical system. In Φ, each point describes a possible state of the system. We denote by P_t $(-\infty < t < \infty)$ the point which describes the state of the system at the instant t, if at the instant 0 this system is in the state P. Thus we have defined in Φ a continuous motion

$$P \to P_t.$$

It can be proved with the aid of the Hamilton-Jacobi differential equations that this motion preserves measure, and thus integrals, that is, that

$$\int_\Phi f(P) dv = \int_\Phi f(P_t) dv$$

for every continuous function $f(P)$ summable on Φ (Liouville's theorem). If Ω is a constant energy surface in Φ, it is transformed by this motion into itself, and we can introduce an invariant surface measure $d\sigma$ on Ω; we have only to set

$$d\sigma = \frac{dS}{|\text{grad } H|}$$

where dS denotes the element of surface and where H is the Hamiltonian of the system.

It was B. O. KOOPMAN's [**1**] idea, in 1931, to consider, instead of the continuous motion $P \to P_t$ on the surface Ω, the linear functional transformation

$$U_t f(P) = f(P_t)$$

of continuous functions which are square-summable with respect to the invariant measure $d\sigma$. Since $|f(P)|^2$ and $|f(P_t)|^2$ have the same integral on Ω, U_t is isometric in terms of the metric of $L^2(\Omega)$, and admits the inverse U_{-t}. The transformation U_t can therefore be extended to a unitary transformation of the entire space $L^2(\Omega)$. The obvious property of motion

$$P_{t+s} = (P_s)_t$$

implies that

$$U_{t+s} = U_t U_s.$$

Finally, since $f(P_t)$ is a continuous function of (P, t) when $f(P)$ is a continuous function of P, we verify easily that for all $f, g \in L^2(\Omega)$, the function $(U_t f, g)$ is a measurable function of t.

Hence we can apply Stone's theorem: we have

$$U_t = \int_{-\infty}^{\infty} e^{it\lambda} dE_\lambda,$$

from which it follows in particular that

$$M(a, T)f = \int_{-\infty}^{\infty} M(a, T; \lambda) dE_\lambda f,$$

where

$$M(a, T; \lambda) = \frac{1}{T} \int_a^{a+T} e^{it\lambda} dt,$$

$$M(a, T)f = \frac{1}{T} \int_a^{a+T} U_t f dt \text{ (limit of sums of Riemann type).}$$

When $T \to \infty$, $M(a, T; \lambda)$ tends to the function $\delta(\lambda)$, equal to 1 for $\lambda = 0$ and to zero for $\lambda \neq 0$. Since, moreover,

$$|M(a, T; \lambda) - \delta(\lambda)| \leq 1,$$

we obtain from Lebesgue's theorem on term-by-term integration.

$$\|M(a, T)f - E(0)f\|^2 = \int_{-\infty}^{\infty} |M(a, T; \lambda) - \delta(\lambda)|^2 \, d\|E_\lambda f\|^2 \to 0$$

for all elements f of $L^2(\Omega)$. For f continuous we have $M(a, T)f = g_{a,T}$, where

$g_{a,T}(P)$ is the mean value with respect to the time interval $a \leq t \leq a + T$:

$$g_{a,T}(P) = \frac{1}{T} \int_a^{a+T} f(P_t) dt.$$

Therefore $g_{a,T}$ tends in the mean to the element $f^* = E(0)f$ of $L^2(\Omega)$:

$$\int_\Omega |g_{a,T}(P) - f^*(P)|^2 \, d\sigma \to 0 \quad \text{when} \quad T \to \infty.$$

Furthermore we have

$$U_t f^* = U_t E(0) f^* = E(0) f^* = f^*,$$

therefore $f^*(P_t) = f^*(P)$ if $f^*(P)$ is continuous; the same relation is valid in the general case, at least for almost all P, as one sees by approximating $f^*(P)$ with continuous functions. We thus arrive, with J. VON NEUMANN (**6**), at the "statistical ergodic theorem" which asserts that as $T \to \infty$ the mean value $g_{a,T}(P)$ of the function $f(P_t)$ with respect to the time t tends "statistically," that is, in square mean, to a limit $f^*(P)$ which is "almost invariant" with respect to the motion $P \to P_t$.

It is easy to extend this theorem to non-continuous functions $f(P)$ belonging to $L^2(\Omega)$, and also to non-continuous, but "measurable," motions $P \to P_t$ which are measure-preserving, on an arbitrary measurable set Ω, cf. HOPF [*].

Several direct proofs of the statistical ergodic theorem have been proposed, and it has been generalized in various directions; we shall return to this in Secs. 144—146.

Moreover, one encounters such families $U_t = e^{itA}$ not only in classical mechanics, but also in quantum mechanics. The SCHRÖDINGER wave function for a system satisfies the differential equation

$$\frac{h}{2\pi i} \frac{\partial \varphi}{\partial t} = H\varphi,$$

where H is the energy differential operator, operating on the coordinates $x = (x_1, x_2, \ldots, x_n)$ (and not on the time t), and generates a self-adjoint transformation of the space L^2 of functions defined in the configuration space of the system; h is PLANCK's constant. Denoting by $\varphi(x, t)$ the wave function corresponding to the system at the instant t, we can derive from the differential equation the relation

$$\varphi(x, t) = e^{\frac{2\pi i t}{h} H} \varphi(x, 0).^5$$

[5] Cf. VON NEUMANN [*] (in particular, p. 108); see also COOPER [**2**].

140. Unitary Representations of More General Groups

The one-parameter group of unitary transformations appearing in Stone's theorem can be considered as a *representation* (not necessarily isomorphic) of the group of translations $x \to x + t$ of the real line into itself. On the other hand, the function $e^{i\lambda t}$ under the integral sign is, for every fixed value of λ, a *character* of this group, that is, a representation by complex values of modulus 1.

This gives rise to the problem of studying analogous representations, by means of *unitary* transformations, of more general abstract groups G. Generalizations of this sort are possible if G is a locally bicompact topological group, where "locally bicompact" means that every element has a neighborhood on which the Borel covering theorem is valid. A profound study of these groups G was made possible by the important discovery of A. HAAR, affirming the existence on G of a left-invariant measure $\mu(E)$, and of another measure invariant from the right, that is, measures such that

$$\mu(a \cdot E) = \mu(E), \text{ respectively } \mu(E \cdot a) = \mu(E),$$

for all elements a of G; the products $a \cdot E$ and $E \cdot a$ denote, respectively, the sets of elements $a \cdot x$, $x \cdot a$ where x runs through E. In other words, there exist functionals $A_1 f$ and $A_2 f$ defined for all continuous functions $f(x)$ on G which vanish in the exterior of a bicompact part of G; these functionals are additive, homogeneous, and such that

$$A_1 f > 0 \text{ and } A_2 f > 0 \text{ if } f(x) \geqq 0,\ f(x) \not\equiv 0,$$

and moreover they are invariant from the left, respectively the right, that is,

$$A_1 f(a \cdot x) = A_1 f(x), \text{ respectively } A_2 f(x \cdot a) = A_2 f(x),$$

for all elements a of G.[6]

If, in particular, G is *Abelian*, PONTRJAGIN and VAN KAMPEN[7] have shown, using the invariant HAAR measure, that the *continuous characters* of G, that is, the continuous functions $\chi(x)$ on G with complex values of modulus 1 satisfying

$$\chi(x)\chi(y) = \chi(x \cdot y)$$

for all elements x, y of G, form a group Γ with respect to the composition

$$(\chi_1 \cdot \chi_2)(x) = \chi_1(x)\chi_2(x);$$

this group Γ is also locally bicompact in the "weak"—or TYCHONOFF—topology, in which a neighbourhood

$$V(\chi_0) = V(\chi_0; x_1, x_2, \ldots, x_n, \varepsilon)$$

[6] HAAR [**2**]; cf. also BANACH's note at the end of SAKS's book [*], [**]; as well as WEIL [*] (pp. 30—45).

[7] Cf., for example, WEIL [*] (pp. 94—109).

of a "point" χ_0 is defined to be the set of χ satisfying the inequalities

$$|\chi(x_k) - \chi_0(x_k)| < \varepsilon \qquad (k = 1, 2, \ldots, n),$$

where $x_1, x_2, \ldots, x_n$ are given elements of G, finite in number, and ε is an arbitrary positive quantity. Furthermore, the relation between G and Γ is reciprocal, G being equal (or rather isomorphic) to the group of continuous characters of Γ. If G is the group of translations of the real line, Γ is isomorphic and homeomorphic to G, but in general this is not the case.

Using these theorems, A. WEIL[8] extended Bochner's theorem to continuous functions $p(x)$ of positive type which are defined on an arbitrary locally bicompact Abelian group G. The hypothesis that $p(x)$ is of positive type means that

$$\sum_{i,k=1}^{m} p(x_k^{-1} \cdot x_i)\varrho_i \bar{\varrho}_k \geqq 0,$$

for every positive integer m and arbitrary elements x_i of G and complex quantities ϱ_i. WEIL showed that in the group of characters Γ there exists a function $m(A) \geqq 0$, defined for all sets $A \subseteq \Gamma$ measurable (B), denumerably additive, such that $m(\Gamma) < \infty$, and such that

$$p(x) = \int_{\Gamma} \chi(x) m(d\chi);$$

moreover, this measure $m(A)$ is uniquely determined. In the language of linear functionals, this means that there exists one and only one positive linear functional $L(f)$ in the space $C(\Gamma)$ of functions $f(\chi)$ which are bounded and continuous on Γ for which

$$p(x) = L(f_x),$$

where

$$f_x(\chi) = \chi(x).$$

This theorem makes possible the following generalization of Stone's theorem:[9]

THEOREM. *Let $\{U_x\}$ be a (weakly) continuous representation of the locally bicompact Abelian group G by unitary transformations in Hilbert space, that is, satisfying*

$$U_e = I, \; U_x U_y = U_{x \cdot y},$$

where e denotes the unit element of G and where x, y are arbitrary elements of G, and $(U_x f, g)$ is a continuous function of x for all elements f and g of Hilbert space. Then there is one and only one spectral family $\{E(A)\}$, distributed on the group

[8] Cf. WEIL [*] (pp. 111—123).
[9] NEUMARK [**2**]; AMBROSE [**1**]; GODEMENT [**1**].

of characters Γ, for which

$$U_x = \int_\Gamma \chi(x) E(d\chi);$$

furthermore, $E(A) \smile\smile \{U_x\}$.

All this is understood in the following sense: $E(A)$ is a projection, a denumerably additive and multiplicative function of the set $A \subseteq \Gamma$ measurable (B), and for every $\varepsilon > 0$ and every element x of G there exists a decomposition of Γ into the sum of a finite number of disjoint Borel sets A_n in such a way that

$$\|U_x - \sum_n \chi_n(x) E(A_n)\| < \varepsilon,$$

where χ_n denotes an arbitrary "point" of A_n.

In recent years the problem of unitary representations of *locally bicompact non-Abelian* groups has been the object of important investigations. The fundamental result obtained by GELFAND and RAIKOV[10] asserts that *each of these groups G admits a complete system of irreducible unitary representations.* We say that a representation of the group G by unitary transformations U_x is *irreducible* if there is no proper subspace (of finite or infinite dimension) of the Hilbert space in question which is invariant with respect to all the transformations U_x. A system of representations is said to be *complete* if to each element x_0 of G distinct from the unit element e, there corresponds at least one representation U_x such that $U_{x_0} \neq U_e = I$. The result cited above generalizes, at least partially, the classical theorems of PETER and WEYL on representations of compact groups by finite unitary matrices.[11]

NON-UNITARY TRANSFORMATIONS

141. Groups and Semi-Groups of Self-adjoint Transformations

We return to the study of one-parameter groups of transformations T_t $(-\infty < t < \infty)$, but we no longer shall assume that these transformations are unitary. We shall also consider one-parameter *semi-groups* of transformations, that is, families of transformations $\{T_t\}$ defined only for $t \geq 0$ and such that

$$T_0 = I, \; T_s T_t = T_{s+t} \quad (s \geq 0, \; t \geq 0).$$

It is clear that if the transformations of a semi-group admit inverses which are defined everywhere, this semi-group can be extended to a group by setting $T_{-t} = T_t^{-1}$.

[10] GELFAND–RAIKOV [1]; cf. also GODEMENT [2], [3]; SEGAL [1]; MAUTNER [1].
[11] A discussion from the modern point of view is found in WEIL [*] (pp. 94–109).

First we consider semi-groups of bounded symmetric transformations $\{A_t\}$. We have $A_t = A^2_{\frac{t}{2}} \geq 0$. We shall show that the functions

$$h_f(t) = \log (A_t f, f) = \log \|A_{\frac{t}{2}} f\|^2$$

are convex in the sense that

$$h_f\left(\frac{s+t}{2}\right) \leq \tfrac{1}{2}[h_f(s) + h_f(t)].$$

In fact,

$$\log (A_{\frac{s+t}{2}} f, f) = \tfrac{1}{2} \log (A_{\frac{s}{2}} f, A_{\frac{t}{2}} f)^2 \leq \tfrac{1}{2} \log (\|A_{\frac{s}{2}} f\|^2 \, \|A_{\frac{t}{2}} f\|^2).$$

Now a convex function $\varphi(t)$ (with finite or infinite values) is continuous in the interior of its interval of definition (including the points where $\varphi(t) = -\infty$) provided this interval contains at least one subinterval in which the function is either bounded from above or measurable.[12] Therefore if we assume, for example, that

1° $\|A_t\| \leq C$ *for* $0 < t < 1$,

or even, alternatively, that

2° *the functions* $(A_t f, f)$ *of* t *are measurable in* $(0, 1)$, it follows automatically that these functions *are continuous for every* $t > 0$.

Let m and M be the greatest lower and least upper bounds of A_1; $m \geq 0$. Let

$$A_1 = \int_{m-0}^{M} \lambda dE_\lambda$$

be the spectral decomposition of A_1 and let us define a semi-group of self-adjoint transformations by the formula

$$B_t = \int_{m-0}^{M} \lambda^t dE_\lambda \qquad (t \geq 0);$$

B_t depends continuously on t (even in the sense of uniform convergence). The equation

$$A_t = B_t$$

holds for $t = 1$ by definition; we verify it successively for

$$t = \tfrac{1}{2}, \tfrac{1}{4}, \tfrac{1}{8}, \ldots$$

by using the fact that the positive square root of a positive self-adjoint transformation is unique (Sec. 104), then we pass to the values

$$t = \frac{k}{2^n} \qquad (k, n = 1, 2, \ldots)$$

[12] Cf., for example, SIERPINSKI [1].

by the semi-group property, and finally to all values of $t > 0$ by using continuity. Moreover, we also have $A_0 = B_0 = I$.

Thus we have arrived at the

THEOREM.[13] *Every semi-group $\{A_t\}$ of bounded self-adjoint transformations satisfying condition 1° or 2° admits the representation*

$$A_t = \int_{m-0}^{M} \lambda^t dE_\lambda \qquad (0 \leq m < M), \tag{10}$$

where $\{E_\lambda\}$ is a spectral family on the interval $[m, M]$.

We observe that

$$\lim_{t \to 0} A_t = I - E_0;$$

hence we do not have continuity at the point $t = 0$ unless $E_0 = 0$, that is, unless the value 0 is not a characteristic value of A_1. When this condition is verified, we can transform representation (10) into the following:

$$A_t = \int_{-\infty}^{\infty} e^{\mu t} dF_\mu, \tag{11}$$

where $\{F_\mu\}$ denotes the spectral family which is derived from $\{E_\lambda\}$ by the relation

$$F_\mu = E_{e^\mu}.$$

Introducing the self-adjoint transformation

$$A = \int_{-\infty}^{\infty} \mu dF_\mu,$$

which is bounded if $m > 0$, and in any case semi-bounded above (since $F_\mu = I$ for $\mu > \log M$), the relation (11) can be put into the form

$$A_t = e^{tA}.$$

The "generating" transformation A is derived from $\{A_t\}$ by the formula

$$A = \lim_{h \to 0} \frac{1}{h} (A_h - I). \tag{12}$$

In fact, if f is an element of the domain of A,

$$\left\| \left[\frac{1}{h} (A_h - I) - A \right] f \right\|^2 = \int_{-\infty}^{\infty} \left| \frac{1}{h} (e^{h\mu} - 1) - \mu \right|^2 d\|F_\mu f\|^2 \to 0,$$

[13] Proved first for groups by Sz.-NAGY [**1**]. For semi-groups, cf. HILLE [**1**], [**2**] and [*] (pp. 373–386); Sz.-NAGY [**2**] and [*] (pp. 73–77). Sz.-NAGY [**1**], [*] has extended the theorem to groups and semi-groups of arbitrary *normal transformations*.

because the function under the integral sign decreases to 0 with h. The transformation in the second member of (12) is therefore an extension of A; furthermore, since it is symmetric, it necessarily coincides with A.

*

One-parameter semi-groups $\{V_t\}$ of (non-unitary) *isometric* transformations of Hilbert space have been studied still more closely. An example of such a semi-group is furnished on the functional space $L^2(0, \infty)$ by the transformations

$$V_t f(x) = f_t(x) = \begin{cases} 0 & \text{for } 0 \leq x < t, \\ f(x-t) & \text{for } x \geq t. \end{cases}$$

Moreover, the space $L^2(0, \infty)$ can be considered as a subspace of the space $L^2(-\infty, \infty)$ by identifying the element $f(x)$ of $L^2(0, \infty)$ with the element of $L^2(-\infty, \infty)$ which arises from it by setting $f(x) = 0$ for $x < 0$. In L^2 $(-\infty, \infty)$ the transformations

$$U_t g(x) = g(x-t) \qquad (-\infty < t < \infty)$$

form a one-parameter group of unitary transformations; for $t \geq 0$ and for $f(x)$ belonging to $L^2(0, \infty)$ we obviously have $V_t f = U_t f$.

It can be shown that the situation is the same in the general case:

THEOREM.[14] *If $\{V_t\}$ is a semi-group of isometric transformations of the Hilbert space $\mathfrak{H}$, we can extend $\mathfrak{H}$ to a larger Hilbert space $\mathfrak{H}_1$ of which $\mathfrak{H}$ will be a subspace and construct a one-parameter group $\{U_t\}$ of unitary transformations of $\mathfrak{H}_1$ such that for $t \geq 0$ and for all elements f of $\mathfrak{H}$ we have $V_t f = U_t f$.*

[14] COOPER [**3**]; this theorem is also obtained by combining an earlier theorem of PLESSNER [**1**] with the results of NEUMARK [**3**] on the "extensions of the second kind" of symmetric transformations. The fundamental result of NEUMARK affirms that: if S is an arbitrary symmetric transformation of Hilbert space $\mathfrak{H}$, we can always embed $\mathfrak{H}$ in a Hilbert space $\mathfrak{H}_1$ of which $\mathfrak{H}$ is a subspace, and construct in $\mathfrak{H}_1$ a self-adjoint transformation which coincides in $\mathfrak{D}_S$ with S. The connection between isometric semi-groups and symmetric transformations is furnished by the result, established by PLESSNER and COOPER, l.c., that every one-parameter semi-group of isometric transformations is generated by an "infinitesimal" maximal symmetric transformation.

We note, in this connection, the following result of Sz.-NAGY [**7**]: Every one-parameter group $\{T_t\}$ of linear transformations in Hilbert space such that $\|T_t\| \leq C(-\infty < t < \infty)$ is "similar" to a group of unitary transformations, that is, there exists a bounded linear transformation Q such that Q^{-1} is also bounded and $Q^{-1}T_tQ$ is unitary $(-\infty < t < \infty)$.

142. Infinitesimal Transformation of a Semi-Group of Transformations of General Type

Semi-groups of bounded linear transformations of general type of a Hilbert space or of a Banach space have been studied by several authors; the results obtained have found important applications in numerous branches of analysis and in the theory of stochastic processes. Since E. HILLE has just published an excellent work on this subject (HILLE [*]), we shall say only a few words about it.

Part of the problem which has been considered is of the following nature. One supposes that the transformations T_t forming the semi-group depend on the parameter t in a continuous (weak or strong) manner, or merely that they depend on t in a measurable manner (where one also distinguishes between weak and strong measurability), and one asks if the hypothesis entails other properties of T_t, in particular its strong continuity and the existence of the derivative $\lim_{\varepsilon\to 0} \frac{1}{\varepsilon}(T_{t+\varepsilon} - T_t)$, on a dense domain of definition. One studies the structural properties of the "infinitesimal" transformation[15] $A = \lim_{\varepsilon\to 0} \frac{1}{\varepsilon}(T_\varepsilon - I)$, and one investigates various possibilities of constructing the semi-group $\{T_t\}$ starting from its given infinitesimal transformation A. In the general case one cannot attain a simple exponential formula $T_t = e^{tA}$, since the transformation A is in general not bounded. For *bounded* linear transformations B of a Banach space one has an obvious definition by the series

$$e^B = \sum_{n=0}^{\infty} \frac{1}{n!} B^n,$$

a series which is majorized in norm by the convergent numerical series

$$\sum_{n=0}^{\infty} \frac{1}{n!} \|B\|^n.$$

If B and C are bounded and permutable, $e^{B+C} = e^B e^C$.[16]

[15] *Translator's footnote.* HILLE [*] calls this transformation the "infinitesimal generator" of the semi-group.

[16] In fact, $e^B e^C - e^{B+C}$ is the limit (as $N \to \infty$) of

$$\sum_{n=0}^{N} \frac{1}{n!} B^n \cdot \sum_{m=0}^{N} \frac{1}{m!} C^m - \sum_{p=0}^{N} \frac{1}{p!} (B+C)^p = \sum \frac{1}{n!m!} B^n C^m,$$

where the last summation is over all m, n such that $0 \le n \le N$, $0 \le m \le N$, and $n + m > N$. Setting $b = \|B\|$, $c = \|C\|$, this sum is majorized in norm by the sum

$$\sum \frac{1}{n!m!} b^n c^m = \sum_{n=0}^{N} \frac{1}{n!} b^n \cdot \sum_{m=0}^{N} \frac{1}{m!} c^m - \sum_{p=0}^{N} \frac{1}{p!} (b+c)^p,$$

which tends, as $N \to \infty$, to $e^b e^c - e^{b+c} = 0$.

We give now a theorem asserting the existence of the infinitesimal transformation on a dense domain:

THEOREM.[17] *Let* $\{T_t\}$ *be a one-parameter semi-group of bounded linear transformations of the Banach space* $\mathfrak{B}$; *we suppose that it is strongly continuous*:

$$T_s \to T_t \text{ when } s \to t \text{ (even at } t = 0).$$

Under these conditions the "infinitesimal" transformation

$$A = \lim_{\varepsilon \to 0} A_\varepsilon \ \left[\text{where } A_\varepsilon = \frac{1}{\varepsilon}(T_\varepsilon - I)\right]$$

has domain $\mathfrak{D}_A$ *dense in* $\mathfrak{B}$; *it is a closed linear transformation. Furthermore, for* $f \in \mathfrak{D}_A$,

$$\lim_{\varepsilon \to 0} \frac{1}{\varepsilon}(T_{t+\varepsilon} - T_t)f = AT_t f = T_t Af. \tag{13}$$

For the proof, we observe first that for given f, $T_t f$ is a continuous function of t, and that consequently

$$f_s = \frac{1}{s}\int_0^s T_t f dt \to f$$

when $s \to 0$, where the integral is defined as the strong limit of sums of Riemann type. In order to prove that $\mathfrak{D}_A$ is everywhere dense in $\mathfrak{B}$, it therefore suffices to prove that it contains all elements of the form f_s, $(s > 0)$. Now

$$A_\varepsilon f_s = \frac{1}{s\varepsilon}\left[\int_s^{s+\varepsilon} T_t f dt - \int_0^\varepsilon T_t f dt\right] = A_s f_\varepsilon \to A_s f$$

for $\varepsilon \to 0$, and therefore Af_s exists; it is equal to $A_s f$.

Equations (13) stem from the obvious equations

$$\frac{1}{\varepsilon}(T_{t+\varepsilon} - T_t) = T_t A_\varepsilon = A_\varepsilon T_t.$$

Therefore there only remains to prove that A is a *closed* transformation.

To do this let us observe first that $\|T_t\|$ is a bounded function of t in every finite interval. For if not there would be a sequence $t_k \to t_0$ such that $\|T_{t_k}\| \to \infty$, which is impossible in view of the fact that the transformations T_{t_k} converge to T_{t_0} and therefore have a common bound (Osgood-Banach theorem, Sec. 31).

It follows that for every f of $\mathfrak{D}_A$

$$\|f_s\| = \left\|\frac{1}{s}\int_0^s T_t f dt\right\| \leq M_s \|f\|,$$

[17] HILLE [*] (Chapt. IX). Cf. also DUNFORD and SEGAL [**1**], YOSIDA [**2**].

where

(14) $$M_s = \sup_{0 \leq t \leq s} \|T_t\| < \infty;$$

consequently

(15) $$f^{(n)} \to f \text{ implies } f_s^{(n)} - f_s = (f^{(n)} - f)_s \to 0,\ f_s^{(n)} \to f_s.$$

Let us recall the relation

(16) $$A_\varepsilon f_s = A_s f_\varepsilon,$$

and observe further that, since $T_t \smile A_\varepsilon$,

(17) $$A_\varepsilon f_s = (A_\varepsilon f)_s.$$

Let f be an arbitrary element of $\mathfrak{D}_A$; for $\varepsilon \to 0$ we have by definition $A_\varepsilon f \to Af$; therefore, in view of (15),

$$(A_\varepsilon f)_s \to (Af)_s.$$

On the other hand, in view of (17), (16), and the fact that $f_\varepsilon \to f$,

$$(A_\varepsilon f)_s = A_s f_\varepsilon \to A_s f.$$

Therefore

(18) $$A_s f = (Af)_s.$$

The fact that A is closed is now proved in the following manner. Let $f^{(n)}$ be an arbitrary sequence of elements of $\mathfrak{D}_A$ such that

$$f^{(n)} \to f, \qquad Af^{(n)} \to g \qquad (n \to \infty).$$

Then on the one hand

$$A_s f^n \to A_s f,$$

and on the other hand, using (18) and (15),

$$A_s f^{(n)} = (Af^{(n)})_s \to g_s,$$

hence

$$A_s f = g_s,$$

which implies that $A_s f \to g$ when $s \to 0$. Therefore f also belongs to $\mathfrak{D}_A$ and $Af = g$, q. e. d.

143. Exponential Formulas

For want of an obvious meaning for an exponential formula $T_t = e^{tA}$, due to the fact that in general A is not bounded, we examine, among others, formulas of the type

$$T_t = \lim_{\varepsilon \to 0} e^{tB_\varepsilon},$$

where B_ε is a bounded linear transformation depending on the parameter $\varepsilon > 0$. Let us prove the following lemma:

LEMMA. *Let $\{T_t\}$ be as in the preceding section, and let B_ε be a bounded linear transformation of the space $\mathfrak{B}$ depending on the parameter ε, $0 < \varepsilon \leq \varepsilon_0$, and such that*

a) $B_\varepsilon \smile T_t$

b) $\sup_{0 \leq t \leq s} \|e^{tB_\varepsilon}\| \leq N_s$ *(a constant independent of ε),*

c) $\lim_{\varepsilon \to 0} B_\varepsilon f = Af$ *for every element f of $\mathfrak{D}_A$,*

where A is the infinitesimal transformation of $\{T_t\}$. One then has

$$T_t f = \lim_{\varepsilon \to 0} e^{tB_\varepsilon} f$$

for every element f of $\mathfrak{B}$; furthermore, for fixed f this convergence is uniform in t in every finite interval $0 \leq t \leq s$.

Let us begin with the formula

$$T_t - e^{tB_\varepsilon} = [\sum_{k=0}^{n-1} T_{k\frac{t}{n}} e^{(n-1-k)\frac{t}{n} B_\varepsilon}] [T_{\frac{t}{n}} - e^{\frac{t}{n} B_\varepsilon}],$$

which is easily verified since a) implies that $e^{tB_\varepsilon} \smile T_t$. Using b), we find that for $0 \leq t \leq s$

$$\|(T_t - e^{tB_\varepsilon})f\| \leq \sum_{k=1}^{n-1} M_s N_s \|(T_{\frac{t}{n}} - e^{\frac{t}{n} B_\varepsilon})f\| \leq C(s) \frac{n}{t} \|(T_{\frac{t}{n}} - e^{\frac{t}{n} B_\varepsilon})f\|,$$

where M_s is defined by (14) and $C(s) = sM_s N_s$. It follows that

$$\|(T_t - e^{tB_\varepsilon})f\| \leq C(s) \limsup_{\tau \to 0} \frac{1}{\tau} \|(T_\tau - e^{\tau B_\varepsilon})f\|.$$

Now

$$\frac{1}{\tau} \|(T_\tau - e^{\tau B_\varepsilon})f\| = \left\| \left(A_\tau - \frac{1}{\tau}(e^{\tau B_\varepsilon} - I)\right) f \right\| \leq$$

$$\leq \|(A_\tau - B_\varepsilon)f\| + \left\| \left(\frac{1}{\tau}(e^{\tau B_\varepsilon} - I) - B_\varepsilon\right) f \right\|$$

and

$$\left\| \frac{1}{\tau}(e^{\tau B_\varepsilon} - I) - B_\varepsilon \right\| = \left\| \sum_{n=2}^{\infty} \frac{\tau^{n-1}}{n!} B_\varepsilon^n \right\| \leq \sum_{n=2}^{\infty} \frac{\tau^{n-1}}{n!} \|B_\varepsilon\|^n =$$

$$= \frac{1}{\tau}(e^{\tau \|B_\varepsilon\|} - 1) - \|B_\varepsilon\|,$$

and the last member tends to 0 with τ. Therefore

$$\|(T_t - e^{tB_\varepsilon})f\| \leq C(s) \limsup_{\tau\to 0} \|A_\tau f - B_\varepsilon f\|$$

for $0 \leq t \leq s$ and for every element f of $\mathfrak{B}$.

If f belongs to the domain $\mathfrak{D}_A$ of the infinitesimal transformation A of $\{T_t\}$, the second member of the last formula is equal to $C(s)\,\|Af - B_\varepsilon f\|$, and therefore in view of hypothesis c) it tends to 0 with ε.

This proves the lemma for the elements f belonging to $\mathfrak{D}_A$, and therefore for a dense set in $\mathfrak{B}$. From here one can pass to the other elements in an obvious manner, by making use of the fact that for $0 \leq t \leq s$ the quantities $\|T_t\|$ and $\|e^{tB_\varepsilon}\|$ are bounded by constants which depend only on s. This completes the proof.

*

An immediate consequence of this lemma is the following theorem of Hille (cf. [*], Chapt. IX).

THEOREM. *For every semi-group $\{T_t\}$ which is strongly continuous (even at $t = 0$) one has*

$$T_t f = \lim_{\varepsilon\to 0} e^{tA_\varepsilon} f \quad \textit{where} \quad A_\varepsilon = \frac{1}{\varepsilon}(T_\varepsilon - I),$$

where for every fixed element f the convergence is uniform in t in every finite interval $0 \leq t \leq s$.

We have only to verify the conditions of the lemma for $B_\varepsilon = A_\varepsilon$. Now a) and c) are obvious. As for b), we first observe, setting $M = M_1 = \sup_{0\leq t\leq 1} \|T_t\|\ (\geq 1)$, that for arbitrary $t \geq 0$ we have $\|T_t\| = \|T_{[t]}T_{t-[t]}\| = \|T_1^{[t]}T_{t-[t]}\| \leq \|T_1\|^{[t]}\,\|T_{t-[t]}\| \leq M^{[t]+1} \leq M^{t+1}$. It follows that

$$\|e^{tA_\varepsilon}\| = \|e^{-\frac{t}{\varepsilon}} e^{\frac{t}{\varepsilon}T_\varepsilon}\| \leq e^{-\frac{t}{\varepsilon}} \sum_{m=0}^{\infty} \left(\frac{t}{\varepsilon}\right)^m \frac{1}{m!} M^{\varepsilon m+1} = Me^{\frac{t}{\varepsilon}(M^\varepsilon - 1)};$$

therefore for $0 < \varepsilon \leq 1$ and for $0 \leq t \leq s$

$$\|e^{tA_\varepsilon}\| \leq Me^{s(M-1)};$$

consequently condition b) is also verified.

We consider an interesting *application* of this theorem.

Let $\mathfrak{B}$ be the space of uniformly continuous and bounded functions $f(x)$ defined in the interval $0 \leq x < \infty$, the norm of $f(x)$ being defined by

$$\|f\| = \max |f(x)|.$$

We consider the following linear transformations of $\mathfrak{B}$ into itself:

$$T_t f(x) = f(x + t) \qquad (t \geq 0);$$

they obviously form a semi-group and when $s \to t$,

$$\|T_s f - T_t f\| = \max |f(x + s) - f(x + t)| \to 0$$

by virtue of the uniform continuity of the function $f(x)$. We can apply the theorem and obtain the following *formula*:

$$f(x + t) = \lim_{h \to 0} \sum_{n=0}^{\infty} \frac{1}{n!} \left(\frac{t}{h}\right)^n \Delta_h^{(n)} f(x), \tag{19}$$

where

$$\Delta_h^{(n)} f(x) = (T_h - I)^n f(x) = \sum_{k=0}^{n} (-1)^{n-k} \binom{n}{k} f(x + kh)$$

is the so-called "n-th difference" of $f(x)$; furthermore, the sum in the second member of (19) converges, when $h \to 0$, uniformly with respect to x in $0 \leq x < \infty$ and with respect to t in every finite interval $0 \leq t \leq \omega$.

This is obviously a sort of generalization of the classical Taylor theorem. Furthermore, we can derive a proof of the Weierstrass approximation theorem from it. In fact, let $f(x)$ be a continuous function in the interval $0 \leq x \leq 1$. We extend it by setting $f(x) = f(1)$ for $x > 1$. Given $\varepsilon > 0$, we fix h so small that

$$\left| f(x + t) - \sum_{n=0}^{\infty} \frac{1}{n!} \left(\frac{t}{h}\right)^n \Delta_h^{(n)} f(x) \right| < \frac{\varepsilon}{2} \quad (0 \leq x < \infty,\ 0 \leq t \leq 1), \tag{20}$$

which is possible in view of the formula obtained above. The series appearing in (20) is uniformly convergent for $x \geq 0$ and for $0 \leq t \leq 1$; this follows from our general results, but can also be verified directly by observing that, denoting the least upper bound of $|f(x)|$ by M, we have the convergent majorizing series

$$\sum_{n=0}^{\infty} \frac{1}{n!} \frac{1}{h^n} 2^n M = M e^{2/h}.$$

Setting $x = 0$ and replacing the infinite series in (20) by a partial sum with sufficiently large index, we therefore obtain a polynomial in t which approaches within ε of $f(t)$ in $0 \leq t \leq 1$.

*

In the above theorem the transformations in the exponent do not have an immediate connection with the infinitesimal transformation of the group. Therefore this theorem does not answer the question of whether the group can be reconstructed from its infinitesimal transformation. There are various possibilities for such a construction, cf. HILLE [*] (Chapt. XI and XII). The following theorem furnishes a construction by an exponential formula, and at the same time it characterizes the transformations which are infinitesimas transformations of groups $\{T_t\}$, at least under the additional hypothesis $\|T_t\| \leq 1$:

THEOREM.[18] (I) *Let* $\{T_t\}$ *satisfy the hypothesis of the preceding theorem and the additional hypothesis*

$$\|T_t\| \leq 1 \qquad (t \geq 0). \tag{21}$$

Then the infinitesimal transformation A *of* $\{T_t\}$ *possesses the following property* (P):

$$(P)\begin{cases} I_\varepsilon = (I - \varepsilon A)^{-1} \text{ exists for every } \varepsilon > 0, \text{ its domain is the entire space } \mathfrak{B}, \\ \text{and } \|I_\varepsilon\| \leq 1. \end{cases}$$

Setting

$$B_\varepsilon = AI_\varepsilon = \frac{1}{\varepsilon}(I_\varepsilon - I)$$

we have

$$T_t f = \lim_{\varepsilon \to 0} e^{tB_\varepsilon} f, \tag{22}$$

where for each fixed element f *of* $\mathfrak{B}$ *the convergence is uniform in every finite interval* $0 \leq t \leq s$.

(II) *Every linear transformation* A *of* $\mathfrak{B}$ *with domain* $\mathfrak{D}_A$ *dense in* $\mathfrak{B}$ *and possessing property* (P) *is the infinitesimal generator of a semi-group of the type considered.*

Let us observe first some facts which are valid for every linear transformation A satisfying the conditions of (II).

For every element f of $\mathfrak{D}_A$ we have

$$(I_\varepsilon - I)f = I_\varepsilon f - I_\varepsilon(I - \varepsilon A)f = \varepsilon I_\varepsilon Af, \quad \|(I_\varepsilon - I)f\| \leq \varepsilon \|Af\|,$$

therefore $(I_\varepsilon - I)f \to 0$ when $\varepsilon \to 0$. Since $\mathfrak{D}_A$ is dense in $\mathfrak{B}$ and $\|I_\varepsilon - I\|$ is bounded by 2, that is to say by a constant independent of ε, this result extends to all elements of $\mathfrak{B}$, therefore $I_\varepsilon \to I$.

Since $I_\varepsilon \smile I_\varepsilon^{-1} = I - \varepsilon A$, and therefore $I_\varepsilon \smile A$, we have for every element f of $\mathfrak{D}_A$

$$B_\varepsilon f - Af = I_\varepsilon Af - Af = (I_\varepsilon - I)Af \to 0 \quad \text{when} \quad \varepsilon \to 0. \tag{23}$$

[18] Cf. HILLE [*] (Theorem 12.2.1) and YOSIDA [**2**]. Formula (22) is due to YOSIDA; HILLE derived, in place of (22), the formula (among others)

$$T_t f = \lim_{k \to \infty} \left(I - \frac{tA}{k}\right)^{-k} f.$$

The proof given in the text differs from that of YOSIDA in that it avoids the use of the dual space $\mathfrak{B}^*$.

Restriction (21) is not essential. In fact, MIYADERA and PHILLIPS [**2**] have just shown that the proof above remains valid without (21) if in property (P) the inequality $\|I_\varepsilon\| \leq 1$ is replaced by the following: *there exists an* $\omega \geq 0$ such that, for $0 < \varepsilon < 1/\omega$:

$$\limsup_{n \to \infty} \| I_\varepsilon^n \|^{1/n} \leq 1/(1 - \varepsilon\omega).$$

Finally we have

$$(24)\qquad \|e^{tB_\varepsilon}\| = \|e^{-\frac{t}{\varepsilon}} e^{\frac{t}{\varepsilon} I_\varepsilon}\| \leq e^{-\frac{t}{\varepsilon}} \sum_{m=0}^{\infty} \left(\frac{t}{\varepsilon}\right)^m \frac{1}{m!} \|I_\varepsilon\|^m \leq e^{-\frac{t}{\varepsilon}} e^{\frac{t}{\varepsilon}} = 1.$$

With this established, part (I) of the theorem is demonstrated as follows. For $\varepsilon > 0$, $t > 0$ we have

$$I - \varepsilon A_t = I - \frac{\varepsilon}{t}(T_t - I) = \frac{t+\varepsilon}{t}\left(I - \frac{\varepsilon}{t+\varepsilon} T_t\right),$$

from which it follows by (21) that

$$I_{t,\varepsilon} = (I - \varepsilon A_t)^{-1}$$

exists, that its domain is the entire space $\mathfrak{B}$, and that for every element f of $\mathfrak{B}$ we have

$$(25)\qquad \|(I - \varepsilon A_t)f\| \geq \frac{t+\varepsilon}{t}\left(\|f\| - \frac{\varepsilon}{t+\varepsilon}\|f\|\right) = \|f\|,$$

therefore

$$(26)\qquad \|I_{t,\varepsilon}\| \leq 1.$$

From (25) it follows that for every element f of the domain $\mathfrak{D}_A$ of the infinitesimal transformation $A = \lim_{t \to 0} A_t$

$$(27)\qquad \|(I - \varepsilon A)f\| \geq \|f\|.$$

This implies in its turn that

$$I_\varepsilon = (I - \varepsilon A)^{-1}$$

exists and that, for every element g of its domain $\mathfrak{D}_{I_\varepsilon}$,

$$(28)\qquad \|I_\varepsilon g\| \leq \|g\|.$$

Since A is closed (cf. Sec. 142) I_ε is also closed and hence (28) implies that $\mathfrak{D}_{I_\varepsilon}$ is a *closed* linear set. Therefore in order to show that $\mathfrak{D}_{I_\varepsilon}$ coincides with the entire space it suffices to show that it contains the domain of A, or in other words that for every fixed ε and arbitrary element g from $\mathfrak{D}_A$ the equation

$$(I - \varepsilon A)h^* = g$$

has a solution h^*.

Now the approximating equation

$$(I - \varepsilon A_t)h = g$$

has the solution $h = I_{t,\varepsilon} g$; we shall show that the latter tends to an element of $\mathfrak{B}$ when $t \to 0$ and that at the same time $(I - \varepsilon A)h$ tends to g; since the transformation $I - \varepsilon A$ is closed it will follow that $\lim_{\varepsilon \to \infty} h$ will be equal to the tension of A since then $(I - \varepsilon\bar{A})^{-1}$ would be a proper extension of $(I - \varepsilon A)^{-1}$, which contradicts the fact that these two transformations are defined in the entire space $\mathfrak{B}$, the first in view of part (I) of the theorem and the second by hypothesis. Therefore $\bar{A} = A$.

This completes the proof of the theorem.

ERGODIC THEOREMS

144. Fundamental Methods

In Sec. 139 we mentioned the "statistical ergodic theorem" of J. VON NEUMANN concerning the motions $P \to P_t$ of a measurable set Ω into itself which preserve measure. The "discrete" counterpart of this theorem (from which we can pass easily to the "continuous" case) concerns the iterates of a single transformation $P \to P'$ of the set Ω into itself which preserves the measure $d\sigma$ on Ω, that is, a transformation such that for every summable function $f(P)$ on Ω,

$$\int_\Omega f(P)d\sigma = \int_\Omega f(P')d\sigma.$$

Setting

$$P'' = (P')', \ldots, P^{(n)} = (P^{(n-1)})', \ldots$$

the theorem can be expressed as follows:

STATISTICAL ERGODIC THEOREM. *For every function $f(P)$ which is square-summable in Ω, the arithmetic means*

$$\varphi_{m,n}(P) = \frac{1}{n-m} \sum_{k=m}^{n-1} f(P^{(k)})$$

tend in the mean, when $n - m \to \infty$, to a square-summable function $f^*(P)$ which is almost invariant with respect to the transformation $P \to P'$, that is, which satisfies

$$f^*(P) = f^*(P')$$

for almost all points P of Ω.

Recalling that the transformation U of the functional space $L^2(\Omega)$, defined by

$$Uf(P) = f(P')$$

is unitary, we can verify that this theorem is included in the following one in which we no longer are dealing with transformations of the set Ω:

desired solution h^*. These propositions follow from the calculation we now make in which we make use of the fact that $I_{t,\varepsilon} \smile T_s$ and that consequently $I_{t,\varepsilon} \smile I_{s,\varepsilon}$, $I_{t,\varepsilon} \smile A$, and also of the boundedness condition (26):

$$\|I_{t,\varepsilon}g - I_{s,\varepsilon}g\| = \|I_{t,\varepsilon}I_{s,\varepsilon}(I - \varepsilon A_s)g - I_{s,\varepsilon}I_{t,\varepsilon}(I - \varepsilon A_t)g\| =$$
$$= \|I_{t,\varepsilon}I_{s,\varepsilon}\varepsilon(A_t - A_s)g\| \leq \varepsilon \|(A_t - A_s)g\| \to 0 \text{ when } t, s \to 0$$
$$\|(I - \varepsilon A)I_{t,\varepsilon}g - g\| = \|I_{t,\varepsilon}(I - \varepsilon A)g - I_{t,\varepsilon}(I - \varepsilon A_t)g\| =$$
$$= \|I_{t,\varepsilon}\varepsilon(A_t - A)g\| \leq \varepsilon \|(A_t - A)g\| \to 0 \text{ when } t \to 0.$$

This completes the proof that A possesses property (P). This property implies (23) and (24), and since $T_t \smile A$ implies $T_t \smile I_\varepsilon$ and hence $T_t \smile B_\varepsilon$, the representation follows from our lemma. Part (I) of the theorem is thus proved.

In order to prove part (II) we consider, for every fixed $\varepsilon > 0$, the semi-group

$$T_t^{(\varepsilon)} = e^{tB_\varepsilon} \text{ where } B_\varepsilon = AI_\varepsilon = A(I - \varepsilon A)^{-1};$$

it is easy to see that this semi-group is strongly continuous, that its infinitesimal transformation is equal to B_ε, and that $T_t^{(\varepsilon)} \smile T_s^{(\eta)}$. One then shows, by an argument which is analogous to that used in the proof of the lemma, that

$$\|T_t^{(\varepsilon)}f - T_t^{(\eta)}f\| \leq t\,\|B_\varepsilon f - B_\eta f\|.$$

If f belongs to $\mathfrak{D}_A$, then $B_\varepsilon f - B_\eta f \to 0$ when $\varepsilon, \eta \to 0$ (cf. (23)), and therefore

$$T_t f = \lim_{\varepsilon \to 0} T_t^{(\varepsilon)} f$$

exists and the convergence is uniform in t in every finite interval $0 \leq t \leq s$. Since $\|T_t^{(\varepsilon)}\| \leq 1$ (cf. (24)) and $\mathfrak{D}_A$ is dense in $\mathfrak{B}$, this proposition extends to all elements f of $\mathfrak{B}$. It follows that $\{T_t\}$ is a strongly continuous semi-group and that $\|T_t\| \leq 1$.

Finally, the equation

$$\frac{1}{s}(T_s^{(\varepsilon)} - I)f = \frac{1}{s}\int_0^s T_t^{(\varepsilon)}B_\varepsilon f\,dt$$

which is only (18) written for the semi-group $\{T_t^{(\varepsilon)}\}$, implies that for f in $\mathfrak{D}_A$

$$\frac{1}{s}(T_s - I)f = \frac{1}{s}\int_0^s T_t A f\,dt$$

(cf. (23)). It follows that

$$\lim_{s \to 0} \frac{1}{s}(T_s - I)f = Af.$$

Denoting the infinitesimal transformation of the semi-group $\{T_t\}$ by $\bar{A}$ we therefore see that $\bar{A} \supseteq A$. But it is impossible that $\bar{A}$ should be a proper e-

THEOREM. *Let U be a unitary transformation of the Hilbert space $\mathfrak{H}$. For every element f of $\mathfrak{H}$ the arithmetic means*

$$\varphi_{m,n} = \frac{1}{n-m} \sum_{k=m}^{n-1} U^k f$$

tend, as $n - m \to \infty$, to an element f^* which is invariant with respect to the transformation U.

This theorem can be deduced from the spectral representation of the transformation U and from its iterates (cf. Sec. 109) just as we derived (in Sec. 139) its "continuous" analogue from Stone's theorem. We give here a direct proof.[19]

Let us consider two subspaces of the space $\mathfrak{H}$. The first, $\mathfrak{L}'$, consists of elements of the type $g - Ug$ and of their limits. When f belongs to $\mathfrak{L}'$, we shall therefore be able to set

$$f = g - Ug + g'$$

with

$$\|g'\| < \varepsilon,$$

where ε is arbitrarily small. From this it follows that

$$\varphi_{m,n} = \frac{U^m g - U^n g}{n-m} + \frac{1}{n-m} \sum_{k=m}^{n-1} U^k g',$$

hence

$$\|\varphi_{m,n}\| \leq \frac{2\|g\|}{n-m} + \varepsilon,$$

and consequently

$$\|\varphi_{m,n}\| \leq 2\varepsilon$$

for $n - m$ sufficiently large. Hence $\varphi_{m,n} \to 0$.

The second subspace, $\mathfrak{L}''$, is made up of the invariant elements of the transformation U, that is, of those elements f such that $Uf = f$. For these f we obviously have $\varphi_{m,n} = f$.

Hence the theorem will be proved if we succeed in decomposing every element f into the sum of an element f' of $\mathfrak{L}'$ and an element f'' from $\mathfrak{L}''$. Now the general identity

$$(g - Tg, h) = (g, h - T^*h),$$

which follows immediately from the definition of adjoint transformation,

[19] It is due to F. RIESZ and is inserted in the discussion by HOPF [*] (§ 8); cf. also RIESZ–SZ.-NAGY [**1**]. The idea involved in this proof was suggested by the paper [**2**] of CARLEMAN.

makes it clear that the set of elements h which are orthogonal to all the $g - Ug$ consists precisely of the invariant elements of U^*, which are, by virtue of the property

$$U^* = U^{-1},$$

which characterizes unitary transformations, the same as the invariant elements of U. Consequently the subspaces $\mathfrak{L}'$ and $\mathfrak{L}''$ are orthogonal and complementary; this completes the proof.

*

This proof applies also to more general transformations. In fact, we have used in it only the following properties of the unitary transformation U:

a) U is everywhere defined and linear,

b) $\|Uf\| = \|f\|$,

c) the invariant elements of U and U^* coincide.

Instead of b) it would obviously have sufficed to know that

b') $\|Uf\| \leq \|f\|$.

The same argument therefore applies to every transformation U satisfying the conditions a), b'), c).

But c) is a consequence of a) and b'). In fact, if the transformation U satisfies the conditions a) and b'), which we express by saying the transformation U is a *contraction*, we have

$$\|U\| \leq 1, \text{ hence also } \|U^*\| \leq 1,$$

from which it follows that for every element f invariant with respect to U,

$$\|f\|^2 = (f, f) = (Uf, f) = (f, U^*f) \leq \|f\| \, \|U^*f\| \leq \|f\|^2,$$

hence that

$$(f, U^*f) = \|f\| \, \|U^*f\|$$

and

$$\|U^*f\| = \|f\|,$$

which implies that

$$\|f - U^*f\|^2 = \|f\|^2 - (f, U^*f) - (U^*f, f) + \|U^*f\|^2 = 0,$$

and therefore that

$$U^*f = f.$$

Interchanging the roles of U and U^*, it follows that *the two classes of invariant elements coincide*, q. e. d.[20]

[20] RIESZ–SZ.-NAGY [**1**].

The ergodic theorem is therefore true not only for unitary transformations, but for all contractions of Hilbert space.

Let us forget for the moment that we have succeeded in proving the identity of the invariant elements of the contraction U and of its adjoint. This proposition would be false even if, instead of restricting ourselves to contractions, we should content ourselves with assuming only that the transformation U to be considered be *linear, everywhere defined, and such that its iterates* U^k *are uniformly bounded, that is*

$$\|U^k\| \leq C \quad \text{for} \quad k = 1, 2, \ldots; \tag{29}$$

nonetheless this hypothesis would suffice in the rest of the above reasoning. But even before having observed this property of contractions, one of the authors succeeded, in 1938,[21] in extending the ergodic theorem to all transformations of this more general type and indeed in passing to functional spaces L^p $(p > 1)$, or even, under certain additional hypotheses, to the space L. In their turn, and at the same time, YOSIDA and KAKUTANI[22] discovered the same method and also formulated it for a class of abstract spaces. Since then, various authors have considered the problem and have found new proofs and generalizations. Let us sketch the method of LORCH [**3**], which seems to be, in a certain sense, the most efficient.

Assuming that U is an arbitrary linear transformation satisfying condition (29), we consider the subspaces $\mathfrak{L}'$ and $\mathfrak{L}''$ corresponding to the transformation U, and the analogous subspaces, say $\mathfrak{M}'$ and $\mathfrak{M}''$, which correspond to the transformation U^*. We show, as above, that the limit φ of the arithmetic means exists when f belongs either to $\mathfrak{L}'$ or to $\mathfrak{L}''$, and that $\varphi = 0$ in the first case and $\varphi = f$ in the second; the same facts hold for $\mathfrak{M}'$ and $\mathfrak{M}''$ with respect to U^*. From this it quickly follows that $\mathfrak{M}'$ and $\mathfrak{M}''$ have no element in common except $f = 0$. Also, as we have seen, the subspaces $\mathfrak{L}'$ and $\mathfrak{M}''$ are orthogonal and complementary; similarly $\mathfrak{M}'$ and $\mathfrak{L}''$. In order to complete the proof, we therefore have only to show that every element f is of the form $f = f' + f''$ where f' and f'' (orthogonal or not, this is of little importance) belong respectively to $\mathfrak{L}'$ and $\mathfrak{L}''$.

Let us show first that the set $\mathfrak{N}$ of elements of the form $f' + f''$ —a set which is clearly linear—is dense in the space $\mathfrak{H}$. If it were not there would exist an element $g \neq 0$ orthogonal to both $\mathfrak{L}'$ and $\mathfrak{L}''$, and consequently contained in $\mathfrak{M}''$ and $\mathfrak{M}'$, contrary to the fact that, as we have just seen, these two subspaces have only the element 0 in common. Therefore $\mathfrak{N}$ is dense in $\mathfrak{H}$. On the other hand, the inequality

$$\|f''\| \leq C\, \|f' + f''\|,$$

[21] F. RIESZ [**19**].

[22] YOSIDA [**1**]; KAKUTANI [**1**], [**2**].

a consequence of (29) and of the fact that f'' is the limit of the arithmetic means formed starting with $f = f' + f''$, implies that if the sequence $\{f'_n + f''_n\}$ is convergent then so are the sequences $\{f'_n\}$ and $\{f''_n\}$ (by writing, in the above inequality, $f'_m - f'_n$ and $f''_m - f''_n$ in place of f', f''). Since $\mathfrak{L}'$ and $\mathfrak{L}''$ are closed, it follows that $\mathfrak{N}$ is also closed, and therefore that $\mathfrak{N} = \mathfrak{H}$. Thus the theorem is proved.

Actually, this was only an adaptation of the proof of LORCH to the case of Hilbert space; by virtue of the particular structure of this space the reasoning becomes simpler than in the general case considered by LORCH. He deals with arbitrary reflexive Banach spaces, that is, with spaces $\mathfrak{B}$ which are the conjugates of their conjugate spaces, $\mathfrak{B}^{**} = \mathfrak{B}$.

145. Methods Based on Convexity Arguments

Here is another way of proving the ergodic theorem for contractions U of Hilbert space $\mathfrak{H}$, suggested to one of the authors[23] by an argument of GARRETT BIRKHOFF [**1**]. It is based on the fact that, given a convex set G in the space $\mathfrak{H}$, and denoting by μ the greatest lower bound of the norms of the elements g of G, *every minimizing sequence* $\{g_n\}$, *that is, every sequence such that*

$$\|g_n\| \to \mu,$$

converges to an element φ *of* $\mathfrak{H}$ *which is independent of the particular sequence.* The convergence of $\{g_n\}$ follows immediately from the characteristic relation of Hilbert space

$$\left\|\frac{f+g}{2}\right\|^2 + \left\|\frac{f-g}{2}\right\|^2 = \tfrac{1}{2}(\|f\|^2 + \|g\|^2) \tag{30}$$

(cf. Sec. 33). As to the uniqueness of the limit element, it follows from the fact that since the union $g_1, g'_1, g_2, g'_2, \ldots$ of two minimizing sequences $\{g_k\}$ and $\{g'_k\}$ is clearly of the same type, the limit φ is necessarily the same for the two sequences.

We consider in particular the convex set G formed by starting with a given element f of $\mathfrak{H}$ and then taking all linear combinations of the form

$$g = \sum_0^\nu c_k f_k, \text{ where } f_k = U^k f, \tag{31}$$

with non-negative coefficients whose sum is 1. The means

$$\varphi_n = \frac{1}{n} \sum_0^{n-1} f_k \text{ and } \varphi_{m,n} = \frac{1}{n-m} \sum_m^{n-1} f_k$$

[23] R. RIESZ [**21**].

clearly belong to G. We shall see that $\{\varphi_n\}$ and $\{\varphi_{m,n}\}$ are minimizing sequences. Since

$$\|\varphi_{m,n}\| = \|U^m\varphi_{n-m}\| \leqq \|\varphi_{n-m}\|,$$

it suffices to consider the sequence $\{\varphi_n\}$. Hence we must show that, for arbitrarily small $\varepsilon > 0$ and for n sufficiently large,

$$\|\varphi_n\| < \mu + \varepsilon.$$

To do this, first let g be an element of G such that

$$\|g\| < \mu + \frac{\varepsilon}{2}, \tag{32}$$

and starting with g, form the analogues g_k and ψ_n of f_k and φ_n. In

$$\psi_n = \frac{1}{n}\sum_0^{n-1} g_k,$$

we write for $g_0 = g$ the second member of (31), and for the other g_k we substitute the expressions which result from this second member when we apply to it the iterates of U. Since $\Sigma c_k = 1$, it is clear that the expression for the ψ_n thus formed and the expression for the φ_n, which are linear combinations of the f_k, agree in their terms from the index ν up to the index $n - 1$, and that consequently these terms cancel in the difference $\varphi_n - \psi_n$. But the other terms in this difference, which are 2ν in number, have coefficients which do not exceed $\frac{1}{n}$ in modulus. Therefore

$$\|\varphi_n - \psi_n\| \leqq 2\nu\,\frac{1}{n}\,\|f\| < \frac{\varepsilon}{2} \tag{33}$$

for n sufficiently large. On the other hand, since U is a contraction, inequality (32) for g implies the analogous inequality for the g_k, hence also for ψ_n, and finally, by (33),

$$\|\varphi_n\| \leqq \|\varphi_n - \psi_n\| + \|\psi_n\| < \mu + \varepsilon,$$

which was to be proved.

Since the sequences $\{\varphi_n\}$ and $\{\varphi_{n+1} = U\varphi_n\}$ are therefore minimizing sequences, they converge to the same limit φ and we have $U\varphi = \varphi$, which completes the proof of the ergodic theorem.

The same method permits us to establish two more general results. The first, to which the method applies in an obvious manner, shows that our theorem remains valid when we replace Hilbert space by a uniformly convex but otherwise arbitrary, Banach space. A *uniformly convex* Banach space is one for which the norm $\|f\|$ satisfies the hypothesis that the assumptions

$$\|f\| \leq 1 + \varepsilon, \ \|g\| \leq 1 + \varepsilon, \text{ and} \left\| \frac{f+g}{2} \right\| > 1$$

imply that $\|f - g\|$ becomes arbitrarily small with ε. In fact, this hypothesis will serve instead of relation (30), which holds only in Hilbert space, to assure the convergence of the minimizing sequences of the convex set G. As was shown by CLARKSON [1], the functional spaces L^p belong to this category of spaces when $p > 1$; but this is not true when $p = 1$. Moreover, we know that *uniformly convex spaces are only a particular case of reflexive spaces.*[24]

The second generalization to which our method applies is that given by DUNFORD [1]; instead of a single contraction, several mutually permutable contractions are considered. For definiteness, we restrict ourselves to the case of *two contractions.* Then we have the

THEOREM. *Let T and U be two permutable contractions of Hilbert space, or, more generally, of a uniformly convex Banach space, and let f be a given element of this space. Setting*

$$f_{i,k} = T^i U^k f \qquad (i, k = 0, 1, \ldots),$$

the arithmetic means

$$\varphi_{m,n;m',n'} = \frac{1}{(n-m)(n'-m')} \sum_{i=m}^{n-1} \sum_{k=m'}^{n'-1} f_{i,k}$$

converge as $n - m \to \infty$, $n' - m' \to \infty$ to an element φ which is invariant with respect to T and U.

The proof differs only slightly from the preceding. We have only to add a single remark. We observe, passing to geometric language, that the above means of the $f_{i,k}$ correspond to the points (i, k) with integral coordinates lying in certain rectangles which we let increase indefinitely. Now the rôle of these rectangles can be taken by other figures, for example, by a sequence of convex figures whose area increases indefinitely, while the quotient of the perimeter by the area converges to 0. The proof remains the same; only the evaluation (33) must be replaced—by the obvious fact that the number of points (i, k) whose distance to the contour is less than a fixed quantity, divided by the total number of points lying in the respective figure, becomes arbitrarity small with $1/n$.

146. Semi-Groups of Non-permutable Contractions

When we wish to proceed further, one of the most interesting tasks is to consider what our theorem becomes when we pass from iterates of a single

[24] Cf. PETTIS [1]; MILMAN [1].

contraction to an arbitrary semi-group $\{U\}$ of contractions of Hilbert space, that is, to a set of contractions, permutable or not, which includes the identity transformation, and which contains, with U_1 and U_2, their product U_1U_2.

The simplest problem in this direction — namely, if the DUNFORD theorem remains valid when T and U are no longer permutable — already shows the difficulties which arise in this way. Here is how these difficulties, without being conquered, can at least be modified by an idea due to ALAOGLU and GARRETT BIRKHOFF [**1**].

Let a semi-group of contractions of Hilbert space be given. We form all the "means" in the general sense of the elements of $\{U\}$, that is, all the transformations of the type

(34) $$T = c_1 U_1 + c_2 U_2 + \dots + c_n U_n,$$

where the U_i are selected from $\{U\}$, $c_i \geq 0$, and $\Sigma c_i = 1$. We shall say that the element g of $\mathfrak{H}$ is a *successor* of the element f if there exists a mean T of this type which transforms f into g. Then the result of ALAOGLU and BIRKHOFF is the

THEOREM. *The successors of each fixed element f_0 of $\mathfrak{H}$ converge to a determined element φ in the following sense (of Moore-Smith): to every successor f_1 of f_0 and to every positive ε we can assign a successor f_2 of f_1 such that*

$$\|\varphi - f\| < \varepsilon$$

for every successor f of f_2.

For consider the subspace $\mathfrak{L}$ of $\mathfrak{H}$ consisting of those elements invariant with respect to all the contractions U belonging to the given semi-group $\{U\}$, or equivalently, invariant with respect to all the adjoint transformations U^* (cf. Sec. 144). The subspace $\mathfrak{M}$ of elements orthogonal to $\mathfrak{L}$ is then transformed into itself by all the U; in fact, if $(f, g) = 0$ for all elements g of $\mathfrak{L}$, we have

$$(Uf, g) = (f, U^*g) = (f, g) = 0.$$

The successors of an element h_1 of $\mathfrak{M}$ obviously form a convex set G. Its closure $\bar{G}$ then admits a unique minimal element ψ. Since the U are contractions, $\|U\psi\| \leq \|\psi\|$, and since $U\psi$ also belongs to $\bar{G}$, we have, by virtue of the uniqueness of the minimal element of $\bar{G}$,

$$U\psi = \psi.$$

Hence the element ψ belongs not only to the subspace $\mathfrak{M}$, but also to the subspace $\mathfrak{L}$ of invariant elements, and this is not possible, since $\mathfrak{M}$ and $\mathfrak{L}$ are orthogonal, unless $\psi = 0$. This means that there is in G—that is, among the successors of h_1—elements of arbitrarily small norm; if h_2 is a successor of h_1 such that $\|h_2\| < \varepsilon$, we also have $\|h\| < \varepsilon$ for every successor $h = Th_2$ of h_2, since the transformations T of type (34) are obviously contractions also.

This being the case, let f_0 be an arbitrary element of $\mathfrak{H}$, and let $f_0 = g_0 + h_0$ be its decomposition into the sum of elements belonging respectively to the subspaces $\mathfrak{L}$ and $\mathfrak{M}$. Since the element g_0 is invariant with respect to all the U, hence also with respect to all the T, all its successors are equal to it; hence the successors of f_0 are also of the type $f_1 = g_0 + h_1$, where h_1 belongs to $\mathfrak{M}$. According to what precedes, every successor of this type admits a successor $f_2 = g_0 + h_2$ all of whose successors $f = g_0 + h$ are such that $\|h\| < \varepsilon$. This proves the theorem with $\varphi = g_0$; the element φ is therefore only the orthogonal projection of the given element f_0 onto the subspace $\mathfrak{L}$ consisting of the invariant elements.

This result also holds for certain more general Banach spaces, in particular for the spaces L^p ($p > 1$), but surprisingly enough, it does not hold in other uniformly convex spaces. A characterization of the spaces admitted, and also a rich collection of other generalizations of the ergodic theorem for linear transformations, are to be found in the works of ALAOGLU and G. BIRKHOFF[25] already cited.

[25] We have considered only the "statistical" ergodic theorem of VON NEUMANN and its generalizations. We refer the reader to HOPF [*] or to F. RIESZ [**20**] for the "individual" ergodic theorem of G. D. BIRKHOFF and its generalizations.

CHAPTER XI

SPECTRAL THEORIES FOR LINEAR TRANSFORMATIONS OF GENERAL TYPE

APPLICATIONS OF METHODS FROM THE THEORY OF FUNCTIONS

147. The Spectrum. Curvilinear Integrals

The spectral theory of non-normal linear transformations in Hilbert space and linear transformations in spaces of more general type is still relatively undeveloped. In the particular case of completely continuous transformations, we have, of course, the theory to which we devoted Chapters IV to VI of this book. Among other things, we have seen that for a completely continuous transformation K, the transformation K_λ defined by the equation $I + \lambda K_\lambda = (I - \lambda K)^{-1}$ is an analytic function of the complex parameter λ which has no singularities other than poles. The transformation

(1) $$R_z = (K - zI)^{-1} = -\frac{1}{z}\left(I + \frac{1}{z} K_{\frac{1}{z}}\right),$$

which is called the *resolvent transformation*, then possesses the same properties except perhaps at the point $z = 0$, which is the only possible accumulation point of the poles of R_z as an analytic function of z.

In the general case of an arbitrary linear transformation T of a Banach space $\mathfrak{B}$, we decompose the complex plane into two complementary sets: the *resolvent set* $\varrho(T)$ and the *spectrum* $\sigma(T)$, where $\varrho(T)$ consists of those points z for which

$$R_z = (T - zI)^{-1}$$

exists and is a bounded transformation with domain dense in $\mathfrak{B}$. If T is closed, R_z is also closed and then it is necessarily defined on the entire space $\mathfrak{B}$. In particular, this is the case if T itself is *defined everywhere* and *bounded*, which, for simplification, we shall assume to be the case; part of the results extend also to unbounded, but closed, linear transformations.[1]

It follows from Sec. 67, or rather we prove directly, that the resolvent set

[1] Cf. TAYLOR [**3**].

$\varrho(T)$ is an open set containing the entire exterior of the circle $|z| = \|T\|$, and that in $\varrho(T)$, R_z is a holomorphic function of z. More precisely,

a) *The resolvent set* $\varrho(T)$ *contains, together with a point* ζ, *the entire neighborhood*

$$|z - \zeta| < \|R_\zeta\|^{-1}, \tag{2}$$

and for these z *there is a development into an entire series which converges in norm,*

$$R_z = R_\zeta + (z - \zeta)R_\zeta^2 + (z - \zeta)^2 R_\zeta^3 + \dots; \tag{3}$$

b) $\varrho(T)$ *contains all the points* z *such that*

$$|z| > \|T\| \tag{4}$$

and for these z *there is a development into a Laurent series which converges in norm,*

$$R_z = -\frac{1}{z} I - \frac{1}{z^2} T - \frac{1}{z^3} T^2 - \dots. \tag{5}$$

In fact, conditions (2) and (4) imply that the respective series are majorized in norm by convergent geometric series, hence they converge in norm, and relations (3) and (5) are verified by multiplying the series in the second member term by term by $T - zI = (T - \zeta I) - (z - \zeta)I$.

If we consider the transformation T^n instead of T, where n is a positive integer, we see that $\varrho(T^n)$ contains all the points $\zeta = z^n$ for which $|\zeta| > \|T^n\|$, that is to say for which $|z| > \|T^n\|^{\frac{1}{n}}$. It follows from the obvious relation

$$T^n - z^n I = (T^{n-1} + zT^{n-2} + z^2 T^{n-3} + \dots + z^{n-1} I)\,(T - zI) \tag{6}$$

that z then belongs to $\varrho(T)$, namely we have

$$(T - zI)^{-1} = (T^n - z^n I)^{-1}\,(T^{n-1} + zT^{n-2} + z^2 T^{n-3} + \dots + z^{n-1} I).$$

Proposition b) can therefore be generalized as follows:

$\varrho(T)$ *contains all points* z *for which*

$$|z| > \|T^n\|^{\frac{1}{n}}$$

for a positive integer n, *that is, all points* z *for which*

$$|z| > \inf_n \|T^n\|^{\frac{1}{n}}. \tag{7}$$

We note further the important relation

$$R_z - R_\zeta = (z - \zeta) R_z R_\zeta, \tag{8}$$

valid for any two points of $\varrho(T)$; this relation follows from formula (1) of

Chapter IV,. or it can be verified directly by multiplying the two members by $(T - zI)(T - \zeta I)$.

The fact that the resolvent transformation T_z behaves like an ordinary holomorphic function suggests that we apply the various methods from the theory of functions to its study, in particular, the *calculus of residues*. One of the authors, in his book published in 1913,[2] has shown how this method permits, in particular, the decomposition of the transformation T in terms of the *disjoint* parts of its spectrum. A decomposition of T in terms of *arbitrary* parts of the spectrum, which would be analogous to the spectral decomposition of self-adjoint or normal transformations, is a much more delicate problem which still awaits a solution. The powerful development and the success of the theory of bounded and unbounded self-adjoint and normal transformations seem to have distracted attention from the problem of transformations of general type. During recent years, mathematicians have begun again to turn to this problem; the method mentioned above has been elaborated and applied with success to several domains of research.[3]

We begin with several almost obvious remarks:

Let $T(z)$ be a bounded linear transformation which is a holomorphic function of the complex parameter z in a certain domain D of the complex plane, in the sense that it can be developed in a neighborhood of every point ζ of D into an entire series analogous to (3) and convergent in norm:

$$T(z) = T(\zeta) + (z - \zeta)T_1(\zeta) + (z - \zeta)^2 T_2(\zeta) + \dots . \tag{9}$$

If C is a rectifiable curve lying in D, we can form the integral

$$\int_C T(z)dz, \tag{10}$$

which is, as usual, the limit (in norm) of sums of the type

$$\sum T(\zeta_k)(z_k - z_{k-1}).$$

What is essential for our purpose is that if the curve C is closed and if we deform it into a curve C' in a continuous manner and without leaving the domain D, the value of the integral (10) does not change.

In fact, we can then pass from C to C' by adding to the oriented curve C a finite number of oriented rectifiable closed curves K_j, each lying in a sufficiently small neighborhood of a point ζ_j of D so that development (9) is valid there; now the integral of $T(z)$ along K_j, evaluated by integrating series (9) term by term, is equal to 0.

In particular, it follows that if the closed curve C can be deformed in D so that it contracts to a single point, integral (10) is equal to O.

[2] F. Riesz [*] (pp. 117–121).

[3] Cf. Gelfand [**1**]; Lorch [**4**]; Dunford [**2**], [**3**]; Taylor [**1**]; Hille [*].

As a first application, we shall prove the

THEOREM.[4] *The spectrum of a bounded linear transformation is never empty unless the space reduces to the single element* 0.

In the contrary case, R_z would be holomorphic in the entire complex plane, and consequently we should have

$$\int_C R_z dz = O$$

for every closed rectifiable curve C. But if C is a circle with radius greater than $\|T\|$ and with center at $z = 0$, we can evaluate this integral by integrating series (5) term by term, and we thus obtain the value $-2\pi i \cdot I$. Now $O \neq I$ except in the obvious case where the space consists of the single element 0, which proves the theorem.

It follows from the results obtained that the spectrum $\sigma(T)$ is a non-empty closed set, and that it lies in the closed disc

(11) $$|z| \leq \inf_n \|T^n\|^{\frac{1}{n}}.$$

148. Decomposition Theorem

An *admissible domain* (with respect to the transformation T) is any bounded open set D of complex numbers, connected or not, whose boundary ∂D consists of a finite number of closed rectifiable curves, lying in the resolvent set $\varrho(T)$ and oriented in conformity with the orientation of D as an open subset of the oriented complex plane.

We denote by σ an *isolated part* of $\sigma(T)$, that is, a part which is at a positive distance from the complementary part $\bar{\sigma} = \sigma(T) - \sigma$; we allow σ or $\bar{\sigma}$ to be empty. There then exists an admissible domain D such that

(12) $$\sigma = \sigma(T) \cap D.$$

This is clear if σ coincides with $\sigma(T)$ or if σ is empty. In the case where σ is at a distance d from $\bar{\sigma}$, we can take for D the union of a finite number of open discs covering σ of radius $\frac{d}{2}$, with centers lying on σ; the existence of such a finite system of discs follows immediately from the Borel covering theorem.

Let D be an arbitrary admissible domain satisfying condition (12); since its boundary ∂D lies in $\varrho(T)$, we can form the integral

(13) $$-\frac{1}{2\pi i} \int_{\partial D} R_z dz.$$

[4] WINTNER [**1**] (p. 242–243); STONE [*] (p. 149).

A simple argument of a topological nature shows that if D' is another admissible domain satisfying (12), we can pass from D to D' by a finite number of cuts and by continuous deformations of the boundary made without leaving the resolvent set $\varrho(T)$, that is, the domain of holomorphy of R_z, which implies that the integral (13) will be the same for D' as for D, and that, consequently it depends only on the part σ of the spectrum in question. We shall express this fact by denoting the value of the integral (13) by P_σ; it is a bounded linear transformation.

If T is a self-adjoint (or normal) transformation of Hilbert space, P_σ is the orthogonal projection onto the proper subspace corresponding to the part σ of the spectrum (see Secs. 135 and 136). We are going to see that in the general case, where we deal with transformations of an arbitrary Banach space $\mathfrak{B}$, P_σ will be a parallel projection (cf. Sec. 78 for terminology).

Since the distance from σ to the closed set complementary to D is positive, there exists an admissible domain D' such that $\sigma \subset D' \subset D$ whose boundary lies in the *interior* of D. We shall have

$$P_\sigma = -\frac{1}{2\pi i}\int_{\partial D} R_z\,dz = -\frac{1}{2\pi i}\int_{\partial D'} R_{z'}\,dz',$$

hence

$$P_\sigma^2 = \frac{1}{(2\pi i)^2}\int_{\partial D}\int_{\partial D'} R_z R_{z'}\,dz dz' = \frac{1}{(2\pi i)^2}\int_{\partial D}\int_{\partial D'} \frac{R_z - R_{z'}}{z - z'}\,dz dz' =$$

$$= \frac{1}{(2\pi i)^2}\int_{\partial D} R_z dz \int_{\partial D'} \frac{1}{z - z'}\,dz' - \frac{1}{(2\pi i)^2}\int_{\partial D'} R_{z'}\,dz' \int_{\partial D} \frac{1}{z - z'}\,dz.$$

But

$$\int_{\partial D'} \frac{1}{z - z'}\,dz' = 0 \quad \text{and} \quad \int_{\partial D} \frac{1}{z - z'}\,dz = 2\pi i,$$

since z lies in the exterior of the domain D' and z' lies in the interior of the domain D.

Consequently $P_\sigma^2 = P_\sigma$, and this is precisely the characteristic property of parallel projections. The space $\mathfrak{B}$ can be decomposed into the vector sum of two linearly independent subspaces $\mathfrak{M}_\sigma$ and $\mathfrak{N}_\sigma$, where $\mathfrak{M}_\sigma$ consists of the elements f of $\mathfrak{B}$ for which $P_\sigma f = f$, and $\mathfrak{N}_\sigma$ of the elements g of $\mathfrak{B}$ for which $P_\sigma g = 0$. P_σ is therefore the "parallel projection onto $\mathfrak{M}_\sigma$ in the direction of $\mathfrak{N}_\sigma$."

Since T is permutable with R_z and hence also with the projection P_σ, it transforms each of the subspaces $\mathfrak{M}_\sigma$ and $\mathfrak{N}_\sigma$ into itself. Thus the study of T in $\mathfrak{B}$ reduces to the study of its parts T' in $\mathfrak{M}_\sigma$ and T'' in $\mathfrak{N}_\sigma$. We denote the respective resolvents by R_z' and by R_z''.

It is clear that the resolvent set $\varrho(T)$ is equal to the intersection of the resolvent sets of T' and of T'':

$$\varrho(T) = \varrho(T') \cap \varrho(T'').$$

But these two sets $\varrho(T')$ and $\varrho(T'')$ are not equal; in fact, we shall show that $\varrho(T')$ contains the entire exterior, and $\varrho(T'')$ the entire interior, of the domain D.

We start from the obvious relation

$$(T - \zeta I)R_z = (T - zI)R_z + (z - \zeta)R_z = I + (z - \zeta)R_z;$$

it follows that

$$(T - \zeta I) \frac{1}{2\pi i} \int_{\partial D} \frac{R_z}{z - \zeta}\, dz = \frac{1}{2\pi i} \int_{\partial D} \frac{dz}{z - \zeta} \cdot I + \frac{1}{2\pi i} \int_{\partial D} R_z\, dz =$$

$$= \begin{cases} 0 \cdot I - P_\sigma = -P_\sigma, & \text{if } \zeta \text{ lies in the exterior of the domain } D, \\ 1 \cdot I - P_\sigma = I - P_\sigma, & \text{if } \zeta \text{ lies in the interior of the domain } D. \end{cases}$$

Noting also the permutability of the transformations which appear, this expresses that for every value of ζ lying in the exterior of D, R'_ζ exists and equals the part of

$$-\frac{1}{2\pi i} \int_{\partial D} \frac{R_z}{z - \zeta}\, dz$$

in $\mathfrak{M}_\sigma$, while for every value of ζ lying in the interior of D, R''_ζ exists and equals the part of

$$\frac{1}{2\pi i} \int_{\partial D} \frac{R_z}{z - \zeta}\, dz$$

in $\mathfrak{N}_\sigma$.

It follows from these considerations that $\sigma(T')$ is equal to the part of $\sigma(T)$ lying in D — that is, to σ — and that $\sigma(T'')$ is equal to the part of $\sigma(T)$ lying in the exterior of D, that is, to $\bar{\sigma}$.

Since the spectrum of a bounded linear transformation is empty only in the obvious case where the space reduces to the single element 0, the set $\bar{\sigma} = \sigma(T'')$ is empty only if $\mathfrak{N}_\sigma = (0)$, hence if $P_\sigma = I$. This means that $P_\sigma = I$ if and only if σ coincides with $\sigma(T)$.

If we had started from the set $\bar{\sigma}$ instead of σ, we would have arrived at the projection $P_{\bar{\sigma}} = I - P_\sigma$, and consequently at the subspaces

$$\mathfrak{M}_{\bar{\sigma}} = \mathfrak{N}_\sigma, \; \mathfrak{N}_{\bar{\sigma}} = \mathfrak{M}_\sigma.$$

In fact, we can choose two *disjoint* admissible domains D and $\bar{D}$ such that $\sigma(T) \cap D = \sigma$, $\sigma(T) \cap \bar{D} = \bar{\sigma}$, and since $D \cup \bar{D}$ is then an admissible domain containing the entire spectrum, we have

$$P_\sigma + P_{\bar{\sigma}} = -\frac{1}{2\pi i}\left(\int_{\partial D} R_z\,dz + \int_{\partial \bar{D}} R_z\,dz\right) = -\frac{1}{2\pi i}\int_{\partial(D\cup\bar{D})} R_z\,dz = P_{\sigma(T)} = I.$$

Hence we have the

DECOMPOSITION THEOREM. *Let T be a bounded linear transformation of the Banach space $\mathfrak{B}$ and let σ and $\bar{\sigma}$ be two complementary isolated parts of its spectrum; we permit σ or $\bar{\sigma}$ to be empty. We can then decompose the space $\mathfrak{B}$ into the vector sum of two linearly independent subspaces, $\mathfrak{M}_\sigma$ and $\mathfrak{M}_{\bar{\sigma}}$, each of which is transformed by T into itself, and with the property that the transformation T restricted to $\mathfrak{M}_\sigma$ or $\mathfrak{M}_{\bar{\sigma}}$ has its spectrum equal to σ or $\bar{\sigma}$, respectively. The parallel projection of $\mathfrak{B}$ onto $\mathfrak{M}_\sigma$ in the direction of $\mathfrak{M}_{\bar{\sigma}}$ is equal to the integral*

$$P_\sigma = -\frac{1}{2\pi i}\int_{\partial D} R_z\,dz$$

taken along the boundary of an arbitrary domain D which is admissible with respect to T and such that $\sigma = \sigma(T) \cap D$. We have $P_\sigma = I$, $P_{\bar{\sigma}} = 0$ if and only if σ coincides with $\sigma(T)$ and $\bar{\sigma}$ is empty.

This theorem can be considered as a generalization of the decomposition theorem which we obtained in Chapter IV for completely continuous linear transformations. We recall that, for T linear and completely continuous, the spectrum $\sigma(T)$ (consisting of the reciprocals of the "singular" values and perhaps also of the point $z = 0$) can have only the single accumulation point $z = 0$. Consequently every point $z_0 \neq 0$ of $\sigma(T)$ is an isolated point of $\sigma(T)$; we can isolate it from the others and from the point $z = 0$, for example, by a small circle C with center z_0. By the obvious relation

$$R_z = \frac{T-(T-zI)}{z} R_z = T\frac{R_z}{z} - \frac{I}{z}$$

we shall have

$$P_{z_0} = -\frac{1}{2\pi i}\int_C R_z\,dz = T\left(-\frac{1}{2\pi i}\int_C \frac{R_z}{z}\,dz\right).$$

The projection P_{z_0} is therefore the product of the completely continuous transformation T by a bounded linear transformation, which implies that this projection itself is completely continuous (cf. Sec. 76). Now P_{z_0} leaves the elements of the corresponding subspace $\mathfrak{M}_{z_0}$ invariant, and consequently every infinite and bounded set of elements of $\mathfrak{M}_{z_0}$ must be compact, hence $\mathfrak{M}_{z_0}$ must be of *finite dimension* (cf. the argument for the subspaces $\mathfrak{M}_n$ in Secs. 77 and 89).

We return to the general case treated in the theorem and deduce several integral formulas which will be useful in the sequel.

The first formula concerns the iterates of the transformation T. It follows from equation (6) that

$$T^n R_z = z^n R_z + z^{n-1} I + \dots + z^2 T^{n-3} + z T^{n-2} + T^{n-1}$$

for every point z of $\varrho(T)$, from which we obtain that

$$(14) \qquad T^n P_\sigma = -\frac{1}{2\pi i} \int_{\partial D} T^n R_z \, dz = -\frac{1}{2\pi i} \int_{\partial D} z^n R_z \, dz,$$

since the integrals of the other terms are zero.

The second formula concerns the iterates of the resolvent transformation R_a, that is, the transformations $(T - aI)^{-n}$ $(n = 1, 2, \dots)$, a being an arbitrary point of the resolvent set. Starting with the obvious relation

$$(\zeta R_a^n + \zeta^2 R_a^{n-1} + \dots + \zeta^n R_a)\,[(T - aI) - \zeta^{-1} I] = \zeta^n I - R_a^n,$$

we obtain, by setting $\zeta = (z - a)^{-1}$, multiplying by R_z, and rearranging,

$$R_a^n R_z = (z - a)^{-n} R_z - (z - a)^{-n} R_a - (z - a)^{-n+1} R_n^2 - \dots - (z - a)^{-1} R_a^n.$$

It follows that

$$(15) \qquad (T - aI)^{-n} P_\sigma = -\frac{1}{2\pi i} \int_{\partial D} R_a^n R_z \, dz = -\frac{1}{2\pi i} \int_{\partial D} (z - a)^{-n} R_z \, dz,$$

at least if the admissible domain D satisfies, in addition to the condition $\sigma(T) \cap D = \sigma$, the condition that the point a be exterior to D; this can be required without loss of generality, since a is in $\varrho(T)$.

If we assign to every rational function

$$u(z) = \sum_{k=0}^{n} c_k z^k + \sum_{k=1}^{n_1} c_{1k} (z - a_1)^{-k} + \dots + \sum_{k=1}^{n_m} c_{mk} (z - a_m)^{-k}$$

whose poles $a_1, \dots, a_m$ belong to $\varrho(T)$ the transformation

$$u(T) = \sum_{k=0}^{n} c_k T^k + \sum_{k=1}^{n_1} c_{1k} (T - a_1 I)^{-k} + \dots + \sum_{k=1}^{n_m} c_{mk} (T - a_m I)^{-k},$$

we obtain from (14) and (15), by addition, the more general formula

$$(16) \qquad u(T) P_\sigma = -\frac{1}{2\pi i} \int_{\partial D} u(z) R_z \, dz,$$

where D is an arbitrary admissible domain such that $\sigma(T) \cap D = \sigma$ which contains the poles $a_1, \dots, a_m$ neither in its interior nor on its boundary.

149. Relations Between the Spectrum and the Norms of Iterated Transformations

We consider the particular case where the two complementary parts of the spectrum $\sigma(T)$ can be separated by the circle C:

$$|z - a| = r$$

lying in the resolvent set $\varrho(T)$. Denoting by σ the part of the spectrum which is interior to C, we have

$$P_\sigma = -\frac{1}{2\pi i} \int_C R_z \, dz = -\frac{1}{2\pi i} \lim_{n\to\infty} \sum_{k=0}^{n-1} R_{z_k}(z_{k+1} - z_k),$$

where $z_k = a + r\varepsilon^k$, $\varepsilon = e^{\frac{2\pi i}{n}}$, and the convergence is convergence in norm. Setting $U = \frac{1}{r}(T - aI)$, it follows that

$$P_\sigma = \lim_{n\to\infty} \frac{\varepsilon - 1}{2\pi i} \sum_{k=0}^{n-1} (I - \varepsilon^{-k}U)^{-1} = \lim_{n\to\infty} n \frac{\varepsilon - 1}{2\pi i} (I - U^n)^{-1} = \lim_{n\to\infty} (I - U^n)^{-1}$$

where we have used the identity

$$\frac{1}{1 - x^n} = \frac{1}{n} \sum_{k=0}^{n-1} \frac{1}{1 - \varepsilon^{-k}x},$$

which clearly remains valid when the indeterminate x is replaced by the transformation U, and the fact that $n(\varepsilon - 1)/2\pi i \to 1$. We obtain in this way the

THEOREM.[5] *If a subset σ of the spectrum of the transformation T can be separated from the rest by the circle $|z - a| = r$ lying in the resolvent set and containing σ in its interior, then*

$$P_\sigma = \lim_{n\to\infty} \left[I - \left(\frac{T - aI}{r} \right)^n \right]^{-1}$$

in the sense of convergence in norm.

Under the same conditions we have, by (16),

$$(T - aI)^n P_\sigma = -\frac{1}{2\pi i} \int_C (z - a)^n R_z \, dz \quad (n = 0, 1, \ldots),$$

from which it follows that

$$\|(T - aI)^n P_\sigma\| \leq \frac{1}{2\pi} 2\pi r \cdot r^n \cdot \max_C \|R_z\|$$

[5] LORCH [**4**].

and consequently that

$$\limsup_{n\to\infty} \|(T - aI)^n P_\sigma\|^{\frac{1}{n}} \leq r.$$

If we vary r subject to C remaining in $\varrho(T)$, the greatest lower bound of r will be equal to the maximum distance from the point a to the point of σ, from which it follows that

$$\limsup_{n\to\infty} \|(T - aI)^n P_\sigma\|^{\frac{1}{n}} \leq \max_{z\in\sigma} |z - a|. \tag{17}$$

This permits us, in this case, to characterize the corresponding subspace $\mathfrak{M}_\sigma$ in the following direct manner: For an element f of $\mathfrak{M}_\sigma$ one has $f = P_\sigma f$ and it follows from (17) that

$$\limsup_{n\to\infty} \|(T - aI)^n f\|^{\frac{1}{n}} < r. \tag{18}$$

Conversely, let f be an arbitrary element of the space $\mathfrak{B}$ for which inequality (18) holds. The series

$$-\sum_{n=0}^{\infty} \frac{1}{(z-a)^{n+1}} (T - aI)^n f$$

is then uniformly convergent on the circle C of radius r. Its sum h_z clearly satisfies the relation $(T - aI)h_z = (z - a)h_z + f$, therefore $(T - zI)h_z = f$, $R_z f = h_z$, and integrating term by term,

$$P_\sigma f = -\frac{1}{2\pi i} \int_C R_z f dz = f.$$

We thus have the following result:

THEOREM. *Under the conditions of the preceding theorem, the elements f of the subspace $\mathfrak{M}_\sigma$ which corresponds to the subset σ of the spectrum can be characterized by inequality* (18).

If, in particular, σ reduces to the single point $z = a$, which is then an isolated point of the spectrum, $\mathfrak{M}_\sigma$ consists of the elements f for which (18) holds for every sufficiently small positive r, hence for which

$$\lim_{n\to\infty} \|(T - aI)^n f\|^{\frac{1}{n}} = 0;$$

$\mathfrak{M}_\sigma$ contains, in particular, all the elements f such that

$$(T - aI)^n f = 0$$

for some exponent $n \geq 0$.

Let us now assume that σ coincides with the entire spectrum $\sigma(T)$. Then

$P_\sigma = I$, and it follows from (17), by writing it for $a = 0$, that

(19) $$\limsup_{n\to\infty} \|T^n\|^{\frac{1}{n}} \leq r_T,$$

where r_T denotes the *spectral radius* of T, that is, the radius of the smallest closed disk centered at 0 which contains all the spectrum of T.

On the other hand, inequality (11) of Sec. 147 shows us that

$$r_T \leq \inf_n \|T^n\|^{\frac{1}{n}} \leq \liminf_{n\to\infty} \|T^n\|^{\frac{1}{n}},$$

and comparing this with inequality (19) we arrive at the following result:

THEOREM.[6] *For every bounded linear transformation T, the limit*

$$\lim_{n\to\infty} \|T^n\|^{\frac{1}{n}}$$

exists and is equal to the spectral radius r_T of T.[7]

It follows that the limit of $\|(T - aI)^n\|^{\frac{1}{n}}$ is equal to the maximum distance from the point $z = a$ to the points of $\sigma(T)$. In particular,

$$\lim_{n\to\infty} \|(T - aI)^n\|^{\frac{1}{n}} = 0$$

means that the spectrum of T consists of the single point $z = a$. An obvious example of a transformation of this type is furnished by $T = aI$. It is, moreover, the only transformation of this type among the normal transformations of Hilbert space. But even in Hilbert space there is an infinite number of such non-normal transformations. For example, every transformation of the space $L^2(a, b)$ of the Volterra type

$$Kf(x) = \int_a^x K(x, y)f(y)dy,$$

where $K(x, y)$ is a bounded measurable function, has the single point 0 for spectrum. In fact, we have shown in Sec. 65 that every complex value is regular with respect to a transformation of this type, hence by relation (1), Sec. 147, every value $z \neq 0$ belongs to the resolvent set.

[6] Cf. GELFAND [**1**] and, for the particular case treated in the following section, BEURLING [**1**].

[7] The fact that $\lim \|T^n\|^{1/n}$ exists already follows from the obvious inequality $\|T^{n+m}\| \leq \|T^n\|\,\|T^m\|$; in fact, for every sequence of positive numbers $\{a_n\}$ such that $a_{n+m} \leq a_n a_m$, the sequence $\{a_n^{1/n}\}$ converges to its greatest lower bound, or, what amounts to the same thing, for every "sub-additive" sequence $\{a_n\}$, i.e. sequence such that $a_{m+n} \leq a_m + a_n$, the sequence $\{a_n/n\}$ converges to its greatest lower bound (see G. PÓLYA-G. SZEGÖ, *Aufgaben und Lehrsätze aus der Analysis.* I (Berlin, 1925), p. 17, problem 98).

The existence of such a variety of linear transformations having the same spectrum concentrated in a single point brings out the difficulties of a characterization of linear transformations of general type by means of their spectra,

Let us now consider two permutable bounded linear transformations. S and T, and seek the relations among the corresponding quantities r_S, r_T, r_{ST} and r_{S+T}.

It follows immediately from the theorem we have just proved that

$$r_{ST} = \lim_{n\to\infty} \|(ST)^n\|^{\frac{1}{n}} = \lim_{n\to\infty} \|S^nT^n\|^{\frac{1}{n}} \leq \lim_{n\to\infty} \|S^n\|^{\frac{1}{n}} \|T^n\|^{\frac{1}{n}} = r_S r_T.$$

As for r_{S+T}, we observe first that

$$\|(S+T)^n\| = \left\| \sum_{\nu=0}^{n} \binom{n}{\nu} S^\nu T^{n-\nu} \right\| \leq \sum_{\nu=0}^{n} \binom{n}{\nu} \|S\|^\nu \|T^{n-\nu}\|. \tag{20}$$

We take arbitrary quantities $p > r_S$ and $q > r_T$; there will be an integer $m > 0$ such that

$$\|S^n\|^{\frac{1}{n}} < p \text{ and } \|T^n\|^{\frac{1}{n}} < q \text{ for every } n \geq m,$$

whereas, for arbitrary n,

$$\|S^n\|^{\frac{1}{n}} \leq \|S\| = s \text{ and } \|T^n\|^{\frac{1}{n}} \leq \|T\| = t.$$

It follows from (20) that, for $n > 2m$,

$$\|(S+T)^n\| \leq \sum_{\nu=0}^{m-1} \binom{n}{\nu} s^\nu q^{n-\nu} + \sum_{\nu=m}^{n-m} \binom{n}{\nu} p^\nu q^{n-\nu} + \sum_{\nu=n-m+1}^{n} \binom{n}{\nu} p^\nu t^{n-\nu} =$$

$$= \sum_{\nu=0}^{m-1} \binom{n}{\nu} p^\nu q^{n-\nu} \left(\frac{s}{p}\right)^\nu + \sum_{\nu=m}^{n-m} \binom{n}{\nu} p^\nu q^{n-\nu} + \sum_{\nu=n-m+1}^{n} \binom{n}{\nu} p^\nu q^{n-\nu} \left(\frac{t}{q}\right)^{n-\nu} \leq$$

$$\leq \left[\sum_{\nu=0}^{n} \binom{n}{\nu} p^\nu q^{n-\nu}\right] \left[\max_{0\leq k\leq m-1} \left(\frac{s}{p}\right)^k + 1 + \max_{0\leq k\leq m-1} \left(\frac{t}{q}\right)^k\right] = (p+q)^n M,$$

where M is a quantity not depending on n; therefore

$$\lim_{n\to\infty} \|(S+T)^n\|^{\frac{1}{n}} \leq \lim_{n\to\infty} (p+q) M^{\frac{1}{n}} = p + q.$$

Letting p and q tend respectively to r_S and r_T, we obtain

$$r_{S+T} \leq r_S + r_T.$$

We summarize:

For permutable S and T we have the relations[8]

$$r_{ST} \leq r_S r_T, \quad r_{S+T} \leq r_S + r_T. \tag{21}$$

[8] Obtained, in the particular case we shall study in the following section, by BEURLING [**1**].

These inequalities give us some information concerning the perturbation of the spectrum caused by the multiplication by or the addition of a permutable transformation.

150. Application to Absolutely Convergent Trigonometric Series

The relations proved in the preceding section admit an interesting application to the theory of continuous functions $f(x)$ defined on the entire real axis, periodic or almost-periodic, and having an absolutely convergent Fourier series, that is, such that

$$f(x) = \sum_{\nu} c(\nu)e^{i\nu x}$$

where

$$\sum_{\nu} |c(\nu)| < \infty, \tag{22}$$

$c(\nu)$ being zero except on a set, at most denumerable, of values ν and the sums being taken over these values.

Considering the sum of the series (22) as a sort of *norm* of f and denoting it by $\|f\|$, we obviously have

$$\|f\| \geq 0, \ \|f\| = 0 \text{ only for } f(x) \equiv 0,$$

$$\|cf(x)\| = |c| \, \|f(x)\|, \ \|f(x) + g(x)\| \leq \|f(x)\| + \|g(x)\|,$$

and we show without difficulty that with this definition of the norm the space $\mathfrak{A}$ of these functions is even *complete*, that is, that $\mathfrak{A}$ is a Banach space.

But there is more: with $f(x)$ and $g(x)$, their *product* also belongs to $\mathfrak{A}$ and

$$\|f(x)g(x)\| \leq \|f(x)\| \, \|g(x)\|;$$

in fact, if

$$f(x) = \sum c(\nu)e^{i\nu x} \text{ and } g(x) = \sum d(\nu)e^{i\nu x},$$

we have

$$f(x)g(x) = \sum_{\nu} (\sum_{\mu} c(\mu)d(\nu - \mu))e^{i\nu x}$$

and

$$\sum_{\nu} |\sum_{\mu} c(\mu)d(\nu - \mu)| \leq \sum_{\nu} \sum_{\mu} |c(\mu)d(\nu - \mu)| = \sum_{\mu} |c(\mu)| \sum_{\nu} |d(\nu)|.$$

It follows that each element f of $\mathfrak{A}$ gives rise to a linear transformation of $\mathfrak{A}$, namely

$$T_f g(x) = f(x)g(x);$$

we obviously have

$$T_{cf} = cT_f, \ T_{f_1+f_2} = T_{f_1} + T_{f_2}, \ T_{f_1 f_2} = T_{f_1} T_{f_2},$$

and consequently

$$T_{f_1} \smile T_{f_2}.$$

Since

$$\|T_f g\| \leq \|f\| \, \|g\|$$

and, for $g(x) \equiv 1$, $T_f g = f$ and $\|g\| = 1$, we have

$$\|T_f\| = \|f\|,$$

from which it also follows that

$$\|T_f^n\| = \|T_{f^n}\| = \|f^n\|.$$

Hence we have

$$r_{T_f} = \lim_{n\to\infty} \|T_f^n\|^{\frac{1}{n}} = \lim_{n\to\infty} \|f^n\|^{\frac{1}{n}}. \tag{23}$$

We shall show, with BEURLING [**1**], that this last limit is equal to

$$M_f = \sup_{-\infty < x < \infty} |f(x)|,$$

and that consequently

$$r_{T_f} = M_f.^9 \tag{24}$$

[9] If we consider the space L^2 of functions $f(x) \sim \sum_{-\infty}^{\infty} c_k e^{2\pi ikx}$ of period 1 which are square-summable and define the norm $\|f\|$ in this space by

$$\|f\| = [\int_0^1 |f(x)|^2 dx]^{\frac{1}{2}} = [\sum_{-\infty}^{\infty} |c_k|^2]^{\frac{1}{2}}$$

(and not by $\Sigma |c_k|$), the transformation $T_f g(x) = f(x)g(x)$ of the space L^2 into itself obviously is bounded if and only if the function $f(x)$ is essentially bounded, and in this case

$$\|T_f\| = \text{true max } |f(x)|.$$

In the case of the space $\mathfrak{A}$ which we are considering we have only the inequality

$$\|T_f\| \geq M_f = \max |f(x)|;$$

and the ratio $\|T_f\| : M_f$ can become as large as we wish. To show this, we have only to consider the FEJÉR means

$$s_n(x) = \sum_{k=-n}^{n} \left(1 - \frac{|k|}{n+1}\right) c_k e^{2\pi ikx}$$

associated with a bounded periodic function

$$h(x) \sim \sum_{-\infty}^{\infty} c_k e^{2\pi ikx}$$

which has at least one discontinuity.

Since

$$|f(x)| = |\sum_{\nu} c(\nu)e^{i\nu x}| \leq \sum_{\nu} |c(\nu)| = \|f\|,$$

we have

(25) $$M_f \leq \|f\|,$$

and since $M_f^n = M_{f^n}$, we also have

(25a) $$M_f = M_{f^n}^{\frac{1}{n}} \leq \|f^n\|^{\frac{1}{n}} \qquad (n = 1, 2, \ldots),$$

hence, by (23),

(26) $$M_f \leq r_{T_f}.$$

Let us assume at first that $f(x)$ is a trigonometric polynomial with p terms (in the extended sense, that is, with arbitrary real quantities ν_k):

$$f(x) = \sum_{k=1}^{p} c(\nu_k)e^{i\nu_k x}.$$

In the development

$$f^n(x) = \sum_{j=1}^{N} d(\mu_j)e^{i\mu_j x}$$

the number N of terms with distinct exponents μ_j will then be less than or equal to $\binom{p+n-1}{p-1}$, that is, to the number of combinations with repetitions of p elements n at a time. Noting that

$$\sum_{j=1}^{N} |d(\mu_j)|^2 = \lim_{a\to\infty} \frac{1}{2a} \int_{-a}^{a} |f(x)|^{2n}dx \leq M_f^{2n}$$

(see Secs. 101—102), we obtain by Cauchy's inequality,

$$\|f^n\|^{\frac{1}{n}} = \left[\sum_{j=1}^{N} |d(\mu_j)|\right]^{\frac{1}{n}} \leq M_f N^{\frac{1}{2n}}.$$

Comparing this result with inequality (25a), it follows that as $n \to \infty$ $\|f^n\|^{\frac{1}{n}}$ converges to the limit M_f, which proves relation (24) in this particular case.

The general case reduces to the case of trigonometric polynomials as follows. It is clear that for every $\varepsilon > 0$ we can decompose $f(x)$ into a sum $g(x) + h(x)$ such that $g(x)$ is a trigonometric polynomial and $\|h(x)\| < \varepsilon$. We then have on the one hand

$$r_{T_g} = M_g \leq M_f + M_h$$

where $M_h \leq \|h\|$ [see (25)], and on the other hand

$$r_{T_h} = \lim_{n} \|h^n\|^{\frac{1}{n}} \leq \|h\|.$$

Since $T_f = T_g + T_h$, we see by applying (21) that

$$r_{T_f} \leq r_{T_g} + r_{T_h} \leq M_f + 2\,\|h\| \leq M_f + 2\varepsilon.$$

Since ε was arbitrary, it follows from this that $r_{T_f} \leq M_f$; taking into account (26), this proves (24)

Relation (24) tells us that the spectrum of T_f is entirely contained in the disc $|z| \leq M_f$. But we can also use it to prove the following more precise proposition.

THEOREM. *The spectrum of the transformation T_f coincides with the closure of the set of values which are assumed by the function $f(x)$.*

Let us suppose first that z is a point of the resolvent set $\varrho(T_f)$. For every element $h(x)$ of $\mathfrak{A}$ there is then an element $g(x)$ of $\mathfrak{A}$ satisfying the equation $(T_f - zI)g(x) = h(x)$; choosing in particular $h(x) \equiv 1$, it follows that the function $\dfrac{1}{f(x) - z}$ also belongs to $\mathfrak{A}$. Consequently z can not be equal to any value taken on by $f(x)$, or to a limit of such values.

Conversely, if z is a point at a distance $\delta_z > 0$ from the set of values taken on by $f(x)$, we set

$$\Delta_z = \sup_x |f(x) - z|$$

and consider the function

$$f^*(x) = (\overline{f(x)} - \bar{z})\,(f(x) - z) - \tfrac{1}{2}(\Delta_z^2 + \delta_z^2);$$

clearly $f^*(x)$ also belongs to $\mathfrak{A}$, and we have $M_{f^*} = \frac{1}{2}(\Delta_z^2 + \delta_z^2)$. Since the point $\zeta = -\frac{1}{2}(\Delta_z^2 + \delta_z^2)$ lies in the exterior of the circle with radius M_{f^*} and center 0, it belongs to the resolvent set of T_{f^*}. The transformation

$$T_{f^*} - \zeta I = (T_{\bar{f}} - \bar{z}I)\,(T_f - zI)$$

therefore admits a bounded inverse, and hence the same is true of the transformation $T_f - zI$; that is, the point z belongs to $\varrho(T_f)$. This completes the proof of the theorem.

An immediate consequence of this theorem is the following theorem, due to N. WIENER.[10]

THEOREM. *Let $f(x)$ be a real or complex-valued function which is continuous, periodic, and does not take on the value zero. If the Fourier series of $f(x)$ is absolutely convergent, the same is true of the Fourier series of the function $1/f(x)$.*

[10] WIENER [*] (§ 12). An ingenious proof of this theorem was given by GELFAND [**2**]; it is based on the theory of *normed rings* which was developed by this author, making special use of the Zermelo axiom; cf. GELFAND [**1**] or HILLE [*].

151. Elements of a Functional Calculus

We have already defined, at the end of Sec. 148, the transformation $u(T)$ for every rational function $u(z)$ no pole of which lies in the spectrum of the transformation T. We started from the canonical decomposition of $u(z)$ into the sum of a polynomial and elementary fractions; in view of the uniqueness of this decomposition, the correspondence between $u(z)$ and $u(T)$ is uniquely determined. This can also be obtained from the formula

$$u(T) = -\frac{1}{2\pi i} \int_{\partial D} u(z) R_z \, dz, \tag{27}$$

where D is an arbitrary domain which is admissible with respect to T and contains the entire spectrum $\sigma(T)$ in its interior, but contains no singularities of $u(z)$. (This is a special case of formula (16) of Sec. 148 for $\sigma = \sigma(T)$; we recall that $P_{\sigma(T)} = I$.)

For the rational functions in question, the algebraic rules of calculation, including the composition of functions, carry over to the corresponding transformations. These facts follow more or less easily from our definition with the aid of the canonical decomposition into the sum of a polynomial and elementary fractions, but we prefer to prove them with the aid of formula (27).

Before doing so, we observe that formula (27) offers us a possibility of extending the correspondence between functions and transformations beyond the field of rational functions. In order to be sure that the value of the integral does not depend on the particular choice of the domain D, it will, however, be necessary to limit ourselves to analytic functions, or at least to "piecewise" analytic functions. More precisely, we shall consider functions $u(z)$ which are defined and differentiable with respect to the complex variable z at all points of some open set E_u containing the spectrum of T. (As the notation indicates, the set E_u can vary with u.) We denote the *class* of these functions by $\mathcal{F}(T)$.

Let D be an admissible domain with respect to T containing the entire spectrum of T, and in its turn contained, along with its boundary ∂D, in E_u. (The existence of a domain of this type is easily deduced with the aid of the Borel covering theorem.) We form the transformation $u(T)$ by the formula (27). This transformation does not depend on the particular choice of D. In fact, we can pass from D to any other domain D' of the same type by means of a finite number of cuts and by continuous deformations of the boundary, all done without leaving the common part of the sets $\varrho(T)$ and E_u; now $u(z)R_z$ is a holomorphic function of z in $\varrho(T) \cap E_u$,[11] which assures that the integrals corresponding to D and D' are equal.

[11] In fact, if z_0 is an arbitrary point of $\varrho(T) \cap E_u$, $u(z)$ and R_z can be developed about z_0 into convergent entire series in $z - z_0$; since the latter are majorized in norm by convergent geometric series, we can multiply them in the Cauchy sense and we thus obtain the entire series for $u(z)R_z$.

It is clear that the correspondence between functions and transformations, thus extended to the class of functions $\mathcal{J}(T)$, is homogeneous and additive; we have only to observe that if $u(z)$ and $v(z)$ are two functions of the class $\mathcal{J}(T)$ their sum is defined and differentiable in the open set $E_u \cap E_v \supset \sigma(T)$, and that the integral (27) is an operation which is homogeneous and additive in u. The correspondence is also multiplicative, which we prove by an argument analogous to that used in Sec. 148 to establish the relation $P_\sigma^2 = P_\sigma$; here we also make use of the classical Cauchy formulas concerning integrals of the type

$$\int_{\partial D} \frac{u(z)}{z - z'}\, dz.$$

The multiplicative property implies that the transformations obtained are all mutually *permutable*. When the function $v(z) = \dfrac{1}{u(z)}$ also belongs to the class $\mathcal{J}(T)$, which is clearly the case if the function $u(z)$ does not take on the value zero in $\sigma(T)$, we have, always by the multiplicative property,

$$u(T)v(T) = v(T)u(T) = I,$$

hence

$$v(T) = [u(T)]^{-1}.$$

Finally, since $\|R_z\|$ is a continuous function of z and therefore bounded on ∂D, it follows from (27) that the correspondence is also *continuous* in the sense that if $u_n(z)$ tends to $u(z)$ in an open set containing $\sigma(T)$ then the transformation $u_n(T)$ tends in norm to $u(T)$.

We observe further that if σ is an isolated part of the spectrum $\sigma(T)$ and if D is an admissible domain such that $\sigma(T) \cap D = \sigma$, the function which is equal to 1 in the interior of D and equal to 0 in the exterior of D belongs to the class $\mathcal{J}(T)$; the corresponding projection P_σ is therefore also a "function" of T, which would not be the case (at least if $\sigma \neq \sigma(T)$) if we had admitted to the functional calculus only functions $u(z)$ holomorphic in the *connected* domains containing $\sigma(T)$.[12]

Between the spectrum of T and its "function" $u(T)$ there is the relation

$$\sigma(u(T)) = u(\sigma(T)), \tag{28}$$

where the second member denotes the set of values assumed by the function $u(z)$ when z runs through the set $\sigma(T)$. This is the *spectral mapping theorem* of N. Dunford.

It is proved as follows. Let ζ be a point of (T). Since the function $v(z) =$

[12] The functional calculus based on formula (27), proposed by F. Riesz [*] (pp. 117–121), was elaborated in all its generality by N. Dunford [**2**], [**3**]. The theorems which follow in this section are due to this author.

$= \dfrac{u(z) - u(\zeta)}{z - \zeta}$ belongs to the class $\mathcal{J}(T)$, $v(T)$ exists and, in view of the multiplicative property of the correspondence,

$$(T - \zeta I)v(T) = u(T) - u(\zeta)I.$$

The transformation appearing in the second member can not have a bounded inverse, since if it did $T - \zeta I$ would also have a bounded inverse, namely, $v(T)\,[u(T) - u(\zeta)I]^{-1}$. Hence $u(\zeta)$ belongs to $\sigma(u(T))$, which proves that $u(\sigma(T)) \subseteq \sigma(u(T))$. On the other hand, every point μ of $\sigma(u(T))$ also belongs to $u(\sigma(T))$. In fact, if μ did not belong to $u(\sigma(T))$, the function $\chi(z) = \dfrac{1}{u(z) - \mu}$ would belong to the class $\mathcal{J}(T)$, hence $\chi(T)$ would exist and we should have, by the multiplicative property,

$$\chi(T)\,[u(T) - \mu I] = I,$$

which would contradict the fact that μ belongs to $\sigma(u(T))$. This completes the proof.

The rule for composite functions is established without difficulty:

If the function $u(z)$ belongs to the class $\mathcal{J}(T)$ and if the function $v(z)$ belongs to the class $\mathcal{J}(u(T))$, the composite function $w(z) = v(u(z))$ belongs to the class $\mathcal{J}(T)$ and $w(T) = v(u(T))$.

In order to prove this, we first choose a domain D^* admissible with respect to $u(T)$ and contained, together with its boundary, in the open set E_v corresponding to the function v. Since $u(\sigma(T)) = \sigma(u(T)) \subset D^*$, we can then choose a domain D admissible with respect to T such that the image of D and of its boundary under u will still be contained in D^*; we can require further that D and its boundary be also contained in the open set E_u corresponding to the function u. Then by the multiplicative property of the correspondence,

$$[u(T) - z^*I]^{-1} = \left[\frac{1}{u - z^*}\right](T) = -\frac{1}{2\pi i}\int_{\partial D} \frac{1}{u(z) - z^*} R_z\,dz$$

for z^* varying on ∂D^* and for $R_z = [T - zI]^{-1}$, and hence

$$v(u(T)) = -\frac{1}{2\pi i}\int_{\partial D^*} v(z^*)\,[u(T) - z^*I]^{-1}dz^* =$$

$$= \left(\frac{1}{2\pi i}\right)^2 \int_{\partial D^*}\int_{\partial D} \frac{v(z^*)}{u(z) - z^*} R_z\,dz^*dz =$$

$$= -\frac{1}{2\pi i}\int_{\partial D}\left[\frac{1}{2\pi i}\int_{\partial D^*} \frac{v(z^*)}{z^* - u(z)}\,dz^*\right] R_z\,dz = -\frac{1}{2\pi i}\int_{\partial D} v(u(z))R_z\,dz = w(T),$$

q. e. d.

152. Two Examples

Let us consider the transformation T_f of the space $\mathfrak{A}$, studied in Sec. 150. Since the spectrum of T_f coincides with the closure of the set of values of the function $f(x)$, the corresponding class $\mathcal{F}(T_f)$ consists of the functions $u(z)$ which are holomorphic in a domain containing the set (which is necessarily connected) of values of $f(x)$. Choosing the domain D in a suitable manner, we shall have for every element h of the space $\mathfrak{A}$:

$$u(T_f)h = -\frac{1}{2\pi i}\int_{\partial D} u(z)\,(T_f - zI)^{-1}\,hdz = -\frac{1}{2\pi i}\int_{\partial D} u(z)\,\frac{h(x)}{f(x) - z}\,dz =$$
$$= u(f(x))h(x).$$

Setting, in particular, $h(x) \equiv 1$, we see that the function $u(f(x))$ belongs to the space $\mathfrak{A}$ also. This fact, first observed by P. LEVY,[13] can be stated for periodic functions, as follows:

THEOREM. *Let $f(x)$ be a continuous periodic function having an absolutely convergent Fourier series. For every function $u(z)$ which is holomorphic in an open domain of the complex plane containing all the values assumed by $f(x)$, the Fourier series of the function $u(f(x))$ is also absolutely convergent.*

*

Let T be a bounded linear transformation of the Banach space $\mathfrak{B}$ and suppose that there exists an entire function $u(z)$ which is zero only for $z = 0$ and such that the transformation $u(T)$ is completely continuous.

Let $\{z_n\}$ be a sequence of points of $\sigma(T)$ tending to z^*, $z_n \neq z^*$. Then $u(z_n) \to u(z^*)$, and since the function $u(z)$ is not constant we have, at least starting with some index n, $u(z_n) \neq u(z^*)$. Since we have, furthermore, the relation $\sigma(u(T)) = u(\sigma(T))$, the points $u(z_n)$ belong to the spectrum of the completely continuous transformation $u(T)$, and consequently they can not have a limit other than 0, hence $u(z^*) = 0$, $z^* = 0$. The spectrum of T can therefore have only the single accumulation point $z^* = 0$. This means that every point $z_0 \neq 0$ of $\sigma(T)$ is an isolated part of $\sigma(T)$; we can isolate it from the rest and from the point 0 by a small circle C with center z_0 which lies in the resolvent set.

Let $v(z)$ be the function equal to $\dfrac{1}{u(z)}$ in the interior of C and zero in the exterior of C. It is clear that $v(z)$ belongs to the class $\mathcal{F}(T)$. Since the function $u(z)v(z)$ is equal to 1 in the interior, and to 0 in the exterior, of C it follows

[13] Cf. WIENER [*].

from the multiplicative property of the correspondence that

$$u(T)v(T) = P_{z_0},$$

where P_{z_0} denotes the (parallel) projection onto the subspace $\mathfrak{M}_{z_0}$ corresponding to the isolated part z_0 of the spectrum of T (Sec. 148). Since it is the product of the completely continuous transformation $u(T)$ and the continuous linear transformation $v(T)$, the projection P_{z_0} is also completely continuous, which implies that the subspace $\mathfrak{M}_{z_0}$ is of *finite dimension*. This fact assures (cf. Secs. 73 and 89) that the Fredholm alternative holds for the transformation T also.

Therefore, in particular, *the Fredholm alternative holds for T if one of its iterates T^k ($k = 2, 3, \ldots$) is completely continuous.*[14]

Moreover, it is clear it would have sufficed to assume that the function $u(z)$ was holomorphic in a connected domain containing the spectrum of T when $z \neq 0$.

VON NEUMANN'S THEORY OF SPECTRAL SETS

153. Principal Theorems

The following proposition is a simple consequence of the results of the preceding sections.

THEOREM. *Let $u(z)$ be a function of the complex variable z which is holomorphic in a domain containing the disc C_r: $|z| \leq r$ in its interior, and let*

$$|u(z)| \leq R$$

in C_r. For every linear transformation T such that

$$\lim_{n\to\infty} \|T^n\|^{\frac{1}{n}} \leq r, \tag{29}$$

$$\lim_{n\to\infty} \|[u(T)]^n\|^{\frac{1}{n}} \leq R. \tag{30}$$

[14] To illustrate the result we have just obtained, we quote a theorem due to PHILLIPS [**1**] which he obtained by making use of certain results of DUNFORD and PETTIS [**1**] on linear functionals of the space L of summable functions. According to this theorem, every linear transformation T of the space L into itself which is "weakly completely continuous," that is, which transforms bounded sets into sets which are compact with respect to weak convergence, has its square T^2 completely continuous in the ordinary sense, that is, with respect to strong convergence. This theorem and the result we have just obtained (in the particular case $u(z) = z^2$) imply that the Fredholm alternative also holds for weakly completely continuous transformations of the space L; cf. DUNFORD [**2**] (p. 208).

In fact, hypothesis (29) means that $r_T \leq r$, that is, that the spectrum of T is contained in the disc C_r. This assures the existence of $u(T)$. By virtue of the spectral mapping theorem (Sec. 151) we have, moreover, $\sigma(u(T)) = u(\sigma(T))$, hence

$$\sigma(u(T)) \subseteq u(C_r) \subseteq C_R,$$

which implies (30).

This theorem brings up the following question: If we require

$$\|T\| \leq r \tag{31}$$

instead of (29), can we conclude that

$$\|u(T)\| \leq R? \tag{32}$$

We observe that hypothesis (31) implies (29), hence also (30), but that (30) does not imply (32).

The answer to the above question is in the negative, at least for arbitrary Banach spaces.

Here is a simple counterexample. We consider the two-dimensional vector space with complex components $x = \{x_1, x_2\}$ with the following definition of the norm:

$$\|x\| = |x_1| + |x_2|;$$

let the addition of vectors and multiplication by complex numbers be defined as usual. The transformation

$$T\{x_1, x_2\} = \{x_2, x_1\}$$

of this space into itself is clearly linear and with norm equal to 1. We consider the function

$$u(z) = \frac{z + a}{1 + \bar{a}z},$$

where a is a fixed complex number with $|a| < 1$. Then $|u(z)| \leq 1$ for $|z| \leq 1$, but the norm of the transformation

$$u(T) = (T + aI)\,(I + \bar{a}T)^{-1}$$

is, in general, greater than unity.

In fact, choosing $a = ti$, $x_1 = ri$, $x_2 = 1$, $0 < t < r < 1$, we have

$$\|(T + aI)x\| - \|(I + \bar{a}T)x\| = |x_2 + ax_1| + |x_1 + ax_2| - |x_1 + \bar{a}x_2| - |x_2 + \bar{a}x_1| =$$
$$= (1 - tr) + (r + t) - (r - t) - (1 + tr) = 2t(1 - r) > 0,$$

hence

$$\|(T + aI)x\| > \|(I + \bar{a}T)x\|.$$

Setting $y = (I + \bar{a}T)x$ it follows that

$$\|u(T)y\| > \|y\|,$$

which proves that

$$\| u(T) \| > 1.$$

It is much more interesting to note that the answer to the above question is in the affirmative if the space in question is Hilbert space. By homogeneity we may limit ourselves to the case $r = R = 1$. The theorem is then stated as follows:

THEOREM A. *Let $u(z)$ be a function of the complex variable z which is holomorphic in a domain containing the disk C_1:*

$$|z| \leq 1,$$

and suppose in addition that

$$|u(z)| \leq 1$$

in C_1. Then for every bounded linear transformation T of the Hilbert space $\mathfrak{H}$ for which

$$\|T\| \leq 1$$

we have

$$\|u(T)\| \leq 1.$$

This theorem is due to J. VON NEUMANN [9]. In his original proof[15] he considered first the linear functions

$$u(z) = \frac{z + a}{1 + \bar{a}z} \qquad (|a| < 1),$$

for which the theorem is demonstrated by an immediate calculation, then he showed, by applying a theorem of I. Schur and by a rather laborious calculation, that the case of functions $u(z)$ of general type can be reduced to the particular case. An alternative proof has since been proposed by E. HEINZ [2]; it is based on the classical formula which expresses the real part of a function which is holomorphic in a circle in terms of the real part of its values on the circle. In this way one first obtains the following theorem of which Theorem A is a consequence:

THEOREM B. *Let $v(z)$ be a function of the complex variable z which is holomorphic in a domain containing the disk C_1 and is such that in this domain*

$$\operatorname{Re} v(z) \geq 0.$$

Then for every bounded linear transformation T of the Hilbert space $\mathfrak{H}$ for which

$$\|T\| \leq 1$$

we have

$$\operatorname{Re}\,(v(T)f, f) \geq 0,$$

for every element f of $\mathfrak{H}$.

[15] The proot was reproduced in the first (French) edition of this book, pp. 427—428.

We first prove theorem B.[16] Let

$$v(z) = \sum_0^\infty c_k z^k \qquad (c_k = a_k + ib_k)$$

be the entire series for $v(z)$; it converges uniformly even on a disk $|z| \leq R$ with $R > 1$, therefore we also have the following series expansion, which converges in norm:

$$v(T) = \sum_0^\infty c_k T^k.$$

Setting, for fixed f,

$$\varrho_k = \text{Re}\,(T^k f, f) \quad \text{and} \quad \sigma_k = \text{Im}\,(T^k f, f),$$

and making use of Parseval's formula for Fourier series, we obtain

$$\text{Re}(v(T)f, f) = \sum_0^\infty (a_k \varrho_k - b_k \sigma_k) = \frac{1}{\pi}\int_0^{2\pi} V_R(x) K_R(x)\,dx \tag{34}$$

where

$$V_R(x) = \sum_0^\infty R^k(a_k \cos kx - b_k \sin kx) = \text{Re}\,v(Re^{ix}) \geq 0 \tag{35}$$

and

$$K_R(x) = \tfrac{1}{2}\varrho_0 + \sum_0^\infty R^{-k}(\varrho_k \cos kx + \sigma_k \sin kx) =$$

$$= \text{Re}\,[\tfrac{1}{2}(f, f) + \sum_1^\infty R^{-k}(T^k f, f) e^{ikx}].$$

Since

$$\tfrac{1}{2}I + \sum_1^\infty R^{-k} e^{ikx} T^k = \left(I - \frac{e^{ix}}{R}T\right)^{-1} - \tfrac{1}{2}I = \tfrac{1}{2}\left(I + \frac{e^{ix}}{R}T\right)\left(I - \frac{e^{ix}}{R}T\right)^{-1},$$

we find, setting

$$g_x = \left(I - \frac{e^{ix}}{R}T\right)^{-1} f,$$

$$2K_R(x) = \text{Re}\left(\left(I + \frac{e^{ix}}{R}T\right)g_x, \left(I - \frac{e^{ix}}{R}T\right)g_x\right) = (g_x, g_x) - \frac{1}{R^2}(Tg_x, Tg_x).$$

Since $\|T\| \leq 1$ and $R > 1$, it follows that $K_R(x) \geq 0$. Together with (34) and (35), this proves that $\text{Re}\,(v(T)f, f) \geq 0$, q. e. d.

Let us now give the proof of Theorem A. We choose the quantity $R > 1$

[16] The proof which follows differs from that of HEINZ in that it is based on Parseval's formula instead of the classical formula cited from the theory of functions.

in such a way that the disk $|z| \leq R$ is contained in the domain of holomorphy of the function $u(z)$, and we denote by M_R the maximum of $|u(z)|$ on this disk.

If there exists a point z_0 such that $u(z_0) = M_R$, $|z_0| < R$, then $u(z)$ must be constant, $u(z) = c$, and therefore $u(T) = cI$; since $u(z) \leq 1$ for $|z| \leq 1$, we necessarily have $|c| \leq 1$ and consequently $\|u(T)\| \leq 1$.

In the contrary case one has $|u(z)| < M_R$ for $|z| < R$, and consequently the function

$$v_R(z) = \frac{M_R + u(z)}{M_R - u(z)}$$

is holomorphic there; moreover $\mathrm{Re}\, v_R(z) \geq 0$. In view of Theorem B this implies that

$$\mathrm{Re}\, (v_R(T)f, f) \geq 0$$

for all f, and consequently

$$\mathrm{Re}\, (v_R(T)\, (M_R g - u(T)g,\ M_R g - u(T)g) \geq 0$$

for all g. Since we clearly have

$$v_R(T)\, (M_R I - u(T)) = M_R I + u(T),$$

this implies that

$$\mathrm{Re}\, (M_R g + u(T)g,\ M_R g - u(T)g) = M_R^2\|g\|^2 - \|u(T)g\|^2 \geq 0,$$

therefore

$$\|u(T)\| \leq M_R.$$

Letting R approach 1, if follows that

$$\|u(T)\| \leq 1, \qquad \text{q. e. d.}$$

154. Spectral Sets

In this and the following section we shall deal with the Hilbert space $\mathfrak{H}$ and its linear transformations T, which we henceforth assume to be everywhere defined and bounded. We wish to investigate the "metric" properties of the correspondence between functions $u(z)$ and transformations $u(T)$, limiting ourselves to *rational* functions which are *regular* at every point of the spectrum of T; as we remarked in Sec. 151, the "algebraic" properties of the correspondence can be established directly for these functions, without recourse to formula (27).

We know that if the transformation T is *symmetric*, the hypothesis that the function $u(z)$ satisfy, at every point of the spectrum of T, the inequality

$$|u(z)| \leq M, \tag{36}$$

implies that

$$\|u(T)\| \leq M. \tag{37}$$

This follows immediately from the spectral decomposition of T (cf. Chapt. VIII, formula (10)). The same fact holds for all *normal* transformations and, in particular, for *unitary* transformations (cf. Sec. 111 and the end of Sec. 132). We recall that the spectrum of a normal transformation is a closed set in the complex plane; in the particular case of a symmetric or unitary transformation, the spectrum lies on the real line or on the unit circle, respectively.

For a *non-normal* transformation T, the spectrum $\sigma(T)$ does not, in general, consist of enough points in order that the validity of inequality (36) in $\sigma(T)$ should imply inequality (37). In fact, there are linear transformations $T \neq 0$ whose spectrum consists of the single point $z = 0$;[17] in this case, the function $u(z) \equiv z$ is zero in $\sigma(T)$ without $\|T\|$ being equal to 0.

The problem arises of characterizing the sets Z of complex numbers which are sufficiently large the that validity of inequality (36), in Z implies inequality (37). It is clear that it suffices to consider closed sets and that, by homogeneity, it suffices to take $M = 1$. Hence the definition:[18]

A set Z of points of the complex plane (completed by the point at infinity) will be called a spectral set of the linear transformation T if it is closed, and if for every rational function $u(z)$ satisfying in Z the inequality

$$|u(z)| \leq 1,$$

the transformation $u(T)$ exists and satisfies the inequality

$$\|u(T)\| \leq 1.$$

First we remark that the condition that $u(T)$ exists means that

$$Z \supseteq \sigma(T);$$

on the other hand, it is clear that every closed set containing a spectral set is itself a spectral set.

We have the following analogue of the spectral mapping theorem:

LEMMA 1. *Let $v(z)$ be a rational function such that $v(T)$ exists. If Z is a spectral set of T, its image under the transformation $z' = v(z)$, which we denote by $Z' = v(Z)$, is a spectral set of $v(T)$.*

In fact, let $u(z')$ be a rational function such that $|u(z')| \leq 1$ in Z'. Since

$$Z' = v(Z) \supseteq v(\sigma(T)) = \sigma(v(T)),^{19}$$

[17] The simplest example is given by the following transformation of two-dimensional space: $T\{x_1, x_2\} = \{x_2, 0\}$.

[18] This definition and the results we shall develop in Sections 154 and 155 are due to J. VON NEUMANN [**9**].

[19] Here we make use of the spectral mapping theorem.

the function $u(z')$ is regular at every point of the spectrum of $v(T)$, hence $u(v(T))$ exists. Since, on the other hand, this transformation equals the transformation $\psi(T)$ corresponding to the function $\psi(z) = u(v(z))$, and since $|\psi(z)| \leq 1$ in Z, we have $\|u(v(T))\| \leq 1$, q. e. d.

We consider, in particular, a linear rational function

$$z' = v(z) \equiv \frac{\alpha z + \beta}{\gamma z + \delta}, \text{ where } \alpha\delta - \beta\gamma \neq 0;$$

its inverse is of the same type:

$$z = v^{-1}(z') \equiv \frac{-\delta z' + \beta}{\gamma z' - \alpha}.$$

Let Z be a closed set which is transformed by the function $z' = v(z)$ into the bounded set $Z' = v(Z)$. If Z is a spectral set of the transformation T, Z' will be a spectral set of the transformation $T' = v(T)$; the latter exists since $v(z)$ is bounded in Z, and consequently also in $\sigma(T)$. Suppose, conversely, that $T' = v(T)$ exists and that Z' is a spectral set of T'; let us show that Z is then a spectral set of T. Since $Z = v^{-1}(Z')$, the problem reduces to showing that the transformation $v^{-1}(T')$ exists and is equal to T. Since $v^{-1}(v(z)) \equiv z$, the second proposition is a consequence of the first, which means that the function $v^{-1}(z)$ is bounded on the spectrum of T', i.e. that the set $v^{-1}(\sigma(T'))$ is bounded. Now $\sigma(T') = v(\sigma(T))$,[19] hence $v^{-1}(\sigma(T') = \sigma(T)$ and the set $\sigma(T)$ is bounded (it is contained in the closed disc $|z| \leq \|T\|$).

Thus we have proved the following proposition:

LEMMA 2. *Let $z' = v(z)$ be a linear rational function which maps the closed set Z onto the bounded set Z'. A necessary and sufficient condition that Z be a spectral set of the linear transformation T is that the transformation $T' = v(T)$ exist and that Z' be a spectral set of T'.*

By the theorem proved in the preceding section, the unit disc C_1 is a spectral set of all linear transformations T such that $\|T\| \leq 1$. Moreover, the condition $\|T\| \leq 1$ is necessary, as we see immediately by considering the function $u(z) \equiv z$. Hence we have the

THEOREM. *In order that the unit disc $|z| \leq 1$ be a spectral set of the linear transformation T, it is necessary and sufficient that*

$$\|T\| \leq 1.$$

Applying Lemma 2 to the closed domains Z characterized respectively by the inequalities

$$|z - a| \leq r, \qquad |z - a| \geq r, \qquad \operatorname{Re} z \geq 0,$$

where a is a complex number and r is a positive number, and to the linear rational functions

$$\frac{z-\alpha}{r}, \qquad \frac{r}{z-\alpha}, \qquad \frac{z-1}{z+1}$$

which map these domains onto the unit disc, we arrive at the following more general theorem:

THEOREM. *A necessary and sufficient condition that a closed domain*

$$|z-\alpha| \leq r, \qquad |z-\alpha| \geq r, \quad \textit{or} \quad \operatorname{Re} z \geq 0$$

be a spectral set of the linear transformation T is that, depending on the case,

$$\|T-\alpha I\| \leq r, \quad \|(T-I\alpha)^{-1}\| \leq \frac{1}{r}, \quad \textit{or} \quad \|(T-I)(T+I)^{-1}\| \leq 1$$

(*the existence of the inverses indicated being understood*).

Every linear transformation T therefore admits the spectral set

$$|z| \leq \|T\|,$$

and if α is a point of the resolvent set of T, T also admits the spectral set

$$|z-\alpha| \geq \|(T-\alpha I)^{-1}\|^{-1}.$$

It follows that the intersection of all the spectral sets of T is equal to the spectrum of T (but this does not imply that the spectrum is itself a spectral set).

Let us show that the condition obtained for the case of the half-plane $\operatorname{Re} z \geq 0$, namely, that $T+I$ admits an inverse (everywhere defined and bounded) and that

$$\|(T-I)(T+I)^{-1}g\| \leq \|g\| \tag{38}$$

for all elements g of $\mathfrak{H}$, is equivalent to the following condition:

$$\operatorname{Re}(Tf, f) \geq 0 \tag{39}$$

for all elements f of $\mathfrak{H}$.

In fact, it follows from the identity

$$\|(T \pm I)f\|^2 = \|Tf\|^2 \pm 2 \operatorname{Re}(Tf, f) + \|f\|^2 \tag{40}$$

that

$$\|(T+I)f\|^2 - \|(T-I)f\|^2 = 4 \operatorname{Re}(Tf, f). \tag{41}$$

Setting $g = (T+I)f$ in (38) we obtain $\|(T-I)f\| \leq \|(T+I)f\|$, which implies, in view of (41), that

$$\|(T+I)f\| \geq \|f\| \tag{42}$$

and

(43) $$\|(T - I)f\| \leq \|(T + I)f\|;$$

since

$$\text{Re}\,(T^*f, f) = \text{Re}\,(Tf, f) \geq 0$$

we also have

(42*) $$\|(T^* + I)f\| \geq \|f\|.$$

From (42) and (42*) it follows that

$$((T^* + I)\,(T + I)f, f) \geq (f, f), \quad ((T + I)\,(T^* + I)f, f) \geq (f, f).$$

The transformations $(T^* + I)\,(T + I)$, $(T + I)\,(T^* + I)$ therefore have greatest lower bound 1; consequently $(T + I)^{-1}$ exists and is everywhere defined and bounded (cf. the end of Sec. 104). It follows finally from (43) that

$$\|(T - I)\,(T + I)^{-1}g\| \leq \|g\|$$

for all elements g of $\mathfrak{H}$, therefore (39) implies (38), q. e. d.

This completes the proof of the equivalence of conditions (38) and (39). Hence we can state the

THEOREM. *A necessary and sufficient condition that the half-plane* $\text{Re}\, z \geq 0$ *be a spectral set of the linear transformation T is that*

$$\text{Re}\,(Tf, f) \geq 0$$

for all elements f of the space $\mathfrak{H}$.

An analogous criterion for the half-plane

$$\text{Re}\,(e^{i\theta}z) \geq 0$$

follows immediately, namely that

$$\text{Re}\,[e^{i\theta}(Tf, f)] \geq 0.$$

155. Characterization of Symmetric, Unitary, and Normal Transformations by Their Spectral Sets

We know that the spectrum of a normal transformation T of a Hilbert space $\mathfrak{H}$ is at the same time a spectral set of T. The unit circle $|z| = 1$ is therefore a spectral set for every unitary transformation, and the real axis is a spectral set for every (bounded) symmetric transformation. We shall show that these properties are even characteristic for unitary and for symmetric transformations.

Let T be a linear transformation of $\mathfrak{H}$ having the unit circle for a spectral set. Since the closed domains

$$|z| \leq 1 \text{ and } |z| \geq 1$$

are, a fortiori, spectral sets of T, it follows from the second theorem of the preceding section that T admits an inverse which is everywhere defined and bounded, and that

$$\|T\| \leq 1 \text{ and } \|T^{-1}\| \leq 1.$$

Thus, for every element f of $\mathfrak{H}$,

$$\|Tf\| \leq \|f\| = \|T^{-1}Tf\| \leq \|Tf\|,$$

hence

$$\|Tf\| = \|f\|.$$

The transformation T is therefore isometric, and since its domain of definition and its range both coincide with the entire space, it is *unitary*.

We now consider a linear transformation T of $\mathfrak{H}$ which has the real axis for a spectral set. Since the half-planes

$$\text{Re}\,(-iz) = \text{Im}\, z \geq 0 \text{ and } \text{Re}\,(iz) = -\text{Im}\, z \geq 0$$

are then also spectral sets of T, it follows from the third theorem of the preceding section that

$$\text{Re}\,[\mp i(Tf, f)] \geq 0,$$

that is, that (Tf, f) is real-valued for all f. This implies that T is *symmetric*.

We summarize:

THEOREM. *Among the bounded linear transformations of Hilbert space, the unitary and the symmetric transformations are characterized by the fact that they admit the unit circle and the real axis, respectively, as spectral sets.*

For *normal* transformations of general type, we still lack an analogous complete characterization. Nevertheless, we can state that a closed set Z of complex values which is sufficiently "sparse" can only be the spectral set of normal transformations. By sparse we mean here that every function which is continuous everywhere in the closed complex plane can be uniformly approximated in Z arbitrarily closely by rational functions. This is the case, for example, if the set Z consists of a *finite* number of points,

$$Z = \{z_1, z_2, \ldots, z_n\}.$$

Hence we have the

THEOREM. *A finite set of complex values can only be the spectral set of a normal transformation.*

Here is a proof. We denote by $l_k(z)$ the polynomial of degree $n - 1$ which takes on the value 1 at the point z_k and which is zero at the other points z_i $(i \neq k)$. Then

$$z = \sum_{k=1}^{n} z_k l_k(z),$$

and consequently

(44) $$T = \sum_{k=1}^{n} z_k l_k(T).$$

Now the spectral set of $l_k(T)$, being equal to $l_k(Z)$, consists of the points 0 and 1 lying on the real axis; in view of the theorem we just proved, $l_k(T)$ is therefore *symmetric*. Since, furthermore, the $l_k(T)$ $(k = 1, 2, \ldots, n)$ are permutable, it follows from (44) that T is *normal*, q. e. d.

In a finite-dimensional space the spectrum of a linear transformation is always a finite set. Hence the above theorem has the following corollary:

A linear transformation T in a finite-dimensional space $\mathfrak{H}$ is normal if and only if the spectrum of T is at the same time a spectral set of T.

BIBLIOGRAPHY

ABEL, N. — [**1**] Solution de quelques problèmes à l'aide d'intégrales définies, *Oeuvres*, **1** (Christiania, 1881), 11—27; [**2**] Résolution d'un problème de mécanique, *ibidem*, 97—101.

ALAOGLU, L.—G. BIRKHOFF — [**1**] General ergodic theorems, *Annals of Math.*, **41** (1940), 293—309.

AMBROSE, W. — [**1**] Spectral resolution of groups of unitary operators, *Duke Math. Journal*, **11** (1944), 589—595.

AMPÈRE, A. M. — [**1**] Recherches sur quelques points de la théorie des fonctions dérivées..., *Ecole Polytechnique*, **6** (1806), fasc. 13.

ARZELÀ, C. — [**1**] Sulla integrazione per serie, *Rendiconti Accad. Lincei Roma*, **1** (1885), 532—537, 566—569; [**2**] Sulle serie di funzioni (parte prima), *Memorie Accad. Sci. Bologna*, **8** (1900), 131—186; [**3**] (parte seconda), *ibidem*, 701—744.

BAIRE, R. — [**1**] Sur les fonctions de variables réelles, *Annali Mat. pura e appl.*, (3) **3** (1899), 1—122.

BANACH, S. — [*] *Théorie des opérations linéaires* (Warszawa, 1932); [**1**] Sur les lignes rectifiables et les surfaces dont l'aire est finie, *Fundamenta Math.*, **7** (1925), 225—237; [**2**] Sur les fonctionnelles linéaires, *Studia Math.*, **1** (1929), 211—216.

BANACH, S.—S. SAKS — [**1**] Sur la convergence forte dans les espaces L^p, *Studia Math.*, **2** (1930), 51—57.

BANACH, S.—STEINHAUS, H. — [**1**] Sur le principe de la condensation de singularités, *Fundamenta Math.*, **9** (1927), 51—57.

BEURLING, A. — [**1**] Sur les intégrales de Fourier absolument convergentes et leur application à une transformation fonctionnelle, *IX. Congrès des Math. Scandinaves, Helsingfors* (1938), 345—366.

BIRKHOFF, G. — [**1**] The mean ergodic theorem, *Duke Math. Journal*, **5** (1939), 635—646; [**2**] An ergodic theorem for general semi-groups, *Proc. Nat. Acad. Sci. USA*, **25** (1939), 625—627.

BLUMBERG, H. — [**1**] The measurable boundaries of an arbitrary function, *Acta Math.*, **65** (1935), 263—282.

BOCHNER, S. — [*] *Vorlesungen über Fouriersche Integrale* (Leipzig, 1932); [**1**] Spektraldarstellung linearer Scharen unitärer Operatoren, *Sitzsber. Preuss. Akad. Wiss.* (1933), 371—376; [**2**] Inversion formulae and unitary transformations, *Annals of Math.*, **35** (1934), 111—115.

BOHNENBLUST, H. F.—A. SOBCZYK — [**1**] Extensions of functionals on complex linear spaces, *Bulletin Amer. Math. Soc.*, **44** (1938), 91—93.

BOREL, E. — [*] *Leçons sur la théorie des fonctions* (Paris, 1898, 2e éd. 1914); [**1**] Le calcul des intégrales définies, *Journal de Math.*, (6) **8** (1912), 159—210.

BOURBAKI, N. — [*] *Intégration* (Éléments de Mathématique, Livre VI, Paris, 1952).

BURKILL, J. C. — [**1**] Functions of intervals, *Proc. London Math. Soc.*, (2) **22** (1924),

275—310; [**2**] The expression of area as an integral, *ibidem*, (2) **22** (1924), 311—336; [**3**] The derivates of functions of intervals. *Fundamenta Math.*, **5** (1924), 321—327.

CALKIN, J. W. — [**1**] Symmetric transformations in Hilbert space, *Duke Math. Journal*, **7** (1940), 504—508.

CARATHÉODORY, C. — [*] *Vorlesungen über reelle Funktionen* (Leipzig-Berlin, 1918, 2nd ed. 1927); [**1**] Entwurf für eine Algebraisierung des Integralbegriffs, *Sitzsber. Bayer. Akad. Wiss.* (1938), 27—68; [**2**] Bemerkungen zum Riesz-Fischerschen Satz und zur Ergodentheorie, *Abhandl. Math. Sem. d. Hansischen Univ.*, **14** (1941), 351—389.

CARLEMAN, T. — [*] *Sur les équations intégrales singulières à noyau réel symétrique* (Uppsala, 1923); [**1**] Zur Theorie der linearen Integralgleichungen, *Math. Zeitschr.*, **9** (1921), 196—217; [**2**] Application de la théorie des équations intégrales linéaires aux équations différentielles non linéaires, *Acta Math.*, **59** (1932), 63—87.

CLARKSON, J. A. — [**1**] Uniformly convex spaces, *Transactions Amer. Math. Soc.*, **40** (1936), 396—414.

COOPER, J. L. B. — [**1**] The spectral analysis of self-adjoint operators, *Quarterly Journal of Math.*, **16** (1945), 31—48; [**2**] Symmetric operators in Hilbert space, *Proc. London Math. Soc.*, **50** (1948), 11—55; (**3**) One-parameter semigroups of isometric operators in Hilbert space, *Annals of Math.*, **48** (1947), 827—842.

COURANT, R. — [**1**] Zur Theorie der linearen Integralgleichungen. *Math, Annalen*, **89** (1923), 161—178; [**2**] Über die Eigenwerte bei den Differentialgleichungen der Mathematischen Physik, *Math. Zeitschr.*, **7** (1920), 1—57.

COURANT, R.—D. HILBERT — [*] *Methoden der Mathematischen Physik*. I (2nd ed. Berlin, 1931).

DANIELL, P. J. — [**1**] A general form of integral, *Annals of Math.*, **19** (1917/18), 279—294; [**2**] Integrals in an infinite number of dimensions, *ibidem*, **20** (1918), 281—288; [**3**] Stieltjes derivatives, *Bulletin Amer. Math. Soc.*, **26** (1920), 444—448.

DAY, M. M. — [**1**] The spaces L^p with $0 < p < 1$, *Bulletin Amer. Math. Soc.*, **46** (1940), 816—823.

DENJOY, A. — [**1**] Mémoire sur les nombres dérivés des fonctions continues, *Journal de Math.*, (7) **1** (1915), 105—240; [**2**] Une extension de l'intégrale de M. Lebesgue, *Comptes Rendus Acad. Sci. Paris*, **154** (1912), 859—862; [**3**] Mémoire sur la totalisation des nombres dérivés non-sommables, *Annales de l'Ecole Normale Sup.*, **33** (1916), 127—222 and **34** (1917), 181—238.

DIEUDONNÉ, J. — [**1**] La dualité dans les espaces vectoriels topologiques, *Annales de l'Ecole Normale Sup.*, (3) **59** (1942), 107—139.

DU BOIS-REYMOND, P. — [**1**] Versuch einer Classification der willkürlichen Functionen reeller Argumente, *Journal für Math.*, **79** (1875), 21—37.

DUFFIN, R. J.—J. J. EACHUS — [**1**] Some notes on an expansion theorem of Paley and Wiener, *Bulletin Amer. Math. Soc.*, **48** (1942), 850—855.

DUNFORD, N. — [**1**] An ergodic theorem for n-parameter groups, *Proc. Nat. Acad. Sci. USA*, **25** (1939), 195—196; [**2**] Spectral theory. I. Convergence to projections, *Transactions Amer. Math. Soc.*, **54** (1943), 185—217; [**3**] Spectral theory, *Bulletin Amer. Math. Soc.*, **49** (1943), 637—651.

DUNFORD, N.—B. J. PETTIS — [**1**] Linear operations on summable functions, *Transactions Amer. Math. Soc.*, **47** (1940), 323—392.

DUNFORD, N.—I. E. SEGAL — [**1**] Semi-groups of operators and the Weierstrass theorem, *Bulletin Amer. Math. Soc.*, **52** (1946), 911—914.

EBERLEIN, W. F. — [**1**] A note on the spectral theorem, *Bulletin Amer. Math. Soc.*, **52**

(1946), 328—331; [2] Closure, convexity and linearity in Banach spaces, *Annals of Math.*, **47** (1946), 688—703.

EGOROFF, D. TH. — [1] Sur les suites des fonctions mesurables, *Comptes Rendus Acad. Sci. Paris*, **152** (1911), 244—246.

FABER, G. — [1] Über stetige Funktionen. II, *Math. Annalen*, **69** (1910), 372—433.

FATOU, P. — [1] Séries trigonométriques et séries de Taylor, *Acta Math.*, **30** (1906), 335—400.

FAVARD, J. — [*] *Leçons sur les fonctions presque-périodiques* (Paris, 1933).

FEJÉR, L. — [1] Über trigonometrische Polynome, *Journal für Math.*, **146** (1915), 53—82.

FISCHER, E. — [1] Sur la convergence en moyenne, *Comptes Rendus Acad. Sci. Paris*, **144** (1907), 1022—1024, 1148—1150; [2] Über quadratische Formen mit reellen Koeffizienten, *Monatshefte für Math. und Phys.*, **16** (1905), 234—249.

FRÉCHET, M. — [1] Sur l'intégrale d'une fonctionnelle étendue à un ensemble abstrait, *Bulletin Soc. Math. France*, **43** (1915), 249—267.

FREDHOLM, I. — [1] Sur une nouvelle méthode pour la résolution du problème de Dirichlet, *Kong. Vetenskaps-Akademiens Förh. Stockholm* (1900), 39—46; [2] Sur une classe d'équations fonctionnelles, *Acta Math.*, **27** (1903), 365—390.

FREUDENTHAL, H. — [1] Über die Friedrichssche Fortsetzung halbbeschränkter Hermitescher Operatoren, *Proceedings Acad. Amsterdam*, **39** (1936), 832—833.

FRIEDRICHS, K. — [1] Spektraltheorie halbbeschränkter Operatoren, *Math. Annalen*, **109** (1934), 465—487, 685—713; **110** (1935), 777—779; [2] Über die ausgezeichnete Randbedingung in der Spektraltheorie der halbbeschränkten gewöhnlichen Differentialoperatoren zweiter Ordnung, *ibidem*, **112** (1935), 1—23; [3] On differential operators in Hilbert spaces, *Amer. Journal of Math.*, **61** (1939), 523—544; [4] Beiträge zur Theorie der Spektralschar, *Math. Annalen*, **110** (1935), 54—62.

FUBINI, G. — [1] Sugli integrali multipli, *Rend. Accad. Lincei Roma*, **16** (1907), 608—614; [2] Sulla derivazione per serie, *ibidem*, **24** (1915), 204—206.

GELFAND, I. — [1] Normierte Ringe, *Math. Collection, Moscow*, **9** (1941), 3—24; [2] Über absolut konvergente trigonometrische Reihen, *ibidem*, **9** (1941), 51—66.

GELFAND, I.—D. RAIKOV — [1] Irreducible unitary representations of locally bicompact groups (in Russian), *Math. Collection, Moscow*, **13** (1943), 301—316.

GODEMENT, R. — [1] Sur une généralisation d'un théorème de Stone, *Comptes Rendus Acad. Sci. Paris*, **218** (1944), 901—903; [2] Les fonctions de type positif et la théorie des groupes, *Transactions Amer. Math. Soc.*, **63** (1948), 1—84; [3] Sur la théorie des représentations unitaires, *Annals of Math.*, **53** (1951), 68—124.

GOURSAT, E. — [*] *Cours d'Analyse Mathématique*. III (3ième éd. Paris, 1923); [1] Sur un cas élémentaire de l'équation de Fredholm, *Bulletin Soc. Math. France*, **35** (1907), 163—173.

HAAR, A. — [1] Über die Multiplikationstabelle der orthogonalen Funktionensysteme, *Math. Zeitschr.*, **41** (1930), 769—798; [2] Der Massbegriff in der Theorie der kontinuierlichen Gruppen, *Annals of Math.*, **34** (1933), 147—169.

HADAMARD, J. — [1] Résolution d'une question relative aux déterminants, *Bulletin Sci. Math.*, (2) **17** (1893), 240—348.

HAHN, H. — [*] *Theorie der reellen Funktionen*. I (Berlin, 1921); [1] Über die Integrale des Herrn Hellinger und die Orthogonalinvarianten der quadratischen Formen von unendlich vielen Veränderlichen, *Monatshefte Math. Phys.*, **23** (1912), 169—224; [2] Über eine Verallgemeinerung der Riemannschen Integraldefinition, *ibidem*, **26** (1915), 3—18; [3] Über lineare Gleichungen in linearen Räumen, *Journal für Math.*, **157** (1927), 214—229.

HALMOS, P. R. — [*] *Introduction to Hilbert space and the theory of spectral multiplicity* (New York, 1951).

HANSON, E. H. — [1] A theorem of Denjoy, Young and Saks, *Bulletin Amer. Math. Soc.*, **40** (1934), 691—694.

HARDY, G. H.—J. E. LITTLEWOOD—G. PÓLYA — [*] *Inequalities* (Cambridge, 1934).

HAUSDORFF, F. — [1] Zur Theorie der linearen metrischen Räume, *Journal für Math.*, **167** (1932), 294—311.

HEINZ, E. — [1] Beiträge zur Störungstheorie der Spektralzerlegung, *Math. Annalen*, **123** (1951), 415—438; [2] Ein v. Neumannscher Satz über beschränkte Operatoren im Hilbertschen Raum, *Göttinger Nachr.*, **1952**, 5—6.

HELLINGER, E. — [1] Die Orthogonalinvarianten quadratischen Formen von unendlich vielen Variablen, *Dissertation Götlingen*, (1907); [2] Neue Begründung der Theorie quadratischer Formen von unendlich vielen Veränderlichen, *Journal für Math.*, **136** (1909), 210—271.

HELLINGER, E.—O. TOEPLITZ — [*] *Integralgleichungen und Gleichungen mit unendlich vielen Unbekannten*, Enzyklopädie d. Math. Wiss., II. C. 13 (Leipzig, 1928); [1] Grundlagen für eine Theorie der unendlichen Matrizen, *Math. Annalen*, **69** (1910), 289—330.

HELLY, E. — [1] Über lineare Funktionaloperationen, *Sitzsber. Akad. Wiss. Wien*, **121** (1912), 265—297.

HERGLOTZ, G. — [1] Über Potenzreihen mit positivem, reellem Teil im Einheitskreis, *Ber. Sächs. Ges. d. Wiss. Leipzig*, **63** (1911), 501—511.

HILB, E. — [1] Über die Auflösung von Gleichungen mit unendlich vielen Unbekannten, *Sitzsber. Phys. Med. Soz. Erlangen* (1908), 84—89.

HILBERT, D. — [*] *Grundzüge einer allgemeinen Theorie der linearen Integralgleichungen*, (Leipzig, 1912).

HILDEBRANT, T. H. — [1] On integrals related to and extensions of the Lebesgue integrals, *Bulletin Amer. Math. Soc.*, **24** (1917), 113—144, 177—202; [2] Definitions of Stieltjes Integrals of the Riemann type, *Amer. Math. Monthly*, **45** (1938), 265—278; [3] Über vollstetige, lineare Transformationen, *Acta Math.*, **51** (1928), 311—318.

HILLE, E. — [*] *Functional analysis and semi-groups* (New York, 1948); [1] On semi-groups of transformations in Hilbert space, *Proc. Nat. Acad. Sci. USA*, **19** (1938), 159—161; [2] Notes on linear transformations. II. Analyticity of semi-groups, *Annals of Math.*, **40** (1939), 1—47.

HOPF, E. — [*] *Ergodentheorie*, Ergebnisse d. Math., V/2 (Berlin, 1937).

JESSEN, B. — [1] The theory of integration in a space of an infinite number of dimensions, *Acta Math.*, **63** (1934), 249—323.

JORDAN, P.—J. VON NEUMANN — [1] On inner products in linear metric spaces, *Annals of Math.*, **36** (1935), 719—723.

KAKUTANI, S. — [1] Iteration of linear operators in complex Banach spaces, *Proc. Imp. Acad. Tokyo*, **14** (1938), 295—300; [2] Mean ergodic theorem in abstract (L)-spaces, *ibidem*, **15** (1939), 121—123.

KARAMATA, J. — [1] Sur certaines limites rattachées aux intégrales de Stieltjes, *Comptes Rendus Acad. Sci. Paris*, **182** (1926), 833.

KELLOGG, O. D. — [1] On the existence and closure of sets of characteristic functions, *Math. Annalen*, **86** (1922), 14—17.

KHINTCHINE, A. — [1] Sur une extension de l'intégrale de M. Denjoy, *Comptes Rendus Acad. Sci. Paris*, **162** (1916), 287—291; [2] Sur le procédé d'intégration de M. Denjoy, *Math. Collection, Moscow*, **30** (1918), 543—557.

Kolmogoroff, A. — [*] *Grundbegriffe der Wahrscheinlichkeitsrechnung*, Ergebnisse d. Math., II/3 (Berlin, 1933); [1] Beiträge zur Masstheorie, *Math. Annalen*, **107** (1932), 351—366.

Koopman, B. O. — [1] Hamiltonian systems and transformations in Hilbert space, *Proc. Nat. Acad. Sci. USA*, **17** (1931), 315—318.

Koopman, B. O.—J. L. Doob — [1] On analytic functions with positive imaginary parts, *Bulletin Amer. Math. Soc.*, **40** (1934), 601—605.

Krasnoselsky, M. A. — [1] On deficiency numbers of closed operators (in Russian). *Reports (Doklady) Acad. Sci. USSR*, **56** (1947), 559—561.

Krein, M. A. — [1] The theory of self-adjoint extensions of semi-bounded Hermitian transformations and its applications, *Math. Collection, Moscow*, **20** (1947), 431—495, **21** (1947), 365—404.

Landau, E. — [1] Ein Satz über Riemannsche Integrale, *Math. Zeitschr.*, **2** (1918), 350—351.

Lebesgue, H. — [*] *Leçons sur l'intégration et la recherche des fonctions primitives* (Paris, 1904; 2e éd. 1928); [$\substack{*\\*}$] *Leçons sur les séries trigonométriques* (Paris, 1906); [1] Intégrale, Longueur, Aire, *Thèse* (Paris, 1902) ou *Annali Mat pura e appl.*, (3) **7** (1902), 231—359; [2] Sur l'intégration des fonctions discontinues, *Annales Ecole Norm. Sup.*, (3) **27** (1910), 361—450; [3] Sur les intégrales singulières, *Annales de Toulouse*, (3) **1** (1909), 25—117; [4] Sur l'intégrale de Stieltjes et sur les opérations linéaires, *Comptes Rendus Acad. Sci. Paris*, **150** (1910), 86—88; [5] Sur les fonctions représentables analytiquement, *Journal de Math.*, (6) **1** (1905), 139—216; [6] Sur la méthode de M. Goursat pour la résolution de l'équation de Fredholm, *Bulletin Soc. Math. France*, **36** (1909), 3—19.

Lengyel, B. — [1] On the spectral theorem of self-adjoint operators, *Acta Sci. Math. Szeged*, **9** (1939), 174—186.

Lengyel, B. A.—M. H. Stone — [1] Elementary proof of the spectral theorem, *Annals of Math.*, **37** (1936), 853—864.

Le Roux, J. — [1] Sur les intégrales des équations linéaires aux dérivées partielles du second ordre à deux variables indépendantes, *Annales Ecole Norm. Sup.*, (3) **12** (1895), 227—316.

Levi, B. — [1] Sopra l'integrazione delle serie, *Rend. Instituto Lombardo di Sci. e Lett.*, (2) **39** (1906), 775—780; [2] Sul principio de Dirichlet, *Rend. Circ. Math. Palermo*, **22** (1906), 293—360.

Liouville, J. — [1] Sur le développement des fonctions ou parties de fonctions en séries dont les divers termes sont assujettis à satisfaire à une même équation différentielle du second ordre contenant un paramètre variable. II, *Journal de Math.*, (1) **2** (1837), 16—35.

Lorch, E. R. — [1] Functions of self-adjoint transformations in Hilbert space, *Acta Sci. Math. Szeged*, **7** (1934), 136—146; [2] On a calculus of operators in reflexive vector spaces, *Transactions Amer. Math. Soc.*, **45** (1939), 217—234; [3] Means of iterated transformations in reflexive Banach spaces, *Bull. Amer. Math. Soc.*, **45** (1939), 945—947; [4] The spectrum of linear transformations, *Transactions Amer. Math. Soc.*, **52** (1942), 238—248; [5] On certain implications which characterize Hilbert space, *Annals of Math.*, **49** (1948), 523—532; [6] Return to the self-adjoint transformations, *Acta Sci. Math. Szeged*, **12 B** (1950), 137—144.

Löwig, H. — [1] Komplexe euklidische Räume von beliebiger endlicher oder unendlicher Dimensionszahl, *Acta Sci. Math. Szeged*, **7** (1934), 1—33.

Lusin, N. — [1] Sur les propriétés des fonctions mesurables, *Comptes Rendus Acad. Sci. Paris*, **154** (1912), 1688—1690.

MAUTNER, F. I. — Unitary representations of locally compact groups, *Annals of Math.*, **51** (1950), 1—25; **52** (1950), 528—556.

MAZUR, S. — [**1**] Über konvexe Mengen in linearen normierten Räumen, *Studia Math.*, **4** (1933), 70—84.

MERCER, T. — [**1**] Functions of positive and negative type and their connection with the theory of integral equations, *Transactions London Phil. Soc.*, (A) **209** (1909), 415—446.

MILMAN, D. — [**1**] On some criteria for the regularity of spaces of type (B), *Reports (Doklady) Acad. Sci. USSR*, **20** (1938), 243—246.

MIMURA, Y. — [**1**] Über Funktionen von Funktionaloperatoren in einem Hilbertschen Raum, *Japan. Journal Math.*, **13** (1936), 119—128.

MIYADERA, I. — [**1**] Generation of a strongly continuous semi-group of operators, *Tôhoku Math. Journal*, (2) **4** (1952), 109—121.

NAKANO, H. — [**1**] Zur Eigenwerttheorie normaler Operatoren, *Proc. Phys.-Math. Soc. Japan*, (3) **21** (1939), 315—339; [**2**] Über Abelsche Ringe von Projektionsoperatoren, *ibidem*, **21** (1939), 357—375; [**3**] Funktionen mehrerer hypermaximaler normaler Operatoren, *ibidem*, **21** (1939), 713—728; [**4**] Unitärinvarianten hypermaximaler normaler Operatoren im Hilbertschen Raum, *Annals of Math.*, **42** (1941), 657—664; [**5**] Unitärinvarianten im allgemeinen euklidischen Raum, *Math. Annalen*, **118** (1941), 112—133.

NEUMANN, C. — [*] *Untersuchungen über das logarithmische und Newtonsche Potential* (Leipzig, 1877).

NEUMANN, J. VON — [*] *Mathematische Grundlagen der Quantenmechanik* (Berlin, 1932); [**1**] Allgemeine Eigenwerttheorie Hermitescher Funktionaloperatoren, *Math. Annalen*, **102** (1929), 49—131; [**2**] Zur Algebra der Funktionaloperationen und Theorie der normalen Operatoren, *ibidem*, **102** (1929), 370—427; [**3**] Über Funktionen von Funktionaloperatoren, *Annals of Math.*, **32** (1931), 191—226; [**4**] Über adjungierte Funktionaloperatoren, *ibidem*, **33** (1932), 294—310; [**5**] Über einen Satz von Herrn M. H. Stone, *ibidem*, **33** (1932), 567—573; [**6**] Zur Operatorenmethode in der klassischen Mechanik, *ibidem*, **33** (1932), 587—648; [**7**] Charakterisierung des Spektrums eines Integraloperators, *Actualités Sci. et Ind.*, **229** (1935); [**8**] On rings of operators. III, *Annals of Math.*, **41** (1940), 94—161; [**9**] Eine Spektraltheorie für allgemeine Operatoren eines unitären Raumes, *Math. Nachrichten*, **4** (1951), 258—281.

NEUMARK, M. — [**1**] On the square of a closed symmetric operator, *Reports (Doklady) Acad. Sci. USSR*, **26** (1940), 866—870; [**2**] Positive definite operator functions on a commutative group, *Bulletin (Izvestiya) Acad. Sci. USSR*, math. series, **7** (1943), 237—244; [**3**] Selfadjoint extensions of the second kind of a symmetric operator, *ibidem*, **4** (1940), 53—104.

NIKODYM, O. — [**1**] Sur une généralisation des intégrales de M. Radon, *Fundamenta Math.*, **15** (1930), 131—179.

OSGOOD, W. F. — [**1**] Non-uniform convergence and the integration of series term by term, *Amer. Journal of Math.*, **19** (1897), 155—190.

PAATERO, V. — [**1**] Über die konforme Abbildung von Gebieten, deren Ränder von beschränkter Drehung sind, *Annales Acad. Sci. Fennicae* **A**, **33** (1931), nº 9, 1—77; [**2**] Über Gebiete von beschränkter Drehung, *ibidem*, **37** (1933), nº 9, 1—20.

PALEY, R. E. A. C. — N. WIENER — [*] *Fourier-transform in the complex domain* (New York, 1934).

PERRON, O. — [**1**] Über den Integralbegriff, *Sitzsber. Heidelberg. Akad. Wiss.*, **16** (1914), 1—16.

PETTIS, B. J. — [**1**] A proof that every uniformly convex space is reflexive, *Duke Math. Journal*, **5** (1939), 249—253.

PHILLIPS, R. S. — [**1**] On linear transformations, *Transactions Amer. Math. Soc.*, **48** (1940), 516—541; [**2**] Perturbation theory for semi-groups of linear operators, *ibidem*, **74** (1953), 199—221.

PLANCHEREL, M. — [**1**] Contribution à l'étude de la représentation d'une fonction arbitraire par des intégrales définies, *Rend. Circ. Math. Palermo*, **30** (1910), 289—335; [**2**] Sur les formules de réciprocité du type de Fourier, *Journal London Math. Soc.*, **8** (1933), 220—226.

PLESSNER. A. — [**1**] Zur Spektraltheorie maximaler Operatoren, *Reports (Doklady) Acad. Sci. USSR.*, **22** (1939), 227—230; [**2**] Über Funktionen eines maximalen Operators, *ibidem*, **23** (1939), 327—330; [**3**] Über halbunitäre Operatoren, *ibidem*, **25** (1939), 710—712.

PLESSNER, A. I.—V. A. ROKHLIN — [**1**] Spectral theory of linear operators. II, *Uspekhi Matem. Nauk*, (N. S.) **1** (1946), 71—191 (in Russian; analyzed by M. H. Stone in *Math. Reviews*, **9** (1948), 43).

RADON, J. — [**1**] Theorie und Anwendungen der absolut additiven Mengenfunktionen, *Sitzsber. Akad. Wiss. Wien.* **122** (1913), Abt. II a, 1295—1438; [**2**] Über lineare Funktionaltransformationen und Funktionalgleichungen, *ibidem*, **128** (1919), 1083—1121; [**3**] Über die Randwertaufgaben beim logarithmischen Potential, *ibidem*, **128** (1919), 1123—1167.

RAJCHMAN, A.—S. SAKS — [**1**] Sur la dérivabilité des fonctions monotones, *Fundamenta Math.*, **4** (1923), 204—213.

REID, W. T. — [**1**] Symmetrizable completely continuous linear transformations in Hilbert space, *Duke Math., Journal*, **18** (1951), 41—56.

RELLICH, F. — [**1**] Spektraltheorie in nicht-separablen Räumen, *Math. Annalen*, **110** (1934), 342—356; [**2**] Über die v. Neumannschen fastperiodischen Funktionen auf einer Gruppe, *ibidem*, **111** (1935), 560—567; [**3**] Störungstheorie der Spektralzerlegung. I, *ibidem*, **113** (1936), 600—619; II, *ibidem*, **113** (1936), 667—685; III, *ibidem*, **116** (1939), 555—570; IV, *ibidem*, **117** (1940), 356—382; V. *ibidem*, **118** (1942), 462—484.

RIESZ, F. — [*] *Les systèmes d'équations linéaires à une infinité d'inconnues* (Paris, 1913); [**1**] Sur les systèmes orthogonaux de fonctions, *Comptes Rendus Acad. Sci. Paris*, **144** (1907), 615—619, 734—736; [**2**] Über orthogonale Funktionensysteme, *Göttinger Nachr.*, **1907**, 116—122; [**3**] Sur les opérations fonctionnelles linéaires, *Comptes Rendus Acad. Sci. Paris*, **149** (1909), 974—977; [**4**] Sur certains systèmes d'équations fonctionnelles et l'approximation des fonctions continues, *ibidem*, **150** (1910), 674—677; [**4a**] Sur certains systèmes singuliers d'équations intégrales, *Annales Ecole Norm. Sup.*, (3) **28** (1911), 33—62; [**5**] Über quadratische Formen von unendlich vielen Veränderlichen, *Göttinger Nachr.*, **1910**, 190—195; [**6**] Untersuchungen über Systeme integrierbarer Funktionen, *Math. Annalen*, **69** (1910), 449—497; [**7**] Démonstration nouvelle d'un théorème concernant les opérations, *Annales Ecole Norm. Sup.*, (3) **31** (1914), 9—14; [**8**] Über Integration unendlicher Folgen, *Jahresber. Deutsch. Math. Ver.*, **26** (1917), 274—278; [**9**] Über lineare Funktionalgleichungen, *Acta Math.*, **41** (1917), 71—98; [**10**] Sur l'intégrale de Lebesgue, *ibidem*, **42** (1919—1920), 191—205; [**1**] Elementarer Beweis des Egoroffschen Satzes, *Monatshefte Math. Phys.*, **35** (1928), 243—248; [**12**] Sur la convergence en moyenne, *Acta Sci. Math. Szeged*, **4** (1928), 58—64, 182—185; [**13**] Über die linearen Transformationen des komplexen Hilbertschen Raumes, *ibidem*, **5** (1930), 23—54; [**14**] Über Sätze von Stone und Bochner, *ibidem*, **6** (1933), 184—198; [**15**] Zur Theorie des Hilbertschen Raumas, *ibidem*, **7** (1934), 34—38; [**16**] Sur les fonctions des transformations hermitiennes dans l'espace de Hilbert, *ibidem*, **7** (1935), 147—159; [**17**] Sur l'existence de la dérivée des fonctions monotones et sur quelques problèmes qui s'y rattachent, *ibidem*, **5** (1932), 208—221; [**18**] Sur l'intégrale de Lebesgue comme l'opération inverse de la dérivation, *Annali Pisa*, (2) **5** (1936), 191—212; [**19**] Some mean ergodic theorems, *Journal London Math. Soc.*, **13** (1938), 274—278; [**20**] Sur la théorie ergodique, *Commentarii Math. Helv.*, **17** (1944/45), 221—239; [**21**] Another proof of the mean ergodic theorem, *Acta Sci. Math. Szeged*, **10** (1941), 75—76; [**22**] Sur la représentation des opérations fonctionnelles linéaires par des intégrales de Stieltjes, *Kung. Fysiografiska Sällskapets i Lund Förhandlingar*, **21** (1952), Nr. 16.

RIESZ, F.—E. R. LORCH — [**1**] The integral representation of unbounded selfadjoint transformations in Hilbert space, *Transactions Amer. Math. Soc.*, **39** (1936), 331—340.

RIESZ, F.—B. SZ.-NAGY — [**1**] Über Kontraktionen des Hilbertschen Raumes, *Acta Sci. Math. Szeged*, **10** (1943), 202—205.

RIESZ, M. — [**1**] Sur le problème des moments. Troisième Note, *Arkiv för Mat., Astronomi och Fysik*, **17** (1923), n° 16.

SAKS, S. — [*] *Théorie de l'intégrale* (Warszawa, 1933); [$^*_*$] *Theory of the integral* (Warszawa—Lwów, 1937); [**1**] Sur les nombres dérivés des fonctions, *Fundamenta Math.*, **5** (1924), 98—104.

SCHAUDER, J. — [**1**] Über lineare, vollstetige Funktionaloperationen, *Studia Math.*, **2** (1930), 183—196.

SCHMIDT, E. — [**1**] Entwicklung willkürlicher Funktionen nach Systemen vorgeschriebener, *Math. Annalen*, **63** (1907), 433—476; [**2**] Auflösung der allgemeinen linearen Integralgleichung, *ibidem*, **64** (1907), 161—174; [**3**] Über die Auflösung linearer Gleichungen mit abzählbar unendlich vielen Unbekannten, *Rend. Circ. Mat. Palermo*, **25** (1908), 53—77.

SCHUR, I. — [**1**] Über Potenzreihen, die im Inneren des Einheitskreises beschränkt sind, *Journal für Math.*, **147** (1917), 205—232.

SEGAL, I. E. — [**1**] Irreducible representations of operator algebras, *Bulletin Amer. Math. Soc.*, **53** (1947), 73—88.

SIERPINSKI, W. — [**1**] Sur les fonctions convexes mesurables, *Fundamenta Math.*, **1** (1920), 125—129.

SMITHIES, F. — [**1**] The Fredholm theory of integral equations, *Duke Math. Journal*, **8** (1941), 107—130.

SOUKHOMLINOFF, G. — [**1**] Über Fortsetzung von linearen Funktionalen in linearen komplexen Räumen und linearen Quaternionräumen, *Recueil* (*Sbornik*) *Math. Moscou*, N. S. **3** (1938), 353—358.

STIELTJES, T. J. — [**1**] Recherches sur les fractions continues, *Annales de Toulouse*, (1) **8** (1894), 1—122 et **9** (1895) 1—47 [= *Oeuvres complètes*. II (Groningen, 1918), 402—566].

STONE, M. H. — [*] *Linear transformations in Hilbert space* (New York, 1932); [**1**] Linear transformations in Hilbert space, *Proc. Nat. Acad. Sci. USA*, **15** (1929), 198—200, 423—425 and **16** (1930), 172—175; [**2**] On one-parameter unitary groups in Hilbert space, *Annals of Math.*, **33** (1932), 643—648; [**3**] Notes on integration, *Proc. Nat. Acad. Sci. USA*, **34** (1948), 336—342, 447—455, 483—490, and **35** (1949), 50—58.

SZ.-NAGY, B. — [*] *Spektraldarstellung linearer Transformationen des Hilbertschen Raumes*, Ergebnisse, d. Math., V/5 (Berlin, 1942); [**1**] Über messbare Darstellungen Liescher Gruppen, *Math. Annalen*, **112** (1936), 286—296; [**2**] Bedingungen für die Multiplikationstabelle eines in sich abgeschlossenen orthogonalen Funktionensystems, *Annali Pisa*, **6** (1937), 211—224; [**3**] On semi-groups of self-adjoint transformations in Hilbert space, *Proc. Nat. Acad. Sci. USA*, **24** (1938), 559—560; [**4**] Perturbations des transformations autoadjointes dans l'espace de Hilbert, *Commentarii Math. Helv.*, **19** (1946/47), 347—366; [**5**] Expansion theorems of Paley-Wiener type, *Duke Math. Journal*, **14** (1947), 975—978; [**6**] Vibrations d'une corde non homogène, *Bull. Soc. Math. France*, **75** (1947), 193—208; [**7**]. On uniformly bounded linear transformations in Hilbert space, *Acta Sci. Math. Szeged*, **11** (1947), 152—157; [**8**] Perturbations des transformations linéaires fermées, *ibidem*, **14** (1951), 125—137.

TAYLOR, A. E. — [**1**] Spectral theory of closed distributive operators, *Acta Math.*, **84** (1950), 189—224.

TITCHMARSH, E. C. — [**1**] A proof of a theorem of Watson, *Journal London Math. Soc.*, **8** (1933), 217—220.

TONELLI, L. — [**1**] Successioni di curve e derivazione per serie. II, *Atti Accad. Lincei*, **25** (1916), 85—91; [**2**] Sul differenziale dell'arco di curve, *ibidem*, **25** (1916) 207—213.

VALLÉE POUSSIN, CH. J. DE LA — [*] *Intégrales de Lebesgue. Fonctions d'ensemble. Classes de Baire* (Paris, 1916; 2e éd. 1936); [**1**] Sur l'intégrale de Lebesgue, *Transactions Amer. Math. Soc.*, **16** (1915), 435—501.

VIGIER, J. P. — [**1**] Etude sur les suites infinies d'opérateurs hermitiens, *Thèse* (Genève, 1946).

VISSER, C. — [**1**] Note on linear operators, *Proc. Acad. Amsterdam*, **40** (1937), 270—272.

VOLTERRA, V. — [**1**] Sulla inversione degli integrali definiti, *Rend. Accad. Lincei* **5** (1896), 177—185, 289—300 and *Annali di Mat.*, (2) **25** (1897), 139—178.

WAERDEN, B. L. VAN DER — [**1**] Ein einfaches Beispiel einer nichtdifferenzierbaren stetigen Funktion, *Math. Zeitschr.*, **32** (1930), 474—475.

WATSON, G. N. — [**1**] General transforms, *Proc. London Math., Soc.*, (2) **35** (1933), 156—199.

WAVRE, R. — [**1**] L'itération directe des opérateurs hermitiens et deux théories qui en dépendent, *Commentarii Math. Helv.*, **15** (1943), 299—317; [**2**] L'itération directe des opérateurs hermitiens, *ibidem* **16** (1944), 65—72.

WECKEN, F. J. — [**1**] Zur Theorie linearer Operatoren, *Math. Annalen*, **110** (1935), 722—725; [**2**] Unitärinvarianten selbstadjungierter Operatoren, *ibidem*, **116** (1939), 422—455; [**3**] Abstrakte Integrale und fastperiodische Funktionen, *Math. Zeitschr.*, **45** (1939), 377—404.

WEIL, A. — [*] *L'intégration dans les groupes topologiques* (Paris, 1940).

WEYL, H. — [**1**] Über gewöhnliche Differenzialgleichungen mit Singularitäten, *Math. Annalen*, **68** (1910), 220—269; [**2**] Über die asymptotische Verteilung der Eigenwerte, *Göttinger Nachr.*, **1911** 110—117; [**3**] Das asymptotische Verteilungsgesetz der Eigenschwingungen eines beliebig gestalteten elastischen Körpers, *Rend. Circ. Mat. Palermo*, **39** (1915), 1—49; [**4**] Über beschränkte quadratische Formen, deren Differenz vollstetig ist, *ibidem*, **27** (1909), 373—392; [**5**] Integralgleichungen und fastperiodische Funktionen, *Math. Annalen*, **97** (1927), 338—356.

WIENER, N. — [*] *The Fourier integral and certain of its applications* (Cambridge, 1933); [**1**] The ergodic theorem, *Duke Math. Journal*, **5** (1939), 1—18.

WINTNER, A. — [*] *Spektraltheorie unendlicher Matrizen* (Leipzig, 1929); [**1**] Zur Theorie der beschränkten Bilinearformen, *Math. Zeitschr.*, **30** (1929), 228—289.

YOSIDA, K. — [**1**] Mean ergodic theorems in Banach spaces. *Proc. Imp. Acad. Japan*, **14** (1938), 292—294; [**2**] On the differentiability and the representation of one-parameter semi-group of linear operators, *Journal of the Math. Soc. Japan*, **1** (1948), 15—21.

YOUNG, G. C. — [**1**] On infinite derivates, *Quarterly Journal of Math.*, **47** (1916), 148—153; [**2**] On the derivates of a function, *Proc. London Math. Soc.*, (2) **15** (1916), 360—384.

YOUNG, G. C.—W. H. YOUNG — [**1**] On the existence of a differential coefficient, *Proc. London Math. Soc.*, (2) **9** (1911), 325—335.

YOUNG, R. C. — [**1**] Functions of Σ defined by addition or functions of intervals in n-dimensional formulation, *Math. Annalen*, **29** (1928), 171—216.

YOUNG, W. H. — [**1**] On a new method in the theory of integration, *Proc. London Math. Soc.*, (2) **9** (1910), 15—50.

ZAANEN, A. C. — [**1**] Über vollstetige symmetrische und symmetrisierbare Operatoren, *Nieuw Archief voor Wiskunde*, (2) **22** (1948), 57—80; [**2**] On linear functional equations, *ibidem*, **22** (1948), 269—282.

APPENDIX

EXTENSIONS OF LINEAR TRANSFORMATIONS IN HILBERT SPACE WHICH EXTEND BEYOND THIS SPACE

Dedicated to Frigyes Riesz
on the occasion of his seventy-fifth birthday
January 22, 1955

1. Introduction

The essential structure of normal, in particular of self-adjoint and unitary, transformations in Hilbert space is known, thanks to the spectral decomposition theorem for these transformations. Much less is known about the structure of non-normal transformations because of the lack of a satisfactory generalization to Hilbert space of either the Jordan canonical form for finite matrices or of the theory of elementary divisors. This is why it is important to find relations between normal and non-normal transformations, which will enable one to reduce certain problems dealing with general linear transformations to the more workable particular case of normal transformations.

The simplest relations of this type are

$$T = A + iB,$$

where the bounded linear transformation T in the Hilbert space $\mathfrak{H}$ is represented by the two self-adjoint transformations

$$A = \operatorname{Re} T = \frac{1}{2}(T + T^*), \quad B = \operatorname{Im} T = \frac{1}{2i}(T - T^*),$$

and

$$T = VR$$

where the bounded linear transformation T in the Hilbert space $\mathfrak{H}$ is represented by the positive self-adjoint transformation

$$R = (T^* T)^{\frac{1}{2}}$$

and V is a partially isometric transformation (which, in certain cases, can be chosen to be unitary, in particular if T is normal or if T is a one-to-one transformation of the space $\mathfrak{H}$ onto itself).[1] The applicability of these relations is

[1] See Sec. 110.

restricted by the fact that neither A and B nor V and R are in general permutable, and there is no simple relation among the corresponding representations of the iterated transformations $T, T^2, \ldots$.

In the sequel, we shall deal with other relations which are connected with extensions of a given transformation. But, contrary to what we usually do, we shall also allow extensions which *extend beyond the given space.*

So, by an *extension* of a linear transformation T of Hilbert space $\mathfrak{H}$, we shall understand a linear transformation $\boldsymbol{T}$ in a Hilbert space $\mathbf{H}$ which contains $\mathfrak{H}$ as a (not necessarily proper) subspace, such that $\mathfrak{D}_{\boldsymbol{T}} \supseteqq \mathfrak{D}_T$ and $\boldsymbol{T}f = Tf$ for $f \in \mathfrak{D}_T$. We shall retain the notation $\boldsymbol{T} \supseteqq T$ which we used for ordinary extensions (where $\mathbf{H} = \mathfrak{H}$).

The orthogonal projection of the "extension" space $\mathbf{H}$ onto its subspace $\mathfrak{H}$ will be denoted by $\boldsymbol{P}_{\mathfrak{H}}$ or simply by $\boldsymbol{P}$.

Among the extensions of a bounded linear transformation T in $\mathfrak{H}$ (with $\mathfrak{D}_T = \mathfrak{H}$) we shall consider in particular those which are of the form $\boldsymbol{PS}$ where $\boldsymbol{S}$ is a bounded linear transformation of an extension space $\mathbf{H}$. We express this relation

$$T \subseteq \boldsymbol{PS}$$

by saying that T is the *projection* of the transformation $\boldsymbol{S}$ onto $\mathfrak{H}$,[2] in symbols

$$T = \mathrm{pr}_{\mathfrak{H}}\, \boldsymbol{S} \text{ or simply } T = \mathrm{pr}\, \boldsymbol{S}. \tag{1}$$

It is obvious that the relations $T_i = \mathrm{pr}\, \boldsymbol{S}_i\ (i = 1, 2)$ imply the relation

$$a_1T_1 + a_2T_2 = \mathrm{pr}\,(a_1\boldsymbol{S}_1 + a_2\boldsymbol{S}_2) \tag{2}$$

(of course, $\boldsymbol{S}_1$ and $\boldsymbol{S}_2$ are transformations in the *same* extension space $\mathbf{H}$). Relation (1) also implies that

$$T^* = \mathrm{pr}\, \boldsymbol{S}^*.\text{[3]} \tag{3}$$

Finally, the uniform, strong, or weak convergence of a sequence $\{\boldsymbol{S}_n\}$ implies convergence of the same type for the sequence $\{T_n\}$ where $T_n = \mathrm{pr}\, \boldsymbol{S}_n$.[4]

If $\mathbf{H}$ and $\mathbf{H}'$ are two extension spaces of the same space $\mathfrak{H}$, $\boldsymbol{S}$ and $\boldsymbol{S}'$ are bounded linear transformations of $\mathbf{H}$ and $\mathbf{H}'$ respectively, then we shall say that the "structures" $\{\mathbf{H}, \boldsymbol{S}, \mathfrak{H}\}$ and $\{\mathbf{H}', \boldsymbol{S}', \mathfrak{H}\}$ are *isomorphic* if $\mathbf{H}$ can be mapped isometrically onto $\mathbf{H}'$ in such a way that the elements of the common subspace $\mathfrak{H}$ are left invariant and that $f \to f'$ implies $\boldsymbol{S}f \to \boldsymbol{S}'f'$.

If $\{\boldsymbol{S}_\omega\}_{\omega \in \Omega}$ and $\{\boldsymbol{S}'_\omega\}_{\omega \in \Omega}$ are two families of bounded linear transformations in $\mathbf{H}$ and $\mathbf{H}'$ respectively, we define the isomorphism of the "structures"

---

[2] The terminology "T is the 'compression' of $\boldsymbol{S}$ in $\mathfrak{H}$, and $\boldsymbol{S}$ is the 'dilation' of T to $\boldsymbol{H}$" was proposed by HALMOS [**1**].

[3] In fact, we have

$$(f, T^*g) = (Tf, g) = (\boldsymbol{PSP}f, g) = (f, \boldsymbol{PS}^*\boldsymbol{P}g) = (f, \boldsymbol{PS}^*g)$$

for $f, g \in \mathfrak{H}$.

[4] In fact, we have $\|T_n - T_m\| \leqq \|\boldsymbol{S}_n - \boldsymbol{S}_m\|$, $\|(T_n - T_m)f\| \leqq \|(\boldsymbol{S}_n - \boldsymbol{S}_m)f\|$ for $f \in \mathfrak{H}$, and $((T_n - T_m)f, g) = ((\boldsymbol{S}_n - \boldsymbol{S}_m)f, g)$ for $f, g \in \mathfrak{H}$.

$\{\mathbf{H}, S_\omega, \mathfrak{H}\}_{\omega\in\Omega}$ and $\{\mathbf{H}', S'_\omega, \mathfrak{H}\}_{\omega\in\Omega}$ in the same manner by requiring that $f \to f'$ imply $S_\omega f \to S'_\omega f'$ for all $\omega \in \Omega$.

It is obvious that, from the point of view of extensions of transformations in $\mathfrak{H}$ which extend beyond $\mathfrak{H}$, two extensions which give rise to two isomorphic "structures" can be considered as identical.

In the sequel, when speaking of Hilbert spaces we shall mean both real and complex spaces. If we wish to distinguish between real and complex spaces, we shall say so explicitly. Of course, an extension space $\mathbf{H}$ of $\mathfrak{H}$ is always of the same type (real or complex) as $\mathfrak{H}$.

2. Generalized Spectral Families. NEUMARK'S Theorem

Extensions which extend beyond the given space were first investigated by M. A. NEUMARK [3, 4]; he investigated self-adjoint extensions of symmetric transformations in particular. If S is a symmetric transformation in the complex Hilbert space $\mathfrak{H}$ (with $\mathfrak{D}_S$ dense in $\mathfrak{H}$), we know that S cannot be extended to a self-adjoint transformation without extending beyond $\mathfrak{H}$ except when the deficiency indices $\mathfrak{m}$ and $\mathfrak{n}$ of S are equal. On the other hand, *there always exist self-adjoint extensions of S if one allows these extensions to extend beyond the space $\mathfrak{H}$.*

This is easily proved: Choose, in a Hilbert space $\mathfrak{H}'$, a symmetric transformation S' whose deficiency indices are $\mathfrak{n}$ and $\mathfrak{m}$, that is equal to the deficiency indices of S, but in the reverse order. One can take for example $\mathfrak{H}' = \mathfrak{H}$ and $S' = -S$. Having done this, we consider the product space $\mathbf{H} = \mathfrak{H} \times \mathfrak{H}'$ whose elements are pairs $\{f, f'\}$ $(f \in \mathfrak{H}, f' \in \mathfrak{H}')$ and in which the vector operations and metric are defined as follows:

$$c\{f, f'\} = \{cf, cf'\};\ \{f_1, f_1'\} + \{f_2, f_2'\} = \{f_1 + f_2, f_1' + f_2'\};$$
$$(\{f_1, f_1'\}, \{f_2, f_2'\}) = (f_1, f_2) + (f_1', f_2').$$

If we identify the element f in $\mathfrak{H}$ with the element $\{f, 0\}$ in $\mathbf{H}$, we embed $\mathfrak{H}$ in $\mathbf{H}$ as a subspace of the latter. The transformation

$$\mathbf{S}\{f, f'\} = \{Sf, S'f'\} \qquad (f \in \mathfrak{D}_S,\ f' \in \mathfrak{D}_{S'})$$

is then, as can easily be seen, a symmetric transformation in $\mathbf{H}$ having deficiency indices $\mathfrak{m} + \mathfrak{n}$, $\mathfrak{n} + \mathfrak{m}$. Consequently, $\mathbf{S}$ can be extended, without extending beyond $\mathbf{H}$, to a self-adjoint transformation $\mathbf{A}$ in $\mathbf{H}$. Since we have

$$S \subseteq \mathbf{S} \subseteq \mathbf{A}$$

(where the first extension is obtained by extension from $\mathfrak{H}$ to $\mathbf{H}$), we obtain a self-adjoint extension $\mathbf{A}$ of S.

Let

$$\mathbf{A} = \int_{-\infty}^{\infty} \lambda\, d\mathbf{E}_\lambda$$

be the spectral decomposition of $\boldsymbol{A}$. We have the relations

$$(Sf, g) = (\boldsymbol{A}f, \boldsymbol{P}g) = \int_{-\infty}^{\infty} \lambda \, d(\boldsymbol{E}_\lambda f, \boldsymbol{P}g) = \int_{-\infty}^{\infty} \lambda \, d(\boldsymbol{P}\boldsymbol{E}_\lambda f, g),$$

$$\|Sf\|^2 = \|\boldsymbol{A}f\|^2 = \int_{-\infty}^{\infty} \lambda^2 \, d(\boldsymbol{E}_\lambda f, f) = \int_{-\infty}^{\infty} \lambda^2 \, d(\boldsymbol{P}\boldsymbol{E}_\lambda f, f)$$

for $f \in \mathfrak{D}_S$, $g \in \mathfrak{H}$. Setting

(4) $$B_\lambda = \operatorname{pr} \boldsymbol{E}_\lambda$$

we obtain a family $\{B_\lambda\}_{-\infty<\lambda<\infty}$ of bounded self-adjoint transformations in the space $\mathfrak{H}$, which have the following properties:

a) $B_\lambda \leqq B_\mu$ for $\lambda < \mu$;

b) $B_{\lambda+0} = B_\lambda$;

c) $B_\lambda \to O$ as $\lambda \to -\infty$; $B_\lambda \to I$ as $\lambda \to +\infty$.

Every one-parameter family of bounded self-adjoint transformations which have these properties will be called a *generalized spectral family*. If this family consists of projections (which are then, as a consquence of a), mutually permutable), then we have an ordinary spectral family.

According to what we just proved, we can assign to each symmetric transformation S in $\mathfrak{H}$ a generalized spectral family $\{B_\lambda\}$ in such a way that the equations

(5) $$(Sf, g) = \int_{\infty}^{\infty} \lambda \, d(B_\lambda f, g), \quad \|Sf\|^2 = \int_{\infty}^{\infty} \lambda^2 \, d(B_\lambda f, f)$$

are satisfied for $f \in \mathfrak{D}_S$, $g \in \mathfrak{H}$ (where the integral in the second equation can also converge for certain f which do not belong to $\mathfrak{D}_S$).

The question arises: Can every generalized spectral family which belongs to S, that is, which satisfies equations (5), be obtained as the projection of the spectral family of a self-adjoint extension $\boldsymbol{A}$ of S? The answer is in the affirmative. Namely, we have the following theorem.

THEOREM I (NEUMARK [4. 5]). *Every generalized spectral family* $\{B_\lambda\}$ *can be represented in the form* (4), *as the projection of an ordinary spectral family* $\{\boldsymbol{E}_\lambda\}$. *One can even require the extension space* $\mathbf{H}$ *to be minimal in the sense that it be spanned by the elements of the form* $\boldsymbol{E}_\lambda f$ *where* $f \in \mathfrak{H}$, $-\infty < \lambda < \infty$; *in this case, the structure* $\{\mathbf{H}, \boldsymbol{E}_\lambda, \mathfrak{H}\}_{-\infty<\lambda<\infty}$ *is determined to within an isomorphism.*

We shall prove this theorem in Sec. 7 as a corollary to the "principal theorem" (Sec. 6).

We observe that if $\mathbf{H}$ is minimal, every interval of constancy of B_λ is also an interval of constancy for $\boldsymbol{E}_\lambda$. In fact, if $a \leqq \lambda < b$ is an interval of constancy of B_λ, we have

$$\|\boldsymbol{E}_\lambda - \boldsymbol{E}_a)\boldsymbol{E}_\mu f\|^2 = \|(\boldsymbol{E}_{\min\{\lambda,\mu\}} - \boldsymbol{E}_{\min\{a,\mu\}})f\|^2 = ((\boldsymbol{E}_{\min\{\lambda,\mu\}} - \boldsymbol{E}_{\min\{a,\mu\}})f, f) =$$
$$= (\boldsymbol{P}(\boldsymbol{E}_{\min\{\lambda,\mu\}} - \boldsymbol{E}_{\min\{a,\mu\}})f, f) = ((B_{\min\{\lambda,\mu\}} - B_{\min\{a,\mu\}})f, f) = 0$$

for $f \in \mathfrak{H}$, $a \leqq \lambda < b$, and μ an arbitrary real number; hence

$$(\boldsymbol{E}_\lambda - \boldsymbol{E}_a)g = 0$$

for every element g of the form $\boldsymbol{E}_\mu f \, (f \in \mathfrak{H})$. Since these elements g span the space $\mathbf{H}$, we have $\boldsymbol{E}_\lambda - \boldsymbol{E}_a = \boldsymbol{O}$, $\boldsymbol{E}_\lambda = \boldsymbol{E}_a$, which completes the proof of the theorem.

The simplest case of the Neumark theorem occurs when the family $\{B_\lambda\}$ is generated by a self-adjoint transformation A such that $O \leqq A \leqq I$, in the following manner:

$$B_\lambda = O \text{ for } \lambda < a, \; B_\lambda = A \text{ for } a \leqq \lambda < b, \; B_\lambda = I \text{ for } \lambda \geqq b.$$

We thus obtain the following corollary.

COROLLARY. *Every self-adjoint transformation A in the Hilbert space $\mathfrak{H}$, such that $O \leqq A \leqq I$, can be represented in the form*

$$A = \operatorname{pr} \boldsymbol{Q}$$

where $\boldsymbol{Q}$ is a projection in an extension space $\mathbf{H}$. *In brief: A is the projection of a projection.*

This corollary can also be proved directly without recourse to the general Neumark theorem. The following construction is due to E. A. MICHAEL.[5]

Consider the product space $\mathbf{H} = \mathfrak{H} \times \mathfrak{H}$; by identifying the element f in $\mathfrak{H}$ with the element $\{f, 0\}$ in $\mathbf{H}$, we embed $\mathfrak{H}$ in $\mathbf{H}$ as a subspace of the latter. If we write the elements $\mathbf{H}$ as one-column matrices $\begin{pmatrix} f \\ g \end{pmatrix}$, then every bounded linear transformation $\boldsymbol{T}$ in $\mathbf{H}$ can be represented in the form of a matrix

$$\boldsymbol{T} = \begin{pmatrix} T_{11} & T_{12} \\ T_{21} & T_{22} \end{pmatrix} \tag{6}$$

whose elements T_{ik} are bounded linear transformations in $\mathfrak{H}$. It is easily verified that the matrix addition and multiplication of the corresponding matrices correspond to the addition and multiplication of transformations. Moreover, relation (6) implies that

$$\boldsymbol{T}^* = \begin{pmatrix} T_{11}^* & T_{21}^* \\ T_{12}^* & T_{22}^* \end{pmatrix}.$$

Finally, we have

$$T = \operatorname{pr} \boldsymbol{T}$$

if and only if

$$T_{11} = T.$$

This done, we consider the transformation

$$\boldsymbol{Q} = \begin{pmatrix} A & B \\ B & I - A \end{pmatrix} \text{ with } B = [A(I-A)]^{\frac{1}{2}}.$$

---

[5] See HALMOS [1].

It is clear that Q is self-adjoint and that $A = \text{pr}\, Q$. It remains only to show that $Q^2 = Q$, which is easily done by calculating the square of the matrix Q.

The following theorem is another, less special, consequence of the Neumark theorem.

THEOREM. *Every finite or infinite sequence $\{A_n\}$ of bounded self-adjoint transformations in the Hilbert space $\mathfrak{H}$ such that*

$$A_n \geqq O, \quad \Sigma A_n = I$$

can be represented in the form

$$A_n = \text{pr}\, Q_n \qquad (n = 1, 2, \ldots),$$

where $\{Q_n\}$ is a sequence of projections of an extension space **H** *for which*

$$Q_n Q_m = O \quad (m \neq n), \quad \Sigma Q_n = I.$$

In fact, one has only to apply the Neumark theorem to the generalized spectral family $\{B_\lambda\}$ defined by

$$B_\lambda = \sum_{n \leqq \lambda} A_n.$$

If $\{E_\lambda\}$ is an ordinary spectral family in a minimal extension space such that $B_\lambda = \text{pr}\, E_\lambda$, the function E_λ of λ increases only at the points n where it has the jumps

$$Q_n = E_n - E_{n-0};$$

these transformations Q_n satisfy the requirements of the theorem.

This theorem in its turn has the following theorem as a consequence.

THEOREM. *Every finite or infinite sequence $\{T_n\}$ of bounded linear transformations in the complex Hilbert space $\mathfrak{H}$ can be represented by means of a sequence $\{N_n\}$ of bounded normal transformations in an extension space* **H** *in the form*

$$T_n = \text{pr}\, N_n \qquad (n = 1, 2, \ldots),$$

where the N_n are pairwise doubly permutable.[6] *If any of the transformations T_n are self-adjoint, the corresponding N_n can also be chosen to be self-adjoint.*

We first consider the case where all the transformations T_n are self-adjoint. If m_n and M_n are the greatest lower and least upper bounds of T_n, we set

$$A_n = \frac{1}{2^n(M_n - m_n + 1)}(T_n - m_n I) \qquad (n = 1, 2, \ldots);$$

then we obviously have

$$A_n \geqq O, \quad \sum_n A_n \leqq I.$$

[6] We shall say that two bounded linear transformations T_1, T_2 are *doubly permutable* if T_1 is permutable with T_2 and with T_2^* (and then T_2 is permutable with T_1 and with T_1^*). Furthermore, for two *normal* transformations, simple permutability implies double permutability; it even suffices to assume that one of the two transformations is normal (FUGLEDE [1]; also see HALMOS [2]).

If we again set

$$A = I - \sum_n A_n$$

we obtain a sequence $A, A_1, A_2, \ldots$ of transformations which satisfies the hypotheses of the preceding theorem, and which, consequently, can be represented in the form

$$A_n = \text{pr } \boldsymbol{Q}_n \qquad (n = 1, 2, \ldots)$$

in terms of the projections $\boldsymbol{Q}_n$, which are pairwise orthogonal (and consequently permutable). It follows that

$$T_n = \text{pr } \boldsymbol{S}_n \qquad (n = 1, 2, \ldots)$$

with

$$\boldsymbol{S}_n = m_n \boldsymbol{I} + 2^n (M_n - m_n + 1) \boldsymbol{Q}_n,$$

where the transformations $\boldsymbol{S}_n$ are self-adjoint and mutually permutable.

The general case is reducible to the particular case of self-adjoint transformations by replacing each transformation T_n in the given sequence by the two self-adjoint transformations Re T_n and Im T_n. In fact, since the representation

$$\text{Re } T_n = \text{pr } \boldsymbol{S}_{2n}, \quad \text{Im } T_n = \text{pr } \boldsymbol{S}_{2n+1} \qquad (n = 1, 2, \ldots)$$

is possible by means of bounded self-adjoint pairwise permutable transformations $\boldsymbol{S}_i$, the representation

$$T_n = \text{pr } \boldsymbol{N}_n \qquad (n = 1, 2, \ldots)$$

follows from this by means of the normal pairwise doubly permutable transformations $\boldsymbol{N}_n = \boldsymbol{S}_{2n} + i\boldsymbol{S}_{2n+1}$. For a self-adjoint T_n, we have T_n = Re T_n, and we can then choose $\boldsymbol{S}_{2n+1} = \boldsymbol{O}$ and hence $\boldsymbol{N}_n = \boldsymbol{S}_{2n}$.

3. Sequences of Moments

1. The following theorem is closely related to Theorem I.

THEOREM II (SZ.-NAGY [9]). *Suppose $\{A_n\}$ $(n = 0, 1, \ldots)$ is a sequence of bounded self-adjoint transformations in the Hilbert space $\mathfrak{H}$ satisfying the following conditions:*

(α_M) *for every polynomial*

$$a_0 + a_1\lambda + a_2\lambda^2 + \cdots + a_n\lambda^n$$

with real coefficients which assumes non-negative values in the interval $-M \leqq \lambda \leqq M$, we have

$$a_0A_0 + a_1A_1 + a_2A_2 + \cdots + a_nA_n \geqq O\,;$$

(β) $A_0 = I.$

Then there exists a self-adjoint transformation $\boldsymbol{A}$ in an extension space **H** *such that*

$$A_n = \text{pr } \boldsymbol{A}^n \qquad (n = 0, 1, \ldots). \tag{7}$$

Furthermore, one can require that **H** *be minimal in the sense that it be spanned by elements of the form $A^n f$ where $f \in \mathfrak{H}$ and $n = 0, 1, \ldots$; in this case, the structure $\{\mathbf{H}, \boldsymbol{A}, \mathfrak{H}\}$ is determined to within an isomorphism, and we have*

$$\|\boldsymbol{A}\| \leqq M.$$

We observe that if $\{B_\lambda\}$ is a generalized spectral family on the interval $[-M, M]$ (that is, $B_\lambda = O$ for $\lambda < -M$ and $B_\lambda = I$ for $\lambda \geqq M$), the transformations

$$(8) \qquad A_n = \int_{-M-0}^{M} \lambda^n \, dB_\lambda \qquad (n = 0, 1, \ldots)$$

satisfy conditions (α_M) and (β). Conversely, if these conditions are satisfied, the sequence $\{A_n\}$ has an integral decomposition of the form (8) with $\{B_\lambda\}$ on $[-M, M]$. This clearly follows from Theorem II if we make use of the spectral decomposition of the transformation $\boldsymbol{A}$. But one can also prove (8) directly, without recourse to Theorem II.

In fact, the correspondence between the polynomials

$$p(\lambda) = a_0 + a_1\lambda + a_2\lambda^2 + \cdots + a_n\lambda^n$$

and the self-adjoint transformations

$$A(p) = a_0 I + a_1 A_1 + a_2 A_2 + \cdots + a_n A_n$$

which is *homogeneous*, *additive*, and *of positive type* with respect to the interval $-M \leqq \lambda \leqq M$, can be extended, with preservation of these properties, to a vaster class of functions which comprises, among others, the discontinuous functions

$$e_\mu(\lambda) = \begin{cases} 1 \text{ for } \lambda \leqq \mu, \\ 0 \text{ for } \lambda > \mu, \end{cases}$$

and then we obtain representation (8) by setting

$$B_\mu = A(e_\mu).$$

We have only to repeat verbatim the line of argument of one of the usual proofs of the spectral decomposition of a bounded self-adjoint transformation A,[7] letting A_n play the role of A^n. The only difference is that now the correspondence $p(\lambda) \to p(A)$ and its extension are no longer multiplicative, and that consequently the relation $e_\mu^2(\lambda) \equiv e_\mu(\lambda)$ does not imply that B_μ^2 is equal to B_μ, and hence that B_μ is in general not a projection.

According to Theorem I, $\{B_\lambda\}$ is the projection of an ordinary spectral family $\{\boldsymbol{E}_\lambda\}$, which one can choose in such a way that it is also on $[-M, M]$, and then (7) follows from (8) by setting

$$\boldsymbol{A} = \int_{-M-0}^{M} \lambda \, d\boldsymbol{E}_\lambda.$$

[7] See Secs. 106, 107.

We shall return to this theorem later (Sec. 8) and prove it as a corollary to the "principal theorem" (Sec. 6).

2. If we replace condition (β) by the less restrictive condition

$$(\beta') \qquad A_0 \leqq I,$$

then representation (7) of the sequence $\{A_n\}$ will still be possible, if only starting from $n=1$.[8] Everything reduces to showing that if the sequence

$$\{A_0, A_1, A_2, \ldots\}$$

satisfies conditions (α_M) and (β'), the sequence

$$\{I, A_1, A_2, \ldots\}$$

satisfies condition (α_M). But, if $p(\lambda) = a_0 + a_1\lambda + \ldots + a_n\lambda^n \geqq 0$ in $[-M, M]$, we have in particular that $p(0) = a_0 \geqq 0$; since by assumption, $a_0A_0 + a_1A_1 + \ldots + a_nA_n \geqq O$ and $I - A_0 \geqq O$, it follows that

$$a_0 I + a_1 A_1 + \cdots + a_n A_n = a_0(I - A_0) + a_0 A_0 + a_1 A_1 + \cdots + a_n A_n \geqq O.$$

One of the most interesting consequences of the representation

$$A_n = \operatorname{pr} \boldsymbol{A}^n \qquad (n = 1, 2, \ldots)$$

is the following. We have

$$(A_2 f, f) = (\boldsymbol{P}\boldsymbol{A}^2 f, f) = (\boldsymbol{A}^2 f, f) = \|\boldsymbol{A}f\|^2 \geqq$$
$$\geqq \|\boldsymbol{P}\boldsymbol{A}f\|^2 = (\boldsymbol{A}\boldsymbol{P}\boldsymbol{A}f, f) = (\boldsymbol{P}\boldsymbol{A}\boldsymbol{P}\boldsymbol{A}f, f) = (A_1^2 f, f)$$

for all $f \in \mathfrak{H}$, where equality holds if, and only if,

$$\boldsymbol{A}f = \boldsymbol{P}\boldsymbol{A}f = A_1 f.$$

If this case occurs for all $f \in \mathfrak{H}$, we have

$$\boldsymbol{A}^2 f = \boldsymbol{A}(\boldsymbol{A}f) = \boldsymbol{A}(A_1 f) = A_1(A_1 f) = A_1^2 f,$$
$$\boldsymbol{A}^3 f = \boldsymbol{A}(\boldsymbol{A}^2 f) = \boldsymbol{A}(A_1^2 f) = A_1(A_1^2 f) = A_1^3 f, \text{ etc.}$$

---

[8] Formula (7) in Theorem II can be replaced by another which is valid under the single condition (α_M) and for $n = 0, 1, \ldots$. Namely, we have that

$$A_n f = A_0^{\frac{1}{2}} \boldsymbol{P}\boldsymbol{A}^n A_0^{\frac{1}{2}} f \qquad (f \in \mathfrak{H};\ n = 0, 1, \ldots),$$

where $\boldsymbol{A}$ is a self-adjoint transformation of a suitable extension space, $\|\boldsymbol{A}\| \leqq M$. (The equality $A_0 \geqq O$ and hence the existence of the square root $A_0^{\frac{1}{2}} \geqq O$ are consequences of the condition (α_M) if we apply it to $p(\lambda) \equiv 1$.) This is obtained immediately when A_0 admits even a positive greatest lower bound, because then the sequence of transformations

$$\tilde{A}_n = A_0^{-\frac{1}{2}} A_n A_0^{-\frac{1}{2}} \qquad (n = 0, 1, \ldots)$$

satisfies conditions (α_M) and (β). The case when A_0 does not have a positive greatest lower bound requires a slightly more refined investigation (see Sz.-Nagy [9]).

and hence

$$A_n f = \boldsymbol{P}\boldsymbol{A}^n f = A_1^n f \qquad (n = 1, 2, \ldots).$$

We have thus obtained the following result.

If the sequence $A_0, A_1, A_2, \ldots$ of bounded self-adjoint transformations in the Hilbert space $\mathfrak{H}$ satisfies hypotheses (α_M) and (β'), then the inequality

$$A_1^2 \leqq A_2 \tag{9}$$

holds, where equality occurs if, and only if, $A_n = A_1^n$ $(n = 1, 2, \ldots)$.

Inequality (9) is due to R. V. KADISON [1] who proved it differently and used it in his researches on algebraic invariants of operator algebras.

Moreover, one can also omit hypothesis (β'), and then the following inequality

$$A_1^2 \leqq \|A_0\| A_2 \tag{10}$$

is obtained; in fact, we have only to apply inequality (9) to the sequence $\{\|A_0\|^{-1} A_n\}$.

4. Contractions in Hilbert Space

1. Whereas the projections of bounded self-adjoint transformations are also self-adjoint, the projections of unitary transformations[9] are already of a more general type. In order that $T = \text{pr}\, \boldsymbol{U}$, with $\boldsymbol{U}$ unitary, it is necessary that

$$\|Tf\| = \|\boldsymbol{P}\boldsymbol{U}f\| \leqq \|\boldsymbol{U}f\| = \|f\|$$

for all $f \in \mathfrak{H}$, that is, $\|T\| \leqq 1$, and hence the transformation T must be a *contraction*.

But, this condition is not only necessary but also sufficient.

THEOREM. *Every contraction T in the Hilbert space $\mathfrak{H}$ can be represented in an extension space* **H** *as the projection of a unitary transformation* $\boldsymbol{U}$ *onto* $\mathfrak{H}$.

The theorem, and the following simple construction of $\boldsymbol{U}$, are due to HALMOS [1]. As in Sec. 2, let us consider the product space $\mathbf{H} = \mathfrak{H} \times \mathfrak{H}$ and the following transformation of **H**:

$$\boldsymbol{U} = \begin{pmatrix} T & S \\ -Z & T^* \end{pmatrix} \text{ where } S = (I - TT^*)^{\frac{1}{2}},\ Z = (I - T^*T)^{\frac{1}{2}}.\text{[10]} \tag{11}$$

The relation $T = \text{pr}\, \boldsymbol{U}$ is obvious. We shall show that $\boldsymbol{U}$ is unitary, or, what

---

[9] It is usual to speak of *unitary* transformations only in the case of a complex Hilbert space; their analogues in the case of real Hilbert space are called *orthogonal*. As a matter of convenience, we agree to say "unitary" in both cases. Hence, the linear transformation T in the Hilbert space $\mathfrak{H}$ is *unitary* if it maps the space $\mathfrak{H}$ isometrically onto itself, or, what amounts to the same thing, if $T^*T = TT^* = I$.

[10] Since $\|T\| \leqq 1$, we have $O \leqq I - TT^* \leqq I$ and $O \leqq I - T^*T \leqq I$.

amounts to the same thing, that $\boldsymbol{U}^* \boldsymbol{U}$ and $\boldsymbol{U} \boldsymbol{U}^*$ are equal to the identity transformation $\boldsymbol{I}$ in $\mathbf{H}$. Since S and Z are self-adjoint, we have

$$\boldsymbol{U}^*\boldsymbol{U} = \begin{pmatrix} T^* & -Z \\ S & T \end{pmatrix} \begin{pmatrix} T & S \\ -Z & T^* \end{pmatrix} = \begin{pmatrix} T^*T + Z^2 & T^*S - ZT \\ ST - TZ & S^2 + TT^* \end{pmatrix},$$

$$\boldsymbol{U}\boldsymbol{U}^* = \begin{pmatrix} T & S \\ -Z & T^* \end{pmatrix} \begin{pmatrix} T^* & -Z \\ S & T \end{pmatrix} = \begin{pmatrix} TT^* + S^2 & -TZ + ST \\ -ZT^* + TS & Z^2 + T^*T \end{pmatrix}.$$

Since $Z^2 = I - T^*T$, $S^2 = I - TT^*$, the diagonal elements of the product matrices are all equal to I. It remains to show that the other elements are equal to O, i.e. that

(12) $$ST = TZ$$

(the equation $T^*S = ZT^*$ follows from this by passing over to the adjoints of both members of (12)).

But, we have

$$S^2T = (I - TT^*)T = T - TT^*T = T(I - T^*T) = TZ^2,$$

from which it follows by complete induction that

$$S^{2n}T = TZ^{2n} \quad \text{for} \quad n = 0, 1, 2, \ldots.$$

Then we also have

$$p(S^2)T = Tp(T^2)$$

for every polynomial $p(\lambda)$. Since S and Z are the positive square roots of S^2 and Z^2 respectively, there exists a sequence of polynomials $p_n(\lambda)$ such that

$$p_n(S^2) \to S, \quad p_n(Z^2) \to Z.$$ [11]

Now (12) follows from the equation

$$p_n(S^2)T = Tp_n(Z^2)$$

by passing to the limit as $n \to \infty$.

This completes the proof of the theorem.

2. The relation between transformations $\boldsymbol{S}$ in an extension space $\mathbf{H}$ of the space $\mathfrak{H}$ and their projections $T = \operatorname{pr} \boldsymbol{S}$ onto $\mathfrak{H}$ is not multiplicative in general, that is, the equations $T_1 = \operatorname{pr} \boldsymbol{S}_1$, $T_2 = \operatorname{pr} \boldsymbol{S}_2$ do not in general imply $T_1T_2 = \operatorname{pr} \boldsymbol{S}_1\boldsymbol{S}_2$. For example, if we consider the transformation $\boldsymbol{U}$ constructed according to formula (11), we have $\operatorname{pr} \boldsymbol{U}^2 = T^2 - SZ$, which in general is not equal to T^2.

The question arises: Is it possible to find, in a suitable extension space, a unitary transformation $\boldsymbol{U}$ such that the powers of the contraction T (which are themselves contractions) are at the same time equal to the projections onto $\mathfrak{H}$ of the corresponding powers of $\boldsymbol{U}$?

---

[11] See Sec. 104.

If we are dealing with only a finite number of powers,

$$T, T^2, \ldots, T^k,$$

then the problem can be solved in the affirmative in a rather simple manner, by suitably generalizing the immediately preceding construction.

Let us consider the product space $\mathbf{H} = \mathfrak{H} \times \cdots \times \mathfrak{H}$, with $k+1$ factors, whose elements are ordered $(k+1)$-tuples $\{f_1, \ldots, f_{k+1}\}$ of elements in $\mathfrak{H}$ and in which the vector operations and metric are defined in the usual way:

$$c\{f_1, \ldots, f_{k+1}\} = \{cf_1, \ldots, cf_{k+1}\},$$
$$\{f_1, \ldots, f_{k+1}\} + \{g_1, \ldots, g_{k+1}\} = \{f_1 + g_1, \ldots, f_{k+1} + g_{k+1}\},$$
$$(\{f_1, \ldots, f_{k+1}\}, \{g_1, \ldots, g_{k+1}\}) = (f_1, g_1) + \cdots + (f_{k+1}, g_{k+1}).$$

We embed $\mathfrak{H}$ in $\mathbf{H}$ as a subspace of the latter by identifying the element f in $\mathfrak{H}$ with the element $\{f, 0, \ldots, 0\}$ in $\mathbf{H}$. The bounded linear transformations $\boldsymbol{T}$ in $\mathbf{H}$ will be represented by matrices (T_{ij}) with $k+1$ rows and $k+1$ columns, all of whose elements T_{ij} are bounded linear transformations in $\mathfrak{H}$. We have $T = \operatorname{pr} \boldsymbol{T}$ if and only if $T_{11} = T$.

Let us now consider the following transformation in $\mathbf{H}$:[12]

$$\boldsymbol{U} = \left(\begin{array}{cccccc} T & S & O & O & \ldots & O \\ O & O & -I & O & \ldots & O \\ O & O & O & -I & \ldots & O \\ \cdot & \cdot & \cdot & \cdot & \cdot & \cdot \\ O & O & O & \ldots & O & -I \\ -Z & T^* & O & \ldots & O & O \end{array} \right) \Bigg\} k+1 \text{ rows and columns,}$$

where S and Z have the same meaning as in the foregoing construction.[13]

The transformation $\boldsymbol{U}$ is unitary. This is proved in the same way as above, by a direct calculation of the matrices $\boldsymbol{U}^*\boldsymbol{U}$, $\boldsymbol{U}\boldsymbol{U}^*$. In order to prove the relations

$$T^n = \operatorname{pr} \boldsymbol{U}^n \qquad (n = 1, 2, \ldots, k),$$

---

[12] The following construction is a modification of that given by EGERVÁRY [1] for the case of a finite-dimensional space.

[13] The $-I$'s could be replaced by $+I$'s, but the choice of the minus sign has the following advantage: Suppose $\mathfrak{H}$ is real and finite-dimensional and represent the transformations T, T^*, S, Z, O and I by their matrices with respect to an orthogonal basis in $\mathfrak{H}$. $\boldsymbol{U}$ will then be an ordinary hypermatrix whose elements are real numbers. Because of the minus signs, the determinant of $\boldsymbol{U}$ is equal to the determinant d of the matrix $\begin{pmatrix} T & S \\ -Z & T^* \end{pmatrix}$. Since the latter matrix is orthogonal, we have $d = \pm 1$. But d depends continuously on T and since the contractions form a convex (and hence connected) set, and since furthermore $d = +1$ for $T = I$, we necessarily have $d = +1$ for all the contractions T. Hence $\boldsymbol{U}$ is an orthogonal transformation which preserves the orientation of the space, that is, it is a *rotation*.

we must calculate the element in the matrix U^n having indices 1, 1 and then note that the latter is equal to T^n for $n = 1, \ldots, k$. We shall even prove more, namely that the first row in the matrix U^n $(n = 1, \ldots, k)$ is the following:

$$(T^n,\ T^{n-1}S,\ -T^{n-2}S^2,\ T^{n-3}S^3,\ -T^{n-4}S^4, \ldots, (-1)^{n-1}S^n,\ \overbrace{O, \ldots, O}^{k-n}).$$

This proposition is obvious for $n = 1$, and we prove it true for $n + 1$, assuming it true for n $(n \leqq k - 1)$ by calculating the matrix U^{n+1} as the matrix product $U^n \cdot U$. We have thus proved the following theorem.

THEOREM. *If T is a contraction in the Hilbert space* $\mathfrak{H}$, *then there exists a unitary transformation* **U** *in an extension space* **H** *such that*

$$T^n = \operatorname{pr} U^n,$$

$n = 0, 1, \ldots, k$ *(the case* $n = 0$ *is trivial), for every given natural number k. The product space* $\mathfrak{H} \times \cdots \times \mathfrak{H}$ *with* $k + 1$ *factors*[14] *can be taken for* **H**.

3. It is important in the above construction that k is a finite number. However, the theorem is also true for $k = \infty$.

THEOREM III (Sz.-Nagy **[10, 11]**). *If T is a contraction in the Hilbert space* $\mathfrak{H}$, *then there exists a unitary transformation* **U** *of an extension space* **H** *such that the relation*

$$T^n = \operatorname{pr} U^n$$

is valid for $n = 0, 1, 2, \ldots$. *Furthermore, one can require that the space* **H** *be minimal in the sense that it is spanned by the elements of the form* $U^n f$ *where* $f \in \mathfrak{H}$ *and* $n = 0, \pm 1, \pm 2, \ldots$; *in this case, the structure* $\{\mathbf{H}, \mathbf{U}, \mathfrak{H}\}$ *is determined to within an isomorphism.*

An analogous theorem is true for *semi-groups* and *one-parameter semi-groups* of contractions, that is, for families $\{T_t\}$ of contractions (where $0 \leqq t < \infty$ or $-\infty \leqq t < \infty$, according to the case at hand) such that

$$T_0 = I, \quad T_{t_1} T_{t_2} = T_{t_1 + t_2},$$

and for which one assumes further that T_t depends strongly or weakly continuously on t; weak continuity means that $(T_t f, g)$ is a continuous numerical-valued function of t for every pair f, g of elements in $\mathfrak{H}$. The theorem in question is the following.

THEOREM IV (Sz.-Nagy **[10, 11]**). *If* $\{T_t\}_{t \geqq 0}$ *is a weakly continuous one-parameter semi-group of contractions in the Hilbert space* $\mathfrak{H}$, *then there exists a one-parameter group* $\{\mathbf{U}_t\}_{-\infty < t < \infty}$ *of unitary transformations in an extension space* **H**, *such that*

$$T_t = \operatorname{pr} U_t \quad \text{for} \quad t \geqq 0.$$

[14] The last proposition is important only in the case where the space $\mathfrak{H}$ is finite-dimensional.

Furthermore, one can require that the space **H** *be minimal in the sense that it is spanned by elements of the form* $U_t f$, *where* $f \in \mathfrak{H}$ *and* $-\infty < t < \infty$; *in this case,* U_t *is strongly continuous and the structure* $\{\mathbf{H}, U_t, \mathfrak{H}\}_{-\infty<t<\infty}$ *is determined to within an isomorphism.*

These two theorems can be generalized to discrete or continuous semi-groups with several generators. We shall formulate only the following generalization of Theorem III.

THEOREM V. *Suppose* $\{T^{(\rho)}\}_{\rho\in R}$ *is a system of pairwise doubly permutable contractions in the Hilbert space* $\mathfrak{H}$. *There exists, in an extension space* **H**, *a system* $\{U^{(\rho)}\}_{\rho\in R}$ *of pairwise permutable unitary transformations such that*

$$\prod_{i=1}^{r} [T^{(\rho_i)}]^{n_i} = \text{pr} \prod_{i=1}^{r} [U^{(\rho_i)}]^{n_i}$$

for arbitrary $\rho_i \in R$ *and integers* n_i, *provided the factor* $[T^{(\rho_i)}]^{n_i}$ *is replaced by* $[T^{(\rho_i)*}]^{-n_i}$ *when* $n_i < 0$. *Moreover, one can require that the space* **H** *be minimal in the sense that it be spanned by the elements of the form* $\prod_{i=1}^{r} [U^{(\rho_i)}]^{n_i} f$ *where* $f \in \mathfrak{H}$; *in this case, the structure* $\{\mathbf{H}, U^{(\rho)}, \mathfrak{H}\}_{\rho\in R}$ *is determined to within an isomorphism.*

We shall prove Theorems III and V in Sec. 9.

3. We now give several applications of these theorems; T will denote a contraction in a complex Hilbert space $\mathfrak{H}$ and $\{T_t\}$ is a weakly continuous one-parameter semi-group of contractions in $\mathfrak{H}$.

a) Invariant elements. *If the element f is invariant with respect to T, then it is also invariant with respect to* T^*.[15]

Proof. We have $T = \text{pr}\, U$, with U unitary, from which it follows that $T^* = \text{pr}\, U^* = \text{pr}\, U^{-1}$. The equations $f = Tf = PUf$, $\|Uf\| = \|f\|$ imply that $Uf = f$. Hence we have $f = U^{-1}f = PU^{-1}f = T^*f$, which completes the proof of the theorem.

b) Ergodic theorems. *For all* $f \in \mathfrak{H}$ *the limits*

$$\lim_{\substack{n > m \geq 0 \\ n-m \to \infty}} \frac{1}{n-m} \sum_{k=m}^{n-1} T^k f$$

and

$$\lim_{\substack{\nu > \mu \geq 0 \\ \nu-\mu \to \infty}} \frac{1}{\nu-\mu} \int_{\mu}^{\nu} T_t f\, dt$$

exist in the sense of strong convergence of elements, where the integral is defined as the strong limit of sums of Riemann type.[16]

Proof. By Theorems III and IV, we have $T^k = \text{pr}\, U^k$ $(k = 0, 1, \ldots)$ and $T_t = \text{pr}\, U_t$ $(t \geq 0)$ with U_t strongly continuous; hence T_t is also strongly continuous. For $f \in \mathfrak{H}$, we have

---

[15] See Sec. 144.

[16] See Sec. 144.

$$\sum_{m}^{n-1} T^k f = P \sum_{m}^{n-1} U^k f$$

and

$$\int_{\mu}^{\nu} T_t f\,dt = P \int_{\mu}^{\nu} U_t f\,dt,$$

respectively, and the propositions thus follow from the ergodic theorems of J. VON NEUMANN on unitary transformations.

DUNFORD'S ergodic theorem on several permutable[17] contractions can be reduced in an analogous manner, by Theorem V, to the particular case of unitary transformations, but this only under the additional condition that these contractions be *doubly* permutable.

c) Theorems of VON NEUMANN and HEINZ.[18] *Suppose*

$$u(z) = c_0 + c_1 z + \cdots + c_n z^n + \cdots$$

is a power series in the complex variable z with

(13) $$|c_0| + |c_1| + \cdots + |c_n| + \cdots < \infty.$$

Set

$$u(T) = c_0 I + c_1 T + \ldots + c_n T^n + \ldots .$$[19]

If the function $u(z)$ satisfies one or the other of the inequalities

$$|u(z)| \leqq 1, \qquad \operatorname{Re} u(z) \geqq 0, \ \textit{with}\ |z| \leqq 1,$$

then we have

$$\|u(T)\| \leqq 1, \qquad \operatorname{Re} u(T) \geqq O,$$

respectively.

Proof: It follows from the representation of powers: $T^k = \operatorname{pr} U^k$ $(k = 0, 1, \ldots)$ that

$$u(T) = \operatorname{pr} u(U).$$

Let

$$U = \int_0^{2\pi} e^{i\lambda}\,dE_\lambda$$

be the spectral decomposition of the unitary transformation U; we then have

$$\|u(T)f\|^2 = \|P\,u(U)f\|^2 \leqq \|u(U)f\|^2 = \int_0^{2\pi} |u(e^{i\lambda})|^2 d(E_\lambda f, f),$$

$$\operatorname{Re}(u(T)f, f) = \operatorname{Re}(P\,u(U)f, f) = \operatorname{Re}(u(U)f, f) = \int_0^{2\pi} \operatorname{Re} u(e^{i\lambda})\,d(E_\lambda f, f)$$

---

[17] See Sec. 145.

[18] See Sec. 153; here they are stated in a slightly generalized form. These theorems are valid in a complex Hilbert space.

[19] This series converges in norm because of (13).

for $f \in \mathfrak{H}$. The above propositions follow in an obvious manner from these formulas.

d) *Let*

$$p(\theta) = \sum_k a_k e^{it_k \theta}$$

be a trigonometric series with arbitrary real t_k *and such that*

$$\sum_k |a_k| < \infty. \tag{14}$$

Set

$$p(T) = \sum a_k T_{t_k}.$$ [20]

If the function $p(\theta)$ *satisfies one or the other of the inequalities*

$$|p(\theta)| \leqq 1, \qquad \text{Re}\, p(\theta) \geqq 0, \text{ for all real } \theta,$$

then

$$\|p(T)\| \leqq 1, \qquad \text{Re}\, p(T) \geqq O$$

respectively.

Proof. The proof proceeds exactly as for c) but now using Theorem IV and Stone's theorem in virtue of which there is a spectral decomposition of $\boldsymbol{U}_t$ of the form

$$\boldsymbol{U}_t = \int_{-\infty}^{\infty} e^{it\lambda}\, d\boldsymbol{E}_\lambda.$$

Analogous theorems could be stated (under suitable hypotheses which assure convergence) for trigonometric integrals.

4. *Isometric* transformations in Hilbert space $\mathfrak{H}$ (into a subspace of $\mathfrak{H}$) are particular cases of contractions. If the isometric transformation T is represented as the projection of a unitary transformation $\boldsymbol{U}$, we have

$$\|f\| = \|Tf\| = \|\boldsymbol{P}\,\boldsymbol{U}f\| \leqq \|\boldsymbol{U}f\|$$

for all f; since, on the other hand, $\|\boldsymbol{U}f\| = \|f\|$, we necessarily have $\boldsymbol{P}\boldsymbol{U}f = \boldsymbol{U}f$, and hence $Tf = \boldsymbol{U}f$; that is, $\boldsymbol{U}$ is an extension of T.

It therefore follows from our theorems on contractions that *every isometric transformation has a unitary extension and that for every weakly continuous one-parameter semi-group of isometric transformations* T_t, *there exists a strongly continuous one-parameter group of unitary transformations* $\boldsymbol{U}_t$ *in an extension space such that* $\boldsymbol{U}_t \supseteq T_t$.

The last theorem was proved earlier by Cooper [3] in an entirely different way.

[20] This series converges in norm because of (14).

5. Normal Extensions

We proved in Sec. 2 in particular that every bounded linear transformation T in the complex Hilbert space $\mathfrak{H}$ can be represented as the projection of a normal transformation in an extension space. The question arises: Does T even have a normal *extension* N?

If a normal extension N of T exists, then a fortiori $T = \text{pr}\, N$, and consequently $T^* = \text{pr}\, N^*$, from which it follows that

$$\|Tf\| = \|Nf\| = \|N^*f\| \geqq \|PN^*f\| = \|T^*f\|$$

for all $f \in \mathfrak{H}$. The inequality

$$\|Tf\| \geqq \|T^*f\| \qquad (\text{for all } f \in \mathfrak{H}) \tag{15}$$

is therefore a necessary condition that T have a normal extension. But, it is easy to construct examples of transformations T which do not satisfy this condition.

Other, less simple, necessary conditions are obtained in the following manner. Suppose $\{g_i\}$ $(i = 0, 1, \ldots)$ is a sequence of elements in $\mathfrak{H}$ almost all of which (that is with perhaps the exception of a finite number of them) are equal to the element 0 in $\mathfrak{H}$. We then have

$$\sum_{i=0}^{\infty} \sum_{j=0}^{\infty} (T^i g_j, T^j g_i) = \sum_i \sum_j (N^i g_j, N^j g_i) = \sum_i \sum_j (N^{*j} N^i g_j, g_i) =$$

$$= \sum_i \sum_j (N^i N^{*j} g_j, g_i) = \sum_i \sum_j (N^{*j} g_j, N^{*i} g_i) = \|\sum_i N^{*i} g_i\|^2 \geqq 0,$$

$$\sum_{i=0}^{\infty} \sum_{j=0}^{\infty} (T^{i+1} g_j, T^{j+1} g_i) = \|\sum_i (N^*)^{i+1} g_i\|^2 \leqq \|N^*\|^2 \|\sum_i N^{*i} g_i\|^2,$$

from which we see that

$$\sum_{i=0}^{\infty} \sum_{j=0}^{\infty} (T^i g_j, T^j g_i) \geqq 0 \tag{16}$$

and

$$\sum_{i=0}^{\infty} \sum_{j=0}^{\infty} (T^{i+1} g_j, T^{j+1} g_i) \leqq C^2 \sum_{i=0}^{\infty} \sum_{j=0}^{\infty} (T^i g_j, T^j g_i) \tag{17}$$

with constant $C > 0$. These two inequalities are therefore *necessary* conditions that T have a bounded normal extension.

But these conditions are also *sufficient*. Namely, the following theorem holds.

THEOREM VI (HALMOS [**1**]). *Every bounded linear transformation T in the Hilbert space $\mathfrak{H}$ which satisfies conditions* (16) *and* (17) *has a bounded normal extension N in an extension space* $\mathbf{H}$. *One can even require that* $\mathbf{H}$ *be minimal in the sense that it is spanned by the elements of the form $N^{*k}f$ where $f \in \mathfrak{H}$ and $k = 0, 1, \ldots$; in this case, the structure $\{\mathbf{H}, N, \mathfrak{H}\}$ is determined to within an isomorphism.*

We shall prove this theorem in Sec. 10 as one of the corollaries to our principal theorem (Sec. 6).

For the present, we shall content ourselves with a remark connecting the problems on extensions with the problems treated in Theorems II-V:

The following three propositions are equivalent for any two bounded linear transformations, T in $\mathfrak{H}$ and $\boldsymbol{T}$ in $\mathbf{H}$ $(\supseteqq \mathfrak{H})$:

a) $T \subseteqq \boldsymbol{T}$;

b) $T = \operatorname{pr} \boldsymbol{T}$ *and* $T^*T = \operatorname{pr} \boldsymbol{T}^*\boldsymbol{T}$;

c) $T^{*i}T^k = \operatorname{pr} \boldsymbol{T}^{*i}\boldsymbol{T}^k$ *for* $i, k = 0, 1, \ldots$.

Proof. a) → c) because

$$(T^{*i} T^k f, g) = (T^k f, T^i g) = (\boldsymbol{T}^k g, \boldsymbol{T}^i g) = (\boldsymbol{T}^{*i} \boldsymbol{T}^k f, g) = (\boldsymbol{P}\boldsymbol{T}^{*i} \boldsymbol{T}^k f, g)$$

for $f, g \in \mathfrak{H}$. c) → b) is obvious. b) → a) is proved as follows: for $f \in \mathfrak{H}$ we have, on the one hand, that

$$\| Tf \|^2 = (T^* Tf, f)) = (\boldsymbol{P}\boldsymbol{T}^* \boldsymbol{T} f, f) = (\boldsymbol{T} f, \boldsymbol{T} f) = \| \boldsymbol{T} f \|^2$$

because

$$T^* T = \operatorname{pr} \boldsymbol{T}^* \boldsymbol{T},$$

and, on the other hand, that

$$\| Tf \| = \| \boldsymbol{P}\boldsymbol{T} f \|$$

because

$$T = \operatorname{pr} \boldsymbol{T}.$$

Hence we have

$$\| \boldsymbol{P}\boldsymbol{T} f \| = \| \boldsymbol{T} f \|$$

which is possible only if $\boldsymbol{T}f = \boldsymbol{P}\boldsymbol{T}f$, that is, if $\boldsymbol{T}f = Tf$; therefore $\boldsymbol{T} \supseteqq T$.

6. Principal Theorem

As we have already stated, Theorems II-VI can be proved as more or less immediate corollaries to a "principal theorem." In order to be able to state this theorem, we must first introduce some concepts of an algebraic nature.

By a *$*$-semi-group* we shall understand a system Γ of elements (which we shall denote by Greek letters) in which two operations are defined: an associative "semi-group operation" $(\xi, \eta) \to \xi\eta$, and a "$*$ operation," $\xi \to \xi^*$, which satisfies the following rules of computation:

$$\xi^{**} = \xi, \quad (\xi \eta)^* = \eta^* \xi^*.$$

We shall assume further that there is a "unit" element ε in Γ such that

$$\varepsilon\xi = \xi\varepsilon = \xi \text{ for all } \xi \in \Gamma, \text{ and } \varepsilon^* = \varepsilon.$$

Any *group* can be considered as a $*$-semi-group if we define the $*$ operation in it as the inverse: $\xi^* = \xi^{-1}$. *In the sequel, when we speak of a group Γ, we shall assume that it is provided with this $*$-semi-group structure.*

By a *representation* of the $*$-semi-group Γ in a Hilbert space $\mathbf{H}$ we shall understand a family $\{\boldsymbol{D}_\xi\}_{\xi \in \Gamma}$ of bounded linear transformations in $\mathbf{H}$ such that

$$D_\varepsilon = I, \quad D_{\xi\eta} = D_\xi D_\eta, \quad D_{\xi^*} = D_\xi^* \text{ for all } \xi, \eta \in \Gamma.$$

The following propositions are obvious: if for an $\xi \in \Gamma$

(i) $\xi^*\xi = \xi\xi^*$, the transformation D_ξ is *normal;*
(ii) $\xi = \xi^*$, the transformation D_ξ is *self-adjoint;*
(iii) $\xi = \xi^* = \xi^2$, the transformation D_ξ is an (*orthogonal*) *projection*;
(iv) $\xi^*\xi = \xi\xi^* = \varepsilon$, the transformation D_ξ is *unitary.*

Suppose $\{D_\xi\}$ is a representation of Γ in $\mathbf{H}$ and let $\mathfrak{H}$ be a subspace of $\mathbf{H}$. Consider the transformations

$$T_\xi = \mathrm{pr}_{\mathfrak{H}} D_\xi .$$

It is obvious that $T_\varepsilon = I$ (the identity transformation in $\mathfrak{H}$), and that $T_{\xi^*} = \mathrm{pr}\, D_{\xi^*} = \mathrm{pr}\, D_\xi^* = T_\xi^*$. Suppose $\{g_\xi\}_{\xi\in\Gamma}$ is a family of elements in $\mathfrak{H}$ such that $g_\xi = 0$ for almost all the ξ.[21] Upon setting $g = \sum_\xi D_\xi g_\xi$,[22] we then have

$$\sum_\xi \sum_\eta (T_{\xi^*\eta} g_\eta, g_\xi) = \sum_\xi \sum_\eta (D_{\xi^*\eta} g_\eta, g_\xi) = \sum_\xi \sum_\eta (D_\xi^* D_\eta g_\eta, g_\xi) = (g, g) \geqq 0$$

and, for all $\alpha \in \Gamma$,

$$\sum_\xi \sum_\eta (T_{\xi^*\alpha^*\alpha\eta} g_\eta, g_\xi) = \sum_\xi \sum_\eta (D_{\xi^*\alpha^*\alpha\eta} g_\eta, g_\xi) = \sum_\xi \sum_\eta (D_\xi^* D_\alpha^* D_\alpha D_\eta g_\eta, g_\xi) =$$
$$= (D_\alpha g, D_\alpha g) \leqq \|D_\alpha\|^2 (g, g).$$

Now the essential content of our "principal theorem" is that these inequalities characterize the families $\{T_\xi\}$ which are obtained by a projection of a representation $\{D_\xi\}$ of Γ.

PRINCIPAL THEOREM. *Let Γ be a $*$-semi-group and suppose $\{T_\xi\}_{\xi\in\Gamma}$ is a family of bounded linear transformations in the Hilbert space $\mathfrak{H}$ which satisfy the following conditions:*

(a) $T_\varepsilon = I, \quad T_{\xi^*} = T_\xi^*$,[23]

(b) *T_ξ, considered as a function of ξ, is of positive type, that is, for every family $\{g_\xi\}_{\xi\in\Gamma}$ of elements in $\mathfrak{H}$ such that $g_\xi = 0$ for almost all ξ, we have*

$$\sum_\xi \sum_\eta (T_{\xi^*\eta} g_\eta, g_\xi) \geqq 0;$$

(c) *for such families $\{g_\xi\}$ and for all $\alpha \in \Gamma$ we have*

$$\sum_\xi \sum_\eta (T_{\xi^*\alpha^*\alpha\eta} g_\eta, g_\xi) \leqq C_\alpha^2 \sum_\xi \sum_\eta (T_{\xi^*\eta} g_\eta, g_\xi)^{24}$$

with constant $C_\alpha > 0$.

---

[21] That is, for all indices with perhaps the exception of a finite number of them. We shall make use of this expression again in the sequel.

[22] The sums extend over all the elements of Γ unless stated expressly otherwise.

[23] In the case of a complex Hilbert space $\mathfrak{H}$ this equation is a consequence of (b).

[24] In particular, the left member is real. This could, after all, have also been deduced from the other hypotheses.

Then there exists a representation $\{\boldsymbol{D}_\xi\}_{\xi\in\Gamma}$ *of* Γ *in an extension space* $\mathbf{H}$ *such that*

$$T_\xi = \operatorname{pr} \boldsymbol{D}_\xi.$$

Furthermore, one can require that $\mathbf{H}$ *be minimal in the sense that it is spanned by elements of the form* $\boldsymbol{D}_\xi f$ *where* $f \in \mathfrak{H}$ *and* $\xi \in \Gamma$. *In this case, the structure* $\{\mathbf{H}, \boldsymbol{D}_\xi, \mathfrak{H}\}_{\xi\in\Gamma}$ *is determined to within an isomorphism and the following propositions are valid:*

1) $\|\boldsymbol{D}_\alpha\| \leqq C_\alpha$;

2) *if the equation* $T_{\xi\alpha\eta} = T_{\xi\beta\eta} + T_{\xi\gamma\eta}$ *is satisfied for fixed* α, β, γ *and for all* $\xi, \eta \in \Gamma$, *we have*

$$\boldsymbol{D}_\alpha = \boldsymbol{D}_\beta + \boldsymbol{D}_\gamma;$$

3) *if*

$$T_{\xi\alpha_n\eta} \rightharpoonup T_{\xi\alpha\eta} \qquad (n \to \infty)$$

for fixed α_n *and* α, *and for all* ξ, η, *and if furthermore* $\lim C_{\alpha_n} < \infty$, *then we have*

$$\boldsymbol{D}_{\alpha_n} \rightharpoonup \boldsymbol{D}_\alpha \qquad (n \to \infty).$$

Remarks. If Γ is a group, we have $\xi^*\xi = \xi\xi^* = \varepsilon$; hence condition (c) is satisfied in an obvious manner by $C_\alpha = 1$ and the representation $\{\boldsymbol{D}_\xi\}$ consists of unitary transformations.[25]

In the case where Γ is a topological group, and $\mathfrak{H}$ is a complex space of dimension 1, our theorem reduces to a theorem of GELFAND and RAIKOV,[26] in virtue of which every continuous complex-valued function $p(\xi)$ of positive type (see (b) in Principal Theorem, above) defined on Γ can be written in the form

$$p(\xi) = (U_\xi f_0, f_0)$$

where $\{U_\xi\}$ is a weakly (and hence also strongly) continuous[27] unitary representation of Γ in a Hilbert space $\mathfrak{H}$, and where f_0 is a fixed element of $\mathfrak{H}$, an

---

[25] The principal theorem was already proved for groups in SZ.-NAGY **[11]**.

[26] GELFAND-RAIKOV **[1]**. Also see GODEMENT **[2]**, in particular pages 21, 22. This theorem is of prime importance in the theory of these authors on irreducible unitary representations of locally bicompact groups.

There is a closely related theorem due to SEGAL **[1, Theorem 1]** on certain complex-valued functions $\omega(A)$, defined on an "operator algebra," that is, on a set $\mathscr{A}$ of bounded linear transformations in a complex Hilbert space, which contains $A + B, AB, cA, A^*$ provided it contains A, B. In the case where $\mathscr{A}$ also contains the transformation I, the SEGAL theorem is also a consequence of our principal theorem.

[27] Since U_ξ is unitary, we have

$$\|U_\xi f - U_\eta f\|^2 = 2\|f\|^2 - 2\operatorname{Re}(U_\xi f, U_\eta f) = 2\operatorname{Re}(U_\eta f - U_\xi f, U_\eta f),$$

for $\xi, \eta \in \Gamma$ and $f \in \mathfrak{H}$, from which it follows that the weak continuity of U_ξ implies its strong continuity.

element whose images under the transformations U_ξ span the space $\mathfrak{H}$; under these conditions the structure $\{\mathfrak{H}, U_\xi, f_0\}$ is determined to within an isomorphism.[28]

Proof. 1) *Extension space.* We denote the set of all the families $\boldsymbol{v} = \{v_\xi\}_{\xi \in \Gamma}$ of elements v_ξ in $\mathfrak{H}$ by $\mathbf{V}$; $\boldsymbol{v}$ can be considered also as a vector whose "component with index ξ" is v_ξ, in symbols:

$$(\boldsymbol{v})_\xi = v_\xi .$$

The addition of these vectors and their multiplication by scalars (that is, by real or complex numbers according as $\mathfrak{H}$ is real or complex) are defined by the corresponding operations on components.

In $\mathbf{V}$ we shall consider in particular two linear manifolds, $\mathbf{G}$ and $\mathbf{F}$. $\mathbf{G}$ consists of vectors almost all of whose components are equal to 0; these vectors will be denoted by the letter $\boldsymbol{g}$. $\mathbf{F}$ consists of vectors $\boldsymbol{f} = \{f_\xi\}$[29] for which there exists a vector $\boldsymbol{g} = \{g_\xi\}$ such that

$$f_\xi = \sum_\eta T_{\xi^* \eta} g_\eta$$

for all $\xi \in \Gamma$; this relation between $\boldsymbol{f}$ and $\boldsymbol{g}$ will be denoted by

$$\boldsymbol{f} = \hat{\boldsymbol{g}}.$$

We define a binary form $[\boldsymbol{f}, \boldsymbol{f}]$ in $\mathbf{F}$ in the following way. If $\boldsymbol{f} = \hat{\boldsymbol{g}}$, $\boldsymbol{f}' = \hat{\boldsymbol{g}}'$, we let

$$\text{(18)} \qquad [\boldsymbol{f}, \boldsymbol{f}'] = \sum_\xi (f_\xi, g'_\xi) = \sum_\xi \sum_\eta (T_{\xi^* \eta} g_\eta, g'_\xi) =$$

$$\text{(19)} \qquad = \sum_\xi \sum_\eta (g_\eta, T_{\eta^* \xi} g'_\xi) = \sum_\eta (g_\eta, f'_\eta).$$

(Here we have made use of the fact that $T^*_{\xi^* \eta} = T_{(\xi^* \eta)^*} = T_{\eta^* \xi}$.) It follows from (18) that this definition does not depend on the particular choice of $\boldsymbol{g}$ in the representation of $\boldsymbol{f}$, and it follows from (19) that it does not depend on the particular choice of $\boldsymbol{g}'$ either; consequently, the form $[\boldsymbol{f}, \boldsymbol{f}']$ is determined uniquely by $\boldsymbol{f}$ and $\boldsymbol{f}'$. It is obviously linear in $\boldsymbol{f}$ and we have $[\boldsymbol{f}', \boldsymbol{f}] = \overline{[\boldsymbol{f}, \boldsymbol{f}']}$. It follows from condition (b) that

$$[\boldsymbol{f}, \boldsymbol{f}] = \sum_\xi \sum_\eta (T_{\xi^* \eta} g_\eta, g_\xi) \geqq 0.$$

We have still to prove that the equality sign holds here only for $\boldsymbol{f} = 0$. But it follows from what has already been proved that the Schwarz inequality is valid for the form $[\boldsymbol{f}, \boldsymbol{f}']$:

---

[28] This means that if $\{\mathfrak{H}', U'_\xi, f'_0\}$ is another structure with the same properties, $\mathfrak{H}$ can be mapped linearly and isometrically onto $\mathfrak{H}'$ in such a way that $f_0 \to f'_0$ and that $f \to f'$ implies $U_\xi f \to U'_\xi f'$ for all $\xi \in \Gamma$.

[29] The parameter of the family will be denoted by ξ ; hence ξ always runs through all the elements of Γ.

$$|[f,f']|^2 \leqq [f,f][f',f'].$$

The equation $[f,f]=0$ for one f therefore implies that $[f,f']=0$ for all $f' \in \mathbf{F}$; but it follows easily from (18) that this is possible only if $f=0$.

Hence the form $[f,f']$ possesses all the properties of a scalar product; therefore, if the scalar product in $\mathbf{F}$ is defined by

$$(f,f')=[f,f'],$$

$\mathbf{F}$ becomes a Hilbert space, which in general is *not complete.* Let $\mathbf{H}$ be the *completion* of $\mathbf{F}$.

The original space $\mathfrak{H}$ can be embedded as a subspace in $\mathbf{H}$, and even in $\mathbf{F}$; this can be done by identifying the element f in $\mathfrak{H}$ with the element

$$\boldsymbol{f}_f=\{T_{\xi^*}f\}$$

in $\mathbf{F}$ (note that $\boldsymbol{f}_f=\hat{\boldsymbol{g}}$ with $(\boldsymbol{g})_\varepsilon=f$ and $(\boldsymbol{g})_\xi=0$ for $\xi \neq \varepsilon$). This identification is justified because we clearly have

$$\boldsymbol{f}_{cf}=c\boldsymbol{f}_f, \quad \boldsymbol{f}_{f+f'}=\boldsymbol{f}_f+\boldsymbol{f}_{f'}, \quad (\boldsymbol{f}_f,\boldsymbol{f}_{f'})=(f,f').$$

Let us now calculate the orthogonal projection $\boldsymbol{Pf}$ of an element $\boldsymbol{f} \in \mathbf{F}$ onto the subspace $\mathfrak{H}$! We should have, for all $h \in \mathfrak{H}$,

$$(\boldsymbol{Pf},h)=(\boldsymbol{f},h),$$

the definition of the scalar product in $\mathbf{F}$ yields

$$(\boldsymbol{Pf},h)=(\boldsymbol{f},\boldsymbol{f}_h)=((\boldsymbol{f})_\varepsilon,h);$$

since $\boldsymbol{Pf}$ and $(\boldsymbol{f})_\varepsilon$ are in $\mathfrak{H}$, this equation is possible for all $h \in \mathfrak{H}$ only if

(20) $$\boldsymbol{Pf}=(\boldsymbol{f})_\varepsilon.$$

2) *The representation* $\{\boldsymbol{D}_\xi\}$. Suppose $\boldsymbol{f}=\hat{\boldsymbol{g}}$, that is,

$$f_\xi=\sum_\eta T_{\xi^*\eta}g_\eta.$$

We then have

$$f_{\alpha^*\xi}=\sum_\eta T_{\xi^*\alpha\eta}g_\eta=\sum_\zeta T_{\xi^*\zeta}g_\zeta^\alpha$$

for arbitrary $\alpha \in \Gamma$, where

(21) $$g_\zeta^\alpha=\sum_{\alpha\eta=\zeta} g_\eta$$

(if there are no η such that $\alpha\eta=\zeta$, then the sum in the second member of (21) is defined to be equal to 0). It is clear that, for given α, $g_\xi^\alpha=0$ for almost all the ξ, and therefore

$$\{g_\xi^\alpha\} \in \mathbf{G}, \quad \{f_{\alpha^*\xi}\} \in \mathbf{F}.$$

Consequently,

$$\boldsymbol{D}_\alpha\{f_\xi\}=\{f_{\alpha^*\xi}\}$$

is a transformation, which is obviously linear, of **F** into **F**. We have

$$(22)\qquad D_\varepsilon\{f_\xi\}=\{f_{\varepsilon^*\xi}\}=\{f_\xi\},$$

$$(23)\qquad D_\alpha D_\beta\{f_\xi\}=D_\alpha\{f_{\beta^*\xi}\}=\{f_{\beta^*\alpha^*\xi}\}=\{f_{(\alpha\beta)^*\xi}\}=D_{\alpha\beta}\{f_\xi\}$$

and, for $\boldsymbol{f}=\hat{\boldsymbol{g}}, \boldsymbol{f}'=\hat{\boldsymbol{g}}'$,

$$(24)\qquad \begin{aligned}(D_\alpha \boldsymbol{f},\boldsymbol{f}') &= \sum_\xi (f_{\alpha^*\xi}, g'_\xi)=\sum_\xi\sum_\eta (T_{\xi^*\alpha\eta}g_\eta, g'_\xi) = \\ &= \sum_\xi\sum_\eta (g_\eta, T_{\eta^*\alpha^*\xi}g'_\xi)=\sum_\eta (g_\eta, f'_{\alpha\eta})=(\boldsymbol{f}, D_{\alpha^*}\boldsymbol{f}');\end{aligned}$$

finally, it follows from (24), (23) and condition (c) of the principal theorem that

$$\begin{aligned}(D_\alpha \boldsymbol{f}, D_\alpha \boldsymbol{f}) &= (D_{\alpha^*}D_\alpha \boldsymbol{f},\boldsymbol{f})=(D_{\alpha^*\alpha}\boldsymbol{f},\boldsymbol{f})=\sum_\xi\sum_\eta (T_{\xi^*\alpha^*\alpha\eta}g_\eta, g_\xi)\leqq \\ &\leqq C_\alpha^2\sum_\xi\sum_\eta (T_{\xi^*\eta}g_\eta, g_\xi)=C_\alpha^2(\boldsymbol{f},\boldsymbol{f}).\end{aligned}$$

Hence D_α is a bounded linear transformation in **F**, $\|D_\alpha\|\leqq C_\alpha$, and consequently it can be extended by continuity to **H**. It follows from (22)-(24) that $\{D_\xi\}$ thus extended will be a representation of Γ in **H**.

Consider in particular an element $\boldsymbol{f}$ which belongs to $\mathfrak{H}$, $\boldsymbol{f}=f$. We then have

$$(25)\qquad D_\alpha \boldsymbol{f}=D_\alpha\{T_{\xi^*}f\}=\{T_{\xi^*\alpha}f\},$$

and hence, by (20),

$$PD_\alpha f=(D_\alpha f)_\varepsilon=T_\alpha f.$$

This proves that

$$T_\alpha=\mathrm{pr}\, D_\alpha.$$

It also follows from (25) that, for $\boldsymbol{f}=\hat{\boldsymbol{g}}\in\mathbf{F}$,

$$(\boldsymbol{f})_\xi=\sum_\eta T_{\xi^*\eta}g_\eta=\sum_\eta (D_\eta g_\eta)_\xi=(\sum_\eta D_\eta g_\eta)_\xi,$$

whence

$$\boldsymbol{f}=\sum_\eta D_\eta g_\eta.$$

This means that **F** consists of finite sums of elements of the form $D_\eta g$ where $g\in\mathfrak{H}$, $\eta\in\Gamma$; then these elements span the space **H**, and therefore the extension space **H** is *minimal.*

The representation $\{D_\xi\}$ of Γ which we have just constructed also satisfies Propositions 2 and 3 of the principal theorem. This follows from the equation

$$(D_\alpha \boldsymbol{f},\boldsymbol{f}')=\sum_\xi\sum_\eta (T_{\xi^*\alpha\eta}g_\eta, g'_\xi)$$

(see formula (24)), valid for arbitrary $\boldsymbol{f}=\hat{\boldsymbol{g}}, \boldsymbol{f}'=\hat{\boldsymbol{g}}'(\in\mathbf{F})$, and from the obvious fact that if the relation

$$(D_\alpha \boldsymbol{f},\boldsymbol{f}')=(D_\beta \boldsymbol{f},\boldsymbol{f}')+(D_\gamma \boldsymbol{f},\boldsymbol{f}'),$$

or the relation

$$(D_{\alpha_n}f, f') \to (D_\alpha f, f') \qquad (n \to \infty)$$

is satisfied for $f, f' \in \mathbf{F}$, and if, moreover, in the second case,

$$\overline{\lim} \| D_{\alpha_n} \| < \infty,$$

the same relation is satisfied for all the elements f, f' in $\mathbf{H}$.

3) *Isomorphism.* It remains to investigate the problem: To what extent is the structure $\{\mathbf{H}, D_\xi, \mathfrak{H}\}$ determined? To this end, let us consider any two representations of Γ, $\{D'_\xi\}$ in $\mathbf{H}'$ and $\{D''_\xi\}$ in $\mathbf{H}''$, where $\mathbf{H}'$ and $\mathbf{H}''$ are two extension spaces of $\mathfrak{H}$, and let us assume that

$$\operatorname{pr} D'_\xi = T_\xi, \quad \operatorname{pr} D''_\xi = T_\xi.$$

Furthermore, we shall assume that each of these extension spaces is minimal, i.e. that $\mathbf{H}'$ is spanned by the elements $D'_\xi g$ and $\mathbf{H}''$ by the elements $D''_\xi g$, where $g \in \mathfrak{H}$ and $\xi \in \Gamma$.

Let

$$f'_1 = \sum_\xi D'_\xi g_{1\xi}, \quad f'_2 = \sum_\xi D'_\xi g_{2\xi}$$

be two elements in $\mathbf{H}'$ (with $\{g_{1\xi}\}, \{g_{2\xi}\} \in \mathbf{G}$), and let

$$f''_1 = \sum_\xi D''_\xi g_{1\xi}, \quad f''_2 = \sum_\xi D''_\xi g_{2\xi}$$

be elements in $\mathbf{H}''$. We have

$$(f'_1, f'_2) = \sum_\xi \sum_\eta (D'_\eta g_{1\eta}, D'_\xi g_{2\xi}) = \sum_\xi \sum_\eta (D'_{\xi^* \eta} g_{1\eta}, g_{2\xi}) = \sum_\xi \sum_\eta (T_{\xi^* \eta} g_{1\eta}, g_{2\xi}),$$

and in an analogous manner

$$(f''_1, f''_2) = \sum_\xi \sum_\eta (T_{\xi^* \eta} g_{1\eta}, g_{2\xi}),$$

hence

$$(f'_1, f'_2) = (f''_1, f''_2).$$

Consequently, if we assign the elements

$$f' = \sum_\xi D'_\xi g_\xi, \quad f'' = \sum_\xi D''_\xi g_\xi \tag{26}$$

to the same $\{g_\xi\} \in \mathbf{G}$, this correspondence $f' \leftrightarrow f''$ will be linear and isometric, and it can then be extended by continuity to a linear and isometric mapping of all the elements of $\mathbf{H}'$ onto $\mathbf{H}''$.

In particular, by setting $g_\varepsilon = g$ and $g_\xi = 0$ for $\xi \neq \varepsilon$, we see that each element g of the common subspace $\mathfrak{H}$ corresponds to itself. For all $\alpha \in \Gamma$, we have

$$D'_\alpha \sum_\xi D'_\xi g_\xi = \sum_\xi D'_{\alpha\xi} g_\xi = \sum_\zeta D'_\zeta g^\alpha_\zeta \leftrightarrow \sum_\zeta D''_\zeta g^\alpha_\zeta = \sum_\xi D''_{\alpha\xi} g_\xi = D''_\alpha \sum_\xi D''_\xi g_\xi$$

(see (21)); hence $f' \leftrightarrow f''$ implies that $D'_\alpha f' \leftrightarrow D''_\alpha f''$ for all f', f'' in the form (26), and then, in virtue of the continuity of the transformations D'_α, D''_α, for all $f' \in \mathbf{H}'$ and $f'' \in \mathbf{H}''$.

Therefore the structures $\{\mathbf{H}', D'_\xi, \mathfrak{H}\}$ and $\{\mathbf{H}'', D''_\xi, \mathfrak{H}\}$ are *isomorphic*. This completes the proof of the theorem.

7. Proof of the Neumark Theorem

Let $\{B_\lambda\}_{-\infty<\lambda<\infty}$ be a generalized spectral family in $\mathfrak{H}$. Set $B_\infty = \lim_{\lambda\to\infty} B_\lambda = I$ and $B_{-\infty} = \lim_{\lambda\to-\infty} B_\lambda = O$. We assign the transformation

$$B_\Delta = B_b - B_a$$

to each half-open interval

$$\Delta = (a,b] \qquad \text{(where } -\infty \leqq a < b \leqq \infty)$$

and the transformation

$$B_\omega = \sum_i B_{\Delta_i};$$

to each set ω which consists of a finite number of disjoint intervals Δ_i; this definition obviously does not depend on the particular choice of the decomposition of ω. For $\Omega = (-\infty, \infty]$ we have $B_\Omega = I$, and for the void set Θ we have $B_\Theta = O$. The family K of these sets ω, including Ω and Θ, is clearly closed with respect to subtraction of any two sets, and with respect to the operation of forming unions and intersections of a finite number of sets. B_ω is a positive additive set function defined on K; more precisely, B_ω is, for all $\omega \in K$, a self-adjoint transformation such that

$$O \leqq B \leqq I, \quad B_\Theta = O, \quad B_\Omega = I, \quad B_{\omega_1 \cup \omega_2} = B_{\omega_1} + B_{\omega_2} \quad \text{si} \quad \omega_1 \cap \omega_2 = \Theta.$$

We shall also consider K as a $*$-semi-group; we do this by setting

$$\omega_1\omega_2 = \omega_1 \cap \omega_2, \quad \omega^* = \omega, \quad \varepsilon = \Omega.$$

We shall see that B_ω, considered as a function defined on this $*$-semi-group, satisfies the conditions of the principal theorem.

Condition (a) is satisfied in an obvious manner. Condition (b) means that

$$s = \sum_i \sum_j (B_{\omega_i \cap \omega_j} g_j, g_i) \geqq 0$$

for arbitrary $\omega_1, \ldots, \omega_n \in K$ and $g_1, \ldots, g_n \in \mathfrak{H}$. In order to prove this inequality, we first consider the intersections

$$\pi = \omega_1^\pm \cap \omega_2^\pm \cap \cdots \cap \omega_n^\pm \qquad (\in K)$$

where each time we can choose one of the signs $+$ or $-$ in an arbitrary manner; ω^+ denotes the set ω itself and ω^- its complement $\Omega - \omega$. Two intersections π corresponding to different variations of sign are obviously disjoint. Each set $\omega_i \cap \omega_j$ $(i \lesseqgtr j)$ is the union of certain of these π, namely of all those obtained by choosing the sign $+$ for i and j, that is, of all those which are contained in $\omega_i \cap \omega_j$. In virtue of the additivity of B_ω as a set function, the sum s then decomposes into a sum of terms of the form

$$(B_\pi g_j, g_i).$$

We combine the terms corresponding to the same π into a partial sum s_π; the latter extends to all the pairs of indices (i, j) for which $\omega_i \cap \omega_j \supseteqq \pi$, that is, for which $\omega_i \supseteqq \pi$ and $\omega_j \supseteqq \pi$ simultaneously. Suppose $i_1, i_2, \ldots, i_r$ are those values of the index i for which ω_i contains the fixed set π; we then have

$$s_\pi = \sum_{h=1}^{r} \sum_{k=1}^{r} (B_\pi g_{i_k}, g_{i_h}) = (B_\pi g, g) \quad \text{with} \quad g = \sum_{h=1}^{r} g_{i_h},$$

and consequently $s_\pi \geqq 0$. Since this is true for all the π, it follows that $s = \sum_\pi s_\pi \geqq 0$, which was to be proved.

Let us now pass on to condition (c). Suppose ω is a fixed element in K and set

$$\omega_i' = \omega_i \cap \omega^+, \quad \omega_i'' = \omega_i \cap \omega^- \qquad (i = 1, 2, \ldots, n).$$

Applying the inequality $s \geqq 0$, which we have just proved, to ω_i' and ω_i'' instead of to ω_i, we obtain

$$s' = \sum_i \sum_j (B_{\omega_i' \cap \omega_j'} g_j, g_i) \geqq 0, \quad s'' = \sum_i \sum_j (B_{\omega_i'' \cap \omega_j''} g_j, g_i) \geqq 0.$$

Since $\omega_i' \cap \omega_j'$ and $\omega_i'' \cap \omega_j''$ are to be contained in the disjoint sets ω^+, ω^-, they are also disjoint; since their union is equal to $\omega_i \cap \omega_j$, it follows from the additivity of B_ω that $s' + s'' = s$. Consequently, we have $0 \leqq s' \leqq s$, that is

$$0 \leqq \sum_i \sum_j (B_{\omega_i \cap \omega \cap \omega \cap \omega_j} g_j, g_i) \leqq \sum_i \sum_j B_{\omega_i \cap \omega_j} g_j, g_i),$$

and hence condition (c) is satisfied with $C_\omega = 1$.

We can then apply the principal theorem. Hence there exists a representation $\{\boldsymbol{E}_\omega\}$ of the $*$-semi-group K in a minimal extension space $\mathbf{H}$ such that

$$B_\omega = \operatorname{pr} \boldsymbol{E}_\omega;$$

here, "minimal" means that the space $\mathbf{H}$ is spanned by the elements of the form $\boldsymbol{E}_\omega f$ where $f \in \mathfrak{H}$, $\omega \in K$. It follows from the structure of K as a $*$-semi-group that $\boldsymbol{E}_\omega$ is a projection, $\boldsymbol{E}_\Omega = \boldsymbol{I}$, and

(27) $$\boldsymbol{E}_{\omega_1 \cap \omega_2} = \boldsymbol{E}_{\omega_1} \boldsymbol{E}_{\omega_2}.$$

We also have

(28) $$\boldsymbol{E}_{\omega_1 \cup \omega_2} = \boldsymbol{E}_{\omega_1} + \boldsymbol{E}_{\omega_2} \quad \text{when} \quad \omega_1 \cap \omega_2 = \Theta;$$

this follows, in virtue of the fact that $\mathbf{H}$ is minimal, from the fact that, B_ω being an additive function of ω, we have

$$B_{(\omega_1 \cup \omega_2) \cap \omega} = B_{\omega_1 \cap \omega} + B_{\omega_2 \cap \omega}$$

for all $\omega \in K$. In particular, we have $\boldsymbol{E}_\Theta = \boldsymbol{E}_{\Theta \cup \Theta} = \boldsymbol{E}_\Theta + \boldsymbol{E}_\Theta$, and hence $\boldsymbol{E}_\Theta = \boldsymbol{O}$. We set

$$\boldsymbol{E}_\lambda = \boldsymbol{E}_{(-\infty, \lambda]} \text{ for } -\infty < \lambda < \infty;$$

since $B_{(-\infty, \lambda]} = B_\lambda - B_{-\infty} = B_\lambda$, we then have

(29) $$B_\lambda = \operatorname{pr} \boldsymbol{E}_\lambda,$$

and by (27) or (28),

$$E_\mu \leqq E_\mu \text{ for } \lambda < \mu.$$

We finally arrive at the relations

$$\begin{array}{ll} E_\lambda \to E_\mu & \text{as } \lambda \to \mu + 0; \\ E_\lambda \to E_\Theta = O & \text{as } \lambda \to -\infty; \\ E_\lambda \to E_\Omega = I & \text{as } \lambda \to \infty,^{30} \end{array}$$

which are consequences, in virtue of the fact that **H** is minimal, of the relations

$$\begin{array}{ll} B_{(-\infty, \lambda] \cap \omega} \to B_{(-\infty, \mu] \cap \omega} & \text{as } \lambda \to \mu + 0, \\ B_{(-\infty, \lambda] \cap \omega} \to O = B_{\Theta \cap \omega} & \text{as } \lambda \to -\infty, \\ B_{(-\infty, \lambda] \cap \omega} \to B_\omega = B_{\Omega \cap \omega} & \text{as } \lambda \to +\infty, \end{array}$$

which are valid for all fixed ω.

Hence $\{E_\omega\}$ is an ordinary spectral family.

Since each of the E_ω is derived from the E_λ by forming differences and sums or by passing to the limit ($\lambda \to \pm \infty$), the space **H** is also minimal with respect to $\{E_\lambda\}$, and the structure $\{\mathbf{H}, E_\lambda, \mathfrak{H}\}$ is determined to within an isomorphism.

This completes the proof of Theorem I. The following theorem, also due to NEUMARK [5], is proved in an analogous manner.

THEOREM. *Suppose K is a family of subsets ω of a set Ω which contains Ω and the void set Θ, and which is closed with respect to subtraction of sets, as well as with respect to forming the union and intersection of a finite number of sets. We assign to each $\omega \in K$ a self-adjoint transformation B_ω in the Hilbert space $\mathfrak{H}$, in such a way that we have*

$$O \leqq B_\omega \leqq I; \quad B_\Theta = O; \quad B_\Omega = I; \quad B_{\omega_1 \cup \omega_2} = B_{\omega_1} + B_{\omega_2} \quad si \quad \omega_1 \cap \omega_2 = \Theta.$$

Then there exists a family $\{E_\omega\}$ of projections in an extension space **H**, *such that the elements of the form $E_\omega f$ ($f \in \mathfrak{H}, \omega \in K$) determine the space* **H**, *and*

$$B_\omega = \operatorname{pr} E_\omega;$$
$$E_\Theta = O; \; E_\Omega = I;$$
$$E_{\omega_1 \cap \omega_2} = E_{\omega_1} E_{\omega_2} \quad \textit{for arbitrary } \omega_1, \omega_2;$$
$$E_{\omega_1 \cup \omega_2} = E_{\omega_1} + E_{\omega_2} \quad \textit{for disjoint } \omega_1, \omega_2.$$

8. Proof of the Theorem on Sequences of Moments

We saw in Sec. 3 how Theorem II is derivable from Theorem I of Neumark (at least with the exception of the last propositions of Theorem II, which however can be proved directly without difficulty). We are now going to see how this theorem is derivable directly from our principal theorem.

Suppose Γ is the $*$-semi-group of non-negative integers n with addition as the "semi-group operation" and with the identity operation $n^* = n$ as the "$*$ operation"; then the "unit" element is the number 0.

---

[30] For a monotone sequence of self-adjoint transformations, the weak limit is at the same time the strong limit, which fact follows from Sec. 104.

Every representation of Γ is obviously of the form $\{A^n\}$ where A is a bounded self-adjoint transformation.

We shall show that the sequence $\{A_n\}$ $(n=0, 1, \ldots)$ visualized in Theorem II, considered as a function of the variable element n in the $*$-semi-group Γ, satisfies the conditions of the principal theorem. Condition (a) is obviously satisfied; as for the other two conditions, one proves them using the integral formula

$$A_n = \int_{-M-0}^{M} \lambda^n \, dB_\lambda$$

established in Sec. 3, where $\{B_\lambda\}$ is a generalized spectral family on the interval $[-M, M]$. In fact, if $\{g_n\}$ $(n=0, 1, \ldots)$ is any sequence of elements in $\mathfrak{H}$, which are almost all equal to 0, we have

$$\sum_{i=0}^{\infty}\sum_{k=0}^{\infty}(A_{i+k}g_k, g_i) = \sum_{i=0}^{\infty}\sum_{k=0}^{\infty}\int_{-M-0}^{M}\lambda^{i+k}\, d(B_\lambda g_k, g_i) = \int_{-M-0}^{M}(B(d\lambda)g(\lambda), g(\lambda)) \geqq 0$$

where we have set

$$g(\lambda) = \sum_{i=0}^{\infty}\lambda^i g_i$$

and where $B(\Delta)$ denotes the positive, additive interval function generated by B_λ, that is, $B(\Delta) = B_b - B_a$ for $\Delta = (a, b]$. Furthermore, for $r = 0, 1, \ldots,$ we have that

$$\sum_{i=0}^{\infty}\sum_{k=0}^{\infty}(A_{i+2r+k}g_k, g_i) = \int_{-M-0}^{M}\lambda^{2r}(B(d\lambda)g(\lambda), g(\lambda)) \leqq$$

$$\leqq M^{2r}\int_{-M-0}^{M}(B(d\lambda)g(\lambda), g(\lambda)) = M^{2r}\sum_{i=0}^{\infty}\sum_{k=0}^{\infty}(A_{i+k}g_k, g_i).$$

Thus we see that conditions (b) and (c) are also satisfied, and one can then apply the principal theorem, which yields Theorem II.

9. Proof of the Theorems on Contractions

Proof of Theorem III. Suppose Γ is the additive group of all integers n. Every representation of Γ is then of the form $\{U^n\}$ where U is a unitary transformation.

Suppose T is a contraction in $\mathfrak{H}$. Set

$$T_n = \begin{cases} T^n \text{ for } n = 0, 1, \ldots, \\ T^{*|n|} \text{ for } n = -1, -2, \ldots; \end{cases}$$

hence $T_0 = I$ and $T_{-n} = T_n^*$. We shall show that T_n, considered as a function defined on Γ, is of positive type, that is,

$$\sum_m \sum_n (T_{n-m}g_n, g_m) \geqq 0 \tag{31}$$

for every sequence $\{g_n\}_{-\infty}^{\infty}$ of elements in $\mathfrak{H}$ almost all of which are equal to 0.

We first consider the case of a *complex* space $\mathfrak{H}$. We set

$$T(r, \varphi) = \sum_{-\infty}^{\infty} r^n e^{in\varphi} T_n \tag{32}$$

for $0 \leqq r < 1$ and $0 \leqq \varphi \leqq 2\pi$; in view of the fact that $\|T_n\| \leqq 1$, this series converges in norm. Setting $z = re^{i\varphi}$ we have

$$T(r, \varphi) = \left(\frac{1}{2} I + \sum_1^\infty z^n T^n\right) + \left(\frac{1}{2} I + \sum_1^\infty \bar{z}^n T^{*n}\right) =$$

$$= 2 \operatorname{Re} \left(\frac{1}{2} I + \sum_1^\infty z^n T^n\right) = \operatorname{Re} (I + zT)(I - zT)^{-1}.$$

Hence, for $f \in \mathfrak{H}$ and $g = (I - zT)^{-1} f$, we have

$$(33) \quad \begin{aligned} (T(r, \varphi)f, f) &= \operatorname{Re}((I + zT)(I - zT)^{-1} f, f) = \operatorname{Re}((I + zT)g, (I - zT)g) = \\ &= \operatorname{Re}[(g, g) + z(Tg, g) - \bar{z}(g, Tg) - z\bar{z}(Tg, Tg)] = \|g\|^2 - |z|^2 \|Tg\|^2 \geqq 0, \end{aligned}$$

since $|z| < 1$, $\|T\| \leqq 1$. Since this result holds for all $f \in \mathfrak{H}$, we have in particular that

$$(34) \qquad p(r, \varphi) = (T(r, \varphi) f(\varphi), f(\varphi)) \geqq 0$$

with

$$f(\varphi) = \sum_{n=-\infty}^{\infty} e^{-in\varphi} g_n$$

where $\{g_n\}$ is the sequence of elements in $\mathfrak{H}$ considered in inequality (31). If in (34) we replace $T(r, \varphi)$ and $f(\varphi)$ by their series expansions, we obtain that

$$p(r, \varphi) = \sum_{k, m, n} r^{|k|} e^{i(k+m-n)\varphi} (T_k g_n, g_m) =$$

$$= \sum_l e^{il\varphi} \sum_{m, n} r^{|l+n-m|} (T_{l+n-m} g_n, g_m) \geqq 0,$$

whence

$$\sum_{m, n} r^{|n-m|} (T_{n-m} g_n, g_m) = \frac{1}{2\pi} \int_0^{2\pi} p(r, \varphi)\, d\varphi \geqq 0.$$

We obtain (31) by letting $r \to 1$.

The case of a *real* space $\mathfrak{H}$ can be reduced to that of a complex space; we do this by introducing the space $\mathfrak{H}_c$ of pairs $\{g, h\}$ of elements in $\mathfrak{H}$, subject to the following fundamental operations:

$$\{g, h\} + \{g', h'\} = \{g + g', h + h'\},$$

$$(a + ib)\{g, h\} = \{ag - bh, bg + ah\} \quad (a \text{ and } b \text{ are real numbers}),$$

$$(\{g, h\}, \{g', h'\}) = (g, g') + (h, h') + i(h, g') - i(g, h'),$$

$$\|\{g, h\}\|^2 = (\{g, h\}, \{g, h\}) = \|g\|^2 + \|h\|^2;$$

hence $\mathfrak{H}_c$ is a complex Hilbert space. The transformation

$$\bar{T}\{g, h\} = \{Tg, Th\}$$

is a contraction in $\mathfrak{H}_c$; in fact, on the one hand, it is linear :

$$\bar{T}\{g + g', h + h'\} = \{T(g + g'), T(h + h')\} =$$

$$= \{Tg, Th\} + \{Tg', Th'\} = \bar{T}\{g, h\} + \bar{T}\{g', h'\},$$

$$\bar{T}(a + ib)\{g, h\} = \bar{T}\{ag - bh, bg + ah\} = \{aTg - bTh, bTg + aTh\} =$$

$$= (a + ib)\{Tg, Th\} = (a + ib)\, \bar{T}\{g, h\},$$

and, on the other hand, we have

$$\|\bar{T}\{g,h\}\|^2 = \|\{Tg, Th\}\|^2 = \|Tg\|^2 + \|Th\|^2 \leqq \|g\|^2 + \|h\|^2 = \|\{g,h\}\|^2.$$

It is also easily seen that

$$\bar{T}^*\{g,h\} = \{T^*g, T^*h\}.$$

It follows that

$$\bar{T}^n\{g,h\} = \{T^n g, T^n h\} \text{ and } \bar{T}^{*n}\{g,h\} = \{T^{*n} g, T^{*n} h\}$$

for $n = 0, 1, \dots$, and hence

$$\bar{T}_n\{g,h\} = \{T_n g, T_n h\}$$

for $n = 0, \pm 1, \pm 2, \dots$.

But since inequality (31) has already been proved for complex spaces, we shall have

$$\sum_m \sum_n (\bar{T}_{n-m}\varphi_n, \varphi_m) \geqq 0 \tag{35}$$

for $\varphi_n = \{g_n, h_n\}$ (where $g_n = 0$ and $h_n = 0$ for almost all n). When $h_n = 0$ for all n, we have

$$(\bar{T}_{n-m}\varphi_n, \varphi_m) = (T_{n-m}g_n, g_m),$$

and hence inequality (35) then reduces to inequality (31), which completes the proof of (31) also in the case of a real space $\mathfrak{H}$.

We can then apply the principal theorem, which yields Theorem III.

Proof of Theorem IV. Now let Γ be the additive group of all real numbers t. Then the representations of Γ are one-parameter groups $\{U_t\}$ of unitary transformations.

Let $\{T_t\}_{t \geqq 0}$ be the one-parameter semi-group of contractions considered in the theorem. We set

$$T_t = T^*_{-t}$$

for $t < 0$; then T_t will be a weakly continuous function of t, $-\infty < t < \infty$, and we shall have

$$T_0 = I \text{ and } T_{-t} = T_t^* \text{ for } -\infty < t < \infty.$$

We shall show that T_t, considered as a function on Γ, is of positive type, that is, that

$$\sum_s \sum_t (T_{t-s}h_t, h_s) \geqq 0 \tag{36}$$

for every family $\{h_t\}$ of elements in $\mathfrak{H}$ such that $h_t = 0$ for almost all values of t.

Suppose $t_1, t_2, \dots t_r$ are those values of t for which $h_t \neq 0$. We assign to each $t_n (n = 1, \dots, r)$ a sequence of rational numbers $t_{n\nu}$ $(\nu = 1, 2, \dots)$ which converges to t in such a manner that the numbers $t_{n\nu}$ $(n = 1, 2, \dots, r)$ are distinct for every fixed index ν. Since T_t is a weakly continuous function of t, setting

$$f_n = h_{t_n} \qquad (n = 1, 2, \ldots, r),$$

we have

(37) $$\sum_s \sum_t (T_{t-s} h_t, h_s) = \sum_{m=1}^{r} \sum_{n=1}^{r} (T_{t_n - t_m} f_n, f_m) =$$

$$= \lim_{\nu \to \infty} \sum_{m=1}^{r} \sum_{n=1}^{r} (T_{t_{n\nu} - t_{m\nu}} f_n, f_m).$$

For every fixed ν, the rational numbers $t_{n\nu}$ $(n = 1, \ldots, r)$ are commensurable, that is, they can be written in the form

$$t_{n\nu} = \tau_{n\nu} d_\nu$$

with a $d_\nu > 0$ and distinct integers $\tau_{n\nu}$. Then we have

$$T_{t_{n\nu} - t_{m\nu}} = T_{(\tau_{n\nu} - \tau_{m\nu}) d_\nu} = \begin{cases} (T_{d_\nu})^{\tau_{n\nu} - \tau_{m\nu}} & \text{when } \tau_{n\nu} \geqq \tau_{m\nu}, \\ (T^*_{d_\nu})^{\tau_{m\nu} - \tau_{n\nu}} & \text{when } \tau_{n\nu} \leqq \tau_{m\nu}, \end{cases}$$

and hence

(38) $$\sum_{m=1}^{r} \sum_{n=1}^{r} (T_{t_{n\nu} - t_{m\nu}} f_n, f_m) = \sum_{m=1}^{r} \sum_{n=1}^{r} (T^{(\nu)}_{\tau_{n\nu} - \tau_{m\nu}} f_n, f_m)$$

where $T^{(\nu)}_n$ is defined in a manner analogous to (30), starting with the transformation $T^{(\nu)} = T_{d_\nu}$. Since the latter is a contraction, inequality (31) holds for it also; choosing the g_n in (31) in such a way that

$$g_n = f_p \ \text{ when } \ n = \tau_{p\nu},$$

$$g_n = 0 \text{ when } n \text{ is not equal to any of the } \tau_{q\nu} \ (q = 1, 2, \ldots, r),$$

the first member of inequality (31) reduces to the second member of equation (38), and hence the latter is $\geqq 0$; and this is true for all fixed values of ν. Inequality (36) follows, in virtue of (37).

Then we can apply the principal theorem and obtain that

$$T_t = \text{pr}\, \boldsymbol{U}_t,$$

and that in the case where the extension space $\mathbf{H}$ in question is minimal, the structure $\{\mathbf{H}, \boldsymbol{U}_t, \mathfrak{H}\}$ is determined to within an isomorphism. In this case, $\boldsymbol{U}_t$ is also a weakly (and hence strongly) continuous function of t, and this in virtue of proposition 3) of the principal theorem and because of the fact that T_{t+t_0} is obviously a weakly continuous function of t for an arbitrary fixed value t_0 of t.

This completes the proof of Theorem IV.

Proof of Theorem V. We now choose Γ to be the group of all the "vectors" $\boldsymbol{n} = \{n^{(\rho)}\}_{\rho \in R}$ whose components are integers, almost all of which are equal to 0. If $\{T^{(\rho)}\}_{\rho \in R}$ is the given system of pairwise doubly permutable contractions, we set

(39) $$T_{\boldsymbol{n}} = \prod_{\rho \in R} T^{(\rho)}_{n(\rho)}$$

where $T^{(\rho)}_n$ is defined in a manner analogous to (30). Since $n^{(\rho)} = 0$ for almost all ρ, almost all the factors in the product (39) are equal to I; therefore, this product has meaning even in the case where the set R is infinite. We note, which is essential, that since the $T^{(\rho)}$ are pairwise doubly permutable the factors in (39) are all permutable.

We obviously have $T_{\mathbf{o}} = I$, $T_{-\mathbf{n}} = T^*_{\mathbf{n}}$, where $\mathbf{o}$ denotes the vector all of whose components equal zero.

It remains to prove that $T_{\mathbf{n}}$, considered as a function on the group Γ, is of positive type, that is

$$\sum_{\mathbf{m}} \sum_{\mathbf{n}} (T_{\mathbf{n}-\mathbf{m}} g_{\mathbf{n}}, g_{\mathbf{m}}) \geqq 0 \tag{40}$$

for every family $\{g_{\mathbf{n}}\}$ of elements in $\mathfrak{H}$ such that $g_{\mathbf{n}} = 0$ for almost all $\mathbf{n} \in \Gamma$.

If one considers only those vectors $\mathbf{n}$ for which $g_{\mathbf{n}} \neq 0$, there is a finite number of indices ρ, say $\rho_1, \rho_2, \ldots, \rho_r$, such that all the components of the vectors $\mathbf{n}$ whose indices are different from these, are equal to 0. Since the factors with $n^{(\rho)} = 0$ in the product (39) can be omitted, it suffices to consider the sums of the type

$$\sum_{n_1=-\infty}^{\infty} \cdots \sum_{n_r=-\infty}^{\infty} (T^{(1)}_{n_1-m_1} \cdots T^{(r)}_{n_r-m_r} g_{n_1, \ldots, n_r}, g_{m_1, \ldots, m_r}) \tag{41}$$

where we have set $T^{(i)}$ in place of $T^{(\rho_i)}$ for simplicity in writing.

In the case of a complex space $\mathfrak{H}$ one can reason as following. We set

$$T(r, \varphi_1, \ldots, \varphi_r) = \sum_{n_1=-\infty}^{\infty} \cdots \sum_{n_r=-\infty}^{\infty} r^{|n_1|+\ldots+|n_r|} e^{i(n_1\varphi_1+\ldots+n_r\varphi_r)} T^{(1)}_{n_1} \cdots T^{(r)}_{n_r} =$$

$$= \prod_{i=1}^{r} T^{(i)}(r, \varphi_i),$$

for $0 \leqq r < 1$ and $0 \leqq \varphi_i \leqq 2\pi$, where the factors in the last member have a meaning analogous to (32). Since these factors are, according to (33), $\geqq O$, and since they are pairwise permutable, their product is also $\geqq O$. Hence, we have in particular that

$$(T(r, \varphi_1, \ldots, \varphi_r) g(\varphi_1, \ldots, \varphi_r), g(\varphi_1, \ldots, \varphi_r)) \geqq 0$$

with

$$g(\varphi_1, \ldots, \varphi_r) = \sum_{n_1=-\infty}^{\infty} \cdots \sum_{n_r=-\infty}^{\infty} e^{-i(n_1\varphi_1+\ldots+n_r\varphi_r)} g_{n_1, \ldots, n_r}.$$

Integrating with respect to each variable φ_i from 0 to 2π, and then letting r tend to 1, we have the result that the sum (41) is $\geqq 0$.

This proves inequality (40) for the case of a complex space. The case of a real space can be reduced to that of a complex space in the same way this was done in the proof of Theorem III.

The principal theorem can then be applied. In order to obtain Theorem V, it only remains to observe that every representation $\{U_{\mathbf{n}}\}$ of the group Γ is of the form

$$U_{\mathbf{n}} = \prod_{\varrho \in R} [U^{(\varrho)}]^{n^{(\varrho)}} \qquad (\mathbf{n} = \{n^{(\varrho)}\})$$

where $\{U^{(\rho)}\}$ is a system of permutable unitary transformations. This follows from the fact that n can be written in the form

$$n = \sum_{\varrho \in R} n^{(\varrho)} e_\varrho$$

where e_ρ denotes the vector all of whose components equal zero except the component with index ρ, which is equal to 1; all one has to do is set

$$U^{(\varrho)} = U_{e_\varrho}.$$

10. Proof of the Theorem on Normal Extensions

Now let Γ be the following $*$-semi-group: its elements are the ordered pairs $\pi = \{i, j\}$ of non-negative integers; the semi-group operation is defined in it by

$$\pi + \pi' = \{i, j\} + \{i', j'\} = \{i + i', j + j'\}$$

and the $*$ operation by

$$\pi^* = \{i, j\}^* = \{j, i\};$$

then the "unit" element is

$$\varepsilon = \{0, 0\}.$$

Let $\{D_\pi\}$ be a representation of Γ. Since the semi-group operation in Γ is commutative, the transformations D_π are *normal* and pairwise doubly permutable. If we set $\eta = \{0, 1\}$, every $\pi = \{i, j\}$ can be written in the form

$$\pi = i\eta^* + j\eta,$$

and consequently we have

$$D_\pi = N^{*i} N^j$$

where

$$N = D_\eta.$$

Let T be a bounded linear transformation in the space $\mathfrak{H}$ which satisfies the conditions of Theorem VI:

$$\sum_{i=0}^{\infty} \sum_{j=0}^{\infty} (T^i g_j, T^j g_i) \geqq 0, \tag{42}$$

$$\sum_{i=0}^{\infty} \sum_{j=0}^{\infty} (T^{i+1} g_j, T^{j+1} g_i) \leqq C^2 \sum_{i=0}^{\infty} \sum_{j=0}^{\infty} (T^i g_j, T^j g_i) \qquad (C > 0). \tag{43}$$

We shall show that the transformation

$$T_{\{i, j\}} = T^{*i} T^j,$$

considered as a function defined on the $*$-semi-group Γ, satisfies the conditions of the principal theorem.

It is first of all obvious that $T_\varepsilon = I$, $T_{\pi^*} = T_\pi^*$. In order to prove that T_π is a function of positive type, we choose any family $\{g_\pi\}$ of elements in $\mathfrak{H}$, almost all of which are equal to 0, and consider the sum

$$s = \sum_\pi \sum_{\pi'} (T_{\pi^*+\pi'} g_{\pi'}, g_\pi)$$

where $\pi = \{i,j\}$ and $\pi' = \{i',j'\}$ run through the elements of Γ. It is easy to see that

$$s = \sum_\pi \sum_{\pi'} ((T^*)^{i'+j} T^{i+j'} g_{\pi'}, g_\pi) = \sum_i \sum_{i'} (T^i h_{i'}, T^{i'} h_i)$$

where

$$h_i = \sum_j T^j g_{\{i,j\}}.$$

In virtue of (42), it follows from this that $s \geqq 0$ and therefore T_π is a function of positive type.

Repeating inequality (43) $(i_0 + j_0)$ times we obtain, in an analogous manner, that for fixed $\pi_0 = \{i_0, j_0\}$, we have

$$\sum_\pi \sum_{\pi'} (T_{\pi^*+\pi_0^*+\pi_0+\pi'} g_{\pi'}, g_\pi) = \sum_i \sum_{i'} (T^{i_0+j_0+i} h_{i'}, T^{i_0+j_0+i'} h_i) \leqq$$

$$\leqq C^{2(i_0+j_0)} \sum_i \sum_{i'} (T^i h_{i'}, T^{i'} h_i) = C^{2(i_0+j_0)} \sum_\pi \sum_{\pi'} (T_{\pi^*+\pi'} g_{\pi'}, g_\pi).$$

Conditions (a) - (c) of the principal theorem are therefore satisfied, and then applying this theorem, it follows that, in an extension space **H**, there exists a bounded normal transformation $\boldsymbol{N}$ (with $\|\boldsymbol{N}\| \leqq C$) such that

$$T^{*i} T^j = \text{pr } \boldsymbol{N}^{*i} \boldsymbol{N}^j \qquad (i, j = 0, 1, \ldots),$$

which is equivalent to $T \subseteqq \boldsymbol{N}$ (see Sec. 5). If **H** is minimal in the sense that it is spanned by the elements $\boldsymbol{N}^{*i} \boldsymbol{N}^j f$ $(f \in \mathfrak{H})$, it is also spanned by the elements $\boldsymbol{N}^{*i} f$, inasmuch as $\boldsymbol{N}^j f \in \mathfrak{H}$ for $f \in \mathfrak{H}$.

This completes the proof of Theorem VI.

SUPPLEMENTARY BIBLIOGRAPHY

EGERVÁRY, E. — [**1**] On the contractive linear transformations of n-dimensional vector space, *Acta Sci. Math. Szeged*, **15** (1954), 178-182.

FUGLEDE, B. — [**1**] A commutativity theorem for normal operators, *Proc. Nat. Acad. Sci. USA*, **36** (1950), 35-40.

HALMOS, P. R. — [**1**] Normal dilations and extensions of operators, *Summa Brasiliensis Math.*, **2** (1950), 125-134; [**2**] Commutativity and spectral properties of normal operators, *Acta Sci. Math. Szeged*, **12 B** (1950), 153-156.

KADISON, R. V. — [**1**] A generalized Schwarz inequality and algebraic invariants for operator algebras, *Annals of Math.*, **56** (1952), 494-503.

NEUMARK, M. A. — [**4**] On spectral functions of a symmetric operator, *ibidem*, **7** (1943), 285-296; [**5**] On a representation of additive operator set functions, *Comptes Rendus* (*Doklady*) *Acad. Sci. URSS*, **41** (1943), 359-361.

RIESZ, F. — [**22**] Sur la représentation des opérations fonctionnelles linéaires par des intégrales de Stieltjes, *Kungl. Fysiografiska Sällskapets i Lund Förhandlingar*, **21** (1952), Nr. 16.

SZ.-NAGY, B. — [**9**] A moment problem for self-adjoint operators, *Acta Math. Acad. Sci, Hung.*, **3** (1952), 285-293; [**10**] Sur les contractions de l'espace de Hilbert, *Acta Sci. Math. Szeged*, **15** (1953), 87-92; [**11**] Transformations de l'espace de Hilbert, fonctions de type positif sur un groupe, *ibidem*, **15** (1954), 104-114.

INDEX

NOTATION AND SYMBOLS